PHYSICS RESEARCH AND TECHNOLOGY

NAVIER-STOKES EQUATIONS

PROPERTIES, DESCRIPTION AND APPLICATIONS

PHYSICS RESEARCH AND TECHNOLOGY

Additional books in this series can be found on Nova's website
under the Series tab.

Additional E-books in this series can be found on Nova's website
under the E-book tab.

MATHEMATICS RESEARCH DEVELOPMENTS

Additional books in this series can be found on Nova's website
under the Series tab.

Additional E-books in this series can be found on Nova's website
under the E-book tab.

PHYSICS RESEARCH AND TECHNOLOGY

NAVIER-STOKES EQUATIONS

PROPERTIES, DESCRIPTION AND APPLICATIONS

R. YOUNSI
EDITOR

Nova Science Publishers, Inc.
New York

For permission to use material from this book please contact us:
Telephone 631-231-7269; Fax 631-231-8175
Web Site: http://www.novapublishers.com

LIBRARY OF CONGRESS CATALOGING-IN-PUBLICATION DATA

Navier-Stokes equations : properties, description, and applications / editors, R. Younsi.
p. cm.
Includes index.
ISBN 978-1-61324-590-3 (hardcover)
1. Navier-Stokes equations. I. Younsi, R.
QA374.N38 2011
515'.353--dc23

2011014548

Published by Nova Science Publishers, Inc. † New York

CONTENTS

PREFACE

It is well known that the Navier–Stokes equations are one of the pillars of fluid mechanics. These equations are useful because they describe the physics of many things of academic and economic interest. They may be used to model the weather behavior, ocean currents, water flow in a pipe and air flow around a wing. The Navier–Stokes equations in their full and simplified forms also help with the design of train, aircraft and cars, the study of blood flow, the design of power stations and pollution analysis. This book presents contributions on the application of Navier-Stokes in some engineering applications and provides a description of how the Navier-Stokes equations can be scaled. (Imprint: Nova)

In Chapter 1 the authors survey some results about partial difference equations (PdEs) in weighted graphs obtained previously by the authors, in the case of finite graphs and countable graphs of finite degree. They discuss the existence of solution of elliptic PdEs via an associated semilinear matrix equation and under growth conditions of the forcing term. As applications of these techniques, they study the discrete analogues of some classical elliptic partial differential equations such as Matukuma, Helmholtz and Lane-Emden-Fowler. Finally the authors discuss their results concerning the discrete Navier- Stokes equation and give some explicit solutions for concrete weighted graphs and discuss possible graphs that simplify the applications in the planar case.

In the first part of Chapter 2, the Czochralski techniques of crystal growth driven by the combined action of buoyancy, surface-tension and electromagnetic forces is considered. This study was done for Prandlt number Pr=0.02 and aspect ratio A=2. The transport equations for continuity, momentum and energy transfer are solved. The numerical results are reported for the effect of Marangoni and Hartmann numbers. A uniform magnetic field is applied in the axial direction. The system of equations governing the growth phenomenon has been treated in steady state). The axisymmetric Navier-Stokes and energy equations, with the Boussinesq approximation, have been discretised by means of finite volume procedure and the SIMPLER algorithm is used to treat the coupled velocity-pressure. The thermocapillary convection and the magnetic field effects on the flow and temperature field are discussed. A parametric study was carried out.

In the second part, the Navier-Stokes and energy equations are numerically solved to investigate two-dimensional convection (originating from the combined effect of buoyancy and surface tension forces) in liquid metals subjected to transverse magnetic field. In particular, a laterally heated horizontal cavity with aspect ratio (height/width =1) and Prandtl number Pr=0.015 is considered (typically associated with the horizontal Bridgman crystal growth process and commonly used for benchmarking purposes). The effect of a uniform magnetic field with different magnitudes and orientations on the stability of the two distinct convective solution branches (with a single-cell or two-cell pattern) of the steady state flows is investigated. The effects induced by increasing values of the Rayleigh and Hartmann numbers on the heat transfer rate are also discussed.

Chapter 3 reviews a decade of research on the use of the Navier-Stokes Equations to study the flow generated by different turbines impellers, conducted by their group at the University of Sfax. For this reason, computational fluid dynamics (CFD) methods are developed to predict the flow field in stirred tanks equipped by various turbines impellers type. The three-dimensional flow of a fluid is numerically analyzed using the Navier-Stokes equations in conjunction with standard k-ε turbulence model. The relevant governing equations are given, and alternative turbulence models are described. The computer method developed permits the numerical analyses of turbines with complex geometries.

Their main objective has been to study and compare the shape effects on the field flow and turbulence characteristics to choose the most effective agitation system for the mixing applications. In these applications, the effect on the local and global flow characteristics have been determined and compared. The numerical results such as flow patterns profile and power numbers variation on Reynolds number were compared to their experimental results and those obtained by other authors. The comparison proved a good agreement.

Chapter 4 is devoted to investigation of the region of instability of the Navier-Stokes equations. This region is of critical impotence for modeling turbulent flows. Stochastic approach to turbulence based upon the Stabilization Principle is introduced. Onset of turbulence is interpreted as loss of stability of solutions to the Navier-Stokes equations *in the class of differentiable functions*. A nonlinear version of the Liouville- Fokker-Planck equation with negative diffusion is proposed for describing postinstability motions of the Navier-Stokes equations. As a result, the flow velocities are represented by a set of stochastic invariants found from the Stabilization Principle. The theory is illustrated by examples.

A recently developed stabilized finite element method is generalised for the incompressible Navier Stokes equations for high Reynolds numbers and free surface flows problems. The proposed method starts by the use of a finite element variational multiscale (VMS) method, which consists in here of decomposition for both the velocity and the pressure fields into coarse/resolved scales and fine/unresolved scales. This choice of decomposition is shown to be favorable for simulating flows at high Reynolds number. However, the main challenge remains in solving turbulent two-phase flows which occurs in a wide range of real life problems, industrial processes including molten metal flow, sloshing in tanks, wave mechanics, flows around structure. A successful approach to deal with such flows, especially in the presence of turbulent behaviour is the use of a local convected level set method coupled to a Large Eddy Simulation method.

In Chapter 5 they assess the behaviour and accuracy of the proposed formulation in the simulation of three test cases. Results are compared with the literature and the experimental data and show that the present implementation is able to exhibit good stability for high Reynolds number flows using unstructured meshes.

The equations which govern the motion of an incompressible viscous fluid are precisely the Navier-Stokes equations. It is a known fact that the Navier-Stokes equations in general are nonlinear partial differential equations and few analytic solutions of these equations are reported in the existing literature.

In Chapter 6 the authors have considered unsteady Navier-Stokes equations for viscous, incompressible three (two) dimensional flow in cylindrical polar coordinates (polar coordinates). The Lie point symmetries for these equations have been obtained by utilizing a direct approach. Using the symmetries obtained via this approach, similarity reductions in order have been performed. A successive similarity reduction has reduced the governing partial differential equations to a system of ordinary deferential equations. The resulting differential equations have been solved analytically and then by process of back substitutions, the solutions of the original system have been obtained. Some interesting discussions and conclusions have been presented at the end of the chapter.

High order low diffusion numerical schemes with the capability of shock and contact discontinuities capturing is essential to study high speed flows, turbulence, acoustics, fluid-structural interactions, combustions, etc. Developing these schemes is numerically challenging. This paper introduces several recent studies of high order accuracy shock capturing numerical algorithms for Navier-Stokes equations. It includes implicit time marching algorithms, high order weighted essentially non-oscillatory(WENO) schemes for the inviscid fluxes, high order conservative schemes for the viscous terms, and preconditioning methods for solving unified compressible and incompressible flows at all speeds.

Chapter 7 is organized as follows. Section 1 gives the full Navier-Stokes equations in the Cartesian coordinates and the generalized computational coordinates, including the form of Reynolds averaged Navier-Stokes equations and spatially filtered Navier-Stokes equations for large eddy simulation. Section 2 presents the study on implicit time marching algorithms, a comparison of three different methods including the unfactored implicit Gauss-Seidel relaxation scheme, the lower-upper symmetric Gauss-Seidel method, and a new hybrid method. Section 3 describes an improvement for the fifth order WENO schemes near shock points, a generalized finite compact scheme for shock capturing, an improved seventh order WENO scheme. Section 4 introduces the high order conservative central difference schemes for viscous terms and applications. Section 5 depicts the preconditioning methods(Roe-type method and E-CUSP scheme) for solving unified compressible and incompressible flows. Each section has its independent sub-section of introduction, results, and conclusions.

Chapter 8 is expository and contains both recent and the latest results of the research devoted to investigation of the transfer process occurring under the fluid flow in inhomogeneous granular medium. When fluid flows around the particles of a granular medium in a longitudinal direction, pulsating components of the velocity occur in the transverse direction, which are ignored by Darcy's law. If the average flow velocity changes in a transverse direction, these pulsations transfer additional momenta from layer to layer, which leads to the occurrence of effective viscosity forces. It is assumed that the fluid flow in a porous space is described by the Navier–Stokes equations. Using the fundamental propositions of Reynolds' averaging theory and Prandtl's mixing path, the structure of the effective viscosity coefficient is determined and hypotheses are formulated which enable it to be assumed to be independent of the flow velocity. It is established by comparison with experimental data that the effective viscosity coefficient can exceed the viscosity coefficient of the flowing fluid by an order of magnitude. The equations of average motion are obtained, which in the case of an incompressible fluid have the form of the Navier–Stokes equations with body forces proportional to the velocity. It is shown that the solution of these equations can be found in boundary layer approximation. It is established that, in addition to the well-known dimensionless flow numbers, there is a new number So which characterizes the ratio of the Darcy porous drag forces to the effective viscosity forces. Structured packing of granular medium is shown to be an effective tool of elimination of velocity profile non-uniformity. It is shown that flow of a fluid in regularly packed granular medium is anisotropic and the medium permeability coefficients form a second-rank tensor. The values of these coefficients may vary in wide range what allows to control the field of velocities meeting certain given criteria of optimization and reduce flow porous drag by 5 – 8 times compared to irregular packing. The proposed equations are extended to the case of the flow of an aerated fluid. The solution of a new problem of the fluid flow in a plane channel with permeable walls is presented using three models: Darcy's law for an incompressible and aerated fluid, and also of an aerated fluid taking the effective viscosity into account. It is established that, for the same pressure drop, the maximum flow rate corresponds to Darcy's law. Compressibility leads to its reduction, but by simultaneously taking into account the compressibility and the effective viscosity one obtains minimum values of the flow rate. The effective viscosity and aeration of the fluid has a considerable effect on the flow parameters. By definition, a fluid is a substance that deforms continuously under the action of a shearing stress.

As explained in Chapter 9, the Navier-Stokes equations govern the motion of fluids in general, and are applicable to Newtonian as well as non-Newtonian fluids. Fluids are classified based on the rate at which they deform in response to an imposed shear stress. In Newtonian fluids there is a linear relation between the sear stress and the strain rate, whereas in non-Newtonian fluids this relation is non-linear. When the complete set of conservation equations governing the flow of a pure fluid, along with its equation of state, are used, the velocity, the pressure, the density, and the temperature (or, equivalently, the enthalpy or internal energy) are the unknown parameters. In the case of a three – dimensional flow, for example, the pressure, three velocity components, the density and temperature constitute the six unknowns in a set of six equations that include the three components of the momentum conservation equation, the mass continuity, the energy conservation, and the fluid equation of state.

The storage of thermal energy as latent heat of fusion presents advantages over sensible heat due to its high storage density and to the isothermal nature of the storage process at melting temperature. Latent heat thermal energy storage systems find application in space craft, solar energy system, greenhouses, heating and cooling of buildings and so on.

The use of phase change material (PCM) for thermal storage in buildings was one of the first applications considered for such materials along with typical sensible heat storage reservoirs and enclosures. Their general objective is to study the thermal behaviour of phase change materials so as to incorporate "bricks" of such materials into passive solar components.

Chapter 10 presents a formulation and an implementation of a numerical method in order to optimize the design of solar passive walls involving phase change materials (PCMs). Particularly, it explores numerically the melting and solidification processes of a PCM. Indeed, the fusion and the solidification of the commercially available PCM (hydrated salts, designed to melt at *27 °C*), engineered by Cristopia™ are investigated.

The mathematical model for the numerical simulations is based on the enthalpy-porosity method, which is traditionally used to track the motion of the liquid-solid front and to obtain the temperature and velocity profiles in the liquid-phase. The governing equations are discretized on a fixed grid by means of a finite volume method. Numerical predictions are obtained with custom one-dimensional and two-dimensional Fortran codes. Several simulation runs were conducted to provide the heat storage during the melting process, as well as the heat recovery during the solidification process. Moreover, these results were compared to experimental data. Numerical simulations provided in 2D important features such as the streamlines in the liquid phase, and the temperature profiles inside the cavity along with the time evolution of the melted fraction.

In: Navier-Stokes Equations
Editor: R. Younsi
ISBN: 978-1-61324-590-3

Chapter 1

PARABOLIC AND ELLIPTIC PARTIAL DIFFERENCE EQUATIONS: TOWARDS A DISCRETE SOLUTION OF NAVIER-STOKES EQUATION

***Rubén A. Hidalgo*[1,*] *and Mauricio Godoy Molina*[2,†]**
[1]Departamento de Matemática,
Universidad Técnica Federico Santa María, Chile
[2]Department of Mathematics, University of Bergen, Norway

Abstract

In the present chapter we survey some results about partial difference equations (PdEs) in weighted graphs obtained previously by the authors, in the case of finite graphs and countable graphs of finite degree. We discuss the existence of solution of elliptic PdEs via an associated semilinear matrix equation and under growth conditions of the forcing term. As applications of these techniques, we study the discrete analogues of some classical elliptic partial differential equations such as Matukuma, Helmholtz and Lane-Emden-Fowler. Finally we discuss our results concerning the discrete Navier-Stokes equation and give some explicit solutions for concrete weighted graphs and discuss possible graphs that simplify the applications in the planar case.

Keywords: Graphs, Partial difference equations, Nonlinear elliptic equations, Laplacian, Navier-Stokes.

AMS Subject Classification: 05C12, 39A12, 35J05.

1. Introduction

Partial differential equations play a central role in the description of physical phenomena. Particularly important examples of this are Laplace equation, Poisson equation, heat and wave equations and of course Navier-Stokes equation. The problem of existence of

*E-mail address: ruben.hidalgo@usm.cl
†E-mail address: mauricio.godoy@math.uib.no

solutions and understanding the behavior of Navier-Stokes equation is nowadays one of the most studied problems in the theory of non-linear partial differential equations. As it is well-known, the smoothness and global definition of the Navier-Stokes equation in dimension 3 is one of the Millenum Problems and this is still far from being solved in full generality.

For the sake of completeness, let us recall that the Navier-Stokes equation is posed on a region $\Omega \subset \mathbb{R}^n$, for $n \geq 2$,

$$\rho \left(\frac{\partial \nu}{\partial t} + \nu \cdot \nabla \nu \right) = -\nabla p + \mu \Delta \nu + f \tag{1}$$

where, if $\nu = (\nu_1, \ldots, \nu_n)$, then

$$\nu \cdot \nabla \nu := (\nu \cdot \nabla \nu_1, \ldots, \nu \cdot \nabla \nu_n)$$

$$\Delta \nu := (\Delta \nu_1, \ldots, \Delta \nu_n).$$

In the above, the parameter μ represents the dynamic viscosity, $\rho = \rho(x,t)$ represents the density of the fluid, i.e. is a real nonnegative function depending on the time variable t and space variable x, $\nu = \nu(x,t)$ is a vector function representing the incompressible fluid speed, $p = p(x,t)$ is a real-valued function which represents the pressure and $f(x,t)$ plays the role of an external force. The factor $\nu \cdot \nabla \nu$ is called the convection acceleration and $\mu \Delta \nu$ the fluid viscosity. The fluid is called Newtonian if it conserves its mass, that is, it satisfies the Newton's conservation equation

$$\nabla \cdot \nu = \sum_{j=1}^{n} \frac{\partial \nu_j}{\partial x_j} = 0 \tag{2}$$

where $x = (x_1, \ldots, x_n)$. A good source for generalities and results on Navier-Stokes equations is [14].

In [12] we proposed a way of discretizing Navier-Stokes equations by using appropriately constructed graphs, in this way we were able to find discrete solutions in some particular cases. The general idea behind discretization of partial differential equations (PDEs) is to find numerical solutions which are sufficiently good approximations of the original solutions to the original PDE. This discrete versions of partial differential equations are usually referred as partial difference equations (PdEs).

Historically one can trace back the use of partial difference equations to the seminal paper by R. Courant, K. Friederichs and H. Lewy [8]. In that article, the authors introduce the finite differences method as a convenient way of dealing numerically with partial differential equations. From the introduction of the aforementioned paper we can see the insight of their philosophy: "Problems involving the classical linear partial differential equations of mathematical physics can be reduced to algebraic ones of a very much simpler structure by replacing the differentials by difference quotients on some (say rectilinear) mesh." In recent papers, e.g [10, 11, 13], the study of partial difference equations has appeared as a subject on its own, dealing with problems of existence and qualitative behavior of solutions.

In the present chapter we survey some results about PdEs in weighted graphs obtained previously by the authors, in the case of finite graphs and countable graphs of finite degree. We discuss the existence of solution of elliptic PdEs via an associated semilinear matrix equation and under growth conditions of the forcing term. As applications of these techniques, we study the discrete analogues of some classical elliptic partial differential equations such as Matukuma, Helmholtz and Lane-Emden-Fowler. Finally we discuss our results concerning the discrete Navier-Stokes equation and give some explicit solutions for concrete weighted graphs.

This chapter is organized as follows. In Section 2. we review some of the necessary preliminaries of discrete calculus and set the notations we will be using throughout the chapter. In Section 3. we will review some of the known results for elliptic PdEs. In Section 4. we discuss a novel application of the techniques to study the Lane-Emden-Fowler equation. Finally in Section 5. we review the discretization of Navier-Stokes equation proposed by us and see some of its applications and conclude in Section 6. by briefly discussing which types of graphs simplify the task of solving explicitly the proposed discretization.

2. Preliminaries

We will denote a graph as a pair $\mathcal{G} = (V, E)$, where V is the set of vertices and E the set of edges. Associated to any graph $\mathcal{G} = (V, E)$ is the real vector space $C^0(\mathcal{G})$ formed by all real functions $\mu : V \to \mathbb{R}$. A particularly interesting vector subspace is the Hilbert space $L^2(\mathcal{G})$ formed by those real functions $\mu : V \to \mathbb{R}$ so that $\sum_{v\in V} \mu(v)^2 < \infty$, with inner product defined by

$$\langle \mu, \nu \rangle = \sum_{v\in V} \mu(v)\nu(v).$$

A derivation on $\mathcal{G} = (V, E)$ is a linear operator $D : C^0(\mathcal{G}) \to C^0(\mathcal{G})$. In many problems, one needs to restrict to derivations on $L^2(\mathcal{G})$, that is, to linear operators $D : L^2(\mathcal{G}) \to L^2(\mathcal{G})$.

Given any function $f : V \times \mathbb{R} \to \mathbb{R}$ and any derivation D on the graph, we have associated a *discrete partial difference equation* on the graph given by

$$D\mu = f(v, \mu) \tag{3}$$

A solution $\mu \in C^0(\mathcal{G})$ of the above partial difference equation is so that for every $v \in V$ it holds that $D\mu(v) = f(v, \mu(v))$. Sufficient conditions for the existence of solutions of this kind of equations may be found in [10, 11, 13].

A metric graph is a pair $(\mathcal{G}, d)$, where $\mathcal{G} = (V, E)$ is a graph and $d : E \to (0, +\infty)$ a function, called a metric on $\mathcal{G}$. Metrics give rise in a natural way to derivations, as we will see in Subsection2.1.

A graph $\mathcal{G} = (V, E)$ is called *admissible* if it is connected, simple (that is, it contains no loops as edges) and its vertex set V being either finite or countable infinite and locally finite (that is, each vertex has finite degree).

Remark 1. Let $(\mathcal{G}, d)$ be a metric graph so that $\mathcal{G} = (V, E)$ is an admissible graph. As consequence of the connectivity, for each pair of vertex, say $v, w \in V$, there is a path between them, that is a finite collection of edges, say $\{v_1 = v, v_2\}, \{v_2, v_3\}, \ldots, \{v_m, v_{m+1} = w\}$. The length of such a path is defined as the sum $\sum_{j=1}^{m} d(\{v_j, v_{j+1}\})$. The requirement that each vertex has finite degree ensures that this provides a metric on the vertex set V by taking the infimum of the lengths of paths connecting two given vertices. In particular, if $\emptyset \neq W \subset V$ is finite, we may consider the distance from $v \in V$ to W, denoted by $d(v, W)$.

We will restrict our attention to admissible graphs only.

2.1. Discrete Laplace Operator

Let us consider a metric graph $(\mathcal{G}, d)$ where $\mathcal{G} = (V, E)$ is admissible. Associated to this pair there is a classical derivation $\Delta_2 : C^0(\mathcal{G}) \to C^0(\mathcal{G})$, called the *discrete Laplace operator*, defined by

$$\Delta_2 \mu(v) = \sum_{w \in N(v)} \frac{\mu(v) - \mu(w)}{d^2(\{v, w\})}, \tag{4}$$

where $N(v)$ denotes the collection of vertices $w \in V$ with the property that $\{v, w\} \in E$; called the neighborhood of v. Note that $v \notin N(v)$ as we are assuming the graph to be simple and that $N(v)$ is finite.

The discrete Laplace operator Δ_2 is symmetric and positive semi-definite on $C^0(\mathcal{G})$. In some situations, for instance if the graph is finite, it happens that $\Delta_2 \mu \in L^2(\mathcal{G})$ for each $\mu \in L^2(\mathcal{G})$. Generalities about the discrete Laplace operator and its relations to the topology of the graph, matrix theory and spectral geometry can be found in [2, 3, 4, 5, 6, 7].

The following simple fact will be used in order to obtain strictly positive solutions once we have non-trivial non-negative ones for the problem we study in Section 4..

Lemma 2. *Let $(\mathcal{G}, d)$ be a metric graph, with $\mathcal{G} = (V, E)$ admissible, and let $f : \mathbb{R} \to \mathbb{R}$ be a function so that $f(t) \geq 0$ if $t \geq 0$ and $f(0) = 0$. If $\mu \in C^0(\mathcal{G})$ is a solution of the partial difference equation $\Delta_2 \mu = f(\mu)$, so that $\mu(v) \geq 0$ for every $v \in V$ and there is some $v_0 \in V$ so that $\mu(v_0) = 0$, then $\mu \equiv 0$.*

Proof. Let $N(v_0) = \{v_1, \ldots, v_m\}$. Then

$$0 = f(\mu(v_0)) = \Delta_2 \mu(v_0) = \sum_{j=1}^{m} \frac{\mu(v_0) - \mu(v_j)}{d^2(\{v_0, v_j\})} = -\sum_{j=1}^{m} \frac{\mu(v_j)}{d^2(\{v_0, v_j\})}.$$

As $\dfrac{\mu(v_j)}{d^2(\{v_0, v_j\})} \geq 0$, then we must have $\mu(v_j) = 0$ for $j = 1, \ldots, m$. Now, as the graph is connected, it follows that $\mu \equiv 0$. □

2.2. Admissible Metric Graphs

If the graph is finite, then $L^2(\mathcal{G}) = C^0(\mathcal{G})$. Clearly, if the graph is infinite, then this is not longer true. In this way, a solution of a partial difference equation as in (3) may not belong to $L^2(\mathcal{G})$ in the infinite situation. We say that d is an *admissible metric* if $\Delta_2(L^2(\mathcal{G})) < L^2(\mathcal{G})$.

An admissible metric graph is a pair $(\mathcal{G}, d)$, where $\mathcal{G}$ is admissible graph and d is admissible metric.

2.3. Directional Derivatives

A *vector field* on the graph $\mathcal{G}$ is a function $X : V \to V$, so that $X(v) \in N(v) \cup \{v\}$. If $v \in V$ and $w \in N(v)$, then an example of a vector field is the following

$$X_{v,w}(\tau) = \begin{cases} \tau, & \tau \neq v \\ w, & \tau = v. \end{cases} \tag{5}$$

If $w \in N(v) \cup \{v\}$, for some $v \in V$ fixed, and $\mu \in C^0(\mathcal{G})$, then the *directional derivative* of μ in v in the direction of w is defined by

$$\partial_w \mu(v) = \begin{cases} \dfrac{\mu(w) - \mu(v)}{d(\{v, w\})}, & v \neq w \\ 0, & v = w. \end{cases} \tag{6}$$

Each vector field X defines a derivation D_X on $\mathcal{G}$ by the rule $D_X(\mu)(v) = \partial_{X(v)}\mu(v)$, for $v \in V$ and $\mu \in C^0(\mathcal{G})$. For instance, $D_{X_{v,w}}\mu(\tau) = 0$ if $\tau \neq v$ and $D_{X_{v,w}}\mu(v) = \partial_w \mu(v)$.

3. Existence of Solutions of PdEs

3.1. Semilinear Equations on ℓ^2

A semilinear equation is an equation on a separable Hilbert space $\mathcal{H}$ of the form

$$Qx = f(x), \tag{7}$$

where $Q : \mathcal{H} \to \mathcal{H}$ is a linear operator, and $f : \mathcal{H} \to \mathcal{H}$ is a map such that $f(0) = 0$. The same problem can be considered in finite dimensional spaces, and it is of course contained in the analysis of this general case.

We want to find conditions for the existence of solutions of the problem (7) assuming

$$\mathcal{H} = \ell^2 = \{x = (x_1, x_2, \ldots) : x_j \in \mathbb{R}, \sum_{j=1}^{\infty} x_j^2 < \infty\}, \quad < x, y >= \sum_{j=1}^{\infty} x_j y_j,$$

and $f(x) := (f_1(x_1), f_2(x_2), \ldots)$ where $f_j : \mathbb{R} \to \mathbb{R}$ satisfy $f_j(0) = 0$ and, for every $x = (x_1, x_2, \ldots) \in \ell^2$, it holds that $f(x) := (f_1(x_1), f_2(x_2), \ldots) \in \ell^2$.

Theorem 3. *Let $Q : \ell^2 \to \ell^2$ be a symmetric and positive semi-definite linear operator and $\{0\} \neq W < \ell^2$ be an invariant subspace under Q. Assume there are continuous maps $f_j : \mathbb{R} \to \mathbb{R}$, for $j = 1, 2, \ldots$, so that they all satisfy the following conditions*

(1) *$f_j(0) = 0$;*

(2) *either,*

(2.1) *for all j, f_j is locally decreasing at 0, or*

(2.2) *for all j, f_j is locally increasing at 0 and, for small $|t|$, it holds $\int_0^t f_j(s)\, ds < t^2$;*

(3) $\lim_{t\to+\infty} \frac{f_j(t)}{t} = +\infty$, *and*

(4) $\left|\int_0^{x_j} f_j(t)\, dt\right| \leq x_j^2$, *for every* $x = (x_1, \ldots) \in \ell^2$,

then the semilinear problem (7) *has non-trivial solutions in* W. *Moreover, if* G *is a group of orthogonal linear automorphisms of* ℓ^2 *so that each transformation in* G *commutes with* Q, *then the semilinear problem* (7) *has non-trivial solutions invariant under the action of* G.

The proof of Theorem 3 and a broad discussion about its consequences in the study of PdEs can be found in [11]. Theorem 3 generalizes results in [13, 10] to the infinite dimensional weighted setting.

3.2. Applications to PdEs

A first problem which is natural to study in this context is the discretization of the Dirichlet problem. In this context, the following result holds.

Corollary 4 (Dirichlet's problem). *Let* $\mathcal{G} = (V, E)$ *be an finite degree simple graph with vertex set* V *either finite or countable infinite and* $W < V$ *so that* $W \neq \emptyset$ *and* $V - W \neq \emptyset$. *Let* $F : (V - W) \times \mathbb{R} \to \mathbb{R}$ *be some continuous function in the real variable.*

(1) *If* W *is finite, then write*

$$V = \{v_{-r+1}, v_{-r+2}, \ldots\}, \quad W = \{v_{-r+1}, \ldots, v_0\} \subset V.$$

Consider real values $a_0, a_{-1}, \ldots, a_{1-r} \in \mathbb{R}$ *and the (finite) column vector* $a = {}^t[a_{1-r} \cdots a_0]$.

(2) *If* W *is infinite, then write*

$$V = \{\ldots, v_{-2}, v_{-1}, v_0, v_1, \ldots.\}, \quad W = \{\ldots, v_{-2}, v_{-1}, v_0\} \subset V.$$

Consider real values $a_j \in \mathbb{R}$, *where* $j \leq 0$, *and the infinite column vector* $a = {}^t[\cdots a_{-1}\ a_0]$.

Assume, with this enumeration of the vertex set V, *that the (infinite size) matrix representation of* D *is*

$$J = \begin{bmatrix} R & {}^tU \\ U & S \end{bmatrix}$$

where R *is the matrix corresponding to* W *(of size* $r \times r$ *in* (1) *and of infinite size in* (2)*). Set* $b = Ua$ *and,* $f_j(t) = F(v_j, t)$. *If each* $f_j : \mathbb{R} \to \mathbb{R}$ *satisfies the conditions* $f_j(0) = b_j$, *for all* j, *and conditions* (2), (3) *and, if* V *is infinite, also condition* (4) *of Theorem* 3, *then the Dirichlet problem*

$$\begin{cases} D\mu = F(v, \mu), & v \in V - W \\ \mu(v_j) = a_j, & v_j \in W \end{cases}$$

has solutions.

Proof. In this case, we need to solve $Sy = g(y)$, where $g(y) = f(y) - Ua$, $f(y)$ is the column vector whose j-th coordinate is $f_j(y_j)$ and S is symmetric and positive semi-definite. The existence is now consequence of Theorem 3 using g instead of f. □

Another consequence of Theorem 3, is the existence of nontrivial solutions of discrete Helmholtz equations. Let $\mathcal{G} = (V, E)$, with V either finite or countable infinite, and a metric $d : E \to (0, +\infty)$. Let $W \subset V$, $W \neq \emptyset$, $V - W \neq \emptyset$, let $\sigma : W \to V - W$ be some function and let real values b_w $(w \in W)$, a_v $(v \in V - W)$ be no necessarily different ones. The classical Helmholtz-Newmann's problem may be adapted to the above metric graph as

$$\left\{ \begin{array}{rcll} \Delta_2 \mu(v) & = & k^2 \mu(v) + a_v, & v \in V - W \\ \mu(w) - \mu(\sigma(w)) & = & b_w, & w \in W \end{array} \right. \tag{8}$$

Let us write the matrix J_2 (representing Δ_2) as

$$J_2 = \begin{bmatrix} A & {}^tC \\ C & D \end{bmatrix}$$

where A and D are positive semi-definite symmetric matrices of sizes $|W|$ and $|V - W|$, respectively. The function σ may be though as a function

$$\sigma : W \to V.$$

Let us set $M_\sigma = [m_{wv}]$ (where $w \in W$ and $v \in V$) by

$$m_{wv} = \left\{ \begin{array}{ll} -1, & \sigma(w) = v \\ 0, & \sigma(w) \neq v \end{array} \right. .$$

Theorem 5. *Helmholtz-Newmann's problem* (8) *has a unique solution if and only if* k^2 *is not an eigenvalue of* $D - CM_\sigma$*. If* k^2 *is an eigenvalue of* $D - CM_\sigma$*, then* (8) *may not have solution and, if it has at least one, then it must have infinitely many solutions.*

Also, as a consequence of Theorem 3, we have the existence of nontrivial solutions of discrete Matukuma-type equations. In fact, in the case of Matukuma-type equations we can even discuss the behavior of their solutions.

Theorem 6. *The equation*

$$-\Delta_2 \mu + f(v)\mu^p = 0, \; v \in V,$$

has non-trivial solutions and all of them must be a sign changing solution, in particular, there is no non-negative nor non-positive non-trivial solution.

The proofs of Theorems 5 and 6 can be found in [11].

4. Discrete Lane-Emden-Fowler Equation

Let $D \subset \mathbb{R}^n$ is a bounded domain, $\Omega = \mathbb{R}^n - D$, $n \geq 3$ and $p > p_0 = \frac{n+2}{n-2}$. If $\Omega \subset \mathbb{R}^n$ is a domain with boundary $\partial\Omega \subset \mathbb{R}^n$, then the Lane-Emden-Fowler equation is given as

$$\begin{cases} \Delta\mu + \mu^p = 0, & \text{in } \Omega \\ \mu > 0, & \text{in } \Omega \\ \mu = 0, & \text{in } \partial\Omega \\ \lim\limits_{|x|\to\infty} \mu(x) = 0 & \end{cases} \tag{9}$$

The existence of solutions of equation (9) was stated in [9].

Next, we adapt Lane-Emden-Fowler equation (9) to the case of metric graphs and study the existence of solutions of such problems (see Section 4.4.). This adaptation may be of interest in studying either Lane-Emden-Fowler equation in other geometries or for the numerical approach of solutions of (9).

4.1. Adaptation of Lane-Emden-Fowler Equation

Let us consider an admissible graph $\mathcal{G} = (V, E)$ and any metric $d : E \to (0, +\infty)$ (non necessarily admissible one for the moment).

Let us fix a finite non-empty subset $W \subset V$, which we identify as the finite boundary vertex set of the graph. With the above definitions and notations, we may adapt equation (9) to the admissible graph $\mathcal{G}$ (and the metric d) as the problem

$$\begin{cases} \Delta_2\mu(v) = \mu^p(v), & v \in V - W \\ \mu(v) > 0, & v \in V - W \\ \mu(w) = 0, & w \in W \\ \lim\limits_{d(v,W)\to+\infty} \mu(v) = 0 & \end{cases} \tag{10}$$

where in the above the condition $\lim\limits_{d(v,W)\to+\infty} \mu(v) = 0$ only is considered for infinite graphs.

The equation (10) is an adaptation of Lane-Emden-Fowler equation (9) to the pair $(\mathcal{G}, d)$.

4.2. Case of Admissible Metric Graphs

Let us assume $(\mathcal{G} = (V, E), d)$ is an admissible metric graph. In this case, the discrete Laplace operator Δ_2 defines a operator $\Delta_2 : L^2(\mathcal{G}) \to L^2(\mathcal{G})$, and we may search for solutions of (10) inside $L^2(\mathcal{G})$, that is, we may consider the *discrete Lane-Emden-Fowler equation*

$$\begin{cases} \Delta_2\mu(v) = \mu^p(v), & v \in V - W \\ \mu(v) > 0, & v \in V - W \\ \mu(w) = 0, & w \in W \\ \mu \in L^2(\mathcal{G}) & \end{cases} \tag{11}$$

As consequences of the results in [10, 11, 13] the following existence results hold.

Theorem 7. *Each of the two problem*

$$\begin{cases} \Delta_2\mu(v) = \mu^p(v), & v \in V - W \\ \mu(w) = 0, & w \in W \end{cases} \tag{12}$$

or

$$\begin{cases} \Delta_2\mu(v) = |\mu(v)|^p, & v \in V - W \\ \mu(w) = 0, & w \in W \end{cases} \tag{13}$$

has non-trivial solutions in $L^2(\mathcal{G})$.

The existence is a consequence of Mountain Pass Lemma [1]. Unfortunately, the solutions of (12) and (13) may be sign-changing and the problem (10) search for positive solutions.

4.3. Matrix Equations

As usual, we set $t^q = |t|^{q-1}t$, for $t \in \mathbb{R}$ and $q > 1$. Let us denote the elements of l^2 as column vectors of real numbers. We identify $\mathbb{R}^k$ with the subspace of l^2 consisting of those $x \in l^2$ whose coordinates $x_j = 0$ if $j \geq k+1$.

Let us consider an admissible metric graph $(\mathcal{G}, d)$, where $\mathcal{G} = (V, E)$. Let $\emptyset \neq W \subset V$ be a finite set, let $p > 1$ and let $r = |W|$. Consider the associated discrete Lane-Emden-Fowler equation (11). Set

$$\begin{cases} V &= \{v_{-r+1}, \ldots, v_0, v_1, \ldots\} \\ W &= \{v_{-r+1}, \ldots, v_0\} \end{cases}$$

The discrete Laplace operator Δ_2 is represented by an symmetric positive semi-definite matrix Q (of infinite size if and only if V is infinite)

$$Q = \begin{bmatrix} R & {}^tU \\ U & S \end{bmatrix}$$

where R is a symmetric matrix of size $r \times r$. As Q is positive semi-definite, the same holds for S. As $\mathcal{G}$ is connected, it follows that $U \neq 0$.

Some of the basic properties of the Laplacian matrix Q are the following:

1. the coefficients of any row of Q are equal to zero, by the exception of a finite number of them;

2. the sum of the coefficients of any row of Q is equal to zero;

3. the coefficients on the diagonal of Q are positive and outside its diagonal are less or equal to zero.

It follows from the above properties on Q that (i) the coefficients of any row of S are equal to zero, by the exception of a finite number of them, (ii) the sum of the coefficients of any row of S is bigger or equal to zero and (iii) the coefficients on the diagonal of S are positive and outside its diagonal are less or equal to zero.

As $U \neq 0$, it also follows that at least for one row of S the sum of its coefficients is strictly positive. As U has only finite number of columns and each column of Q has only finite number of coefficients different from zero, there is some index $I > 1$ so that the sum of the coefficients of the rows of S (starting at the row I) is equal to zero.

Note that our assumptions on admissibility of d assert that Qx, $Sx \in l^2$ for every $x \in l^2$. In particular, S defines a positive semi-definite quadratic form on l^2

$$\langle x, y \rangle_S = {}^t x S y, \quad x, y \in l^2.$$

Remark 8. In many cases one may have that the graph whose vertices are $V - W$ and its edges are those edges in E which do not contain a vertex in W is still connected, but this property it is not assumed.

4.3.1. Infinite Graphs

If $z = {}^t[z_1 \cdots z_r]$, $x = {}^t[x_1\, x_2 \cdots] \in l^2$, and $x^p := {}^t[x_1^p\, x_2^p \cdots]$, then (11) is equivalent to solve the following matrix problem in the case of infinite graphs

$$\begin{cases} \begin{bmatrix} I & 0 \\ U & S \end{bmatrix} \begin{bmatrix} z \\ x \end{bmatrix} = \begin{bmatrix} 0 \\ x^p \end{bmatrix} \\ \\ x_j > 0, \ j \geq 1, \end{cases}$$

which is equivalent to have $z = 0$ and to solve the problem

$$\begin{cases} Sx = x^p \\ \\ x_j > 0, \ j \geq 1. \end{cases}$$

4.3.2. Finite Graphs

If $z = {}^t[z_1 \cdots z_r]$, $x = {}^t[x_1 \cdots x_m] \in \mathbb{R}^m$, and $x^p := {}^t[x_1^p \cdots x_m^p]$, then (11) is equivalent to solve the following matrix problem in the case of finite graphs (with $|V - W| = m$)

$$\begin{cases} \begin{bmatrix} I & 0 \\ U & S \end{bmatrix} \begin{bmatrix} z \\ x \end{bmatrix} = \begin{bmatrix} 0 \\ x^p \end{bmatrix} \\ \\ x_j > 0, \ j \in \{1, \ldots, m\} \end{cases}$$

which is equivalent to have $z = 0$ and to solve the problem

$$\begin{cases} Sx = x^p \\ \\ x_j > 0, \ j \in \{1, \ldots, m\}. \end{cases}$$

Let us note that in the finite case, the symmetric matrix S is in fact positive definite as seen in the following.

Lemma 9. *Let $\mathcal{G} = (V, E)$ be a finite admissible graph and $d : E \longrightarrow (0, +\infty)$. Set*

$$V = \{v_1, \ldots, v_r, v_{r+1}, \ldots, v_{r+m}\}$$

where $m, r > 0$. Assume the matrix associated to Δ_2 is

$$Q = \begin{bmatrix} R & {}^tU \\ U & S \end{bmatrix}$$

where R and S are square matrices of sizes $r \times r$ y $m \times m$, respectively. Then, S is invertible.

Proof. Let $\mathcal{L}$ be the real vector space of functions $\mu : V \longrightarrow \mathbb{R}$ which are discrete harmonics in $\{v_{r+1}, \ldots, v_{r+m}\}$ and $\mu(v_1) = \cdots = \mu(v_r) = 0$. As at a maximum or a minimum of μ it cannot be discrete harmonic except if μ is locally constant, the connectivity of the graph asserts that $\mathcal{L} = \{0\}$. If we identify μ with the column vector ${}^t[x\ y]$, where $x_j = \mu(v_j)$, $j = 1, \ldots, r$, and $y_k = \mu(v_{r+k})$, $k = 1, \ldots, m$, it follows that $\mathcal{L}$ corresponds to the solution space of

$$\begin{bmatrix} I & 0 \\ U & S \end{bmatrix} \begin{bmatrix} x \\ y \end{bmatrix} = 0$$

As, by the above, such space is the trivial one, it follows that the previous matrix should be invertible, in particular, S should be invertible. □

4.4. Positive Solutions

Let S be as in Subsection 4.3.. If S is infinite, we consider the functional

$$\begin{aligned} H \ : \ l^2 &\longrightarrow \mathbb{R} \\ x &\longmapsto \frac{1}{2}{}^txSx - \sum_{j=1}^{\infty} \int_0^{x_j} t^p dt = \frac{1}{2}{}^txSx - \frac{1}{p+1}\sum_{j=1}^{\infty} |x_j|^{p+1} \end{aligned} \tag{14}$$

and if S is of finite, say $m \times m$, we consider

$$\begin{aligned} H \ : \ \mathbb{R}^m &\longrightarrow \mathbb{R} \\ x &\longmapsto \frac{1}{2}{}^txSx - \sum_{j=1}^{m} \int_0^{x_j} t^p dt = \frac{1}{2}{}^txSx - \frac{1}{p+1}\sum_{j=1}^{m} |x_j|^{p+1} \end{aligned} \tag{15}$$

As, for $x \neq 0$, it holds that

$$\nabla H(x) = Sx - x^p, \tag{16}$$

the non-zero critical points of H corresponds to the non-trivial solutions of the equation

$$Sx = x^p \tag{17}$$

The existence of non-zero critical points of H is a consequence of the Mountain Pass Lemma [1].

As consequence of Lemma 2, in order to ensure the existence of positive solutions of the previous problem (then existence of solutions of (4.3.1.) and so of (11)) we only need to find non-trivial non-negative solutions. We will proceed as in [13] in order to find non-trivial non-negative solutions.

4.5. The Nehari Variety

Let us start with a natural object associated to elliptic equations, called the Nehari variety, but adapted to the case of partial difference equations on graphs as done in [13] for the case of constant metric and in [10] for any metric.

The Nehari variety S_H is defined, in case (14), by

$$\begin{aligned} S_H &= \{x \in l^2 - \{0\} : \langle \nabla H(x), x \rangle = 0\} \\ &= \left\{ x \in l^2 - \{0\} : {}^t x S x = \sum_{j=1}^{\infty} x_j^{p+1} \right\}, \end{aligned} \tag{18}$$

where $x = {}^t[x_1, \ldots] \in l^2$; and, in case (15), by

$$\begin{aligned} S_H &= \{x \in \mathbb{R}^m - \{0\} : \langle \nabla H(x), x \rangle = 0\} \\ &= \left\{ x \in \mathbb{R}^m - \{0\} : {}^t x S x = \sum_{j=1}^{m} x_j^{p+1} \right\}, \end{aligned} \tag{19}$$

where $x = {}^t[x_1, \ldots, x_m] \in \mathbb{R}^m$.

By the definition of S_H, all non-zero critical points of H belong to S_H.

Proposition 10. *The Nehari variety S_H is a differentiable submanifold of co-dimension* 1.

Proof. We proceed to prove in the case of l^2, but the arguments for $\mathbb{R}^m$ are exactly the same.

If we consider $G : l^2 \to \mathbb{R}$, where

$$G(x) = \sum_{j=1}^{\infty} x_j \frac{\partial H}{\partial x_j}(x), \tag{20}$$

then $S_H \subset G^{-1}(0)$.

Next, we proceed to observe that 0 is a regular value for G, obtaining that S_H is in fact a submanifold of l^2 of co-dimension 1.

Let us assume there is a point $\widehat{x} \in S_H$ for which $\nabla G(\widehat{x}) = 0$. Set $\widehat{x} = {}^t[x_1\, x_2 \cdots]$.

As

$$\frac{\partial G}{\partial x_k}(\widehat{x}) = \frac{\partial H}{\partial x_k}(\widehat{x}) + \sum_{j=1}^{\infty} x_j \frac{\partial^2 H}{\partial x_k \partial x_j}(\widehat{x}),$$

the condition

$$\frac{\partial G}{\partial x_k}(\widehat{x}) = 0, \quad \forall k \geq 1,$$

ensures

$$0 = \sum_{k=1}^{\infty} x_k \frac{\partial G}{\partial x_k}(\widehat{x}) = \sum_{k=1}^{\infty} x_k \frac{\partial H}{\partial x_k}(\widehat{x}) + \sum_{k=1}^{\infty} x_k \sum_{j=1}^{\infty} x_j \frac{\partial^2 H}{\partial x_k \partial x_j}(\widehat{x}),$$

or equivalently

$$\sum_{k=1}^{\infty} x_k \frac{\partial H}{\partial x_k}(\widehat{x}) + {}^t\widehat{x}\; Hess(H)(\widehat{x})\; \widehat{x},$$

where

$$Hess(H)(\widehat{x}) = S - p \begin{bmatrix} x_1^{p-1} & 0 & 0 & \cdots \\ 0 & x_2^{p-1} & 0 & \cdots \\ \vdots & \vdots & \vdots & \ddots \end{bmatrix}$$

As $\widehat{x} \in S_H$, then all the above ensure that

$${}^t\widehat{x}\; Hess(H)(\widehat{x})\; \widehat{x} = 0.$$

Since

$${}^t\widehat{x}\; Hess(H)(\widehat{x})\; \widehat{x} = {}^t\widehat{x}S\widehat{x} - p\sum_{j=1}^{\infty} x_j^{p+1},$$

and $p > 1$, it holds

$$0 = {}^t\widehat{x}S\widehat{x} - p\sum_{j=1}^{\infty} x_j^{p+1} < {}^t\widehat{x}S\widehat{x} - \sum_{j=1}^{m} x_j^{p+1} = \langle \nabla H(\widehat{x}), \widehat{x}\rangle = 0,$$

a contradiction. □

Proposition 11. (i) *$\forall x \in l^2 - \{0\}$ (respectively $x \in \mathbb{R}^m - \{0\}$) there exists $\lambda_x > 0$ (unique) so that $\lambda_x x \in S_H$. Moreover, if $\mu \neq \lambda_x$, then $H(\mu x) < H(\lambda_x x)$.*

(ii) *Every non-trivial solution of* (17) *corresponds to a critical point of the restriction of H to S_H.*

Proof. We only provide the proof for the case of l^2, but it is similar for the case $\mathbb{R}^m$.

Let $x \in l^2 - \{0\}$ and set $U : (0, +\infty) \to \mathbb{R}$ by

$$U(t) = \langle \nabla H(tx), tx\rangle = t(t\,{}^t xSx - {}^t xt^p x^p).$$

We need to see the existence of a unique zero of U. As $t \neq 0$, it is enough to check the existence of a unique zero of $T : (0, +\infty) \to \mathbb{R}$, where

$$T(t) = \langle \nabla H(tx), x\rangle = t\,{}^t xSx - {}^t xt^p x^p = {}^t x(S(tx) - t^p x^p).$$

Since ${}^t xS(tx) = t\,{}^t xSx \geq 0$ grows linearly with t, one obtains that

$$\lim_{t\to+\infty} T(t) = -\infty.$$

As

$$\lim_{t\to 0} T(t) = 0$$

and $T(t) > 0$ for $t > 0$ close to 0, we obtain the existence of zeros for T. Now, if we define $R : (0, +\infty) \to \mathbb{R}$ by $R(t) = H(tx)$, then $R'(t) = T(t)$; so the zeros of T are exactly the critical points of R. On the other hand,

$$R''(t) = T'(t) = {}^t x S x - p \sum_{j=1}^{\infty} x_j^{p+1} t^{p-1} = \frac{1}{t^2}\left({}^t(tx) Hess(H)(tx)(tx)\right). \tag{21}$$

Clearly, as $p > 1$, we also have

$$R''(t) = T'(t) < {}^t x S x - \frac{1}{t} \sum_{j=1}^{\infty} x_j^{p+1} t^p.$$

If $\sigma > 0$ is zero of T, then $\sigma x \in S_H$ and

$$\begin{aligned} R''(\sigma) &= T'(\sigma) < {}^t x S x - \frac{1}{\sigma} \sum_{j=1}^{\infty} x_j^{p+1} \sigma^p \\ &= \frac{1}{\sigma^2} \left({}^t(\sigma x) S(\sigma x) - p \sum_{j=1}^{\infty} \sigma^p x_j^p \right) = 0, \end{aligned}$$

as $\sigma x \in S_H$.

The above asserts that each critical point of R is a local maximum, from which we obtain the uniqueness as desired, and that it is a global maximum for $H(tx)$. □

Remark 12. Fix $x \in l^2 - \{0\}$ (respectively, for $x \in \mathbb{R}^m - \{0\}$) and consider, as in the previous proof, $R_x(t) = H(tx)$, for $t > 0$. Let $\lambda_x > 0$ be the unique value so that $\lambda_x x \in S_H$ provided by the above Proposition. The last part of the proof above asserts that R_x is a strictly concave function with λ_x as its unique maximum. The concavity of the function R_x and the fact that $H(y) > 0$ for $y \neq 0$ close to 0 asserts that $R_x(t) > 0$ for $0 < t < \lambda$ (see Figure 1).

4.6. Existence of Positive Solutions

Let us start with the following simple fact.

Lemma 13. *Let $M = [m_{ij}]$ be so that $M = {}^t M$, $m_{ij} \leq 0$ for $i \neq j$ and $m_{jj} \geq 0$, and that for every $x \in l^2$ it holds that $Mx \in l^2$. Then,*

$${}^t x^+ M x^- = {}^t x^- M x^+ \geq 0,$$

where for $x = {}^t[x_1, \ldots] \in l^2$, we set $x_j^+ = Max\{0, x_j\}$ and $x_j^- = Min\{0, x_j\}$.

Proof. The finiteness of the ${}^t x^+ M x^-$ and ${}^t x^- M x^+$ are consequence of the fact that for every $x \in l^2$ it holds that $Mx \in l^2$. The equality ${}^t x^+ M x^- = {}^t x^- M x^+$ is a consequence of ${}^t M = M$. As

$${}^t x^+ M x^- = \sum_{r=1}^{\infty} \sum_{j=1}^{\infty} m_{rj}\, x_r^- x_j^+ \geq 0,$$

$x_j^+ x_j^- = 0$, $x_r^+ x_j^- \leq 0$ and $m_{rj} \leq 0$ for $r \neq j$, the non-negativity follows. □

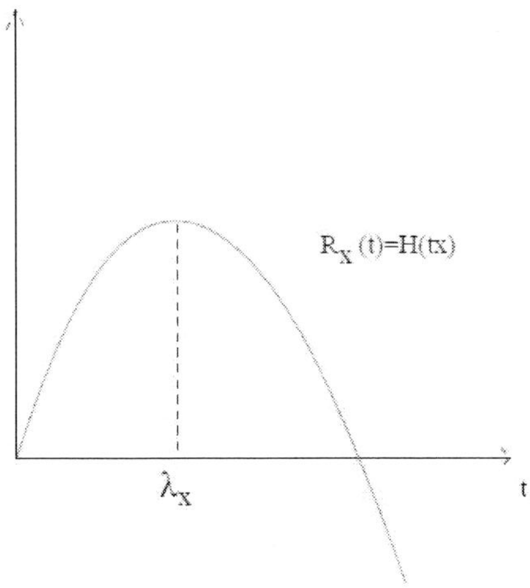

Figure 1.

Theorem 14. *If H has a global maximum or a local maximum on its Nehari variety S_H, then the discrete Lane-Emden-Fowler equation* (11) *has solution.*

Proof. Let us assume that z is a global minimum of H over S_H. By Proposition 11 such a point z provides a solution of (17). As $0 \notin S_H$, clearly $z \neq 0$. We proceed to see that all non-zero coordinates of z must have the same sign. By contradiction, let us assume z has coordinates with different sign and consider $z^+, z^- \in l^2 - \{0\}$. Part (i) of Proposition 11 ensures the existence of unique positive values, say λ_+, λ_-, so that $\lambda_+ z^+, \lambda_- z^- \in S_H$. We may assume $\lambda_+ \leq \lambda_-$. By the concavity of the function $R_z(t) = H(tz)$ we have that

$$H(z) \geq H(\lambda_+ z).$$

By Lemma 13

$$H(\lambda_+ z) \geq H(\lambda_+ z^+) + H(\lambda_+ z^-).$$

As $H(\lambda_+ z^-) > 0$, the above asserts

$$H(z) > H(\lambda_+ z^+),$$

a contradiction to the fact that z is global minimum of H over S_H. Now, we have the existence of a non-trivial solution whose non-zero coordinates have the same sign. As we are using the convention that $t^p = |t|^{p-1}t$, $w = -z$ is also a solution of (17). It follows the existence of a non-trivial non-negative solution of problem (17) so, by Lemma 2, a positive solution of (17). We have obtained a positive solution of our discrete problem (17), in particular, a solution of problem (11).

The above arguments can also be carry out using global maximum of H on the Nehari variety S_H. □

As consequence of Theorem 14, in order to obtain solutions of equation (11) we need to see the existence of global minima (or global maxima) of H over its Nehari variety S_H.

In the case that H is in (14) we define $G : l^2 \to \mathbb{R}$, as

$$G(x) = \sum_{j=1}^{\infty} x_j \frac{\partial H}{\partial x_j}(x), \tag{22}$$

and in the case that H is in (15) we define $G : \mathbb{R}^m \to \mathbb{R}$, as

$$G(x) = \sum_{j=1}^{m} x_j \frac{\partial H}{\partial x_j}(x), \tag{23}$$

Note that in both cases the function G is a continuous function. It follows, from the continuity of G, that $S_H \cup \{0\} = G^{-1}(0)$ is a closed subset of the corresponding space (l^2 in case (22) and $\mathbb{R}^m$ in case (23)).

4.7. Finite Case

Let us assume H is as in (15). Then we have .

Proposition 15. *The Nehari variety is a compact submanifold.*

Proof. We already know that $S_H \cup \{0\}$ is closed. Let us consider the equality

$$^t x S x = \sum_{j=1}^{m} x_j^{p+1}, \tag{24}$$

whose left hand side grows quadratically, but the right hand side grows faster than quadratically. In this way, there is no sequence in S_H going to neither 0 nor ∞ and keeping such an equality. It follows that S_H is closed and bounded. □

As a consequence of the above, the existence of global minima or maxima for H on S_H is provided. As a consequence, we have the following.

Corollary 16. *Let $(\mathcal{G}, d)$ be so that $\mathcal{G} = (V, E)$ is a finite admissible graph and $d : E \to (0, +\infty)$ is a metric. Then, for every $\emptyset \neq W \subset V$ and $p > 1$, the discrete Lane-Emden-Fowler equation*

$$\begin{cases} \Delta_2 \mu(v) = \mu^p(v), & v \in V - W \\ \mu(v) > 0, & v \in V - W \\ \mu(w) = 0, & w \in W \end{cases} \tag{25}$$

has a nontrivial solution.

The infinite case seems to be require a more delicate analysis and it will be addressed in a forthcoming paper.

5. Discrete Navier-Stokes Equation

Let us consider a finite simple connected weighted graph $(\mathcal{G}, d)$, where $\mathcal{G} = (V, E)$ and $V = \{v_1, \ldots, v_m\}$. Let $0 < n < m$ be so that the degree at each vertex is at least n.

5.1. The n-gradient

Let us consider n permutations in $\mathfrak{S}_m$, the permutation group on m letters, say

$$\sigma_1, \ldots, \sigma_n : \{1, \ldots, m\} \to \{1, \ldots, m\}$$

so that, for each $j \in \{1, \ldots, m\}$, the following properties hold

1. $\sigma_r(j) \neq \sigma_s(j)$, if $r \neq s$;
2. $v_{\sigma_r(j)} \in N(v_j)$.

We define the n-gradient of a function $\alpha \in C^0(\mathcal{G})$, associated to the choice of the above permutations, as

$$\nabla_n \alpha(v_j) := \left(\partial_{v_{\sigma_1(j)}} \alpha(v_j), \ldots, \partial_{v_{\sigma_n(j)}} \alpha(v_j)\right) . \tag{26}$$

5.2. The Adaptation

The idea is now to adapt equation (1) to the case of finite admissible graphs. We assume fixed a n-gradient ∇_n defined by a collection of n permutations $\sigma_1, \ldots, \sigma_n$. Let us consider functions

$$\rho, p : V \times \mathbb{R} \to \mathbb{R}$$

both of them differentiable with respect to the real variable, and a function

$$f : V \times \mathbb{R} \to \mathbb{R}^n : (v, t) \mapsto (f_1(v, t), \ldots, f_n(v, t))$$

continuous in the real variable. The corresponding Discrete Navier-Stokes Equation is given by

$$\begin{cases} \rho(v_j, t)\left(\dfrac{\partial \nu}{\partial t}(v_j, t) + \nu(v_j, t) \cdot \nabla_n \nu(v_j, t)\right) \\ \qquad\qquad \| \\ -\nabla_n p(v_j, t) + \mu \Delta \nu(v_j, t) + f(v_j, t), \\ \qquad\qquad j = 1, \ldots, n. \end{cases} \tag{27}$$

A solution of (27) is a function

$$\nu : V \times \mathbb{R} \to \mathbb{R}^n : (v, t) \mapsto (\nu_1(v, t), \ldots, \nu_n(v, t)),$$

differentiable with respect to the real variable, and where

$$\nu \cdot \nabla_n \nu := (\nu \cdot \nabla_n \nu_1, \ldots, \nu \cdot \nabla_n \nu_n)$$

$$\Delta \nu := (-\Delta_2 \nu_1, \ldots, -\Delta_2 \nu_n) .$$

5.3. The Associated Ordinary Differential Equation

The discrete Navier-Stokes equation (27) is equivalent to an ordinary differential equation. In order to state such an equation, we need some notations.

Let J_2 be the Laplacian matrix (representing Δ_2) and let us define, for $j = 1, \ldots, m$ and $k = 1, \ldots, n$, the following:

$$\begin{array}{lllllll} \rho_j(t) & := & \rho(v_j, t), & & p_j(t) & := & p(v_j, t) \\ f_j^k(t) & := & f_k(v_j, t), & & \nu_j^k(t) & := & \nu_k(v_j, t) \\ \rho & := & {}^t[\rho_1 \cdots \rho_m], & & p & := & {}^t[p_1 \cdots p_m] \\ f^k & := & {}^t[f_1^k \cdots f_m^k], & & \nu^k & := & {}^t[\nu_1^k \cdots \nu_m^k] \end{array}$$

$$L_k = [l_{ij}^k], \quad k = 1, \ldots, n,$$

$$l_{ij}^k = \begin{cases} \dfrac{-1}{d(\{v_i, v_{\sigma_k(i)}\})}, & i = j \\ \dfrac{1}{d(\{v_i, v_{\sigma_k(i)}\})}, & \sigma_k(i) = j \\ 0, & \text{otherwise} \end{cases}$$

$$M = \mathrm{diag}(\rho_1, \rho_2, \ldots, \rho_m)$$

$$S_k = \mathrm{diag}(\nu_1^k, \nu_2^k, \ldots, \nu_m^k)$$

With the previous notations, Equation (27) may be written as the following system of ordinary differential equations

$$\dot{\nu}^k + \left(\left(\sum_{l=1}^{n} S_l L_l \right) + \mu M^{-1} J_2 \right) \nu^k = M^{-1}(f^k - L_k p), \qquad k = 1, \ldots, n. \tag{28}$$

or equivalently, in matrix notation, as

$$\dot{\nu} + (SL + \mu M^{-1} J_2)\, \nu = M^{-1}[(f^1 - L_1 p) \cdots (f^n - L_n p)] \tag{29}$$

where $S = [S_1 \cdots S_n]$ and $L = {}^t[{}^tL_1 \cdots {}^tL_n]$.

The Newton conservation equation is given, in this way, by

$$\sum_{k=1}^{n} L_k \nu^k = 0\ . \tag{30}$$

5.4. Examples

In the case of the complete graph K_m, $m \geq 3$ with $\sigma_1(j) = j+1$ (j modulo m), and $\sigma_k = \sigma_1^k$, for $k = 2, \ldots, n = m-1$, we see that

$$L_1 = \cdots = L_n = \begin{bmatrix} -w & w & \cdots & w & w \\ w & -w & w & \cdots & w \\ \vdots & \vdots & \vdots & \vdots & \vdots \\ w & w & \cdots & w & -w \end{bmatrix},$$

$$J_2 = \begin{bmatrix} nw^2 & -w^2 & \cdots & -w^2 & -w^2 \\ -w^2 & nw^2 & -w^2 & \cdots & -w^2 \\ \vdots & \vdots & \vdots & \vdots & \vdots \\ -w^2 & -w^2 & \cdots & -w^2 & nw^2 \end{bmatrix},$$

Newton conservation equation (30) asserts that

$$\nu^n = -\sum_{j=1}^{n-1} \nu^j$$

and, in particular,

$$S_n = -\sum_{j=1}^{n-1} S_j$$

At this point we note that, from the above,

$$\sum_{l=1}^{n} S_l L_l = L_1 \sum_{l=1}^{n} S_l = 0,$$

therefore the equation to solve is

$$\dot{\nu}^k + \mu M^{-1} J_2 \nu^k = M^{-1}(f^k - L_k p), \qquad k = 1, \ldots, n. \tag{31}$$

and a necessary condition to have a solution is

$$\sum_{k=1}^{n} (f^k - L_k p) = 0.$$

Now assuming the above necessary condition, the original system of ordinary differential equations has solutions if and only if the following one has solutions

$$\dot{\nu}^k + \mu M^{-1} J_2 \nu^k = M^{-1}(f^k - L_k p), \qquad k = 1, \ldots, n-1, \tag{32}$$

which can be explicitly solved as it is a linear system of degree one, uniquely determined by the initial conditions for $\nu^k(0)$, $k = 1, \ldots, n-1$.

Then, the following Theorem takes place.

Theorem 17. *In the case of K_m, with the previous notations, the condition*

$$\sum_{k=1}^{n}(f^k - L_k p) = 0$$

is necessary and sufficient in order for the discrete Navier-Stokes equation (28) *under the conservation law* (30) *to have solution.*

On the other hand, if we are interested in the problem of approximation of the solutions to Navier-Stokes equation in the case of a circle, it may be useful to consider the case of a weighted cyclic graph C_m, $m \geq 3$, $n = 1$ and $\sigma(j) = j + 1$ (modulo m). First, observe that incompressibility takes the form

$$\begin{bmatrix} -w & w & 0 & \cdots & 0 & 0 \\ 0 & -w & w & & 0 & 0 \\ 0 & 0 & -w & & 0 & 0 \\ \vdots & \vdots & & \ddots & & \vdots \\ 0 & 0 & 0 & & -w & w \\ w & 0 & 0 & \cdots & 0 & -w \end{bmatrix} \begin{bmatrix} \nu_1 \\ \nu_2 \\ \nu_3 \\ \vdots \\ \nu_{m-1} \\ \nu_m \end{bmatrix} = \begin{bmatrix} 0 \\ 0 \\ 0 \\ \vdots \\ 0 \\ 0 \end{bmatrix}$$

which implies $\nu = \nu_i = \nu_j$ for all $i, j = 1, \ldots, m$. It is easy to see that this restriction forces the discrete Laplace operator to vanish identically. More explicitly

$$\Delta_2 \nu_j = \sum_{w \in N(v_j)} \frac{\nu(w) - \nu(v_j)}{d(w, v_j)^2} = \frac{\nu(v_{j-1}) - 2\nu(v_j) + \nu(v_{j+1})}{d^2} = 0.$$

Thus, Navier-Stokes equation (27) in this case is simply

$$\rho \dot{\nu} = -w(p_{j+1} - p_j) + f \tag{33}$$

from which we have the following result.

Theorem 18. *In the case of C_m, with the previous notations, the discrete Navier-Stokes equation* (28) *under the conservation law* (30) *has solution if and only if $p^+ = p_{j+1} - p_j$ is independent of j. In such case, there is a unique solution with given initial conditions and it is*

$$\nu(t) = \nu(0) + \int_0^t \frac{f - wp^+}{\rho} d\tau.$$

6. Convenient Graphs

In this section, we discuss general ideas of how the adapted Navier-Stokes equation may be of use to produce numerical solutions of Navier-Stokes equation in a region $\Omega \subset \mathbb{R}^n$. Let us consider the case $n = 2$. Let us assume we have a bounded region $\Omega \subset \mathbb{R}^2$, with boundary Γ, which we assume to be a finite collection of rectifiable Jordan curves.

6.1. First Step: Replacing Ω by a Weighted Graph

We may start with a cellular decomposition Σ of $\mathbb{R}^2$ and define a graph $\mathcal{G}_\Sigma$ as follows. The vertices of this graph are the 0-dimensional components of Σ contained in $\overline{\Omega}$. The edges are the 1-dimensional components of Σ contained inside $\overline{\Omega}$ and extra edges connecting vertices at the boundary Γ (we need to take care in this part to avoid intersections of these extra edges with the previous ones). In Figures 2 and 3 it is shown such a procedure for two different cellular decompositions of $\mathbb{R}^2$ (the first one corresponds to a cellular decomposition for which every 2-dimensional component is a hexagon and in the second one every 2-dimensional component is a rectangle). The weight of the graph is defined by the rule that the weight of an edge e is equals the Euclidean distance between the end vertices.

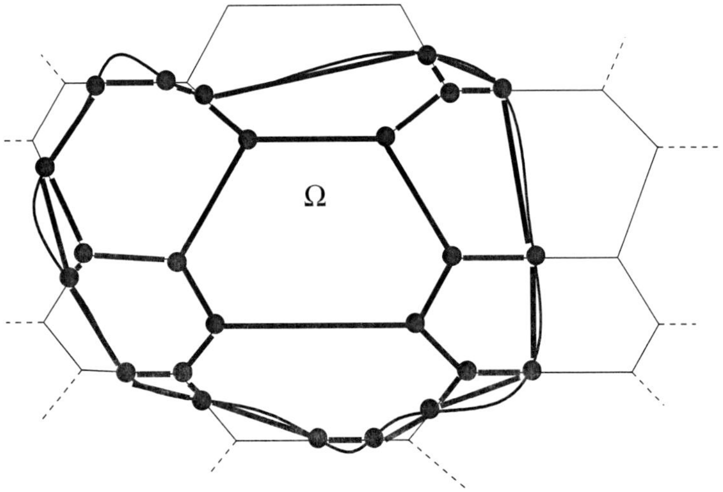

Figure 2. The graph in thicker lines.

6.2. Second Step: Choosing Correct Graphs

A problem that arises in this construction is the fact that the discrete Laplace operator on the graph $\mathcal{G}_\Sigma$ may not approximate (a multiple of) the classical Laplace operator. More precisely, assume we have a weighted graph, say $\mathcal{G} = (V, E)$ with metric d, as above. If $\mu \in C^0(\mathcal{G})$, $v \in V \cap \Omega$ and $N(v) = \{w_1, \ldots, w_{r_v}\}$, then

$$\Delta_2\mu(v) = \sum_{j=1}^{r_v} \frac{\mu(v) - \mu(w_j)}{d(\{v, w_j\})^2}.$$

Set $w_j = v + (a_{j,v}, b_{j,v})d_{j,v}$, where $a_{j,v}^2 + b_{j,v}^2 = 1$ and $d(\{v, w_j\}) = d_{j,v}$, for each $j = 1, \ldots, r_v$.

If we keep the unit vectors $(a_{j,v}, b_{j,v})$ fixed, then

$$\lim_{(d_{1,v},\ldots,d_{r_v,v})\to(0,\ldots,0)} \sum_{j=1}^{r_v} \frac{\mu(v) - \mu(w_j)}{d_{j,v}^2}$$

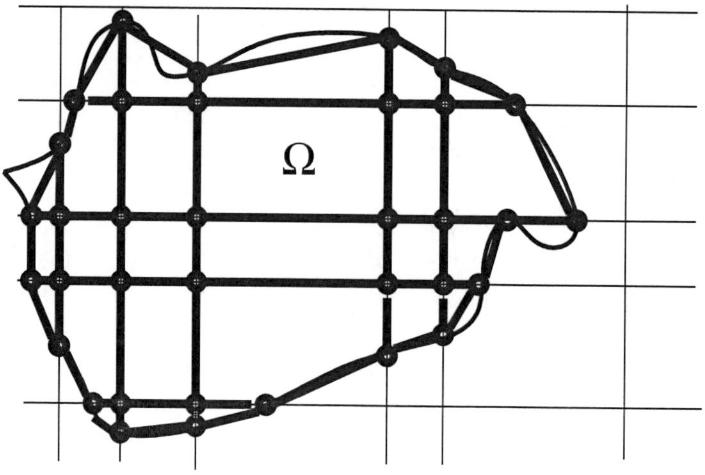

Figure 3. The graph in thicker lines.

$$= \frac{1}{2}\left\{\left(\sum_{j=1}^{r_v} a_{j,v}^2\right)\frac{\partial^2\mu(v)}{\partial x^2} + \left(\sum_{j=1}^{r_v} b_{j,v}^2\right)\frac{\partial^2\mu(v)}{\partial y^2} + 2\left(\sum_{j=1}^{r_v} a_{j,v}b_{j,v}\right)\frac{\partial^2\mu(v)}{\partial x\partial y}\right\}.$$

In this way, if we want to use Δ_2 to approach $-\Delta$ (up to a multiple), then we need, for the graph $\mathcal{G}$, the existence of some $D > 0$ with the property that at each vertex $v \in \Omega \cap V$ it holds the following

(P1) $\sum_{j=1}^{r_v} a_{j,v}^2 = D = \sum_{j=1}^{r_v} b_{j,v}^2,$

(P2) $\sum_{j=1}^{r_v} a_{j,v}b_{j,v} = 0.$

Thus, if (P1) and (P2) are satisfied, the discrete operator $\frac{2}{D}\Delta_2$ will approximate weakly the differential operator $-\Delta$. Next, we proceed to describe a nice family of equiangular graphs we may use.

6.3. Equiangular Planar Graphs

Assume we have a weighted graph, say $\mathcal{G} = (V, E)$ with metric d, as defined in Subsection 6.1.. As noted in Subsection 6.2., we need to assume some extra geometric conditions on such a graph. We assume that at each vertex $v \in V$ the angles between two consecutive edges are the same, that is, if the degree of v is $n_v \geq 3$, then the angle under consideration is $2\pi/n_v$. We say that $\mathcal{G}$ is an *equiangular planar graph*. We first note that (P2) is satisfied for such an equiangular planar graph; at a vertex v it holds that

$$\sum_{j=1}^{r_v} a_{j,v}^2 = \frac{n_v}{2} = \sum_{j=1}^{r_v} b_{j,v}^2$$

In order, for that equiangular planar graph, to satisfies (P1), we should have $n_v = n$, that is, the degree at each vertex is the same. With this restriction, $D = n/2$. But in this situation, the bounded connected components of the complement of the graph $\mathcal{G}$ are polygons whose interior angles are all equal to $2\pi/n$. Let $L \geq 3$ be the number of sides of one of these polygons. Clearly, $(L-2) = 2L/n$, from which $(n, L) \in \{(3,6),(4,4),(6,3)\}$. In the case $(n, L) = (3,6)$, so $2/D = 4/3$, we obtain *hexagonal graphs* (see Figure 2); in the case $(n, L) = (4,4)$, so $2/D = 1$, we obtain *rectangular graphs* (see Figure 3); and in the case $(n, L) = (6,3)$, so $2/D = 2/3$, we obtain *equilateral triangle graphs* (these graphs also define regular hexagonal graphs). Up to translations and rotations, the rectangular and hexagonal graphs depend on infinite number of parameters.

The case $(n, L) = (6,3)$ seems to be to expensive in computations, in comparison with the other two cases. The case $(n, L) = (4,4)$ is the classical types of graphs used in the method of finite differences (in the planar setting), but at each vertex we need to take care of 4 computations, one per each neighbor vertex. The case $(n, L) = (3,6)$ seems to be the less expensive in terms of number of computations of discrete Laplacian.

6.4. The Modified Ordinary Differential Equation

Let us choose an equiangular planar graph (as above), for some fixed (n, L). First, for each vertex $v \in V$, let us enumerate the edges about v counterclockwise starting at one of them. We use the trivial permutation to define the n-gradient in (27). Secondly, we replace Δ_2 by $\frac{4}{n}\Delta_2$ in (27), which is equivalent to keep Δ_2 but to change the dynamical viscosity μ by $4\mu/n$. With these modifications, (29) is given by

$$\dot{\nu} + \left(SL + \frac{4\mu}{n}M^{-1}J_2\right)\nu = M^{-1}[(f^1 - L_1 p)\cdots(f^n - L_n p)] \tag{34}$$

In [12], equation (34) is employed to find discrete solutions to the well-known case of the Poiseuille flow. More concrete examples of this point of view together with generalizations of the convenient graphs to higher dimensions, seems to be a problem worth studying and we will address them in forthcoming research articles.

Acknowledgements

The first author is partially supported by projects Fondecyt 1070271 and UTFSM 12.09.02. The second author is partially supported by the grant of the Norwegian Research Council # 177355/V30, by the grant of the European Science Foundation Networking Programme HCAA and NordForsk Research Network "Analysis and applications".

References

[1] Ambrosetti, A. and Rabinowitz, P. Dual Variational Methods in Critical Point Theory and Applications. *J. Functional Analysis* **14** (1973), 349-381.

[2] W.N. Anderson, T.D. Morley. Eigenvalues of the Laplacian of a graph. *Lin. Multilin. Algebra* **18** (1985) 141-145.

[3] Bapat, R.B. The Laplacian Matrix of a Graph. *Math. Student* **65** (1996), 214-223.

[4] N. L. Biggs, B. Mohar, and J. Shawe-Taylor. The spectral radius of infinite graphs. *Bull. London Math. Soc.* **20** (1988) 116-120.

[5] R.A. Brualdi, J.L. Goldwasser. Permanent of the Laplacian matrix of trees and bipartite graphs. *Discrete Math.* **48** (1984) 1-21.

[6] Colin de Verdiere, Y. Spectre d'operateurs différentiels sur les graphes. In *Random walks and discrete potential theory*, Cortona, June 22-28 (1997), 1-26.

[7] Colin de Verdiere, Y. *Spectres de graphes.* Societé Mathématique de France (1998).

[8] Courant, R.; Friedrichs, K. and Lewy, H. On the partial difference equations of mathematical physics. *Math. Ann.* **100**, 32–74 (1928).

[9] J. Dávila, M. del Pino, M. Musso, J. Wei. Fast and slow decay solutions for supercritical elliptic problems in exterior domains. *Calc. Var. Partial Differential Equations* **32**, No. 4 (2008), 453–480.

[10] Hidalgo, R.A. Zeros of semilinear systems with applications to nonlinear partial difference equations on graphs. *Journal of Difference Equations and Applications* **14**, No. 9 (2008), 953–969.

[11] R. A. Hidalgo, M. Godoy Molina Existence of solutions of Semilinear systems in ℓ^2. *Revista Proyecciones* **27**, No 2 (2008), 149–161.

[12] R. Hidalgo, M. Godoy Molina. Navier-Stokes equations on weighted graphs. *Complex Analysis and Operator Theory* **4** (2008), 525–540.

[13] Neuberger, John M. Nonlinear Elliptic Partial Difference Equations on Graphs. *Experimental Mathematics* **15** (2006), 91–107.

[14] Salvi, R. *The Navier-Stokes Equations. Theory and Numerical Methods*. Proceedings of the International Conference held in Varenna, 2000. Edited by Rodolfo Salvi. Lecture Notes in *Pure and Applied Mathematics*, **223**. Marcel Dekker, Inc., New York, 2002.

In: Navier-Stokes Equations
Editor: R. Younsi

ISBN: 978-1-61324-590-3

Chapter 2

NUMERICAL SOLUTION OF NAVIER-STOKES EQUATIONS IN LIQUID METALS UNDER MAGNETIC FIELD

***S. Nouri*[1], *S. Hamimid*[1], *A. Harkati*[3], *M.Guellal*[2], *D. Ouadjaout*[1] and *R. Younsi*[4]**

[1]Simulation's Laboratory, UDTS, 2Bd Frantz Fanon, Alger Gare, Algiers, (Algeria)
[2]Laboratoire de Génie des Procédés Chimiques (LGPC), UFAS -Setif, (Algeria)
[3]Theoretical Physics Laboratory, Physics institute, USTHB B.P.32 El-Alia, 16111 Algiers, (Algeria)
[4]University of Quebec at Chicoutimi, G7H2B1 Chicoutimi (QC), Canada

PART I
NUMERICAL COMPUTATION OF CZOCHRALSKI TECHNIQUE UNDER STATIC MAGNETIC FIELD EFFECT

S. Nouri, A. Harkati, D. Ouadjaout and R. Younsi

ABSTRACT

In this chapter, the Czochralski techniques of crystal growth driven by the combined action of buoyancy, surface-tension and electromagnetic forces is considered. This study was done for Prandlt number Pr=0.02 and aspect ratio A=2. The transport equations for continuity, momentum and energy transfer are solved. The numerical results are reported for the effect of Marangoni and Hartmann numbers. A uniform magnetic field is applied in the axial direction. The system of equations governing the growth phenomenon has been treated in steady state). The axisymmetric Navier-Stokes and energy equations, with the Boussinesq approximation, have been discretised by means of finite volume procedure and the SIMPLER algorithm is used to treat the coupled velocity-pressure. The thermocapillary convection and the magnetic field effects on the flow and temperature field are discussed. A parametric study was carried out.

Keywords: Natural convection, thermocapillary convection, magnetic field, finite volume.

I.1. INTRODUCTION

The most utilized crystal growth techniques from the melt are: the directional solidification (Bridgman technique); the crystal pulling technique named after Czochralski; the zone melting method, the floating zone method is one of variants of its vertical configuration. Fig.I.1 shows techniques commonly used for the crystal growth from the melt [1].

Convection in the bulk liquid is an important parameter in crystal growth from the melt. A typical driving force for convection in the melt is the buoyancy force induced by thermal and solutal gradients in the liquid, besides buoyancy, is the surface-tension gradient that results from temperature or concentration gradients on a free surface. Surface tension typically decreases with increasing temperature. In a free melt surface with a temperature gradient, surface tension forces drive the flow from regions of low surface tension (hot) to areas of high surface tension (cold). Surface tension forces are balanced by viscous shear which transfers momentum to neighboring liquid layers because of fluid viscosity. Similar to the buoyancy driven flow, continuity causes the development of a bulk flow in the whole melt volume. This convective motion is referred to as Marangoni or thermocapillary convection and occurs if free melt surfaces exist in the growth configuration [2-3]

The study of Marangoni convection in crystal growth from the melt has proven to be of particular interest with respect to its influence on the distribution of impurities in the obtained crystal structure [4-5]. The role of Marangoni convection becomes quite dominant in reduced gravity environments. Even under normal gravity, the thermocapillary effect can be very important in the overall transport pattern. A review of various numerical and experimental studies of the interaction of the surface tension driven flow and the buoyancy driven flow in crystal growth melts can be found in [6-7].

Studies of surface tension effects have concentrated largely on flows without phase change [8-9]. In [10], the solidification of pure materials was considered during which liquid convection was driven by a combination of buoyancy and surface tension forces. The action of thermocapillary convection on crystal growth processes is generally undesirable because the transition to unsteady behavior is hard to control [11].

Application of magnetic fields is known to stabilize both flow and temperature oscillations in the melt and thereby represents a promising opportunity to obtain an improved crystal quality. The effects of a magnetic on melt convection have been previously investigated by several authors [12-13]. The effects of a strong vertical magnetic field on convection and segregation in the vertical Bridgman crystal growth process were considered in [14].

In this paper, a computational model was developed to study the interaction between convection and magnetic field effect for pure metal fluid that is crystallized in a cylindrical container with an open boundary. The finite volume method was adopted for the discretization of the equation governing the studied physical phenomenon. The numerical solution of the system was done via the SIMPLER algorithm. The numerical results of the fluid flow and temperature field are analyzed and discussed.

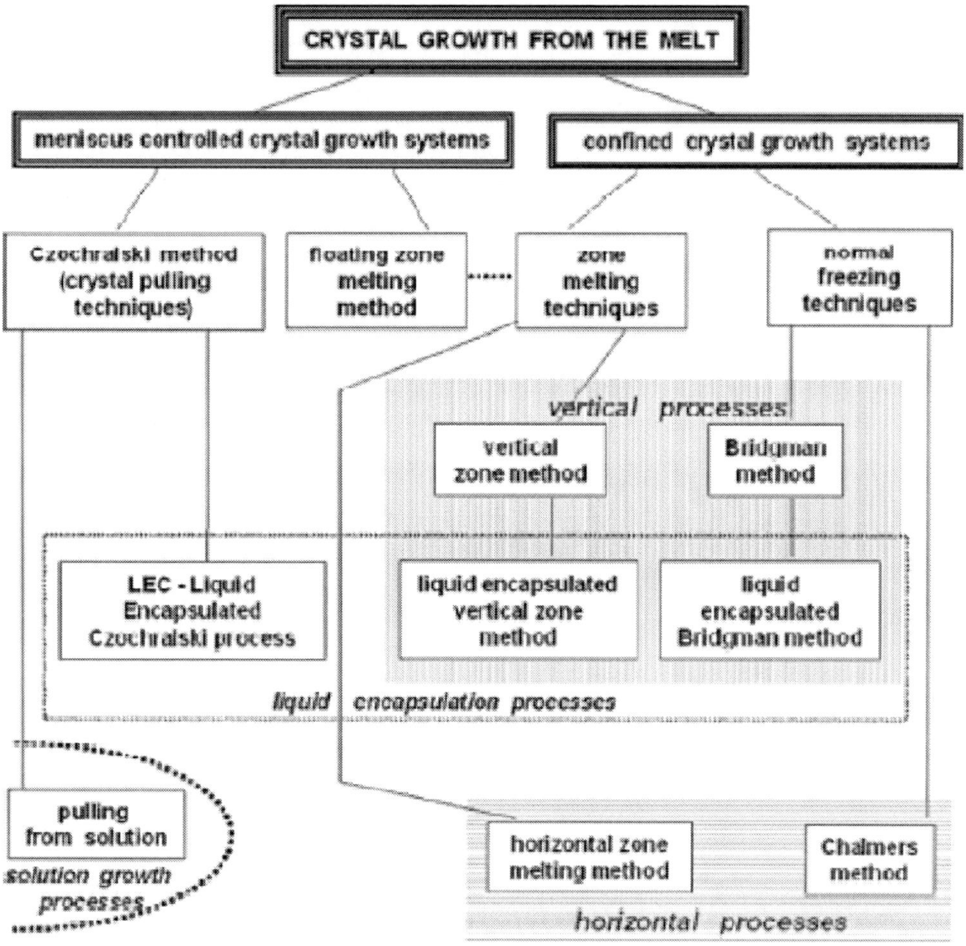

Figure I.1. Main categories of crystal growth techniques from the melt [1].

I.2. Mathematical Model

The geometry of the problem is shown in Fig.I.1. The crucible consists of a right cylinder of radius rc filled with a melt to a depth d. The crucible rotates with an angular velocity w_c about its axis of symmetry, and a constant axial magnetic field $\vec{B} = B_0 i_z$ is applied. The melt is bounded by a coaxial crystal of radius r_s rotating with an angular velocity w_s.

In the solidification system considered in this work, the following assumptions are introduced regarding the transport of heat and momentum:

- The thermal and physical properties are constant.
- The system is axisymmetric.
- The melt flow is assumed to be a laminar flow induced by thermal and surface tension gradients.
- The Boussineq approximation can be invoked.

- The free –surface deformation is negligible.
- A macroscopically stable solid – liquid interface exists between the solid and the liquid regions.

The governing equations the system are now introduced. The crucible consists of a right cylinder of radius r_c filled with a melt to a depth d. The crucible rotates with an angular velocity w_c about its axis of symmetry, and a constant axial magnetic field $\vec{B} = B_0 i_z$ is applied. The melt is bounded by a coaxial crystal of radius r_s rotating with an angular velocity w_s.

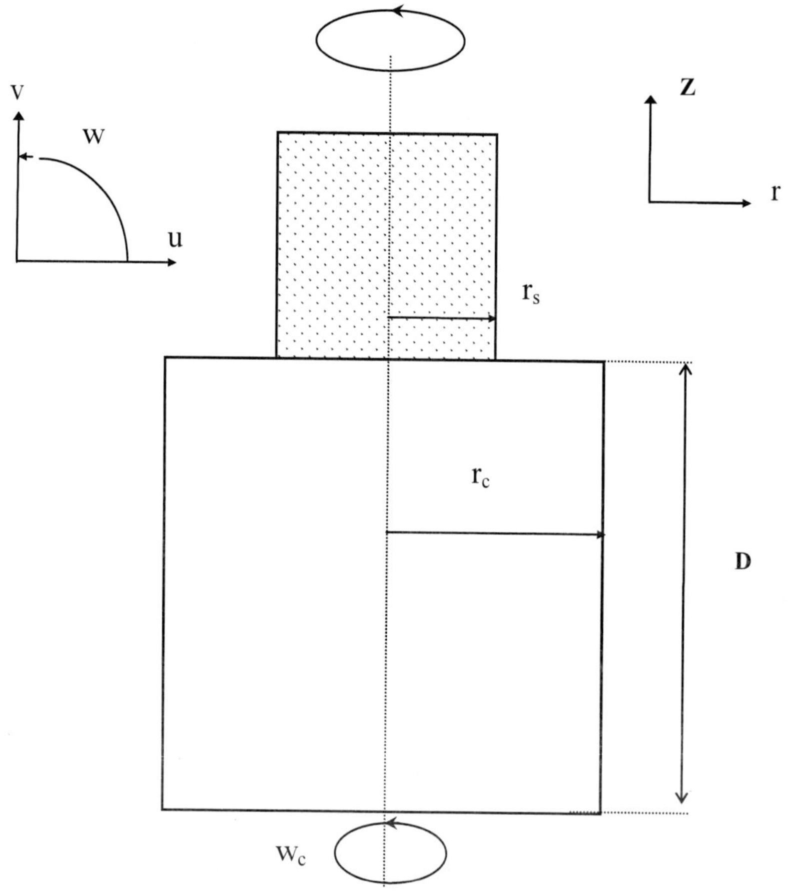

Figure I.2. Geometry of the Czochralski bulk flow and system of grid.

The basic equations used in the simulation of the melt flow are the incompressible Navier-Stokes equations, including the Lorentz force. The Boussinesq approximation is also used for defining buoyancy. The other equation governing the fluid flow in the liquid domain is the energy equation. These equations are written as:

$$\frac{\partial u}{\partial z} + \frac{\partial (rv)}{r \partial r} = 0 \tag{I.1}$$

$$u\frac{\partial u}{\partial z}+v\frac{\partial(u)}{\partial r}=-\frac{\partial p}{\partial z}+\left[\frac{\partial}{r\partial}\left(r\frac{\partial u}{\partial r}\right)+\frac{\partial^2 u}{\partial z^2}\right]-Ha^2u \tag{I.2}$$

$$u\frac{\partial v}{\partial z}+v\frac{\partial(v)}{\partial r}-\frac{w^2}{r}=-\frac{\partial p}{\partial z}+\left[\frac{\partial}{r\partial}\left(r\frac{\partial v}{\partial r}\right)+\frac{\partial^2 v}{\partial z^2}-\frac{v}{r^2}\right]+Gr\left[T-1\right] \tag{I.3}$$

$$u\frac{\partial w}{\partial z}+v\frac{\partial(w)}{\partial r}+\frac{vw}{r}=-\frac{\partial p}{\partial z}+\left[\frac{\partial}{r\partial}\left(r\frac{\partial w}{\partial r}\right)+\frac{\partial^2 w}{\partial z^2}-\frac{w}{r^2}\right]-Ha^2w \tag{I.4}$$

$$u\frac{\partial T}{\partial z}+v\frac{\partial(T)}{\partial r}=-\frac{\partial p}{\partial z}+\frac{1}{\Pr}\left[\frac{\partial}{r\partial}\left(r\frac{\partial T}{\partial r}\right)+\frac{\partial^2 T}{\partial z^2}\right] \tag{I.5}$$

where the governing dimensionless groups are the Prandlt number $\left(\Pr=\nu/\alpha\right)$, the thermal Grashof number $\left(Gr=g\,\beta_T\Delta T\,r_c^3/\nu^2\right)$ and the Hartmann number $\left(Ha=\sqrt{\sigma_e/\rho\nu}\,B_0\,r_c\right)$ where g is the gravity constant, β_T is the thermal coefficient of expansion, B_0 is the magnetic induction and σ_e is the electrical conductivity.

A no-slip and no penetration boundary condition on the velocity field is imposed on all the liquid boundaries.

$$u=v=0 \tag{I.6}$$

The surface tension on the free surface is approximated as

$$\frac{\partial v}{\partial z}=\frac{Ma}{\Pr}\frac{\partial T}{\partial r} \tag{I.7}$$

In order to complete the mathematical model, we still use the appropriate boundary conditions. They are as follows:

<u>at the crucible bottom</u>: $z=0;\ 0\leq r<1$

$$u=v=0,\quad w=Re_c\,r,\quad \frac{\partial T}{\partial z}=0 \tag{I.8}$$

<u>at the crucible side</u>: $r=1;\quad 0\leq z\leq\frac{d}{r_c}$

$$u = v = 0, \quad w = Re_c, \quad T = \frac{T_c}{T_f} \tag{I.9}$$

at the melt-crystal interface : $z = \frac{d}{r_c}; \quad 0 \le r \le \frac{r_s}{r_c}$

$$u = v = 0, \quad w = Re_s\, r, \quad T = 1 \tag{I.10}$$

at the symmetry axis: $r = 0\ ; \quad 0 \le z \le \frac{d}{r_c}$

$$v = w = 0, \quad \frac{\partial u}{\partial r} = 0, \quad \frac{\partial T}{\partial r} = 0 \tag{I.11}$$

at the melt free surface: $z = \frac{d}{r_c}\ ; \quad \frac{r_s}{r_c} < r < 1$

$$u = 0, \quad \frac{\partial w}{\partial z} = 0, \quad \frac{\partial v}{\partial z} = \frac{Ma}{Pr}\frac{\partial T}{\partial r} \tag{I.12}$$

where $\left(Ma = (\partial\delta/\partial T)\Delta T\, r_c \,/\, (\rho \nu \alpha)\right)$ is the thermal Marangoni number and δ is the surface tension.

A control volume based discretisation method, described by Patankar [15] has been applied for solving the system of equation (1), together with the boundary and interfacial conditions (I.2-I.6). In evaluating convective and diffusive fluxes, a power-law scheme and the SIMPLER algorithm were considered.

I.3. Results and Discussion

I.3.1. Effect of Buoyancy and Thermocapillary Forces

The surface tension varies from one place to another of the surface. This case involves the tangential forces at the interface, involving explicitly the gradient along the surface of the surface tension $\partial\delta/\partial T$. Therefore a flow is induced. The movement at the surface is transmitted to the adjacent fluid layer by the viscosity effect. This case of capillary movement is known Marangoni effect. Our numerical problem allowed us to analyse the Marangoni number effect on flow and thermal fields where convection is well pronounced $\left(Gr = 2.10^6\right)$. Fig.I.3 demonstrates that at low Marangoni number $Ma \le 10^3$, recirculation zones appear and the flow becomes bi cellular when the capillary convection is important $Ma \ge 10^4$.

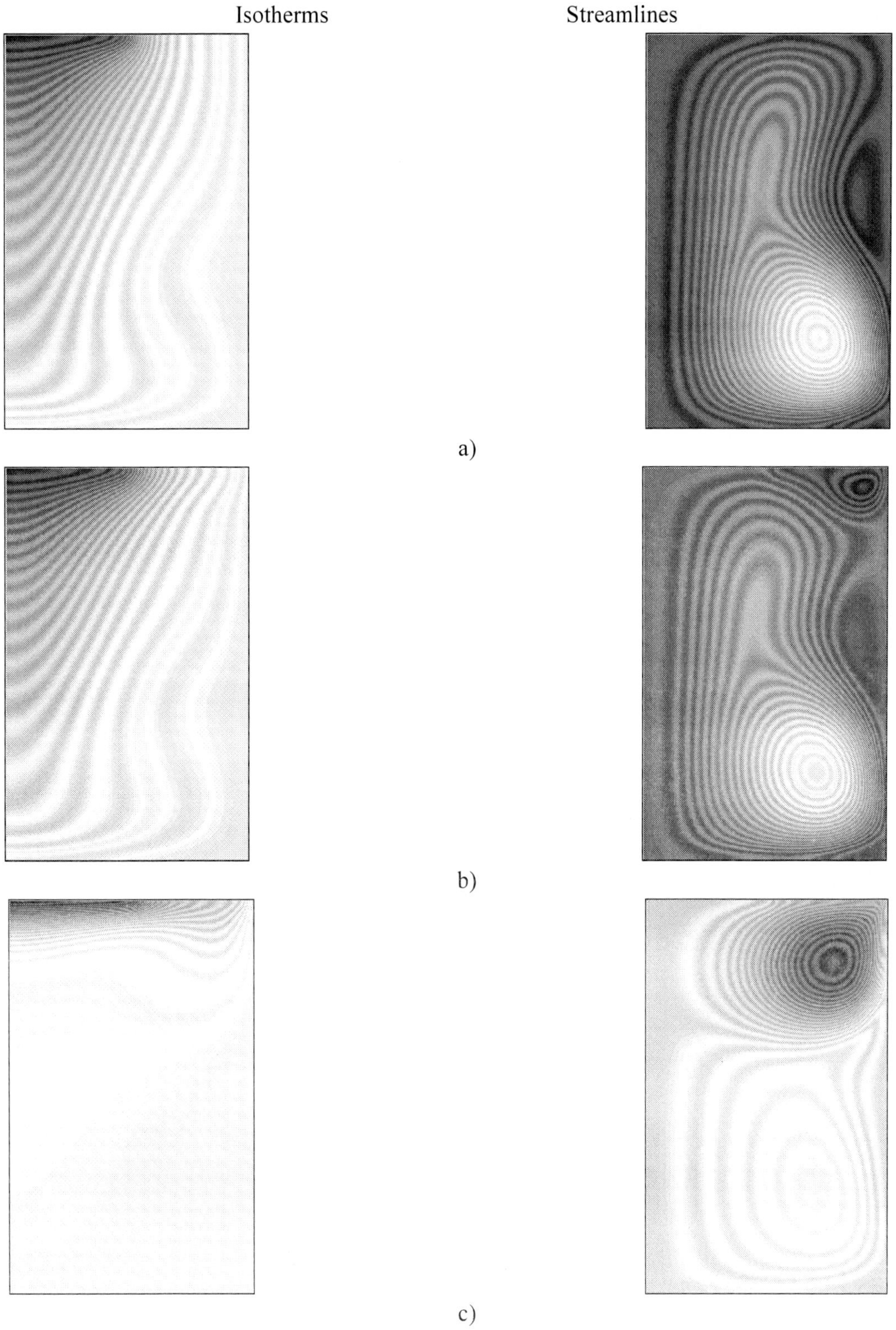

Figure I.3. Combined Buoyancy and Thermocapillary forces $Gr=2.10^6$: a) Ma = 0, b) Ma = 1000, c) Ma = 10000.

I.3.2. Effect of an Externally Applied Magnetic Field

In recent years, strong magnetic fields have been used in crystal growth of semiconductor materials in order to reduce macroscopic inhomogeneity in the crystal by suppression of buoyancy driven convection. In addition, static magnetic fields have also been used to avoid occurrence of unsteady convection which is considered to be the origin of micro-inhomogeneity in crystals. It is becoming increasingly evident that the homogeneity of crystals can be drastically influenced by imposed magnetic fields. Furthermore, recent studies indicate that there is tremendous potential in effectively using magnetic fields in order to control the solidification process. In an attempt to explore this effect numerically, we consider the influence of an externally applied magnetic field on the above Czochralski problem.

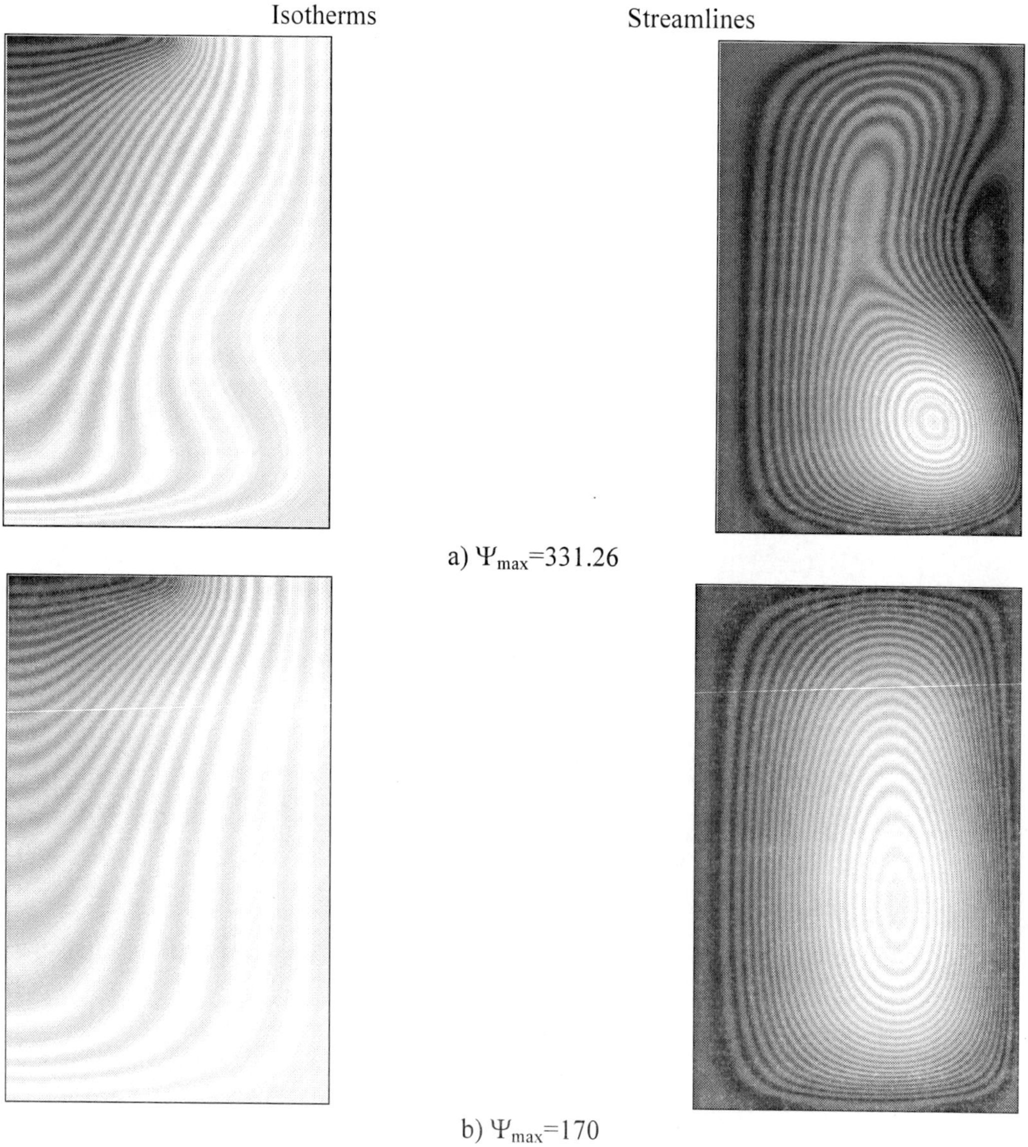

Figure I.4. Continued on next page.

c) $\Psi_{max}=18$

Figure I.4. Effect of Magnetic field: a) Ha=0, b) Ha=50, c) Ha=150.

It is well known that application of a magnetic field damps the melt flow. Fig.I.4 shows that for Ha=0, the flow is multicellular with recirculation zones at the right of the cavity (fig. I.4a). This flow structure is resulting from buoyancy forces. For Ha=150, convection in the bath becomes weak and the multicellular flow disappears.

The magnetic field affects directly the temperature field. This can be clearly seen by comparing the isotherm in Fig. I.4.

I.4. Conclusion

The Czochralski crystal growth with an externally applied magnetic field was considered in this work. Thermocapillary convection analyzed via Marangoni number affects the flow structure which becomes more complex. However, the applied magnetic field analyzed via Hartmann number decreases the convection in the melt and the flow at single cell appears.

Nomenclature

B_0:	intensity of magnetic field
Gr:	Grashof number, $g\ \beta\ T_f\ r_c^3\ /\ \nu^2$
d:	height of the crucible
Ma:	Marangoni number,$(\partial\gamma/\partial T).(r_c T_f/\mu\alpha)$
P:	dimensionless pressure
Pr:	Prandtl number, $\mu\ C_{pm}\ /\ k_m$
r_c:	radius of the crucible
Re:	Reynolds number, $w\ r_c\ /\ \nu$
T:	dimensionless temperature
u, v, w:	dimensionless velocity components

Subscripts

m: melt
c: crucible
f: fusion
s: crystal

Greek symbols

β: expansion coefficient of the melt
ν: kinematics viscosity
Ψ: stream function
γ: surface tension for interface

REFERENCES

[1] Hiroyuki, O.; Janusz, S. S.; Toshio, T. In *Historical Evolution and Trends*; S. Molokov et al.; Eds; Springer, NETHERLANDS, 2007, pp 375–390.
[2] Zeng, Z.; Mizuseki, H.; Higashino, K.; Shimamura, K.; Fukuda, T. and Kawazoe, Y. *Mater. Trans.* 2001, 42, 2322-2331.
[3] Z. Zeng, H. Mizuseki, J.-Q.; Chen, K. Ichinoseki and Y. Kawazoe. *Mater. Trans*. 2004, 45, 1522.1527.
[4] Chen, B.; Zeng, Z.; Mizuseki, H.; Kawazoe, Y. *Mater. Trans.* 2008, 49, 2566-2571.
[5] Braescu, L.; George, T. F. *J. of math. Mod. and meth. app. sci.* 2008, 2(3).
[6] Zhong, Z.; Chen, J.; Mizuseki, H.; Shishido,T.; Ichinoseki,K.; Kawazoe, Y. *J .of thermal science.* 2002, 11(4), 348-352.
[7] K.H. Lie, D.N. Riahi **and** Walker, **J.S.** *J. of heat and mass transfer,* 1989, 32(12) 2409-2420.
[8] Mokhtari, F.; Bouabdallah, A.; Zizi, M.; Hanchi, S.; Alemani, A. *Magneto-hydrodynamics.* 2009, 45(3), 269-274.
[9] Mokhtari, F.; Bouabdallah, A.; Zizi, M.; Hanchi, S.; Alemani, A. *Crystal Research and Technology,* 2009, 44(8), 787–799.
[10] Sampath, **R.**; Zabaras, *J. of Computational Physics*, 2001, 168(2), 384-411.
[11] Younsi, R.; Harkati, A.; Ouadjaout, D. *AMO - Advanced Modeling and Optimization,* 2007, 9(2).
[12] Testsuo, M.; Satoshi, S.; Ichiro, T. *Int. J. of heat and mass transfer,* 2004, 47(21), 4525-4533.
[13] Kader, Z., Nathalie, M.N.; René, M. *Comptes rendus. Mécanique.* 2007, 335(5-6), 330-335.
[14] Nouri, S.; Benzeghiba, M.; Benzaoui, A. *Defect and Diffusion Forum.* 2010, 297-301.
[15] Patankar S.V. *Numerical Heat Transfer and Fluid Flow;* Hemisphere, Washington DC, USA, 1980.

In: Navier-Stokes Equations
Editor: R. Younsi

ISBN: 978-1-61324-590-3

PART II
MARANGONI-NATURAL CONVECTION IN LIQUID METALS IN THE PRESENCE OF A TILTED MAGNETIC FIELD

S. Hamimid and M.Guellal

ABSTRACT

The Navier-Stokes and energy equations are numerically solved to investigate two-dimensional convection (originating from the combined effect of buoyancy and surface tension forces) in liquid metals subjected to transverse magnetic field. In particular, a laterally heated horizontal cavity with aspect ratio (height/width =1) and Prandtl number Pr=0.015 is considered (typically associated with the horizontal Bridgman crystal growth process and commonly used for benchmarking purposes). The effect of a uniform magnetic field with different magnitudes and orientations on the stability of the two distinct convective solution branches (with a single-cell or two-cell pattern) of the steady state flows is investigated. The effects induced by increasing values of the Rayleigh and Hartmann numbers on the heat transfer rate are also discussed.

Keywords: Numerical modeling, semiconductor melt, magnetic field suppression, thermocapillary convection, solidification.

II.1. INTRODUCTION

When a free surface is present in free convection liquid flow, the variation in the surface tension at the free surface due to temperature gradients can induce motion within the fluid. Such flow is known either as thermocapillary flow or Marangoni convection.

Thermocapillary convection is of importance in a wide variety of materials processes associated with unbalanced surface tension, in particular, benefiting; from the reduction of buoyancy convection and hydrostatic pressure in low gravity.

Marangoni convective flows are encountered in many technological processes involving free surfaces of a liquid with a nonuniform temperature distribution. Widespread interest in these flows is related with manufacturing of semiconductor monocrystals in a microgravity environment [Ostrach (1982) and Favier (1990)].

[Hadid and Roux (1990)] have also investigated numerically the influence of thermocapillary forces on natural convection flow in a shallow cavity. Numerous studies have been already devoted to the numerical modeling of buoyancy flows in cavities [e.g., Achobir et al. (2008), Bcchignani (2009), Djebali et al. (2009), Mezrhab and Naji (2009)] and the electromagnetic stabilization of the convective flows in several different configurations [e.g., Weiss (1981), Ben Hadid and Henry (1997)].

Such magnetoconvective flows in bounded domains have been solved through two- and three-dimensional numerical simulations by several researchers in recent years [Ozoe and Maruo (1987); Ozoe and Okada (1989); Armour and Dost (2009): Mechighel et al. (2009)].

Rudraiah and Venkatachalappa (1995), in particular, investigated the effect of surface tension gradients on buoyancy driven flow of an electrically conducting fluid in a square cavity in the presence of a vertical transverse magnetic field. The purpose of the investigation was to see how this force damps hydrodynamic movements (since, this is required to enhance crystal purity, increase compositional uniformity and reduce defect density).

In the present investigation, we have considered the problem of combined buoyancy and thermo-capillary convection flow of an electrically conducting fluid within a square enclosure under an externally imposed constant uniform magnetic field. We have chosen a fluid that is characterized by a small Prandtl number (Pr = 0.015, which is appropriate for liquid metal and semi-conductor melts, (Tab. 1)) and a Marangoni number (Ma =1000). The transport equations describing the momentum and heat transfer have been discretized using the finite volume method (FVM) with staggered grids. Solutions of the problem in terms of streamlines, isotherms as well as heat transfer rate from the heated surface have been obtained for values of the Rayleigh number, Ra, equal to 10^4, 10^5 and 10^6, the Hartmann number, Ha, which depends on the transverse magnetic field magnitude, ranges from 0.0 to 150.

Table 1. Physical properties of silicon melt

Melting point temperature $T_m\,[K]$	1685
Density $\rho\left[g/cm^3\right]$	2.42
Thermal diffusivity $\alpha\left[cm^2/s\right]$	2.44×10^{-1}
Kinematic viscosity $\nu\left[cm^2/s\right]$	2.45×10^{-3}
Prandtl number	0.015
Thermal conductivity $\lambda\left[w/cmk\right]$	0.64
Surface tension $\sigma\left[dyne/cm\right]$	7.33×10^{2}

II.2. Mathematical Model and Analysis

The liquid is assumed as a Newtonian fluid filling a square enclosure as shown in Fig. 1. The right and the left walls are maintained at uniform temperatures T_H and T_C, respectively, and are such that $T_H \succ T_C$. The upper and lower boundaries are considered to be adiabatic.

We further assume that the cavity is permeated by a uniform magnetic field

$$\overrightarrow{B} = B_x \overrightarrow{ex} + B_y \overrightarrow{ey} \tag{II.1}$$

The free surface is idealized as non-deformable and adiabatic from the environmental gas. The surface tension is considered to be a linearly decreasing function of the temperature, as:

$$\sigma(T) = \sigma(T_0) - \gamma(T - T_0) \tag{II.2}$$

Fluid flow in the system is described by the equation of continuity and by the Navier–Stokes equation for momentum.

Assuming the model fluid to be incompressible, the equation of continuity can be expressed as

$$\frac{\partial u}{\partial x} + \frac{\partial v}{\partial y} = 0 \tag{II.3}$$

Using the Boussinesq approximation, the Navier–Stokes equation for momentum for a unsteady, laminar flow can be written as

$$\frac{\partial u}{\partial t} + u\frac{\partial u}{\partial x} + v\frac{\partial u}{\partial y} = -\frac{1}{\rho}\frac{\partial P}{\partial x} + \nu(\frac{\partial^2 u}{\partial x^2} + \frac{\partial^2 u}{\partial y^2}) + \frac{\sigma_e B_0^2}{\rho}(v\sin\varphi\cos\varphi - u\sin^2\varphi) \tag{II.4}$$

$$\frac{\partial v}{\partial t} + u\frac{\partial v}{\partial x} + v\frac{\partial v}{\partial y} = -\frac{1}{\rho}\frac{\partial P}{\partial y} + \nu(\frac{\partial^2 v}{\partial x^2} + \frac{\partial^2 v}{\partial y^2}) + g\beta(T - T_0) + \frac{\sigma_e B_0^2}{\rho}(u\sin\varphi\cos\varphi - v\sin^2\varphi) \quad (\text{I}$$

Assuming negligible viscous heat dissipation, the differential thermal energy balance equation may be expressed as

$$\frac{\partial T}{\partial t} + u\frac{\partial T}{\partial x} + v\frac{\partial T}{\partial y} = \alpha(\frac{\partial^2 T}{\partial x^2} + \frac{\partial^2 T}{\partial y^2}) \tag{II.6}$$

The effect of the induced electric current on the external magnetic field and the Joule heating are neglected in the formulation (II.2)–(II.). This is justified by the estimation of non-dimensional parameters characteristic for liquid metals and semiconductors (some details are given in Ref. [Gelfgat and Bar-Yoseph]).

The scales L, $\frac{L^2}{\alpha}$, $\frac{\alpha}{L}$, $\rho\frac{\alpha^2}{L^2}$, are adopted for length, time, velocity, and pressure with α indicating thermal diffusivity. The normalized temperature is $\theta = \frac{T - T_0}{\Delta T}$ with $T_0 = \frac{T_C + T_F}{2}$ and $\Delta T = T_C - T_F$

Introducing the above dimensionless dependent and independent variables into the governing equations (II.3)–(II.6) yields the following equations:

$$\frac{\partial U}{\partial X} + \frac{\partial V}{\partial Y} = 0 \tag{II.7}$$

$$\frac{\partial U}{\partial \tau} + U\frac{\partial U}{\partial X} + V\frac{\partial U}{\partial Y} = -\frac{\partial P}{\partial X} + \Pr\left(\frac{\partial^2 U}{\partial X^2} + \frac{\partial^2 U}{\partial Y^2}\right) + \Pr Ha^2 (V \sin\varphi \cos\varphi - U \sin^2\varphi) \tag{II.9}$$

$$\frac{\partial V}{\partial \tau} + U\frac{\partial V}{\partial X} + V\frac{\partial V}{\partial Y} = -\frac{\partial P}{\partial Y} + \Pr\left(\frac{\partial^2 V}{\partial X^2} + \frac{\partial^2 V}{\partial Y^2}\right) + Ra \Pr\theta + \Pr Ha^2 (U \sin\varphi \cos\varphi - V \cos^2\varphi) \tag{II.10}$$

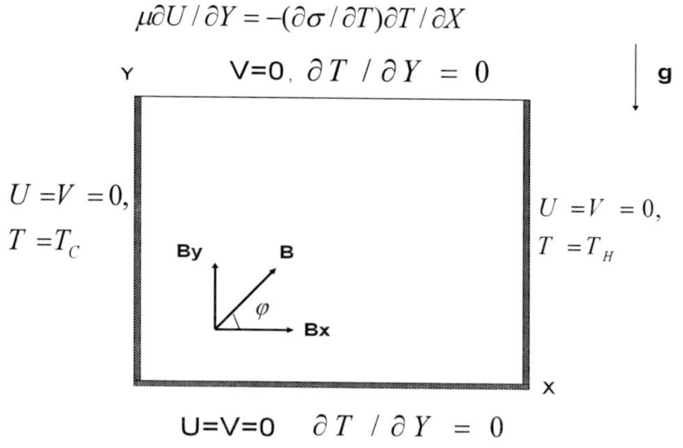

Figure II.1. The flow configuration and coordinate system.

$$\frac{\partial \theta}{\partial \tau} + U\frac{\partial \theta}{\partial X} + V\frac{\partial \theta}{\partial Y} = \left(\frac{\partial^2 \theta}{\partial X^2} + \frac{\partial^2 \theta}{\partial Y^2}\right) \tag{II.11}$$

The dimensionless initial and boundary conditions are:

$$U = V = \theta = 0 \text{ for } \tau = 0$$

$$U = V = 0,\ \theta = \theta_H = 0.5 \text{ for } 0 \le Y \le 1 \text{ at } X = 1$$

$$U = V = 0,\ \theta = \theta_C = -0.5 \text{ for } 0 \le Y \le 1 \text{ at } X = 0$$

$$U = V = 0,\ \frac{\partial \theta}{\partial Y} = 0 \text{ for } 0 \le X \le 1 \text{ at } Y = 0$$

$$V = 0,\ \frac{\partial \theta}{\partial Y} = 0,\ \frac{\partial U}{\partial Y} = Ma\frac{\partial \theta}{\partial X} \text{ for } 0 \le X \le 1 \text{ at } Y = 1$$

In the above equations Ra, $\Pr$, Ha and Ma are, respectively, the Rayleigh number, Prandtl number, Hartmann number and the Marangoni number which are defined as follows:

$$Ra = \frac{g\beta\Delta TL^3}{\nu^2},\ \Pr = \frac{\nu}{\alpha},\ Ha = B_0 L(\sigma_e / \rho\nu)^{1/2},\ Ma = \frac{\gamma\Delta TL}{\mu\alpha}.$$

The non-dimensional heat transfer rate in terms of local Nusselt number, Nu, from the right vertical heated surface is given by

$$Nu(Y) = -\left.\frac{\partial\theta(X,Y)}{\partial X}\right|_{X=1} \tag{II.12}$$

The corresponding value of the average Nusselt number, denoted by Nu_{av}, may be calculated from the following relation

$$Nu_{av} = \int_0^1 Nu(y)dy = -\int_0^1 \left(\frac{\partial\theta(x,y)}{\partial x}\right)dy \tag{II.13}$$

II.3. Numerical Solution Methodology

The governing equations are discretized using the control volume approach of Patankar. In addition, the power law formulation is employed to determine the combined advective and diffusive fluxes across the boundaries of each control volume.

The unsteady term is treated with backward difference. The buoyancy and Lorentz forces in the x and y-momentum equations are treated as source terms. The conventional staggered grid system used originally in the SIMPLE scheme is adopted. The discretized equations are solved iteratively with the line-by-line procedure of tri-diagonal matrix algorithm. In this study, no uniform mesh sizes are used and 140 X 140 grids are chosen for the grid arrangement and it is found that this grid solution is enough.

II.4. Results and Discussion

Numerical results are presented in order to determine the effects of the presence of a magnetic field, buoyancy and thermocapillary forces on the natural convection flow of an electrically conducting fluid in a square cavity. Values of the magnetic field parameter Ha range between 0 to 250, the Rayleigh number Ra, between 0 and 10^5 but for the Marangoni number, Ma, equal to 1000. Typical value of direction of the external magnetic field with the horizontal considered to be:

$\varphi = 0, \pi/6, \pi/4, \pi/3, \pi/2, 4\pi/6, 5\pi/6, \pi, 7\pi/6, 8\pi/6, 9\pi/6, 10\pi/6, 11\pi/6$ and 2π.

In order to obtain grid independent solution, a grid refinement study is performed for a square cavity ($A = 1$) with $Ha = 100$, $Ra = 20$ and $\varphi = 0$. Fig. II.2 shows the convergence of the average Nusselt number, *Nu*, at the heated surface, Fig. 2(a), the maximum ψ_{max}, Fig. II.2(b), and the minimum ψ_{min}, Fig. II.2(c), values of the stream function with grid refinement. It is observed that grid independence is achieved with a 140×140 grid beyond which there is insignificant change in Nu_{av}, ψ_{max} and ψ_{min}. This grid resolution is therefore used for all subsequent computations.

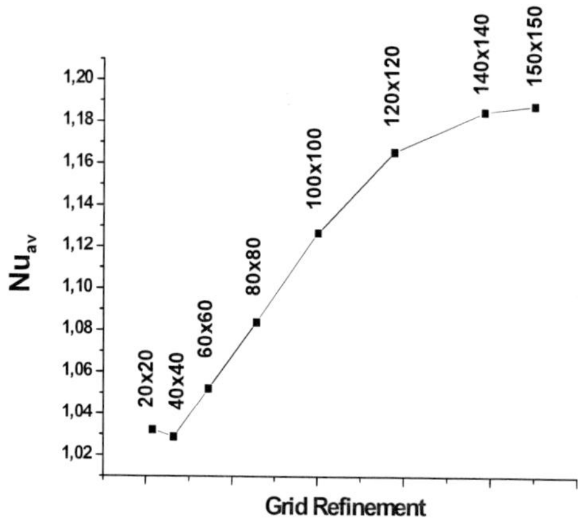

(a)

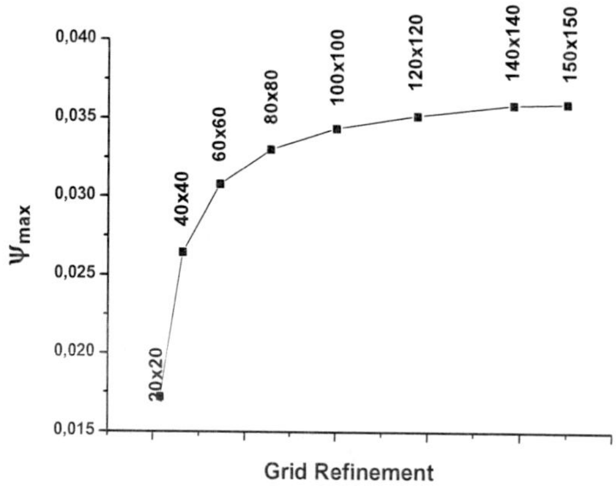

(b)

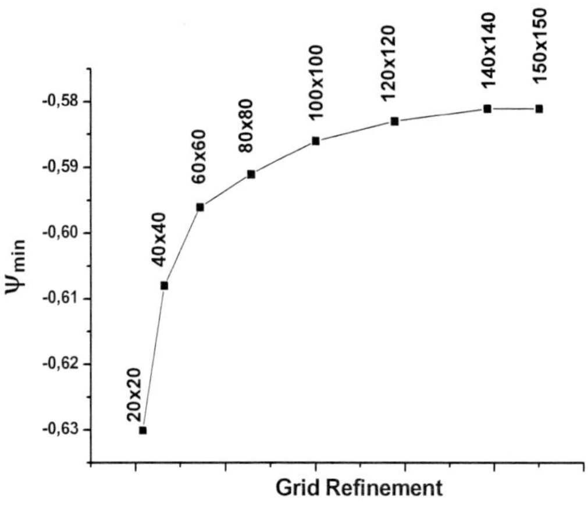

(c)

Figure II.2. Convergence of the (a) average Nusselt number, (b) the maximum values of the stream function and (c) the minimum values of the stream function with grid refinement.

II.4.1. Effects of Varying the Rayleigh Number on the Flow Field and the Heat Transfer

The resulting flow and temperature distributions are depicted in Fig. 3 where the convective flow streamlines for increasing values of the Rayleigh number. In this figure the presence of two cells flow patterns is noticed (Fig. II.3a); one cell on the upper region of the cavity induced by of thermocapillary forces. From the streamlines one may also see that the size of the upper cell gradually decreases with the increase of the buoyancy parameter, *Ra*. This is possible, since an increase in the value of *Ra* will increase the dominancy of the buoyancy force over the magnetic field effect and the thermo-capillary force.

The corresponding effect of the increasing buoyancy forces on the isotherms is shown in Fig. II.3b. From the figure we can ascertain that the increase in the buoyancy force causes the isotherms to deform significantly, and thin thermal boundary layers form near both the heated and cooled surfaces, which have enhanced the heat transfer rate as displayed in fig. II.4.

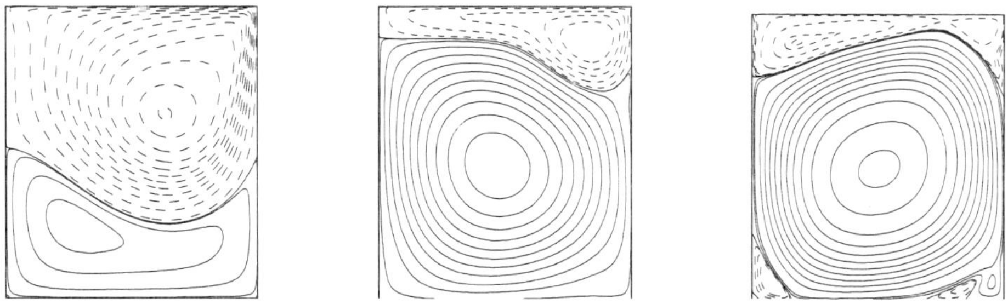

$\psi_{max}=$ 0.121 1.941 2.217

$\psi_{min}=$ -1.018 -1.003 -0.719

(a)

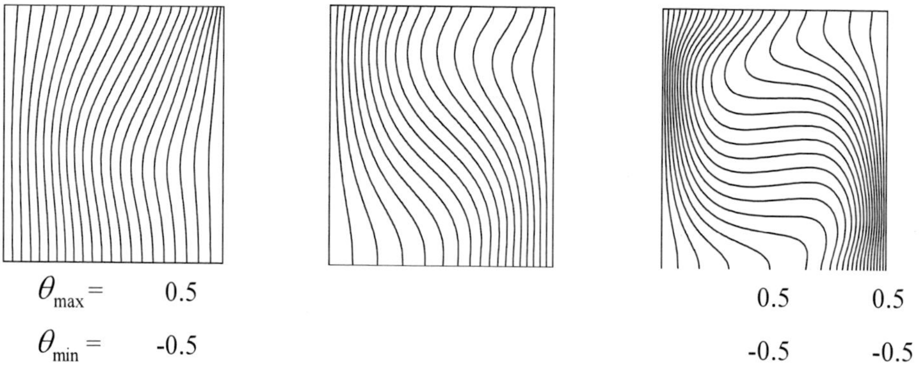

(b)

Figure II.3. Streamlines (top) and isotherms (bottom) for Ra = 10^3,10^4,10^5 while Ma =10^3, Ha =20 and $\varphi = 0$.

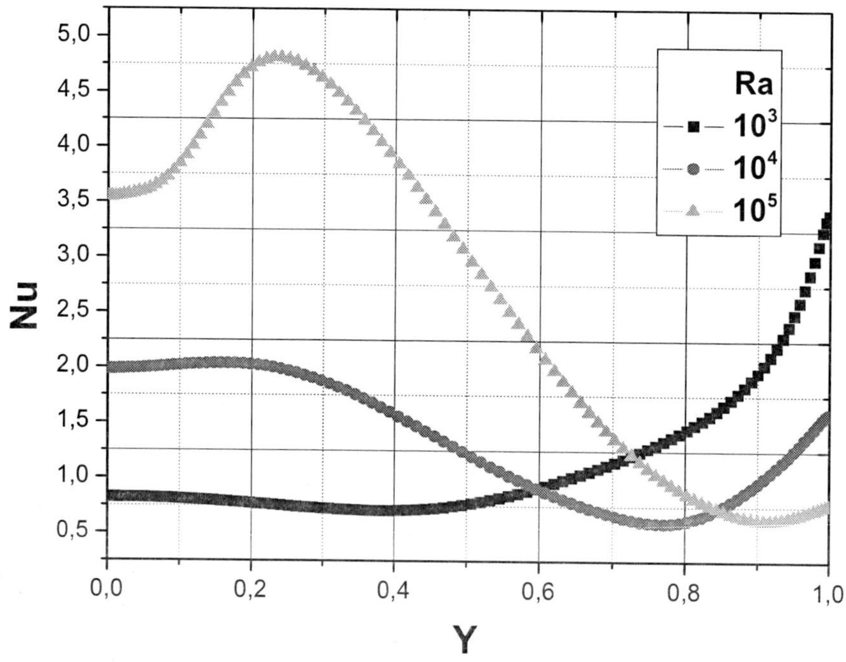

Figure II.4. Numerical values of the local Nusselt number, Nu , on the right heated surface for different Ra and with $Ma = 10^3$, $Ha = 20$ and $\varphi = 0$.

II.4.2. Effects of Varying the Hartmann Number on the Flow Field and the Heat Transfer

Fig. 5a illustrates the effect of increasing values of the magnetic field parameter on the flow patterns when the magnetic field is applied horizontally ($\varphi = 0$). As the magnetic field increases forces, the convective flow in the cavern slows down and the vortex shifts to the right and upwards, thus, concentrating near the free surface of the melt. At Ha = 100, the Lorentz forces dominate and, as a result the distortion of the isotherms caused by the melt flow is very small, Fig. II.5b. In the case of the horizontal magnetic field, it is seen that the streamlines in the upper part of the maps and the isotherms are equidistant. The main vortex shifts noticeably towards the upper surface, and the flow as a whole becomes multilayered. At Ha = 100 and 150, four and five vortices, respectively, are observed. From the value Ha =175, the structure of streamlines do not have a big change in their form, and remains the same for values of $Ha > 200$, Fig. II.6. It may be seen also, that the isotherms become more vertical and straighten out owing to the increase of the magnetic field strength, which is expected; since the magnetic field tend to weaken the flow, as observed above. The flow can be considered to be dominated by conduction phenomenon, as the convective motion has been totally inhibited. This effect can also be seen in Fig. II.7.

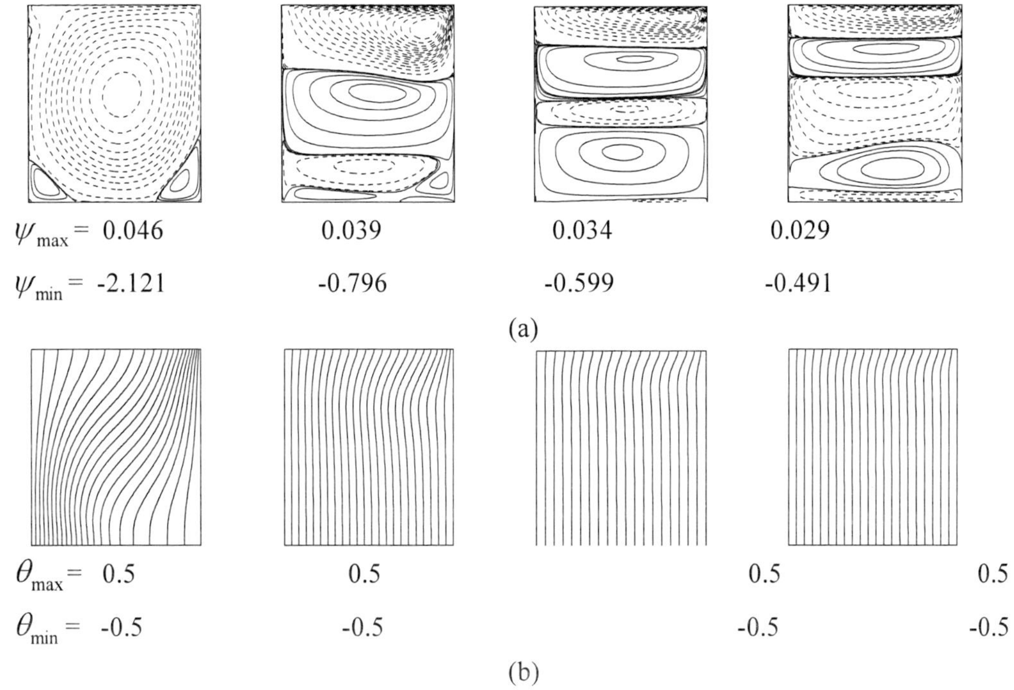

Figure II.5. Streamlines (top) and isotherms (bottom) for Ha =0, 50,100,150 with Ma =10^3, Ra =20 and $\varphi = 0$.

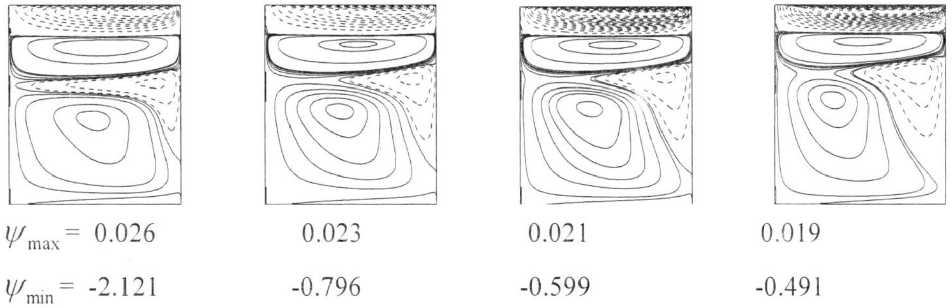

Figure II.6. Streamlines for Ha =175, 200, 220,250 with Ma =10^3, Ra =20 and $\varphi = 0$.

II.4.3. Effect of the Direction of the External Magnetic Field on the Flow and the Temperature Distribution

The effect of the direction of the external magnetic field, on the flow and the temperature distribution, is now discussed. Fig.8a, represents the streamlines obtained for $\varphi = 0, \pi/6, \pi/4, \pi/3$ and $\pi/2$ at $Ra = 10^2$, $Ma = 10^3$ and $Ha = 10^2$. As the direction of the external magnetic changes from horizontal to vertical, the flow rate in both the primary and the secondary cells decreases which causes an increase in the effect of the thermocapillary force and the flow becomes unicellular for $\varphi = \pi/2$. The corresponding isotherms are depicted in Fig. 8b. In this figure, one can see that as the direction of the external magnetic field changes from 0 to $\pi/4$, the isotherms near the heated surface become parabolic; whereas a further change of the direction to $\pi/2$, that is when the magnetic field acts in the vertical direction, the isotherms near the heated and cold surface become more vertical and straighten out. All these because change in the direction reduces the flow rate in the cells which results in heat transfer rate reduction on the heated surface Fig. II.9.

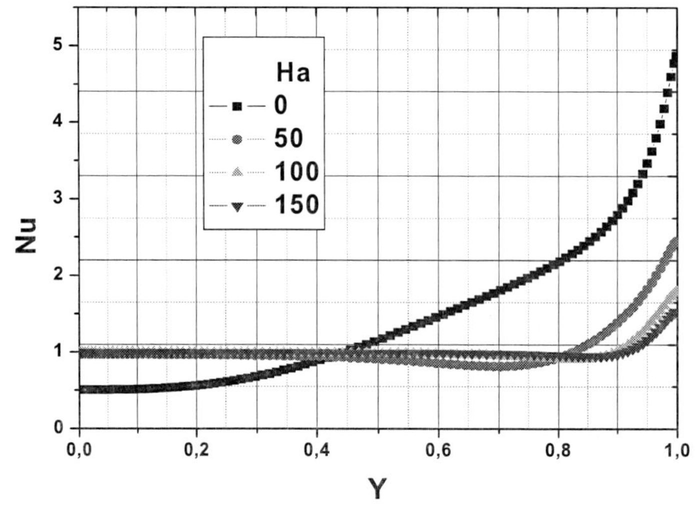

Figure II.7. Numerical values of the local Nusselt number, Nu, on the right heated surface for different Ha with $Ma = 10^3$, $Ra = 20$ and $\varphi = 0$.

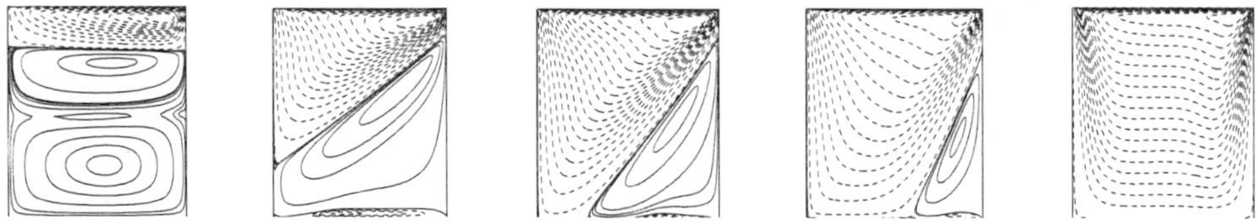

ψ_{max} = 0.035 0.0137 0.005 0.0025 3.69×10^{-6}

ψ_{min} = -0.600 -0.319 -0.193 -0.134 -0.098

(a)

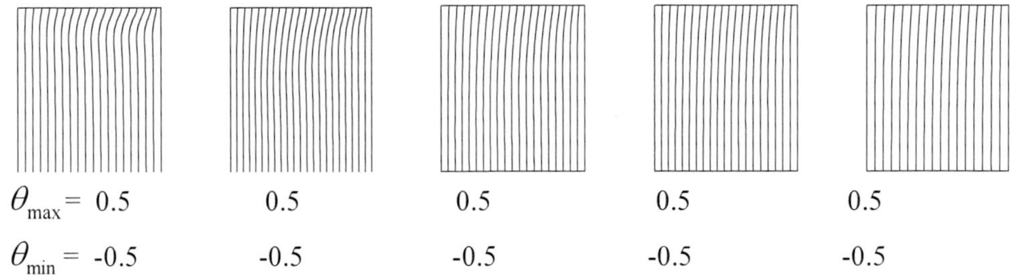

(b)

Figure II.8. Streamlines (top) and isotherms (bottom) $\varphi = 0, \pi/6, \pi/4, \pi/3$ and $\pi/2$ while $Ra = 10^2$, $Ma = 10^3$ and $Ha = 10^2$.

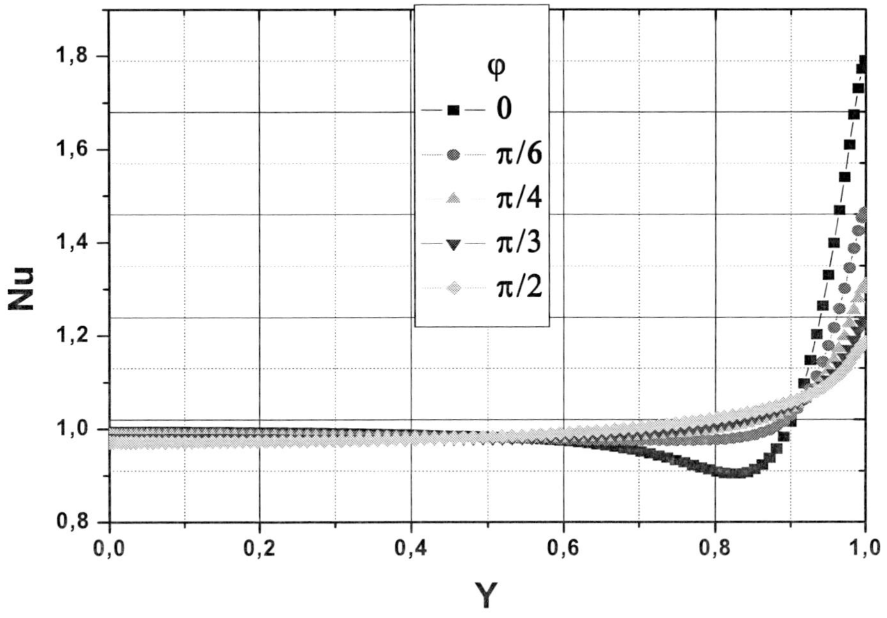

Figure II.9. Numerical values of local Nusselt number, Nu, at the right heated surface for different φ at $Ra = 10^2$, $Ma = 10^3$ and $Ha = 10^2$.

The variation of the average Nusselt number along the heated surface for different Hartmann number is shown in Fig. II.10. It can be seen that the mean Nusselt number for Case $Ra = 10^5$ is higher than that of Case $Ra = 20$. Globally, the mean Nusselt number is a decreasing function of Hartmann number on nonisothermal wall. Differences between Nusselt numbers decrease with the increasing of Hartmann number due to increasing of domination of conduction mode of heat transfer, due to the domination of the Lorentz forces. Variation of mean Nusselt numbers becomes constant for $Ha > 100$ in case that $Ra = 20$, and for $Ha > 150$ in case that $Ra > 10^5$. This result is valid for all two cases considered for $\varphi = 0$ or $\varphi = \pi/2$.

Fig. II.11 depicts Variation of the average Nusselt number at the heated surface for different φ, for $Ra = 20$ and 10^5, the inclination angle φ of the magnetic field have an influence on Nu_{av} . As can be further seen from this figure, Nu_{av} tends to decrease when the magnetic field is inclined between $\varphi = 0°$ and $\varphi = 90°$ (and between $\varphi = 180°$ and $\varphi = 270°$), and tends to increase when the magnetic field is inclined between $\varphi = 90°$ and $\varphi = 180°$ (and between $\varphi = 270°$ and $\varphi = 360°$), and everything is in the case of the dominancy of the thermo-capillary force over the buoyancy force. A minimal value of Nu_{av} has been observed at $\varphi = 180°$ (and φ=270). The physical mechanism behind this conclusion is that both the convective effect of the hot wall and the unstable effect of the cold wall are reduced by the inclination angle φ of magnetic field. But higher values of Nu_{av} are obtained in the case of φ = 0°, 180° (and 360°).

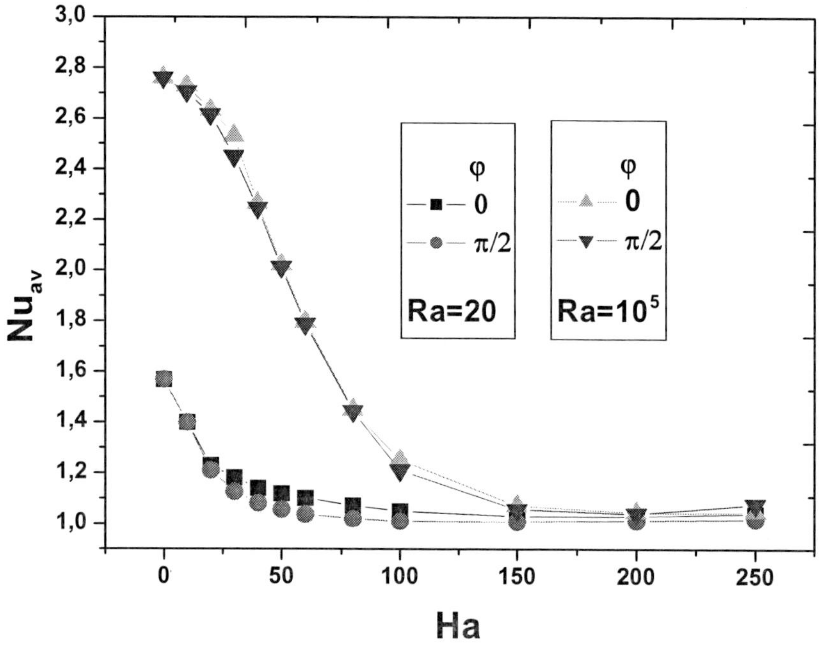

Figure II.10. Variation of the average Nusselt number at the heated surface with Hartmann number Ha while $Ra = (10^2, 10^5)$, $Ma = 10^3$, $Ha = 10^2$ and $\varphi = (0, \pi/2)$.

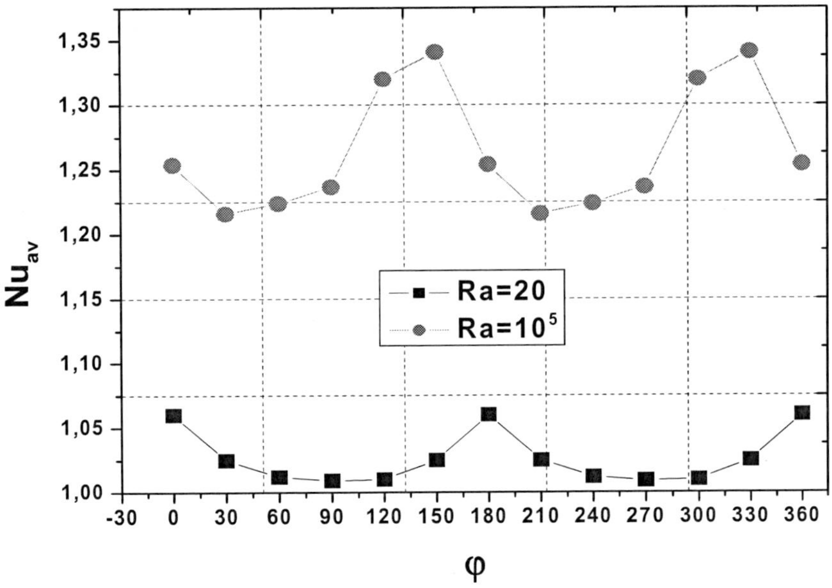

Figure II.11. Numerical values of average Nusselt number, Nu_{av}, at the right heated surface for different φ while $Ra = (10^2, 10^5)$, $Ma = 10^3$ and $Ha = 10^2$.

In the case of the dominancy of the buoyancy force over the thermo-capillary force ($Ra = 10^5$), we have a sinusoidal change of the average Nusselt number, Nu_{av} tends to decrease when the magnetic field is inclined between $\varphi = 0°$ and $\varphi = 30°$ and between $\varphi = 150°$ and $\varphi = 180°$, and tends to increase when the magnetic field is inclined between $\varphi = 30°$ and $\varphi = 150°$. The same change is observed in the other semi-circle (between $\varphi = 180°$ and $\varphi = 360°$). A minimal value of Nu_{av} has been observed at $\varphi = 30°$ (and $\varphi = 210°$) and a maximal value at $\varphi = 150°$ (and $\varphi = 330°$).

Conclusion

The following conclusions may be drawn from the present investigations in which a laterally heated horizontal cavity with aspect ratio (height/width) =1 and Pr=0.015 subjected to both surface tension and buoyancy forces was considered:

In the case of the vertical magnetic field, both the isotherms and the streamlines are equidistant. The flow becomes unicellular; the flow can be considered to be dominated by conduction phenomenon. In the case of the horizontal magnetic field, as the intensity of the magnetic field grows, the flow becomes multilayered with the main vortex shifted towards the free surface of the melt. The maximum absolute values of the stream function appear to be higher than the corresponding values obtained for the vertical field.

NOMENCLATURE

$\vec{B}$	Uniform magnetic field, $B_x \overrightarrow{ex} + B_y \overrightarrow{ey}$ T
B_x, B_y	Space independent components of $\vec{B}$ T
B_0	Magnitude of $\vec{B}$ T
Cp	Specific heat at constant pressure. $J.kg^{-1}.k^{-1}$
g	Gravitational acceleration $m.s^{-2}$
Ha	Hartmann number $= B_0 L/(\sigma_e \rho \nu)^{1/2}$
H	enclosure height m
k	Effective thermal conductivity $w.m^{-1}.k^{-1}$
Ma	Marangoni number $= \gamma \Delta T / \mu\alpha$.
Nu	Nusselt number
Nu_{avg}	Average Nusselt number
P	Fluid pressure Pa
Pr	Prandtl number $= \nu / \alpha$.
Ra	Rayleigh number $= \dfrac{g \beta \Delta T L^3}{\nu\alpha}$.
t	Time s
T	Temperature K
T_0	Reference temperature $= \dfrac{T_C - T_H}{2}$ K
u	Velocity in x-direction $m.s^{-1}$
v	Velocity in y-direction $m.s^{-1}$
$\vec{V}$	Field velocity $(u\overrightarrow{ex} + v\overrightarrow{ey})$ $m.s^{-1}$
x,y	m Cartesian coordinates m
X, Y	Dimensionless coordinates

Greek symbols

α	Thermal diffusivity $m^2.s^{-1}$
β	Coefficient of thermal expansion of fluid k^{-1}
ρ	Fluid density at reference temperature (T_0)
σ	Surface tension $N.m^{-1}$
σ_e	Electrical conductivity $S.m^{-1}$
μ	Effective dynamic viscosity $Kg.m^{-1}s^{-1}$
ν	Effective kinematic viscosity $m^2.s^{-1}$

γ	Temperature coefficient of the surface Tension $N.m^{-1}.K^{-1}$
ΔT	Difference in temperature = $T_C - T_F$ K
Ψ	Streamfunction $m^2.s^{-1}$
θ	Dimensionless temperature $= \frac{T - T_0}{\Delta T}$.
φ	The orientation of the magnetic field with horizontal axis ^{0}
τ	Dimensionless time

Subscripts

max	Maximum value
min	Minimum value
avg	Average value
0	Reference state
C	Cold
H	Hot

REFERENCES

Achoubir, K.; Bennacer, R.; Cheddadi, A.; El Ganaoui, M.; Semma, E. *Fluid Dyn. Mater. Process.*, 2008, 4(3), 199-210.

Armour, N.; Dost, S. *Fluid Dyn. Mater. Process.* 2009, 5(4), 331-344.

Baumgartl, J.; Hubert, A.; Muller, G. *Phy. of Fluids.* 1993, A-5(12), 3280- 3289.

Ben Hadid, H.; Henry, D.; Kaddeche, H. *J. Fluid Mech.* 1997, 333 23–56.

Ben Hadid, H.; Roux, B. *J. Fluid Mech.* 1992, 235, 1.

Bucchignani, E. *Fluid Dyn. Mater. Process.* 2009, 5(1), 37-60.

Djebali, R.; El Ganaoui, M.; Sammouda, H.; Bennacer, R. *Fluid Dyn. Mater. Process.* 2009, 5(3), 261-282.

Favier, J. J. *J. Cryst. Growth.* 1990, 99, 18.

Gelfgat, A. YU.; Bar-Yoseph, P. Z., *Phys. Fluids*, submitted for publication.

Mechighel, F.; El Ganaoui, M.; Kadja, M.; Pateyron, B.; Dost, S. *Fluid Dyn. Mater. Process.* 2009, 5(4), 313-330.

Mezrhab, A.; Naji H. *Fluid Dyn. Mater. Process.* 2009, 5(3), 283-296.

Ostrach, S. (1990): Low-gravity fluid flows, *Annu. Rev. Fluid Mech.* 1990, 14, 313.

Ozoe, H.; Maruo, E. *JSME Int. J.*, 1987, 30, 774–784.

Ozoe, H.; Okada, K. *Int. J. Heat Mass Transfer.* 1989, 32, 1939–1954.

Rudraiah, N.; Venkatachalappa, M.; Subbaraya, C. K. *Internat. J. Non-Linear Mech.* 1995, 30(5), 759.

Weiss, N. O. *J. Fluid Mech.* 1981, 108, 247–272.

In: Navier-Stokes Equations
Editor: R. Younsi

ISBN: 978-1-61324-590-3

Chapter 3

USE OF THE NAVIER-STOKES EQUATIONS TO STUDY OF THE FLOW GENERATED BY TURBINES IMPELLERS

Z. Driss* and M.S. Abid
Laboratory of Electro-Mechanic Systems (LASEM),
National Engineering School of Sfax (ENIS), University of Sfax (US),
B.P. 1173, km 3.5 Soukra, 3038 Sfax, Tunisia

ABSTRACT

The present chapter reviews a decade of research on the use of the Navier-Stokes Equations to study the flow generated by different turbines impellers, conducted by our group at the University of Sfax. For this reason, computational fluid dynamics (CFD) methods are developed to predict the flow field in stirred tanks equipped by various turbines impellers type. The three-dimensional flow of a fluid is numerically analyzed using the Navier-Stokes equations in conjunction with standard k-ε turbulence model. The relevant governing equations are given, and alternative turbulence models are described. The computer method developed permits the numerical analyses of turbines with complex geometries.

Our main objective has been to study and compare the shape effects on the field flow and turbulence characteristics to choose the most effective agitation system for the mixing applications. In these applications, the effect on the local and global flow characteristics have been determined and compared. The numerical results such as flow patterns profile and power numbers variation on Reynolds number were compared to our experimental results and those obtained by other authors. The comparison proved a good agreement.

Keywords: Navier-Stokes, laminar; turbulent, flow, Modeling, CFD, Finite volume, Meshing, Turbines Impellers, Stirred tanks.

1. INTRODUCTION

The Navier-Stokes equations of fluid dynamics are a formulation of Newton's laws of motion for a continuous distribution of matter in the fluid state, characterized by an inability

* E-mail address: Zied.Driss@enis.rnu.tn

to support shear stresses. They may be derived from the microscopic description with some additional fundamental assumptions. They are a set of nonlinear partial differential equations which are thought to describe fluid motions for gases and liquids, from laminar to turbulent flows, on scales ranging from below a millimeter to astronomical lengths. They are exactly soluble only for the simplest examples, usually corresponding to laminar flows. The Navier-Stokes equations form the basis upon which the entire science of viscous flow theory has been developed. Strictly speaking, the term Navier-Stokes equations refers to the components of the viscous momentum equation. However, it is common practice to include the continuity equation and the energy equation in the set of equations referred to as the Navier-Stokes equations. In many important applications, including turbulence, they must be modified and matched, truncated and closed, or otherwise approximated analytically or numerically in order to extract any predictions. On its own this is not a fundamental barrier, for a good approximation can sometimes be of equal or greater utility than a complicated exact result. In fact, many interesting quantities vary rapidly in time in a turbulent flow and cannot be readily measured. In practice, all that can be measured in laboratory experiments are averages. These averages are well-defined, reproducible quantities. This leads to the concept of ensemble averages underlying the conventional theory of turbulence, and to the concept of statistical solutions of the Navier-Stokes equations.

In this chapter, we will restrict our attention to the incompressible Navier-Stokes equations for a Newtonian fluid. In this context, a mathematical model has been developed to simulate the flow processes occurring in the vessel tanks. In fact, it's essential to understand the basics of mathematical modeling of single phase flow processes before one attempt to model complex flow processes occurring in industrial vessel tanks. The unifying processes, which form part of the mechanical unit operations in process technology, have been used by man since time immemorial in preparing food and drink and in constructing his dwelling. Since the emergence of manufacturing and the advent of industrial production, stirring has been used in almost all branches of industry like metallurgy, building materials, glass, paper, chemicals, food and pharmaceuticals.

A general prediction of the mixing process is then possible for laminar and turbulent flow by applying the Navier-Stokes equations. For this reason, a specific computational fluid dynamics (CFD) code are developed to predict the flow field in stirred tanks equipped by various turbines impellers type.

The computer method developed permits the numerical analyses of turbines with complex geometries. Our main objective has been to study and compare the shape effects on the field flow and turbulence characteristics to choose the most effective agitation system for the mixing applications. In these applications, the effect on the local and global flow characteristics have been determined and compared. The numerical results such as flow patterns profile and power numbers variation on Reynolds number were compared to our experimental results and those obtained by other authors. The comparison proved a good agreement.

This chapter represents a brief summary of the state of the art in the field of stirring technology from the viewpoint of the author. It particularly, we focus on recent research results account being taken of scientific literature published up to the 2011 [1-36].

2. Model Formulation

2.1. Conservation Laws

The fundamental equations of Navier-Stokes Equations are based on the universal laws of conservation of mass, momentum and energy. The equation that results from applying the conservation of mass law to a fluid flow is called the continuity equation. The conservation of momentum law is the Newton's second law. When this law is applied to a fluid flow it yields a vector equation known as the momentum equation. The conservation of energy law is identical to the first law of thermodynamics and the resulting fluid dynamic equation is named the energy equation. They can be written as mathematical equations once a representation for the state of a fluid is chosen. In the context of mathematics, there are two classical representations. One is the so-called Lagrangian representation, where the state of a fluid particle at a given time is described with reference to its initial position. The other representation, adopted throughout this chapter is the so-called Eulerian representation. In this representation, the state of the fluid particle at that position and time is given. In the Eulerian representation of the flow, the density is represented as a function of the position and time.

In these conditions, the Navier-Stokes equations can be derived in integral formulation, in accordance with the conservation laws. Applying Gauss's theorem, equations can be rewritten in differential form. In fact, the differential form is often found in literature [37-50].

2.1.1. Continuity Equation

The equation of continuity is developed by making a mass balance over a small element of volume through which the fluid is flowing. Then the size of this element is allowed to go to zero and the desired partial differential equation is generated. The law of mass conservation expresses the fact that mass cannot be created in such a fluid system, nor can disappear from it. There is also no diffusive flux contribution to the continuity equation, since for a fluid at rest, any variation of mass would imply a displacement of fluid particles. In order to derive the continuity equation, consider the model of a finite control volume fixed in space, as presented in figure 1. The arbitrary finite region of the flow, bounded by the closed surface $\partial\Omega$ and fixed in space, defines the control volume Ω.

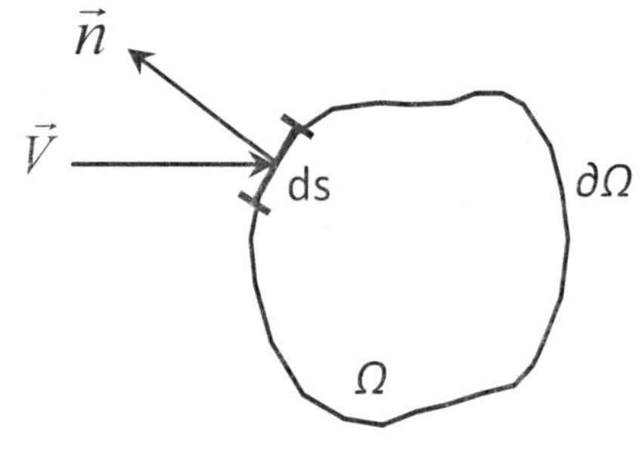

Figure 1. Finite control volume fixed in space.

At a point on the control surface, the flow velocity is $\vec{V}$, the unit normal vector is $\vec{n}$ and dS denotes an elemental surface area. The conserved quantity in this case is the density ρ. The time rate of change of the total mass inside the finite volume Ω have this expression: $\frac{\partial}{\partial t}\int_{\Omega} \rho \, d\Omega$.

The mass flow of a fluid through some surface fixed in space equals to the product of density, surface area and velocity component perpendicular to the surface. Therefore, the contribution from the convective flux across each surface element dS becomes equal to ($\rho \, \vec{V} \, \vec{n} \, dS$).

Since by convection $\vec{n}$ always points out of the control volume, we speak of inflow if the product ($\vec{V} \, \vec{n}$) is negative, and of outflow if it is positive and hence the mass flow leaves the control volume.

As stated above, there are no volume or surface sources present. Thus, by taking into account the general formulation, we can write:

$$\frac{\partial}{\partial t}\int_{\Omega} \rho \, d\Omega + \oint_{\partial\Omega} \rho \vec{V} \, \vec{n} \, ds = 0 \quad (1)$$

This represents the integral form of the continuity equation or the conservation law of mass.

In the case of a compressible Newtonian fluid and in absence of source terms in coordinate invariant, we can write equation (1) in differential form after applying Gauss's theorem:

$$\frac{\partial \rho}{\partial t} + \frac{\partial}{\partial x_i}(\rho V_i) = 0 \quad (2)$$

V_i denotes the velocity component and x_i stands for a coordinate direction.

A flow in which the density of each fluid element remains constant is called incompressible. Mathematically, this implies that:

$$\frac{\partial \rho}{\partial t} = 0 \quad (3)$$

Which reduce Equation (2) to the form:

$$\frac{\partial V_i}{\partial x_i} = 0 \quad (4)$$

2.1.2. Momentum Equations

We may start the derivation of the momentum equation by recalling the particular form of Newton's second law which states that the variation of momentum is caused by the net force acting on a mass element. For the momentum of an infinitesimally small portion of the control volume Ω, we have $\rho \vec{V}\, d\Omega$ (Figure 1).

The variation in time of momentum within the control volume Ω is equals to $\frac{\partial}{\partial t}\int_{\Omega} \rho \vec{V}\, d\Omega$.

Here, the product of density times the velocity $\rho \vec{V}$ is the conserved quantity.

The contribution of the convective flux tensor to the conservation of momentum is then given by: $-\oint_{\partial\Omega} \rho \vec{V}\left(\vec{V}\,\vec{n}\right) ds$.

In this case, two kinds of forces acting on the control volume can be identified: the external forces and the Surface forces. The external forces act directly on the mass of the volume. These are for example gravitational, Coriolis or centrifugal forces. The Surface forces act directly on the surface of the control volume. They result from only two sources: the pressure distribution and the shear and normal stresses. The pressure distribution is imposed by the outside fluid surrounding the volume. However, the shear and normal stresses, are resulted from the friction between the fluid and the surface of the volume.

The body force per unit volume denoted by $\rho \vec{f}_e$ corresponds to the volume sources. Thus, the contribution of the external force to the momentum conservation is equal to $\int_{\Omega} \rho \vec{f}_e\, d\Omega$.

The effect of the surface sources on the control volume is presented in figure 2. In fact, the surface sources consist of two parts: an isotropic pressure component $-p\bar{\bar{I}}$ and a viscous stress tensor $\bar{\bar{\tau}}$:

$$\bar{\bar{Q}}_s = -p\bar{\bar{I}} + \bar{\bar{\tau}} \tag{5}$$

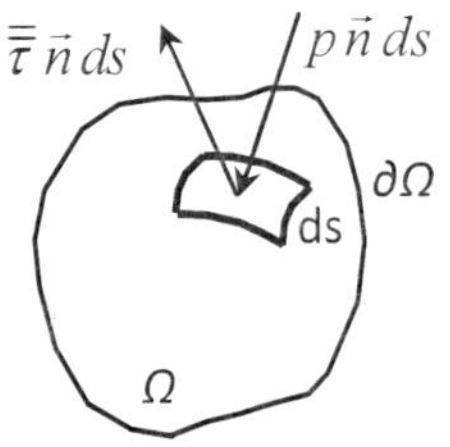

Figure 2. Surface forces acting on a surface element of the control volume.

The viscous stresses are originated from the friction between the fluid and the surface of an element. They depend on the dynamical properties of the medium. For fluids like air or water, Isaac Newton stated that the shear stress is proportional to the velocity gradient. Therefore, medium of such a type is designated as Newtonian fluid. On the other hand, fluids like for example melted plastic or blood behave in a different manner are non-Newtonian fluids.

The notation τ_{ij} means by convention that the particular stress component affects a plane perpendicular to the i-axis, in the direction of the j-axis. In Cartesian coordinates, the components τ_{xx}, τ_{yy} *and* τ_{zz} represent the normal stresses, the other components τ_{xy}, τ_{yx}, τ_{xz}, τ_{zx}, τ_{yz} *and* τ_{zy} stand for the shear stresses. Figure 3 shows the stresses for a quadrilateral fluid element. One can notice that the normal stresses try to displace the faces of the element in three mutually perpendicular directions, whereas the shear stresses try to shear the element.

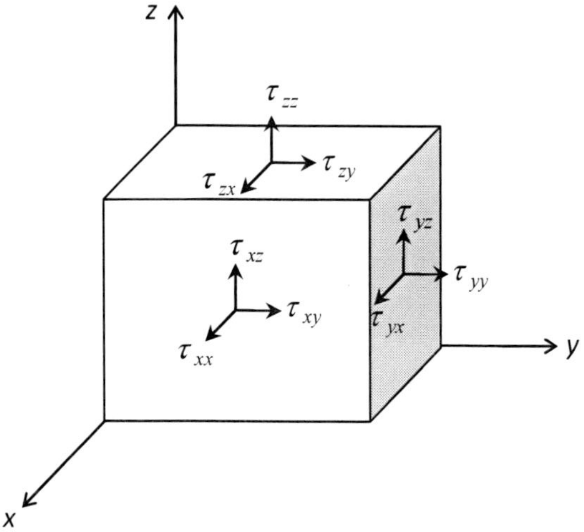

Figure 3. Normal and shear stresses acting on a fluid element.

For the vast majority of practical problems, where the fluid can be assumed to be Newtonian, the components of the viscous stress tensor are defined by the relations:

$$\tau_{xx} = \lambda\left(\frac{\partial U}{\partial x} + \frac{\partial V}{\partial y} + \frac{\partial W}{\partial z}\right) + 2\mu\frac{\partial U}{\partial y} \quad (6)$$

$$\tau_{yy} = \lambda\left(\frac{\partial U}{\partial x} + \frac{\partial V}{\partial y} + \frac{\partial W}{\partial z}\right) + 2\mu\frac{\partial V}{\partial y} \quad (7)$$

$$\tau_{zz} = \lambda\left(\frac{\partial U}{\partial x} + \frac{\partial V}{\partial y} + \frac{\partial W}{\partial z}\right) + 2\mu\frac{\partial W}{\partial y} \quad (8)$$

$$\tau_{xy} = \tau_{yx} = \mu\left(\frac{\partial U}{\partial y} + \frac{\partial V}{\partial x}\right) \quad (9)$$

$$\tau_{xz} = \tau_{zx} = \mu\left(\frac{\partial U}{\partial z} + \frac{\partial W}{\partial x}\right) \quad (10)$$

$$\tau_{yz} = \tau_{zy} = \mu\left(\frac{\partial V}{\partial z} + \frac{\partial W}{\partial y}\right) \tag{11}$$

λ and μ represent respectively the second viscosity coefficient and the dynamic viscosity coefficient. Also, we can define the so-called kinematic viscosity coefficient, which is given by:

$$\nu = \frac{\mu}{\rho} \tag{12}$$

The expressions were derived by the Englishman George Stokes in the middle of the 19th century. The term $\mu\frac{\partial V}{\partial x}$ in the normal stresses represent the rate of linear dilatation a change in shape.

On the other hand, the term $\lambda\, div\vec{V}$ represents volumetric dilatation rate of change in volume, which is in essence a change in density. In order to close the expressions for the normal stresses, Stokes introduced the hypothesis that:

$$\lambda + \frac{2}{3}\mu = 0 \tag{13}$$

With the exception of extremely high temperatures or pressures, there is so far no experimental evidence that Stokes's hypothesis does not hold. It is therefore used in general to eliminate λ from equations. Hence, we obtain for the normal viscous stresses:

$$\tau_{xx} = 2\,\mu\left(\frac{\partial U}{\partial x} - \frac{1}{3}div\vec{V}\right) \tag{14}$$

$$\tau_{yy} = 2\,\mu\left(\frac{\partial V}{\partial y} - \frac{1}{3}div\vec{V}\right) \tag{15}$$

$$\tau_{zz} = 2\,\mu\left(\frac{\partial W}{\partial z} - \frac{1}{3}div\vec{V}\right) \tag{16}$$

For an incompressible fluid, it should be noted that the expressions for the normal stresses simplify for an incompressible fluid because of $div\vec{V} = 0$.

According to the general conservation law, the expression for the momentum conservation inside an arbitrary control volume Ω is obtained as-follows:

$$\frac{\partial}{\partial t}\int_{\Omega} \rho \vec{V}\, d\Omega + \oint_{\partial\Omega} \rho \vec{V}\left(\vec{V}\,\vec{n}\right) ds = \int_{\Omega} \rho \vec{f}_e\, d\Omega - \oint_{\partial\Omega} p\,\vec{n}\, ds + \oint_{\partial\Omega} \tau\,\vec{n}\, ds \quad (17)$$

Applying Gauss's theorem, we can write in differential form and in absence of source terms:

$$\frac{\partial}{\partial t}\left(\rho V_i\right)+\frac{\partial}{\partial x_j}\left(\rho V_j V_i\right)=-\frac{\partial p}{\partial x_i}+\frac{\partial \tau_{ij}}{\partial x_j} \quad (18)$$

According to the Stokes's hypothesis, the components of the viscous stress tensor τ_{ij} are defined as-follows:

$$\tau_{ij}=2\,\mu\, S_{ij}+\lambda\frac{\partial V_k}{\partial x_k}\delta_{ij}=2\,\mu\, S_{ij}-\frac{2\,\mu}{3}\frac{\partial V_k}{\partial x_k}\delta_{ij} \quad (19)$$

In this equation, the first term corresponds to the components of the strain-rate tensor defined as-follows:

$$S_{ij}=\frac{1}{2}\left(\frac{\partial V_i}{\partial x_j}+\frac{\partial V_j}{\partial x_i}\right) \quad (20)$$

The antisymmetric part of the velocity gradient tensor defines the rotation-rate tensor:

$$A_{ij}=\frac{1}{2}\left(\frac{\partial V_i}{\partial x_j}-\frac{\partial V_j}{\partial x_i}\right) \quad (21)$$

For incompressible flows, the term of the divergence of the velocity disappears and we can reduce equation (11) to the form:

$$\frac{\partial V_i}{\partial t}+V_j\frac{\partial V_i}{\partial x_j}=-\frac{1}{\rho}\frac{\partial p}{\partial x_i}+\nu\nabla^2 V_i \quad (22)$$

2.1.3. Energy Equation

The first law of thermodynamics applied to the control volume (Figure 1) is used to obtain the energy equation. It states that any changes in time of the total energy inside the volume are caused by the rate of work of forces acting on the volume and by the net heat flux into it. The total energy per unit mass E of a fluid is obtained by adding its internal energy per unit mass e to its kinetic energy per unit mass $\left|\vec{V}\right|^2/2$. Thus, we can write for the total energy:

$$E = e + \frac{\left|\vec{V}\right|^2}{2} = e + \frac{U^2 + V^2 + W^2}{2} \tag{23}$$

Also, the total enthalpy per unit mass H of a fluid is obtained by adding its internal enthalpy per unit mass h to its kinetic energy per unit mass $\left|\vec{V}\right|^2/2$. Thus, we can write:

$$H = h + \frac{\left|\vec{V}\right|^2}{2} = h + \frac{U^2 + V^2 + W^2}{2} \tag{24}$$

The relation between the total enthalpy, the total energy and the pressure is given by:

$$H = E + \frac{p}{\rho} \tag{25}$$

In this case, the conserved quantity is the total energy per unit volume ρE. Its variation in time within the control volume Ω is equals to $\frac{\partial}{\partial t}\int_{\Omega} \rho\, E\, d\Omega$.

The contribution of the convective flux is given by: $-\oint_{\partial\Omega} \rho\, E\left(\vec{V}\,\vec{n}\right) ds$.

In contrast to the continuity and the momentum equation, there is now a diffusive flux. The Fick's law announces that it is proportional to the gradient of the conserved quantity per unit mass. Since the diffusive flux $\vec{F}_D$ is defined for fluid at rest, only the internal energy becomes effective:

$$\vec{F}_D = -\gamma\,\rho\,\kappa\,\nabla e \tag{26}$$

$\gamma = \frac{C_p}{C_v}$ is the ratio of specific heat coefficients and κ is the thermal diffusivity coeficient.

The diffusion flux represents one part of the heat flux into the control volume, namely the diffusion of heat due to molecular thermal conduction - heat transfer due to temperature gradients. Therefore, the equation (26) is in general written in the form of Fourier's law of heat conduction:

$$\vec{F}_D = -k\,\nabla T \tag{27}$$

k is the thermal conductivity coefficient and T is the absolute static temperature.

The other part of the net heat flux into the finite control volume consists of volumetric heating due to absorption or emission of radiation, or due to chemical reactions. We will denote the heat sources or the time rate of heat transfer per unit mass as $\dot{q}_h$. Together with the rate of work done by the body forces $\vec{f}_e$, it completes the volume sources, expressed as follows:

$$Q_v = \rho \vec{f}_e \vec{V} + \dot{q}_h \tag{28}$$

The surface sources Qs contribute also to the conservation of energy. As seen in the figure 2, they correspond to the time rate of work done by the pressure as well as the shear and normal stresses on the fluid element:

$$\vec{Q}_s = -p\vec{V} + \bar{\bar{\tau}}\vec{V} \tag{29}$$

Thus, we can write the energy conservation equation:

$$\frac{\partial}{\partial t}\int_{\Omega} \rho E \, d\Omega + \oint_{\partial\Omega} \rho H \left(\vec{V}\,\vec{n}\right) ds$$
$$= \oint_{\partial\Omega} k\left(\nabla T\,\vec{n}\right) ds + \int_{\Omega} \left(\rho \vec{f}_e \vec{V} + \dot{q}_h\right) d\Omega + \oint_{\partial\Omega} \left(\bar{\bar{\tau}}\vec{V}\right)\vec{n}\, ds \tag{30}$$

Applying Gauss's theorem, we can write in differential form and in absence of source terms:

$$\frac{\partial}{\partial t}(\rho E) + \frac{\partial}{\partial x_j}(\rho V_j H) = -\frac{\partial\left(V_i \tau_{ij}\right)}{\partial x_j} + \frac{\partial}{\partial x_j}\left(k\frac{\partial T}{\partial x_j}\right) \tag{31}$$

For incompressible flows, we can obtain the form:

$$\frac{\partial T}{\partial t} + V_j \frac{\partial T}{\partial x_j} = k\nabla^2 T \tag{32}$$

In the absence of buoyancy effects, the equation for the temperature T becomes decoupled from the mass conservation and momentum equations.

2.1.4. Complete System of the Navier-Stokes Equations

The conservation laws of mass, momentum and energy can be collected into one system of equations in order to obtain a better overview of the various terms involved. In these conditions, we introduce two flux vectors, namely $\vec{F}_c$ and $\vec{F}_v$. The first one is related to the convective transport of quantities in the fluid. It is usually termed vector of convective fluxes, although for the momentum and the energy equation it also includes the pressure terms. The second flux vector contains the viscous stresses as well as the heat diffusion. In addition, the source term $\vec{Q}$

comprises all volume sources due to body forces and volumetric heating. With all this in mind and conducting the scalar product with the unit normal vector $\vec{n}$, we obtain:

$$\frac{\partial}{\partial t}\int_{\Omega} \vec{W}\, d\Omega + \oint_{\partial\Omega} \left(\vec{F}_c - \vec{F}_v\right) ds = \int_{\Omega} \vec{Q}\, d\Omega \tag{33}$$

$\vec{W}$ is the vector of the conservative variables and it consists in three dimensions of the following five components:

$$\vec{W} = \begin{bmatrix} \rho \\ \rho U \\ \rho V \\ \rho W \\ \rho E \end{bmatrix} \tag{34}$$

The vector of convective fluxes $\vec{F}_c$ is given by:

$$\vec{F}_c = \begin{bmatrix} \rho\left(\vec{V}\,\vec{n}\right) \\ \rho U\left(\vec{V}\,\vec{n}\right) \\ \rho V\left(\vec{V}\,\vec{n}\right) \\ \rho W\left(\vec{V}\,\vec{n}\right) \\ \rho H\left(\vec{V}\,\vec{n}\right) \end{bmatrix} \tag{35}$$

The contravariant velocity is defined as the scalar product of the velocity vector and the unit normal vector to the surface element dS:

$$\left(\vec{V}\,\vec{n}\right) = n_x U + n_y V + n_z W \tag{36}$$

The vector of viscous fluxes $\vec{F}_v$ is given by:

$$\vec{F}_v = \begin{bmatrix} 0 \\ n_x\,\tau_{xx} + n_y\,\tau_{xy} + n_z\,\tau_{xz} \\ n_x\,\tau_{yx} + n_y\,\tau_{yy} + n_z\,\tau_{yz} \\ n_x\,\tau_{zx} + n_y\,\tau_{zy} + n_z\,\tau_{zz} \\ n_x\,\Theta_x + n_y\,\Theta_y + n_z\,\Theta_z \end{bmatrix} \tag{37}$$

The terms describing the work of viscous stresses and the heat conduction in the fluid are given by:

$$\Theta_x = U\,\tau_{xx} + V\,\tau_{xy} + W\,\tau_{xz} + k\frac{\partial T}{\partial x} \tag{38}$$

$$\Theta_y = U\,\tau_{yx} + V\,\tau_{yy} + W\,\tau_{yz} + k\frac{\partial T}{\partial y} \tag{39}$$

$$\Theta_z = U\,\tau_{zx} + V\,\tau_{zy} + W\,\tau_{zz} + k\frac{\partial T}{\partial z} \tag{40}$$

Finally, the source term is presented as-follows:

$$\vec{F}_c = \begin{bmatrix} 0 \\ \rho\, f_{e,x} \\ \rho\, f_{e,y} \\ \rho\, f_{e,z} \\ \rho\, \vec{f}_e\,\vec{V} + \dot{q}_h \end{bmatrix} \tag{41}$$

In the case of a Newtonian fluid, the above system of equations is called the Navier-Stokes equations. They describe the exchange of mass, momentum and energy through the boundary $\partial\Omega$ of a control volume $\partial\Omega$, which is fixed in space (Figure 1). The Navier-Stokes equations are derived in integral formulation, in accordance with the conservation laws. Applying Gauss's theorem, equations can be re-written in differential form. In fact, the differential form is often found in literature [38].

The Navier-Stokes equations represent in three dimensions a system of five equations for the five conservative variables: ρ, ρU, ρV, ρW and ρE. But they contain seven unknown flow field variables: ρ, U, V, W, E, p and T. Therefore, it's necessary to supply two additional equations, which have to be thermodynamic relations between the state variables, like for example the pressure as a function of density and temperature, and the internal energy or the enthalpy as a function of pressure and temperature. Beyond this, we have to provide the viscosity coefficient μ and the thermal conductivity coefficient k as a function of the state of the fluid, in order to close the entire system of equations. Clearly, the relationships depend on the kind of fluid being considered. In addition to the equation developed from these universal laws, it is necessary to establish relationships between fluid properties in order to close the system of equations. An example of such a relationship is the equation of state which relates the thermodynamic variables pressure p, density ρ and temperature T.

2.2. Turbulent Flow Modelling

Viscous fluids flow in a laminar or in a turbulent state. There are, however, transition regimes between them where the flow is intermittently laminar and turbulent. Laminar flow is smooth and quiet without lateral motions. Turbulent flow is irregular, random, chaotic and has lateral motions as a result of eddies superimposed on the main flow, which results in random or irregular fluctuations of velocity, pressure, and, possibly, temperature. The flow consists of a spectrum of different scales where largest eddies are of the order of the flow geometry. At the other end of the spectra, we have the smallest eddies which are by viscous forces (stresses) dissipated into internal energy. Even though turbulence is chaotic, it is deterministic and is described by the Navier-Stokes equations. In turbulent flow, the diffusivity increases. This means that the spreading rate of boundary layers increases as the flow becomes turbulent. The turbulence increases the exchange of momentum in boundary layers and reduces or delays there by separation at bluff bodies. The increased diffusivity also increases the resistance (wall friction) in internal flows. Turbulent flow occurs at high Reynolds number and is always three-dimensional. Also, it is dissipative, which means that kinetic energy in the small (dissipative) eddies are transformed into internal energy. The small eddies receive the kinetic energy from slightly larger eddies. The slightly larger eddies receive their energy from even larger eddies and so on. The largest eddies extract their energy from the mean flow. This process of transferred energy from the largest turbulent scales (eddies) to the smallest is called cascade process.

An entire range of turbulence models has been developed. In these models there is a clear trade-off between complexity and representation of the underlying physics. The various theoretical approaches can be formulated in wave number space or physical space; can use long time averages, averages over specific structures, or no averages at all; and will usually involve some closure approximation based on statistical reasoning, dimensional analysis, experimental evidence, or simplified conceptual modeling. Many facets of the physics need to be addressed to accurately represent any process of industrial importance. First, accurate models of the physics based on fundamental understanding are needed. Second, the inherent dynamics of turbulence must be addressed.

Fluid motion is described as turbulent if it is irregular, rotational, intermittent, highly disordered, diffusive and dissipative. Turbulent motion is inherently unsteady and three-dimensional. Visualizations of turbulent flows reveal rotational flow structures with a wide range of length scales. Such eddy motions and interactions between eddies of different length scales lead to effective contact between fluid particles which are initially separated by a long distance. As a consequence, heat, mass and momentum are very effectively exchanged. The rate of scalar mixing in turbulent flows is greater by orders of magnitude than that in laminar flows. Heat and mass transfer rates are also significantly higher in turbulent flows. Turbulent flows are also associated with higher values of friction drag and pressure drop. However, more often than not, advantages gained with the enhanced transport rates are more valuable than the costs of higher frictional losses. It can be concluded that for many (if not most) engineering applications, turbulent flow processes are necessary to make the desired operation realizable and more efficient. It is, therefore, essential to develop suitable methods to predict and control turbulent flow processes.

2.2.1. Time-Resolved Simulations

Both direct numerical simulation (DNS) and large eddy simulations (LES) use the governing equations directly without time averaging. These equations are the Navier–Stokes equations, the continuity equation and the energy balance equations. In such an approach there are as many equations as unknowns, so the problem is deterministic and the equations are closed. However, the partial differential equations are nonlinear, higher order and coupled. Problems in numerical resolution can be extreme, especially when DNS calculations are used. The problem is complicated by the large range of length scales which are relevant to the process results and by the highly three dimensional nature of the flow field. Therfore, simulation results that are grid and time step-independent can be extremely difficult to attain. At the time of writing, time-resolved simulations are still in the province of the expert user. Despite this, a good deal of insight into modeling issues can be gained from a brief explanation of this approach to turbulence modelling.

2.2.1.1. Direct Numerical Simulation

The Navier-Stokes equations describe a momentum balance on a differential control volume at any instant in time. They are exactly correct, at any instant in time, so in principle all that is needed to solve turbulent flow is a transient solution of the Navier–Stokes equations with appropriate initial and boundary conditions. This is the approach used in DNS. In a high Reynolds number turbulent flow, the changes with time can be very rapid, and the range of scales is extreme. A full computational domain would be 27 10^9 cells big, a number that is far too large for used computers. The task is even more impressive when one realizes that thesimulation must be transient with adequate resolution in time. It takes a very large computer indeed to do such modeling, even at low turbulent Reynolds numbers. Present computations can only be applied to low Reynolds numbers in somewhat simple geometries. Despite the lack of ability to do extremely detailed space and time resolution calculations, calculations in more modest grid structures (still fine when compared to LES) can be of use. In particular, when the geometry is complex and local conditions are not as critical, such calculations could be very helpful in design. In an ideal world, one could use DNS to reproduce the experimental flow field that controls mixing, then obtain measures of the individual terms in the Navier–Stokes equations on scales down to a small multiple of the grid size. These terms determine the coupling between mixing with the instantaneous flow field. The results of these detailed, fully coupled calculations could then be used to test and develop models for subgrid scales in LES, and for other computational fluid dynamics (CFD) calculations where average forms of the equations are used [39].

2.2.1.2. Large Eddy Simulation

The second approach is to use large eddy simulations. The limitations of this method are much less severe. The large scale motions are computed in a manner similar to DNS but on a much coarser grid. The scale might be as coarse as 1/30. The computational domain would then be 27 10^3, which would not be difficult with current machines [39]. A grid several times as fine as this would not be out of the question, and initial LES simulations in stirred tanks have recently been reported [51-54]. This modeling approach computes the larger scales of turbulence directly as they vary in time and models the finer scales of turbulence. The LES modelling technique has few assumptions, all of which can be modified to provide a match between the experimental statistical measures and the more detailed large scale results. The advantage of LES is that it is far less computationally demanding than DNS, so that the computations can be pushed to higher Reynolds number flows. The problem is to decide which, if any, of the subgrid models and filtering techniques are adequate to represent the data. As in the DNS effort, one cannot expect to match the data on an instantaneous basis, since any instantaneous velocity record is expected to be unique. However, by tracking the statistics that are important to the mixing process, the critical information can (in principle) be extracted. Initial results are promising, showing excellent agreement between experiment and simulation for the trailing vortices associated with a Rushton turbine [53] and macroinstabilities associated with a pitched blade impeller in its resonant geometry [54].

2.2.2. Reynolds-Averaged Navier-Stokes Equations

For incompressible flows, the Navier-Stokes equations can be rewrite in differential form. This is used very often in literature on turbulence modelling and it allows for a compact and clear notation:

$$\frac{\partial V_i}{\partial x_i} = 0 \tag{42}$$

$$\frac{\partial V_i}{\partial t} + V_j \frac{\partial V_i}{\partial x_j} = -\frac{1}{\rho}\frac{\partial p}{\partial x_i} + \nu \nabla^2 V_i \tag{43}$$

To reduce the modeling problem to a single steady solution, Reynolds formulated time-averaging rules. Application of these rules yields a time-averaged form of the Navier–Stokes and other equations, known as the Reynolds averaged, or RANS, equations. These equations now relate time-averaged quantities, not instantaneous time-dependent values. The first approach for the approximate treatment of turbulent flows was presented by Reynolds in 1895 [37]. The methodology is based on the decomposition of the instantaneous variables into a mean value and a fluctuating value. The governing equations are then solved for the mean values, which are the most interesting for engineering applications. Thus, considering incompressible flows, the velocity components and the pressure are substituted by:

$$V_i = \overline{V_i} + v_i' \tag{44}$$

$$p = \overline{p} + p' \tag{45}$$

The mean value is denoted by an overbar and the turbulent fluctuations by a prime. The mean values are obtained by an averaging procedure. Three different forms of the Reynolds averaging can be used: time averaging, spatial averaging and ensemble averaging. One reason why we decompose the variables is that when we measure flow quantities we are usually interested in the mean values rather that the time histories. Another reason is that when we want to solve the Navier-Stokes equation numerically it would require a very fine grid to resolve all turbulent scales and it would also require a fine resolution in time.

The time averaging is appropriated for stationary turbulence or statistically steady turbulence:

$$\overline{V_i} = \lim_{T \longrightarrow \infty} \frac{1}{T} \int_t^{t+T} V_i \, dt \tag{46}$$

$$\overline{p} = \lim_{T \longrightarrow \infty} \frac{1}{T} \int_t^{t+T} p \, dt \tag{47}$$

In these conditions, the mean value does not vary in time, but only in space. In practice, $T \longrightarrow \infty$ means that the time interval T should be large as compared to the typical timescale of the turbulent fluctuations.

The spatial averaging is appropriated for homogeneous turbulence. Inside the control volume Ω, we have:

$$\overline{V_i} = \lim_{\Omega \longrightarrow \infty} \frac{1}{\Omega} \int_\Omega V_i \, d\Omega \tag{48}$$

$$\overline{p} = \lim_{\Omega \longrightarrow \infty} \frac{1}{\Omega} \int_\Omega p \, d\Omega \tag{49}$$

In this case, $\overline{V_i}$ is uniform in space, but it is allowed to vary in time.

The ensemble averaging is appropriated for general turbulence. The mean value is a function of time and of space coordinates:

$$\overline{V_i} = \lim_{N \longrightarrow \infty} \frac{1}{N} \sum_{n=1}^{N} V_i \tag{50}$$

$$\overline{p} = \lim_{N \longrightarrow \infty} \frac{1}{N} \sum_{n=1}^{N} p \tag{51}$$

For all three approaches, the average of the fluctuating part is zero:

$$\overline{v_i'} = 0 \tag{52}$$

However, it can be easily seen that:

$$\overline{v_i' v_i'} \neq 0 \tag{53}$$

The same is true if both turbulent velocity components are correlated:

$$\overline{v_i' v_j'} \neq 0 \tag{54}$$

In cases where the turbulent flow is both stationary and homogeneous, all three averaging forms are equivalent. This is called the ergodic hypothesis [37].

If we apply either the time averaging or the ensemble averaging to the incompressible Navier-Stokes equations, we obtain the following relations for the mass and momentum conservation:

$$\frac{\partial \overline{V}_i}{\partial x_i} = 0 \tag{55}$$

$$\rho \frac{\partial \overline{V}_i}{\partial t} + \rho \overline{V}_j \frac{\partial \overline{V}_i}{\partial x_j} = -\frac{\partial \overline{p}}{\partial x_i} + \frac{\partial}{\partial x_j}\left(\overline{\tau}_{ij} - \rho \overline{V_i' V_j'}\right) \tag{56}$$

These are known as the Reynolds-Averaged Navier-Stokes equations (RANS). The equations are formally identical to the Navier-Stokes equations with the exception of the additional term which constitutes the so-called Reynolds-stress tensor. The new term $\overline{V_i' V_j'}$ appears on the right-hand side. The tensor is symmetric and represents correlations between fluctuating velocities. It is an additional stress term due to turbulence (fluctuating velocities) and it is unknown. We need a model for $\overline{V_i' V_j'}$ to close the equation system. This is called the closure problem: the number of unknowns is larger than the number of equations. In fact, there are ten unknowns (three velocity components, pressure and six stresses) and four equations (the continuity equation and three components of the Navier-Stokes equations). For this simplification, we pay a dear price in that there are now more unknowns than equations. The additional unknowns are the six Reynolds stresses, which are the normal or mean-squared values (autocorrelations) and cross-correlations of the three components of fluctuating velocity. The Reynolds-stress tensor represents the transfer of momentum due to turbulent fluctuations and consists in 3D of nine components:

$$\tau_{ij}^{R} = -\rho \overline{V_i' V_j'} = -\rho\left(\overline{V_i \; V_j} - \overline{V}_i \overline{V}_j\right) \tag{57}$$

$$\rho \overline{V_i' V_j'} = \begin{bmatrix} \rho\overline{\left(V_1'\right)^2} & \rho\overline{V_1' V_2'} & \rho\overline{V_1' V_3'} \\ \rho\overline{V_2' V_1'} & \rho\overline{\left(V_2'\right)^2} & \rho\overline{V_2' V_3'} \\ \rho\overline{V_3' V_1'} & \rho\overline{V_3' V_2'} & \rho\overline{\left(V_3'\right)^2} \end{bmatrix} \tag{58}$$

The terms on the diagonal are the normal stresses or variances, and these squared terms will always be positive. In an idealized flow with no directional preferences, they will all be equal. The off-diagonal elements are symmetric. So, only three of them are unique. If the turbulence has no directional preference and there are no velocity gradients in the flow, the individual fluctuations will be completely random and the covariances will be equal to zero. This assumption of "no directional preference" or "isotropic turbulence" is an important concept for understanding the different classes of time-averaged turbulence models [39].

In the correlations, $\overline{V}_i$ and $\overline{V}_j$ can be interchanged. Thus, the Reynolds-stress tensor contains only six independent components. The sum of the normal stresses divided by density defines the turbulent kinetic energy:

$$k = \frac{1}{2}\overline{V_i' V_i'} = \frac{1}{2}\left[\overline{(V_1')^2} + \overline{(V_2')^2} + \overline{(V_3')^2}\right] \tag{59}$$

The laminar viscous stresses are evaluated using Reynolds-averaged velocity components:

$$\overline{\tau}_{ij} = 2\mu \overline{S}_{ij} = \mu\left(\frac{\partial \overline{V}_i}{\partial x_j} + \frac{\partial \overline{V}_j}{\partial x_i}\right) \tag{60}$$

Then, the fundamental problem of turbulence modelling based on the Reynolds-averaged Navier-Stokes equations is to find six additional relations in order to close the equations [38].

It turns out that this degree of simplification is too severe, and some way of treating the cross-correlations must also be considered. Although, complete texts and regular review articles are written on the subject of turbulence modelling [55-56]. The simplest approach makes an initial assumption that the Reynolds stresses can be modelled using k and its rate of dissipation ε. These are the two-equation isotropic models, including the k–ε model. A more general, but more complex approach models each of the Reynolds stresses separately, allowing the development of anisotropy, or orientation of eddies, in the flow.

3. Numerical Models

This section is to develop a mathematical model for simulating flow processes occurring in vessel tank. The flow processes occurring in these industrial vessel tanks may involve more than one phase. It is, however, essential to understand the basics of mathematical modeling of single phase flow processes before one attempt to model complex flow processes occurring in industrial vessel tanks.

3.1. Governing Equations

In agitated tanks applications, the control volume is rotating with the constant angular velocity $\vec{N}$ around a fixed axis. In this case, the Navier-Stokes equations are transformed into a rotating frame of reference. As a consequence, the source term has to be extended by the effects due to the Coriolis and the centrifugal force for the momentum equations. Furthermore, the energy equation has to be modified because of the centrifugal force. In fact, the Coriolis force does not contribute to the energy balance. Only the continuity equation stays unchanged since the mass balance is invariant to system rotation [37].

The Coriolis force per unit mass and the centrifugal force per unit mass are defined as follows:

$$\vec{f}_{Cor} = -2\left(\vec{N} \wedge \vec{V}\right) \tag{61}$$

$$\vec{f}_{cen} = N^2 \vec{r} \tag{62}$$

Thus, the momentum equations can be written as follows:

$$\frac{\partial\left(\rho \vec{V}\right)}{\partial t} + div\left(\rho \vec{V} \otimes \vec{V} + p \bar{\bar{I}}\right) = div\left(\bar{\bar{\tau}} + \bar{\bar{\tau}}_R\right) + \rho\left(N^2 \vec{r} - 2\vec{N} \wedge \vec{V}\right) \tag{63}$$

The Navier-Stokes equations are often put into nondimensional form. The advantage in doing this is that the characteristic parameters such as Reynolds number, Froud number and Prandtl number can be varied independently. Furthermore, the flow variables are normalised and their values fall between certain prescribed limits such as 0 and 1. Many different nondimensionalizing procedures are possible. An example of one such procedure is:

$$r^* = \frac{r}{d} \tag{64}$$

$$z^* = \frac{z}{d} \tag{65}$$

$$t^* = \frac{t}{2\pi N} \tag{66}$$

$$p^* = \frac{p}{\rho\left(2\pi N d\right)^2} \tag{67}$$

$$U^* = \frac{U}{2\pi N d} \tag{68}$$

$$V^* = \frac{V}{2\pi N d} \tag{69}$$

$$W^* = \frac{W}{2\pi N d} \tag{70}$$

$$k^* = \frac{k}{(2\pi N d)^2} \tag{71}$$

$$\varepsilon^* = \frac{\varepsilon}{(2\pi N)^3 d^2} \tag{72}$$

The nondimensional variables are denoted by an asterisk and the diameter d of the turbine impeller is the reference length used in the Reynolds number Re given as follows:

$$Re = \frac{\rho N d^2}{\mu} \tag{73}$$

As well, the Froude number Fr is defined as follows:

$$Fr = \frac{(2\pi N)^2 d}{g} \tag{74}$$

If this nondimensionalizing procedure is applied to the Navier-Stokes equations, the following nondimensional equations are obtained in cylindrical coordinates (r,θ,z). It's clear that the nondimensional form of the equations is identical to the dimensional form. For convenience, the asterisks can be dropped from the nondimensional equations and this is usually done as follows:

$$\frac{\partial U}{\partial t} + div\left[\vec{V}U - \frac{2}{\pi}\left(\frac{d}{D}\right)^2 \frac{1}{Re} \upsilon_e \overrightarrow{grad}\, U\right]$$

$$= -\frac{\partial p}{\partial r} + \frac{2}{\pi}\left(\frac{d}{D}\right)^2 \frac{1}{Re}\left[\begin{array}{l} -2\upsilon_e\left[\frac{U}{r^2} + \frac{1}{r^2}\frac{\partial V}{\partial \theta}\right] + \frac{1}{r}\frac{\partial}{\partial r}\left[r\,\upsilon_e \frac{\partial U}{\partial r}\right] \\ + \frac{\partial}{r\,\partial \theta}\left[\upsilon_e\, r\frac{\partial}{\partial r}\left(\frac{V}{r}\right)\right] + \frac{\partial}{\partial z}\left[\upsilon_e \frac{\partial W}{\partial r}\right] \end{array}\right] + \frac{V^2}{r} + r + 2V \tag{75}$$

$$\frac{\partial V}{\partial t}+div\left[\vec{V}V-\frac{2}{\pi}\left(\frac{d}{D}\right)^2\frac{1}{Re}\upsilon_e\overrightarrow{grad}V\right]$$
$$=-\frac{\partial p}{r\partial\theta}+\frac{2}{\pi}\left(\frac{d}{D}\right)^2\frac{1}{Re}\left[\begin{array}{l}\upsilon_e\left[\frac{1}{r^2}\frac{\partial u}{\partial\theta}+\frac{\partial}{\partial r}\left(\frac{V}{r}\right)\right]+\frac{1}{r}\frac{\partial}{\partial r}\left[\upsilon_e\left(\frac{\partial U}{\partial\theta}-V\right)\right]\\+\frac{\partial}{r\,\partial\theta}\left[\upsilon_e\left(\frac{\partial V}{r\,\partial\theta}+\frac{2U}{r}\right)\right]+\frac{\partial}{\partial z}\left[\upsilon_e\frac{\partial W}{r\,\partial\theta}\right]\end{array}\right]-\frac{UV}{r}-2U \tag{76}$$

$$\frac{\partial W}{\partial t}+div\left[\vec{V}W-\frac{2}{\pi}\left(\frac{d}{D}\right)^2\frac{1}{Re}\upsilon_e\overrightarrow{grad}\,W\right]$$
$$=-\frac{\partial p}{\partial z}+\frac{2}{\pi}\left(\frac{d}{D}\right)^2\frac{1}{Re}\left[\begin{array}{l}\frac{1}{r}\frac{\partial}{\partial r}\left[r\,\upsilon_e\frac{\partial U}{\partial z}\right]+\frac{\partial}{r\partial\theta}\left[\upsilon_e\frac{\partial V}{\partial z}\right]\\+\frac{\partial}{\partial z}\left[\upsilon_e\frac{\partial W}{\partial z}\right]\end{array}\right]+\frac{1}{Fr} \tag{77}$$

The standard k-ε turbulence equations are given in the following form:

$$\frac{\partial k}{\partial t}+div\left[\vec{V}\,k-\frac{2}{\pi}\left(\frac{d}{D}\right)^2\frac{1}{Re}\left(1+\frac{\upsilon_t}{\sigma_k}\right)\overrightarrow{grad}\,k\right]=\frac{2}{\pi}\left(\frac{d}{D}\right)^2\frac{1}{Re}G-\varepsilon \tag{78}$$

$$\frac{\partial\varepsilon}{\partial t}+div\left[\vec{V}\,\varepsilon-\frac{2}{\pi}\left(\frac{d}{D}\right)^2\frac{1}{Re}\left(1+\frac{\upsilon_t}{\sigma_\varepsilon}\right)\overrightarrow{grad}\,\varepsilon\right]=\frac{\varepsilon}{k}\left[C_{1\varepsilon}\frac{2}{\pi}\left(\frac{d}{D}\right)^2\frac{1}{Re}G-C_{2\varepsilon}\,\varepsilon\right] \tag{79}$$

The turbulent kinetic energy production is given in the following form:

$$G=\upsilon_t\left[2\left[\left(\frac{\partial U}{\partial r}\right)^2+\left(\frac{\partial V}{r\partial\theta}+\frac{U}{r}\right)^2+\left(\frac{\partial W}{\partial z}\right)^2\right]+\left[\frac{\partial V}{\partial r}-\frac{V}{r}+\frac{\partial U}{r\partial\theta}\right]^2+\left[\frac{\partial W}{r\partial\theta}+\frac{\partial V}{\partial z}\right]^2+\left[\frac{\partial U}{\partial z}+\frac{\partial W}{\partial r}\right]^2\right] \tag{80}$$

The viscosity equations [12] are given in the following form:

$$\upsilon_t=C_\mu\frac{\pi}{2}\left(\frac{D}{d}\right)^2 Re\,\frac{k^2}{\varepsilon} \tag{81}$$

$$\upsilon_e=1+\upsilon_t \tag{82}$$

The constants of the standard k-ε model were presented in table 1.

Table 1. Standard *k*-*ε* model constants

$C_{1\varepsilon}$	$C_{2\varepsilon}$	C_{μ}	σ_k	σ_{ε}
1.44	1.92	0.09	1	1.3

The appropriate equations in the range of turbulent flow for an incompressible Newtonian fluid are expressed in the general conservation form which can be written as follows:

$$\frac{\partial}{\partial t}\iiint_D \Phi \, dv = -\iiint_D div\vec{J}_\Phi \, dv + \iiint_D S_\Phi \, dv \tag{83}$$

With:

$$\vec{J}_\Phi = \Phi\,\vec{V} - \Gamma_\Phi\,\overrightarrow{grad}\,\Phi \tag{84}$$

Using vector notation and cylindrical coordinates (r,θ,z), which of course are the most natural choice for cylindrical stirred vessels, the definition of Φ, $\vec{J}_\Phi$ and S_Φ for each equation is given, in dimensionless form, in table 2. The discretisation techniques are suitable for the treatment of the key transport phenomena, convection and diffusion as well as for the source terms and the rate of change with respect to time. The underlying physical phenomena are complex and non-linear so an iterative solution approach is required. The solution to the flow problem is defined at nodes inside each cell. Both the accuracy of the solution and its costs in terms of necessary computer hardware and calculation time are dependent on the fitness of the grid. The residuals for both velocity and pressure during the solution process were required to converge below 10^{-6}.

Table 2. Dimensionless form of flux and sink terms

Φ	Flux terms $\vec{J}_\Phi$	Sink terms S_Φ
1	$\vec{V}$	0
U	$-\frac{2}{\pi}\left(\frac{d}{D}\right)^2\frac{1}{Re}\left(\upsilon_e\overrightarrow{grad}\,U\right)+\vec{V}U$	$-\frac{\partial p}{\partial r}+\frac{2}{\pi}\left(\frac{d}{D}\right)^2\frac{1}{Re}\left[\begin{array}{l}-2\upsilon_e\left[\frac{U}{r^2}+\frac{1}{r^2}\frac{\partial V}{\partial\theta}\right]+\frac{1}{r}\frac{\partial}{\partial r}\left[r\upsilon_e\frac{\partial U}{\partial r}\right]\\+\frac{\partial}{r\partial\theta}\left[\upsilon_e r\frac{\partial}{\partial r}\left(\frac{V}{r}\right)\right]+\frac{\partial}{\partial z}\left[\upsilon_e\frac{\partial W}{\partial r}\right]\end{array}\right]+\frac{V^2}{r}+r+2V$
V	$-\frac{2}{\pi}\left(\frac{d}{D}\right)^2\frac{1}{Re}\left(\upsilon_e\overrightarrow{grad}\,V\right)+\vec{V}V$	$-\frac{\partial p}{r\partial\theta}+\frac{2}{\pi}\left(\frac{d}{D}\right)^2\frac{1}{Re}\left[\begin{array}{l}\upsilon_e\left[\frac{1}{r^2}\frac{\partial u}{\partial\theta}+\frac{\partial}{\partial r}\left(\frac{V}{r}\right)\right]+\frac{1}{r}\frac{\partial}{\partial r}\left[\upsilon_e\left(\frac{\partial U}{\partial\theta}-V\right)\right]\\+\frac{\partial}{r\partial\theta}\left[\upsilon_e\left(\frac{\partial V}{r\partial\theta}+\frac{2U}{r}\right)\right]+\frac{\partial}{\partial z}\left[\upsilon_e\frac{\partial W}{r\partial\theta}\right]\end{array}\right]-\frac{UV}{r}-2U$
W	$-\frac{2}{\pi}\left(\frac{d}{D}\right)^2\frac{1}{Re}\left(\upsilon_e\,\overrightarrow{grad}\,W\right)+\vec{V}\,W$	$-\frac{\partial p}{\partial z}+\frac{2}{\pi}\left(\frac{d}{D}\right)^2\frac{1}{Re}\left[\frac{1}{r}\frac{\partial}{\partial r}\left(r\upsilon_e\frac{\partial U}{\partial z}\right)+\frac{\partial}{r\partial\theta}\left(\upsilon_e\frac{\partial V}{\partial z}\right)+\frac{\partial}{\partial z}\left(\upsilon_e\frac{\partial W}{\partial z}\right)\right]+\frac{1}{Fr}$
T	$-\frac{2}{\pi}\left(\frac{d}{D}\right)^2\frac{1}{Pe}\left[\left(1+\upsilon_t\frac{Pr}{Pr_t}\right)\overrightarrow{grad}\,T\right]+\vec{V}T$	0
k	$-\frac{2}{\pi}\left(\frac{d}{D}\right)^2\frac{1}{Re}\left[\left(1+\frac{\upsilon_t}{\sigma_k}\right)\overrightarrow{grad}\,k\right]+\vec{V}k$	$\frac{2}{\pi}\left(\frac{d}{D}\right)^2\frac{1}{Re}\,G-\varepsilon$
ε	$-\frac{2}{\pi}\left(\frac{d}{D}\right)^2\frac{1}{Re}\left[\left(1+\frac{\upsilon_t}{\sigma_\varepsilon}\right)\overrightarrow{grad}\,\varepsilon\right]+\vec{V}\varepsilon$	$\frac{\varepsilon}{k}\left(C_{1\varepsilon}\frac{2}{\pi}\left(\frac{d}{D}\right)^2\frac{1}{Re}\,G-C_{2\varepsilon}\,\varepsilon\right)$

3.2. Computer Methods

3.2.1. CFD Code Presentation

Modelling a stirred tank using our specific CFD code requires consideration of many aspects of the process. The motion of the impeller in the tank must be treated in a special way. The special treatment employed impacts both the construction of the computational grid as well as the solution method used to numerically obtain the flow field. The basic principle is to split the domain under investigation into small elements. For each one of these elements, a set of partial differential equations is solved, which approximates a solution for the flow in order to achieve the conditions of conservation of mass and momentum. The basic equations are solved simultaneously with any additional equations implemented in a particular model to obtain the flow patterns, pressure, characteristics of turbulence and any derived quantities [1]. For easy access, our specific CFD code contains three main elements: pre-processor, processor and post-processor (Figure 4).

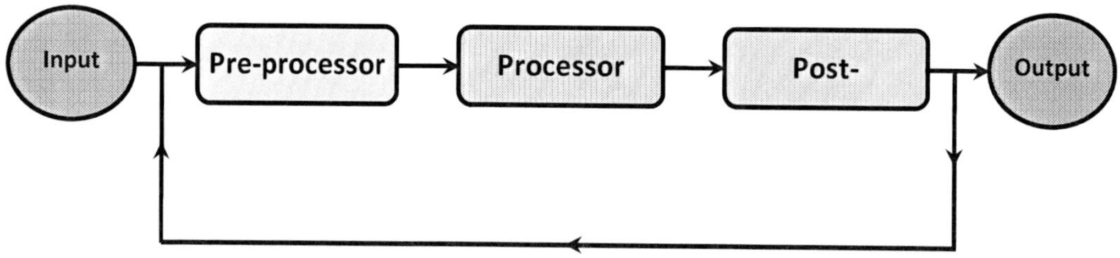

Figure 4. Schematic presentation of our specific CFD code.

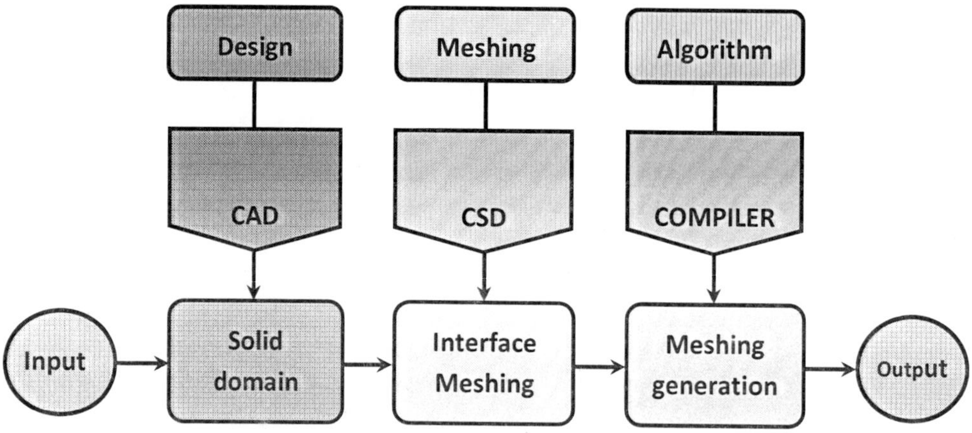

Figure 5. Schematic presentation of the pre-processor.

3.2.1.1. Pre-processor

The pre-processor serves as a source of input for the CFD code by means of an operator interface (Figure 6). This is transformed into a suitable form for the processor. The user activities at the pre-processing stage involve definition of the geometry, mesh generation and specification of the appropriate boundary conditions. The computational model requires that the volume occupied by the fluid inside the vessel to be described by a computational grid or cells. In these cells, variables are computed and stored. The computational grid must fit the contours of the vessel and its internals, even if the components are geometrically complex.

In our CFD applications, a new meshing method is developed to study the impellers turbines with complex geometries. This method permits the numerical analyses of turbines with simple and complex geometries [1]. In this method, the computer aided design (CAD) is used at first to construct the turbine shape. After that, a list of nodes is defined to belong to the interface separating the solid domain from the fluid domain passing through the computational structure dynamics (CSD) code (Figure 5). Using this list, the mesh in the flow domain is automatically generated for the three-dimensional simulations (Figure 6).

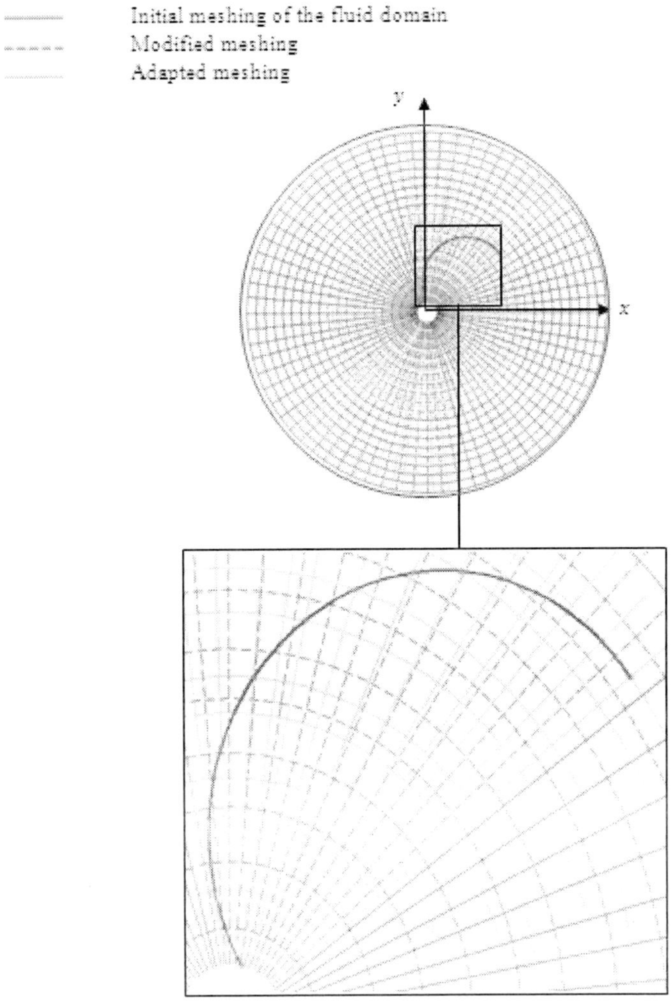

Figure 6. Method of the meshing adaptation.

After that, a staggered mesh is used in such a way that four different control volumes are defined for a given node point, one for each of the three vector components and one for the scalar variable (Figures 7 and 8). Therefore, the region to be modeled is subdivided into a number of control volumes defined on a cylindrical coordinates system [3].

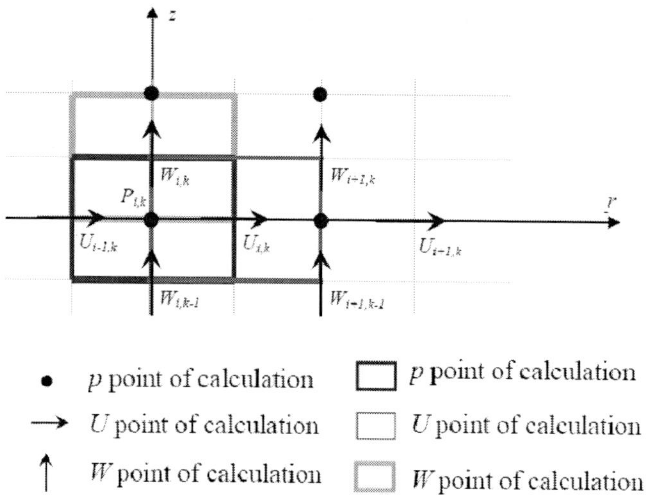

Figure 7. Staggered mesh in the *r-z* plane.

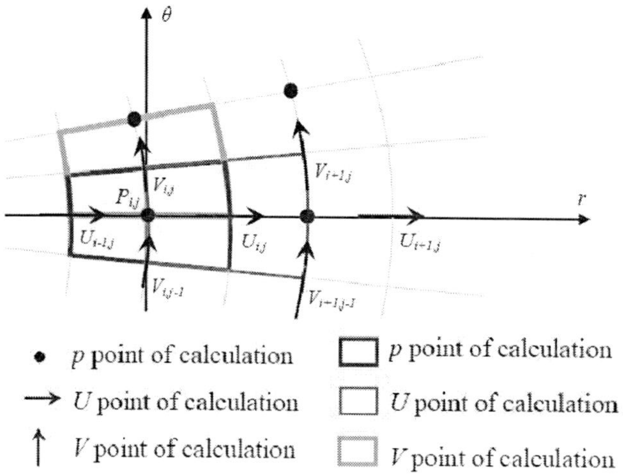

Figure 8. Staggered mesh in the *r-θ* plane.

3.2.1.2. Processor

Solutions of the Naviers-Stokes equations in conjunction with the standard k–ε turbulence model are developed using a control volume discretization method. Numerical methods consist on an approximation of the unknown flow variables by means of simple functions and a discretisation by substitution of the approximation into the governing flow equations and subsequent mathematical manipulations. The finite volume method consists on the formal integration of the governing equations of fluid flow over all the control volumes of the solution domain. The transport equations are integrated over its own control volume using the hybrid scheme discretization method. The pressure-velocity coupling is handled by the SIMPLE algorithm of Patankar [57]. The algebraic equations solutions are obtained in reference to the fundamental paper published by Douglas and Gunn [58]. The control volume integration, distinguishes the finite volume method from all other computers techniques. The resulting statements express the conservation of relevant properties for each finite size cell.

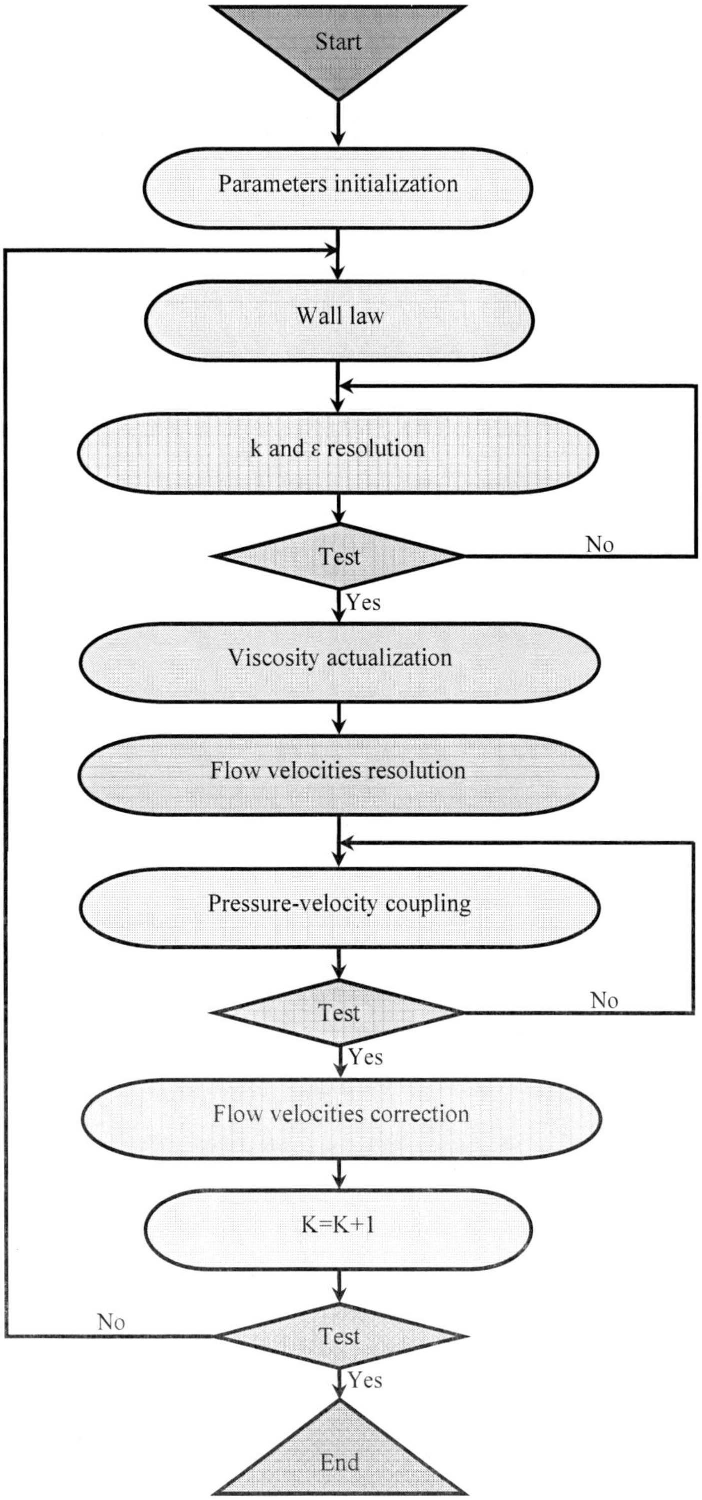

Figure 9. Numerical algorithm.

This clear relationship between the numerical algorithm and the underlying physical conservation principle forms one of the main attractions of the finite volume method and makes its concepts much simpler to understand by engineers than the finite element (Figure 9). The conservation of a general flow variable within a finite control volume can be expressed as a balance between the various processes tending to increase or decrease it.

3.2.1.3. Post-processor

A huge resource has been devoted to the development post-processing techniques and display. Owing to the increased popularity of engineering workstations, many of which have outstanding graphics capabilities. These include geometry domain, grid display, vector plots, surface plots and view manipulation. Several results can be gotten such as the velocity field, the viscous dissipation rate, the characteristics of turbulence and the power number. Also, it's possible to exchange data through the post-processor with other codes or other interfaces [5].

3.2.2. Boundary Conditions

The rotation of the turbine induced a periodic flow in a fixed reference frame. To obtain a steady state mode and to simplify the writing of the boundary conditions, a rotating reference frame fixed on the turbine shaft has been chosen. It is similar to a fixed impeller and a rotating tank which turns at the same rotating velocity but in opposed direction, by taking into account the centrifugal and Coriolis accelerations [1].

3.2.2.1. Velocity Components

On all rigid walls, a setting no-slip condition has been prescribed. Thus, in order to take into account the presence of the turbine, all velocity components mesh nodes, which intersected with turbine, were taken equal to zero.

$$U\,(I, J, K) = 0 \tag{85}$$

$$V\,(I, J, K) = 0 \tag{86}$$

$$W\,(I, J, K) = 0 \tag{87}$$

At the internal wall tank, the angular velocity component has been set equal to the rotating speed because of the rotating frame reference. Therefore, in dimensionless variables:

$$V\,(NR\text{-}1, J, K) = -1 \tag{88}$$

Otherwise, all radial as well as axial velocity mesh nodes were taken equal to zero:

$$U\,(NR\text{-}1,J,K) = 0 \tag{89}$$

$$W\,(NR\text{-}1,J,K) = 0 \tag{90}$$

In the closed tank, the surface of the liquid is always considered as plane that is to say that the vortex phenomenon is considered to be absent. This is warranted because Coriolis forces are reduced due to a decrease of the tangential motion. The liquid surface corresponds to the following conditions:

$$\frac{\partial U}{\partial z}(\text{I, J, NZ-1}) = 0 \tag{91}$$

$$\frac{\partial V}{\partial z}(\text{I, J, NZ-1}) = 0 \tag{92}$$

$$\text{W (I, J, NZ-1)} = 0 \tag{93}$$

At the lateral surface of the computational domain, zero velocity gradients are assumed.

3.2.2.2. Characteristics of Turbulence

Near of the walls, the relations known as laws of the wall are introduced to connect the values of k and ε to the wall with their values on the first point of adjacent grid to the wall. The values of the turbulent kinetic energy k and the dissipation rate of the turbulent kinetic energy ε are given in the following form:

$$k = \frac{\left(V_0\right)^2}{\sqrt{C_\mu}} \left(l_m \frac{dV_R}{dy^+}\right)^2 \tag{94}$$

$$\varepsilon = \frac{\left(V_0\right)^4}{\sqrt{C_\mu}} \left(l_m\right)^2 \left(\frac{dV_R}{dy^+}\right)^3 \tag{95}$$

Then, the k and ε boundary conditions were deduced while knowing the velocity neighbour to the wall V_R and the distance y of the adjacent node to the solid wall. The necessary stages for this calculation are presented as follows:

Calculate the product:

$$\frac{\rho V_R}{\mu} y \tag{96}$$

Determine the value of y^+ with the interpolation help,

Determine the rubbing velocity on the wall:

$$V_0 = \frac{\mu y^+}{\rho y} \tag{97}$$

Calculate the mixing length:

$$l_m = \kappa y^+ \left[1 - exp\left(-\frac{y^+}{A} \right) \right] \tag{98}$$

Determine:

$$\frac{d V_R}{d y^+} = \frac{2}{1+\sqrt{1+4\left(l_m\right)^2}} = \frac{2}{1+\sqrt{1+4\left[\kappa y^+\left[1-exp\left(-\frac{y^+}{A}\right)\right]\right]^2}} \tag{99}$$

Finally, deduct the values of the turbulent kinetic energy k and the dissipation rate of the turbulent kinetic energy ε.

For the flow parallel to the tank wall, the velocity neighbour to the wall V_R is given in the following form:

$$V_R = \sqrt{W_p^2 + V_p^2} \tag{100}$$

In this case, the axial W_p and the tangential V_p velocity compounds are considered:

$$W_p = \frac{W\left(NR-1,J,K\right)+W\left(NR-1,J,K-1\right)}{2} \tag{101}$$

$$V_p = \frac{V\left(NR-1,J,K\right)+V\left(NR-1,J-1,K\right)}{2} \tag{102}$$

On the tank bottom, the same procedure can be used. In this case, the radial U_p and the tangential V_p velocity compounds are considered:

$$V_R = \sqrt{U_p^2 + V_p^2} \tag{103}$$

$$U_p = \frac{U\left(I,J,2\right)+U\left(I-1,J,2\right)}{2} \tag{104}$$

$$V_p = \frac{V\left(I,J,2\right)+V\left(I,J-1,2\right)}{2} + r\left(I\right) \tag{105}$$

At the level of the turbine, the radial U_p and the axial W_p velocity compounds are considered:

$$V_R = \sqrt{U_p^{\,2} + W_p^{\,2}} \tag{106}$$

On the blade downstream, the velocity compounds are written as follows:

$$U_p = a\,U(I,J,K) + (1-a)\,U(I-1,J,K) \tag{107}$$

$$W_p = b\,W(I,J,K) + (1-b)\,W(I,J,K-1) \tag{108}$$

On the blade upstream, the velocity compounds are written as follows:

$$U_p = a\,U(I,J+1,K) + (1-a)\,U(I-1,J+1,K) \tag{109}$$

$$W_p = b\,W(I,J+1,K) + (1-b)\,W(I,J+1,K-1) \tag{110}$$

With:

$$a = \frac{\Delta R(I-1)}{\Delta R(I) + \Delta R(I-1)} \tag{111}$$

$$b = \frac{\Delta Z(K-1)}{\Delta Z(K) + \Delta Z(K-1)} \tag{112}$$

3.2.3. Global Characteristics

Power consumption is usually used in the design of equipment. Therefore, a detailed knowledge of power of stirred tank configurations is required. The dimensional analysis enables us to characterize power consumption through the power number N_p defined as follows:

$$N_p = \frac{P}{\rho N^3 d^5} \tag{113}$$

The power P consumed by the turbine in the stirred tanks should be equal to the power dissipated in the liquid. Hence, the power consumption was calculated from the volume integration predicated from the CFD code [59].

In turbulent flow:

$$P_t = \rho \iiint_D \varepsilon\, dv \tag{114}$$

In laminar flow:

$$P_l = \iiint_D \mu \, \Phi_v \, dv \tag{115}$$

The viscous dissipation rate can be expressed as follows:

$$\Phi_v = \left[2\left[\left(\frac{\partial U}{\partial r}\right)^2 + \left(\frac{\partial V}{r\partial\theta} + \frac{U}{r}\right)^2 + \left(\frac{\partial W}{\partial z}\right)^2 \right] + \left[\frac{\partial V}{\partial r} - \frac{V}{r} + \frac{\partial U}{r\partial\theta}\right]^2 + \left[\frac{\partial W}{r\partial\theta} + \frac{\partial V}{\partial z}\right]^2 + \left[\frac{\partial U}{\partial z} + \frac{\partial W}{\partial r}\right]^2 \right] \tag{116}$$

The pumping flow number was calculated using the following equation:

$$NQp = \frac{Q_p}{N\, d^3} \tag{117}$$

The numerical data of mean axial velocity below the impeller, in the radial direction were used for the calculation of pumping flow Q_p:

$$Q_p = \int_{z_B}^{z_B+h} \pi d \left(U\right)_{r=d/2} dz + \int_0^{d/2} 2\pi r \left(W\right)_{z=z_B} dr \tag{118}$$

The pumping efficiency can be defined by dividing the pumping flow number by the power number:

$$Ep = \frac{NQp}{N_p} \tag{119}$$

However, the energy efficiency can be obtained as follows:

$$Ee = \frac{NQp^3}{N_p} \tag{120}$$

4. Applications

In this section, applications of computational flow modeling tools to simulate flow generated by different impellers turbines are discussed in detail. The three-dimensional flow of fluid is numerically analysed using the continuity and the incompressible Navier-Stokes equations in conjuction with the standard k-ε turbulence model. Our computer simulations results offer local information about the laminar and turbulent mixing in vessel tanks. Our computer simulations results, such as the flow patterns, the viscous dissipation rate and the evolution of the pumping flow number NQp, the pumping efficiency number Ep, the energy efficiency Ee and the power number Np, offer local and global information about the laminar and turbulent mixing in vessel tanks. These results give a more precise understanding of the hydrodynamic mechanism than those obtained by experimental studies. The comparisons of the power number and the velocity profile have been compared with ones found by other researchers [60]. The good agreement between the numerical results and the experimental data validate the numerical method.

4.1. Simulation of Laminar Flow

Simulation of laminar flow is explored by many authors because this situation arises in many practical applications. The steady of the flow characteristics greatly facilitates by modeling. Thus provide an excellent starting point for solid progress. Many reserches haves developed in laminar flows. For exemple, Zalc et al. [61] explore laminar flow in an impeller stirred tank using CFD tools. They extend the analysis to include short and long time mixing performance as a function of the impeller speed. The simulated flow fields are validated extensively by particle image velocimetry (PIV). Also, they used planar laser induced fluorescence (PLIF) to compare the experimental and computed mixing patterns. Alvarez et al. [62] studied a stirred tank system with a single Rushton impeller mounted in a central shaft. Using UV visualization techniques, they illustrated the 3D mechanism by which fluorescent dye is dispersed within the chaotic region of the tank. Also, they compared a system with three Rushton impellers with a system having three discs at the same locations. Wang et al. [63] used electrical resistance tomography (ERT) to understand the mixing fundamentals and to validate advanced CFD models for gas liquid mixing in stirred vessels. Bakker et al. [64] used the large eddy simulation (LES) to predict these large-scale chaotic structures in stirred tanks. The numerical simulations were conducted using FLUENT 5. The predicted flow patterns for the pitched-blade turbine are compared well with digital particle image velocimetry data reported in the literature. Stitt [65] notes that multiphase reactor designs from larger scale and non-catalytic processes are now being considered. These include trickle beds, bubble columns, and jet or loop reactors. Fabrice Guillard and Christian Trägardh [66] have designed and tested a new model for estimating mixing times in aerated stirred tanks with three reactors which were equipped with two, three and four Rushton impellers, respectively. The results showed that the analogy model developed is independent of the scale, the geometry of the tank, the number of impellers used, the distance between impellers and the degree of homogeneity considered. Only the region in which the pulses were added was found to affect the results. Also they have defined loading and flooding regimes, which correspond to different types of interactions between the bubbles and the agitator. Kaneko et al. [67] analysed the three-dimensional motion of particles in a single helical ribbon agitator by the Discrete Element Method (DEM). To validate the computed results experiments were carried out with a cold scale model of 0.3 m inside diameter. Circulation time of particles in the agitator and the horizontal particle velocity distribution in the core region predicted by the simulation agreed well with those obtained by experiments. Based on DEM simulation, the particle circulation and mixing characteristics in the agitator vessel were investigated. Vertical mixing of particles was found rather poor during upward and downward flows through the blade and core regions, respectively. Espinosa-Solares et al. [68] have carried an ungassed power measurements in a dual coaxial mixer composed of an helical ribbon and a Rushton turbine in laminar mixing conditions for Newtonian and non Newtonian shear thinning fluids. For the Newtonian case, the power draw constant for the hybrid geometry was not the sum of the individual. This was explained by considering the radial discharge flow in the turbine region as well as the top-to-bottom circulation pattern of the helical ribbon impeller. For the non-Newtonian fluids, the results showed that, at a given Reynolds number, power consumption decreases as the shear thinning behaviour increases.

Yao et al. [69] analyzed the local and total dispersive mixing performance of large type impellers, a standard type of MAXBLEND and double helical ribbons impellers using two indices, local dispersive mixing efficiency and NPD function. The results indicated that a standard type of MAXBLEND has a satisfactory local dispersive mixing performance, especially in the grid region where the local dispersive mixing efficiency is high to near 1. However, when the Re is low, the total dispersive mixing performance is not as satisfactory as that operated under a moderate Reynolds number. The double helical ribbon impeller can not provide a promising local mixing performance. Although, it can induce a good total circulation throughout the stirring tank. Tanguy et al. [70] investigated the mixing performance of a new dual impeller mixer composed of a disc turbine and an helical ribbon impeller mounted on the same axis but rotating at different speeds. The methodology is based on a blend of experimental measurements and 3D numerical simulations in the case of Newtonian and non-Newtonian shear-thinning fluids. It is shown that the dual impeller mixer outperforms the standard helical ribbon in terms of top-to-bottom pumping when the fluid rheology evolves during the process. The power consumption of this mixer is also studied which allows to derive a generalized power curve. Bertrand *et al.* [71] elucidated the role of elasticity on power draw, though studies with helical ribbon impellers indicate that elasticity increases torque, in the case of viscoelastic fluids. The objective is to show that in the case of second-order fluids, the use of a simple constitutive equation derived from a second-order retarded-motion expansion succeeds in predicting a rise in power draw owing to elasticity. The equations of change governing fluid flow are solved using a finite element method combined with an augmented Lagrangian method for the treatment of the non-linear constitutive equation. They presented how the underlying non-linear tensor equations can be solved directly using a spectral decomposition of the related matrix operator. Niedzielska and Kuncewicz [72] examined the effect of impeller diameter modification and the pitch ratio on power consumption. In the course of experiments of power consumption, rotary frequence of impeller was changed and for this frequence value of moment of rotary was read. A change of liquid viscous was gained in result of water addition. Experiments were continued in range, in which liquid movement was laminar. Wang et al. [73] proposed a simple correlation to predict the Metzner-Otto constant. Through a comparison of different methods for predicting of Metzner-Otto constant from the viewpoint of numerical analysis, they introduced a new algorithm for estimating this constant. On the basis of the previous studies, we are studied the laminar flow of different impellers. Particularly, we are interested to the Flat blade turbine (FBT8) [7-11], the Retreated blade turbine (RBT8) [12-15], the Ancre impeller (AI) [16-24], the Maxblend impeller (MI), the Rushton turbines (RT) [25-29], the Double helical ribbons (DHR) and the Double helical ribbons screw (DHRS) [30-34].

4.1.1. Flat and Retreated Blade Turbines (FBT8 and RBT8)

In this application, computer simulations are carried out to compare the fundamental mechanisms of laminar mixing of the flat-blade turbine (FBT8) with the retreated-blade turbines (RBT8) in a vessel tank. The system under investigation is a stirred tank equipped with eight-blades. The tank is a vertical cylindrical vessel with a height-to-diameter ratio H/D of 1. The shaft is placed concentrically with a diameter ratio s/D of 0.04. The tip to tip impeller diameter ratio d/D is a 0.5 (Figure 10.a). Tow scraping angle equal to $\xi=0°$ and $\xi=40°$ were considered (Figure 10.b). For this impeller, the two rotation senses have tested. The impeller attack surface is a convex type for the first sense, whereas it is a concave type for the second sense. The computational results further elucidate the mechanism of laminar mixing: the convex retreated-blade paddle cause reorientation of fluid elements and the radial movement appears more intense and more active. The results of our computer simulations have compared with Nagata applications [60]. The flow conditions are represented by the Reynolds number Re=14 and the Froude number Fr=0.19.

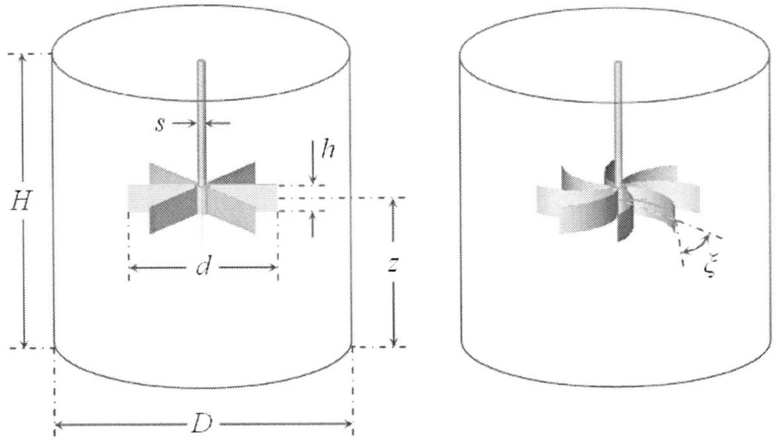

(a) FBT8 (b) RBT8

Figure 10. Stirred vessels configurations.

Figure 11. Flows patterns induced at the mid-height in r-θ plane.

4.1.1.1. Flow Patterns in r-θ Plane

Figure 11 shows a velocity vector plot of the primary flow in r-θ plane situated at the mid-height of the turbine. It appears that the flow is strongly dominated by the tangential component. Far from the region swept by the turbine, the rotating movement is no longer transmitted to the fluid, which remains quasi motionless. Also, it's noted that the velocity vector is directly affected by the blade type and the turbine rotation sense.

4.1.1.2. Flow Patterns in r-z Plane

Figures 12 and 13 show the secondary flow in two different r-z planes. The angular coordinates are equal to θ=20° and θ=40° respectively. In these conditions, the retreated-blade paddle intercepts the presentation plan on a segment with a length equal to the blade height. The blade intersection points with these presentation planes are defined by the radial position equal to r=0.2 and r=0.04 respectively. However, with a flat-blade paddle, these presentation plans are situated respectively in the upstream and in the downstream of the blade plane. These figures show the turbine stream impinging on the outer wall and being deflected in both directions in a wall jet motion producing the upper and lower vortices. Positions of the centre of these vortices are significantly affected by the turbine type. In fact, with a flat-blade paddle the flow movement coming from the turbine is a centrifugal type. The same fact is observed within a convex retreated-blade paddle. But, we note a light displacement of the vortice centre. If the turbine rotation sense is changed, the flow movement becomes a centripetal type. Also, the radial movement appears more intense and more active in the tank bottom within a concave retreated-blade paddle.

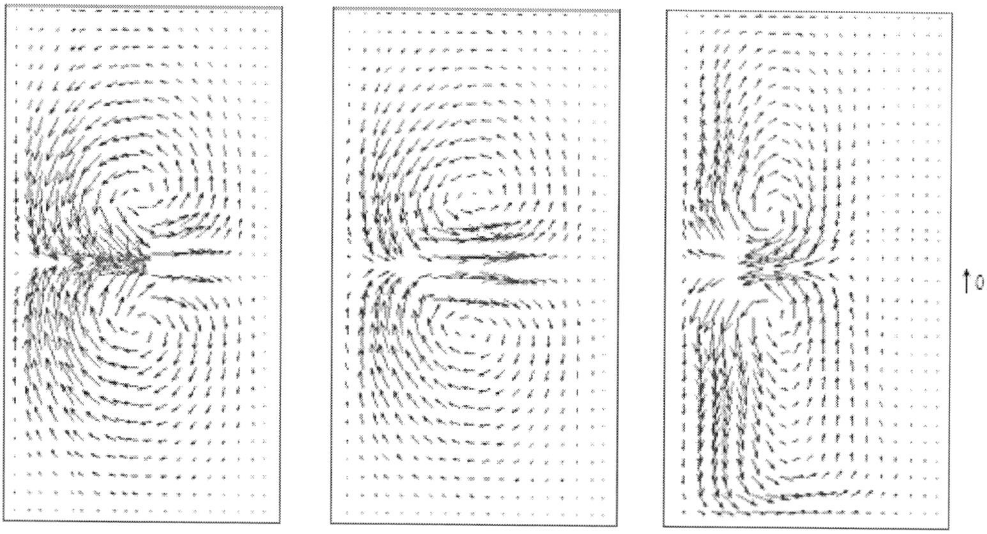

Figure 12. Flows patterns induced in two different r-z planes with angular coordinate θ=20°.

4.1.1.3. Viscous Dissipation Rate

Figures 14 and 15 show the viscous dissipation rate. The presentation planes are situated at the mid-height of the turbine (Figure 14) and just in the turbine bottom (Figure 15). Globally, it's noted a maximal value in the region localized in the blades tip. The biggest rate is reached within a concave retreated-blade paddle (Figure 14.c). This rate is divided by two within a convex retreated-blade paddle (Figure 14.b). It is even weaker within a flat-blade paddle (Figure 14.a). However, the viscous dissipation rate becomes quickly very weak out of the domain swept by blades (Figures 15.a, 15.b and 15.c).

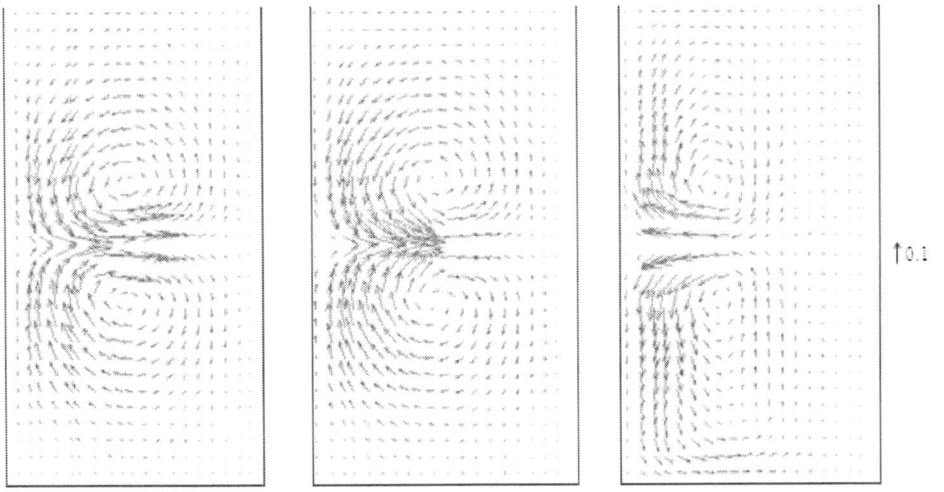

Figure 13. Flows patterns induced in two different r-z planes with angular coordinate θ=40°.

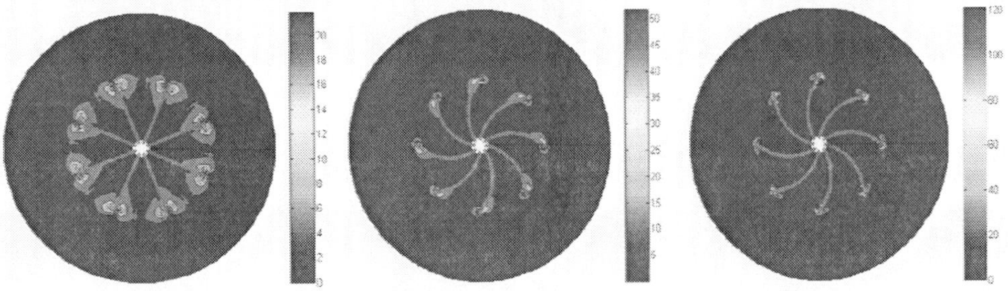

(a) FBT8 (b) Convex RBT8 (c) Concave RBT8

Figure 14. Viscous dissipation rate at the mid-height of the turbine.

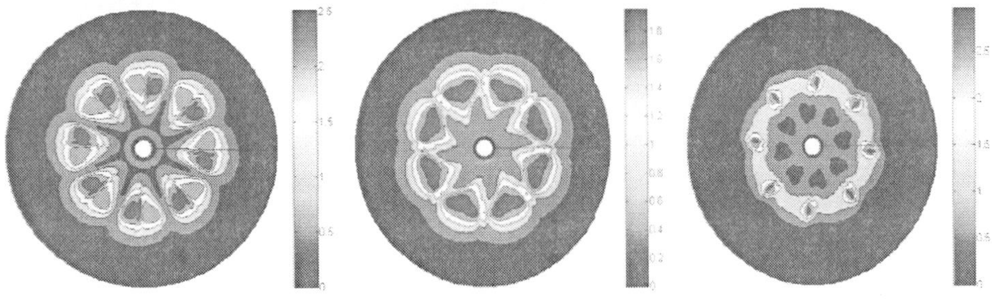

(a) FBT8 (b) Convex RBT8 (c) Concave RBT8

Figure 15. Viscous dissipation rate in the turbine bottom.

4.1.1.4. Axial Profiles of the Radial Velocity Component

Figure 16 illustrates the predicted axial profiles of the dimensionless radial velocity component *U(z)*. These profiles are defined in the flat-blade downstream. The dimensionless radial coordinates are equal to r=0.33 (figure 16.a), r=0.5 (figure 16.b) and r=0.66 (figure 16.c).

These figures adequately portray the swirling radial jet character. Within a flat-blade paddle, the axial profile shows a parabolic pace. The extremer is defined at the mid-height of the blade. The same fact is observed within a convex retreated-blade paddle. But, the radial velocity component value decreases. Within a concave retreated-blade paddle the pace is extensively modified and reverses itself completely. In these figures, we superimposed the experimental results founded within a flat-blade paddle. The good agreement between the Nagata application [60] and the numerical results confirms the validity of the analysis method.

4.1.1.5. Global Characteristics

Figure 17 show the power number *Np* variation on Reynolds number Re of the Flat blade turbine. In these conditions, the numerical value of the power number *Np* is slightly inferior to the experimental value when compared at the same Reynolds number *Re*. These experimental results founded by Nagata [60] were superposed over an average error of 7%. The good agreement between the experimental results and the numerical results quietly confirmed the analysis method.

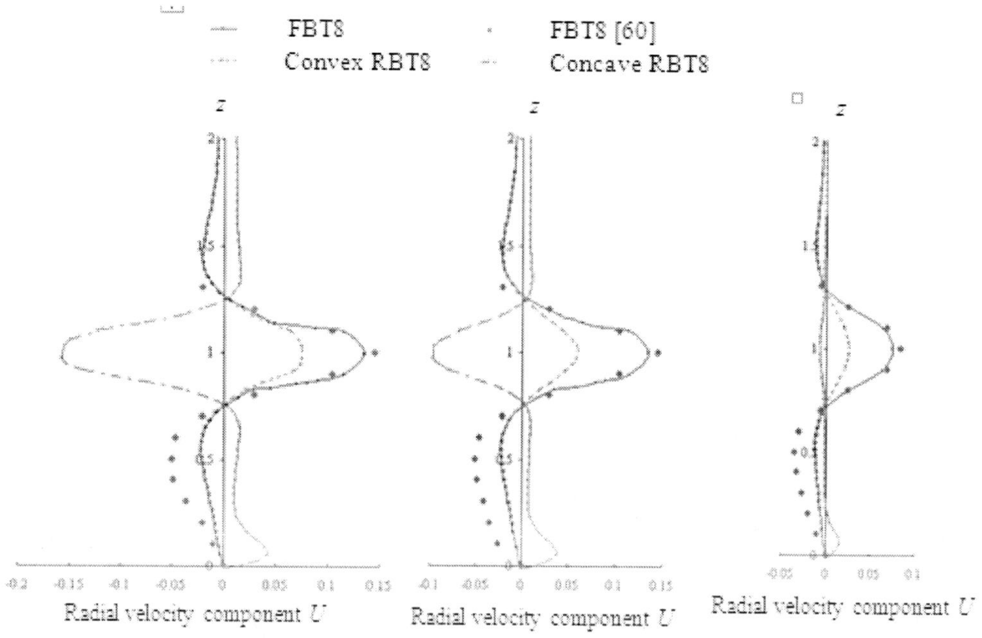

(a) Radial position: r=0.33 (b) Radial position: r=0.5 (c) Radial position: r=0.66

Figure 16. Comparison with Nagata applications [60].

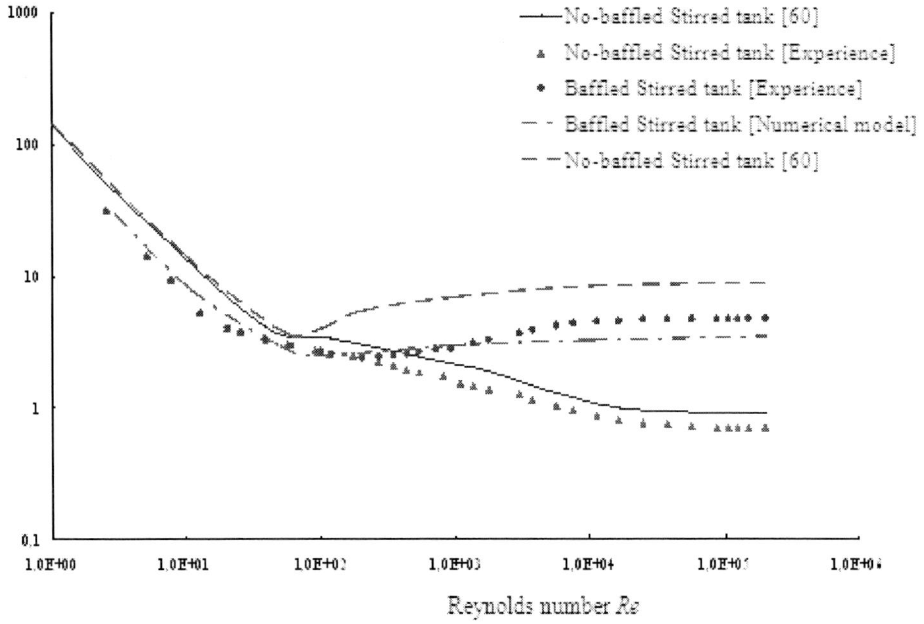

Figure 17. Comparison with experimental results.

4.1.2. Ancre and Maxblend Impellers

The notations used to describe the geometry of the agitated vessel are presented in Figure 18. The geometry system was made up of a flat bottom cylindrical vessel of diameter D and the liquid height in the vessel was $H=D$. The impellers are located in a cylindrical tank of height $h=0.95\ D$. The shaft is placed concentrically with a diameter ratio s/D of 0.04. The Ancre impeller has two arms of diameter $d_A=0.9\ D$, its width is equal to $w_A=0.1\ D$. The off-bottom clearance is equal to $e_A=0.1\ D$ and has an horizontal blade situated in the height equal to $h_{Ab}=0.6\ D$. The geometry of this system resembles to the one already studied by Pedrosa and Nunhez [74]. However, the Maxblend impeller diameter is equal to $d_M=0.9\ D$. Its width is equal to $w_M=$s. The height of the horizontal blade situated in the tank bottom is equal to $h_{Mb}=0.25\ D$. The flow is laminar, where the Reynolds number and the Froude number are equal respectively to Re = 14 and Fr = 0.19.

4.1.2.1. Flow Patterns

4.1.2.1.1. Flow Patterns in r-θ Plane

Figures 19, 20 and 21 show a velocity vector plot of the primary flow (U, V) presented in different r-θ planes situated respectively in the top, in the middle and in the bottom of the tank. For the symmetry reason, the velocity fields are only presented in the sector equal to 180°. In these presentation planes, it's observed that the flow was strongly dominated by the tangential component. In the bottom and in the middle of the tank, the velocity fields of the flow are relatively weak. But in the top, the velocity components improve a lot. Nearby the lateral surface of the tank, it has been noted a progressive slowing of the flow. Also, it has been noted that the velocity vector was directly affected by the proximity impellers type. In fact, the velocity field of the Maxblend impeller is more active than the ancre impeller.

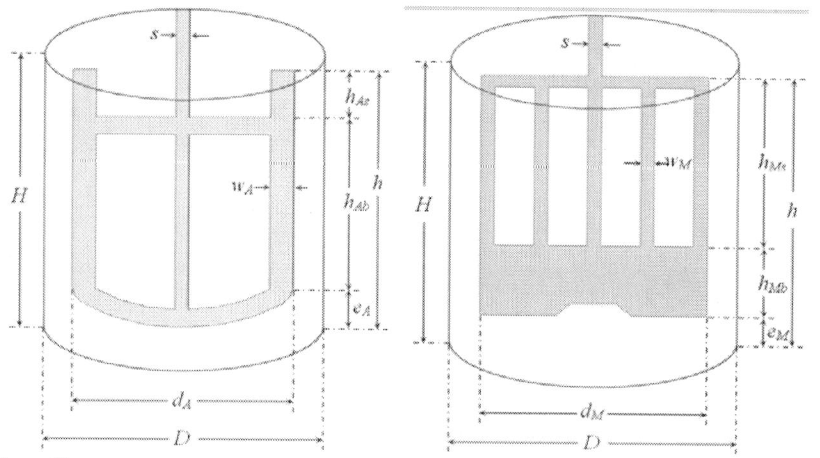

(a) Ancre impeller (b) Maxblend impeller

Figure 18. Stirred vessels configurations.

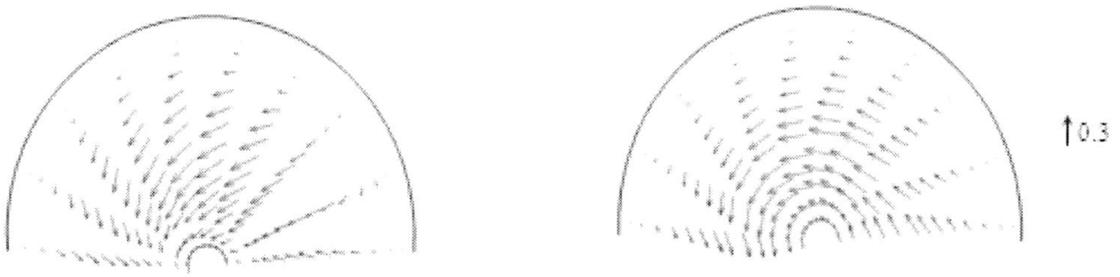

(a) Ancre impeller (b) Maxblend impeller

Figure 19. Viscous dissipation rate in the tank top.

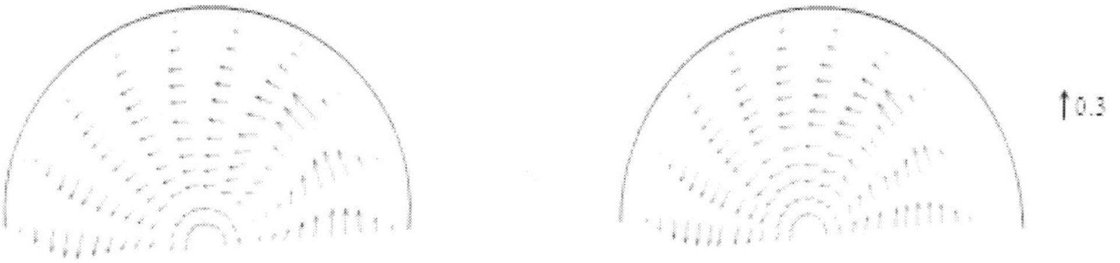

(a) Ancre impeller (b) Maxblend impeller

Figure 20. Viscous dissipation rate in the tank middle.

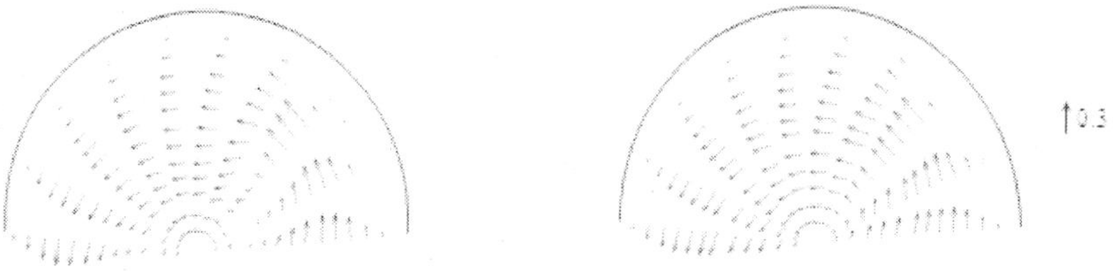

(a) Ancre impeller (b) Maxblend impeller

Figure 21. Viscous dissipation rate in the tank bottom.

4.1.2.1.2. Flow Patterns in r-z Planes

Figures 22, 23 and 24 show a velocity vector plot of the secondary flow (U,W) presented in r-z planes. These presentation planes have been defined respectively by the angular coordinate equal to θ=45°, θ=70° and θ=140°. These positions have been chosen in order to show the velocity field evolution. The presentation plane defined by θ=45° are situated in the ancre and Maxblend blade. However, the two other planes are defined in the downstream. In the first plane, the found result translated well the no slip condition that supposes that the all velocity components mesh nodes, which intersected with impellers shape, were taken equal to zero (Figure 22). In the second plane, it's noted that the Ancre impeller flow have a centripetal radial movement in the tank middle. However, within the Maxblend impeller the flow is more active and has an ascending oblique movement. The same fact is observed in the tank bottom. But, in the tank top, the flow has a descending oblique movement. While approaching to the lateral surface of the tank, the flow becomes weak (Figure 23). In the third plane, a circulation zone appears in the tank top. The axial position of the centre of these vortices are equal to z=0.75 for the Ancre impeller and z=0.82 for the Maxblend impeller. However, the radial position r=0.62 is the same for the two impellers. For this reason, the circulation zone of the Ancre impeller is more developed than the Maxblend impeller. For this impeller, a second circulation zone appears in the tank bottom.

The centre of this vortices is defined by the radial position r=0.48 and the axial position z=0.11. This fact proof that the mixing operation is more effective within the maxblend impeller than the ancre impeller (Figure 24).

4.1.2.2. Viscous Dissipation Rate

Figures 25, 26 and 27 show the viscous dissipation rate in different r-θ planes defined respectively by the axial positions equal to z=0.1, z=0.95 and z=1.15. Figures 28, 29 and 30 show the viscous dissipation rate in r-z planes defined respectively by the angular positions equal to θ=45°, θ=70° and θ=140°. These presentation planes have been chosen in order to show the viscous dissipation rate evolution in the stirred tank. Seen the symmetry of the problem, two symmetry wake shape are observed in these r-θ planes. Globally, the maximal value of the viscous dissipation rate is reached in the meeting of the impeller with these presentation planes.

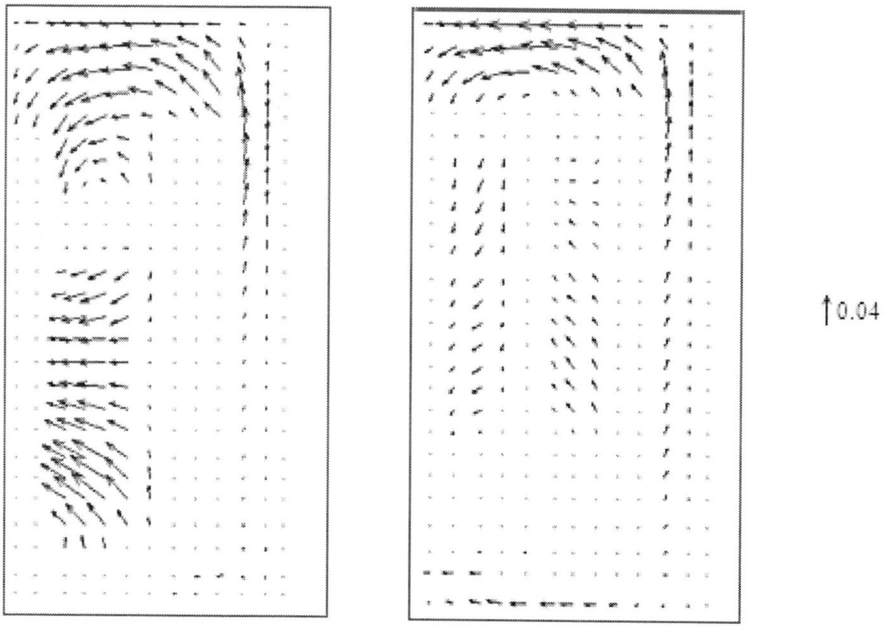

(a) Ancre impeller (b) Maxblend impeller

Figure 22. Flows patterns induced in r-z plane defined by θ=45°.

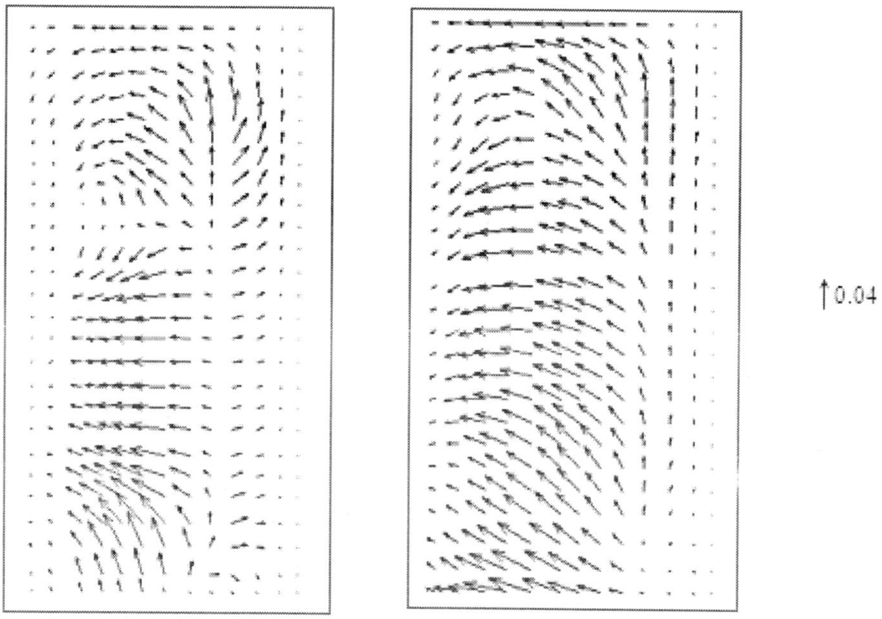

(a) Ancre impeller (b) Maxblend impeller

Figure 23. Flows patterns induced in r-z plane defined by θ=70°.

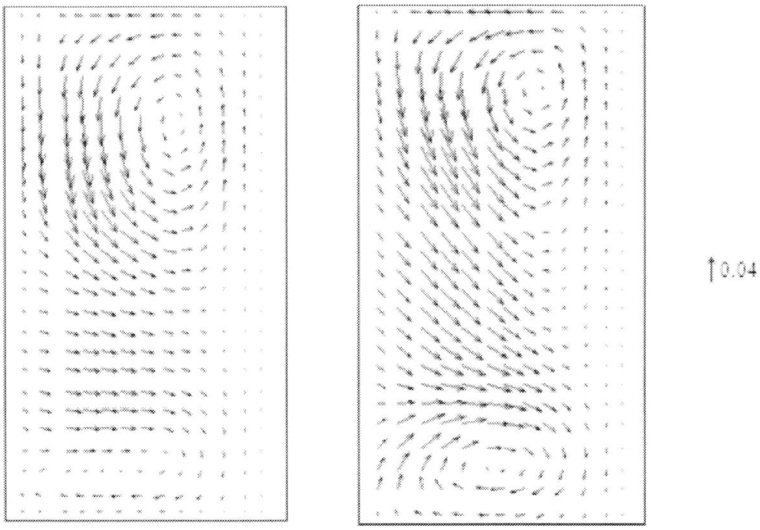

(a) Ancre impeller (b) Maxblend impeller

Figure 24. Flows patterns induced in r-z plane defined by θ=140°.

Out of this domain, the viscous dissipation rate becomes rapidly very weak. All of these observations are available for the two impellers. But, it's noted that the viscous dissipation rate for the Ancre impeller is more important than the Maxblend impeller. Thus confirm that the dissipation effect is more important within the Ancre impeller.

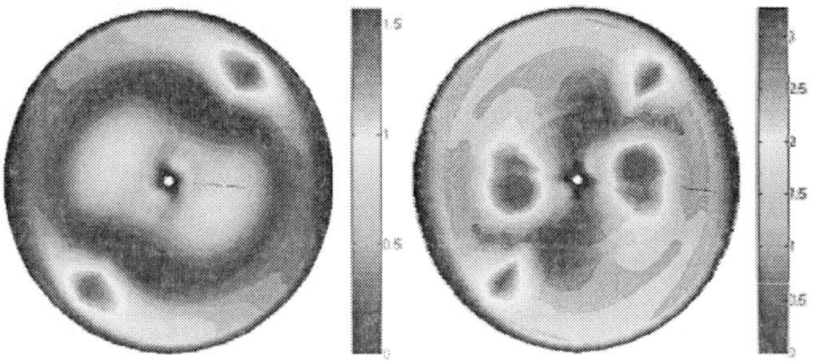

(a) Ancre impeller (b) Maxblend impeller

Figure 25. Viscous dissipation rate in the tank top.

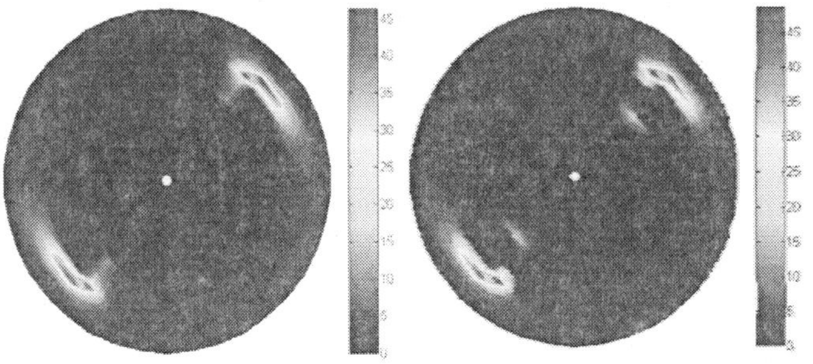

(a) Ancre impeller (b) Maxblend impeller

Figure 26. Viscous dissipation rate in the tank middle.

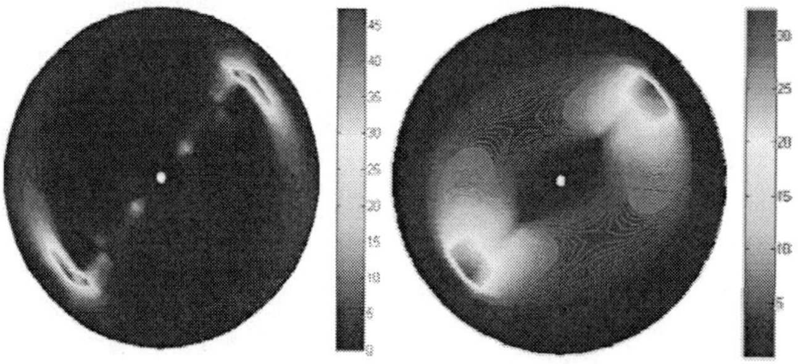

(a) Ancre impeller (b) Maxblend impeller

Figure 27. Viscous dissipation rate in the tank bottom.

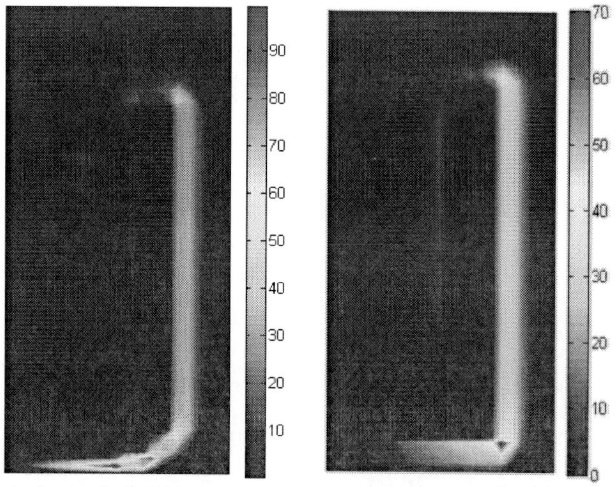

(a) Ancre impeller (b) Maxblend impeller

Figure 28. Viscous dissipation rate in r-z plane defined by θ=42°.

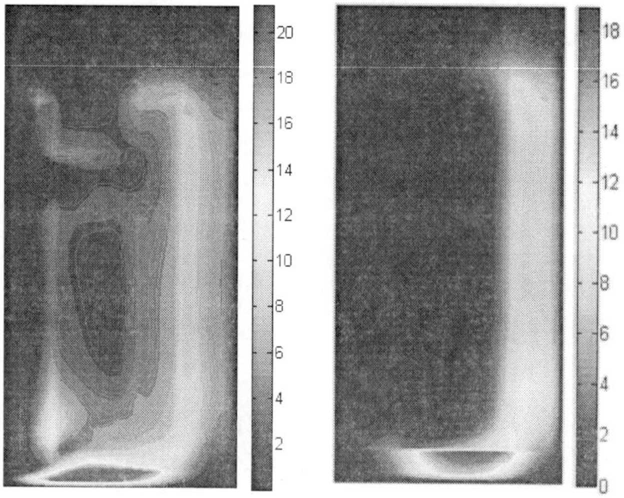

(a) Ancre impeller (b) Maxblend impeller

Figure 29. Viscous dissipation rate in r-z plane defined by θ=70°.

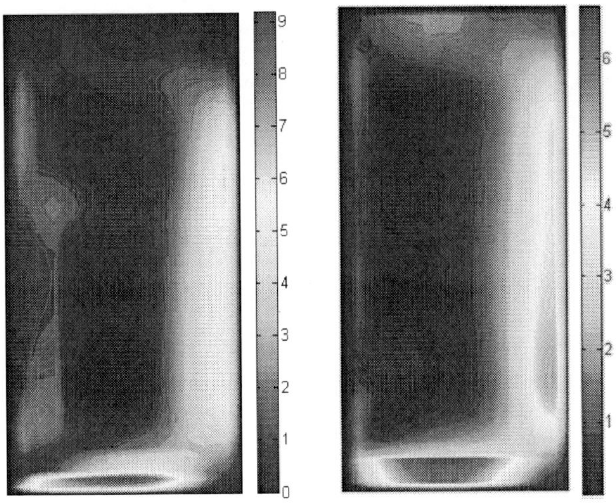

(a) Ancre impeller (b) Maxblend impeller

Figure 30. Viscous dissipation rate in r-z plane defined by θ=140°.

4.1.3. Rushton Turbines (RT4)

CFD modelling is carried out to compare the fundamental mechanisms of laminar mixing with multiple Rushton impellers in a vessel tank. The investigation was carried out in a cylindrical tank with a diameter D equal to height (D=H). The shaft is placed concentrically with a diameter ratio s/D of 0.02. The blade is defined by a height and a width equal to w=0.06 D and l=0.09 D respectively. Agitation was provided with three Rushton turbines of diameter d=D/3 placed at the distance h_1 from the vessel base. Distances between Rushton turbines discs are equal to h_2 and h_3 (Figure 31). These configurations are defined respectively by:

$$h_1=H/3 \tag{121}$$

$$h_1=h_2=H/3 \tag{122}$$

$$2\ h_1=h_2=h_3=H/3 \tag{123}$$

In this application, three different configurations equipped by one two and three impellers have compared (Figure 32). The flow conditions are defined by the Reynolds number Re=14 and the Froude number Fr=0.19. The gotten results confirmed that the multi-impellers systems appear necessary to reduce the amount of unreached zones in the stirred tanks. Thus, we can obviously conclude that the two Rushton impellers system is much more enough to generate an efficient mechanical agitation rather than the multiple Rushton impellers.

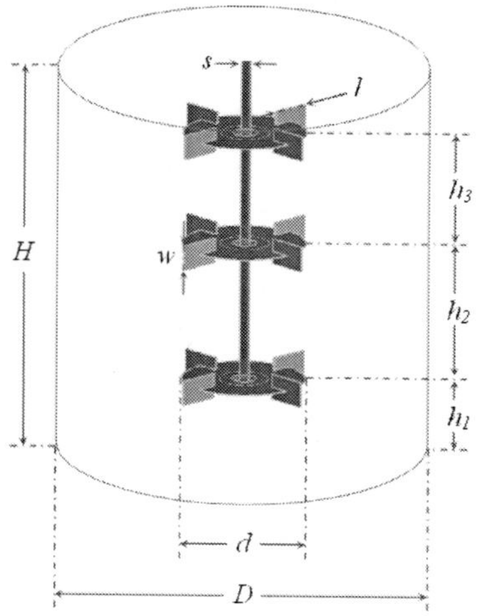

Figure 31. Stirred vessel configurations.

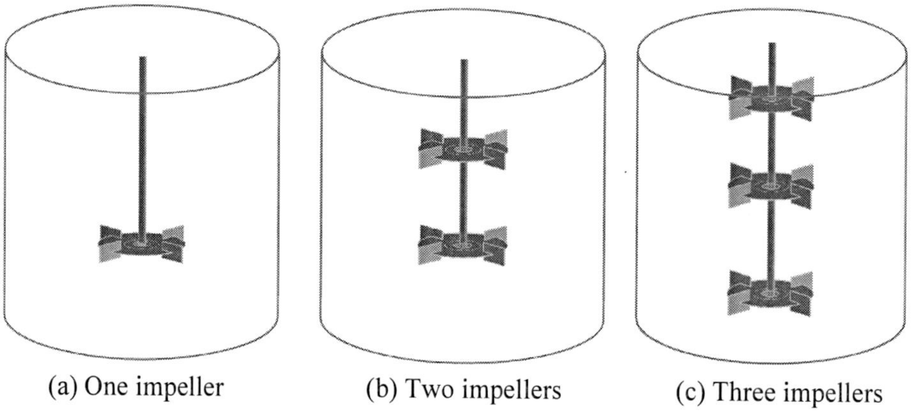

Figure 32. Multiple Rushton impellers.

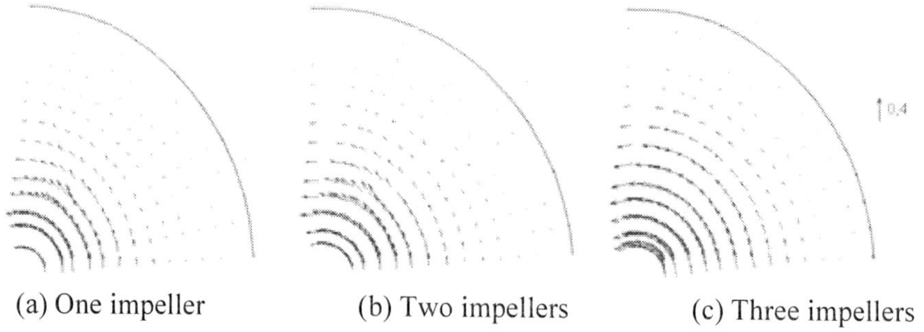

Figure 33. Flows patterns induced in r-θ plane defined by z=0.6.

4.1.3.1. Flow Patterns in r-θ Plane

Figure 33 shows a velocity vector plot of the primary flow in r-θ plane defined by the axial coordinate equal to z=0.6. These results, relative to the three configurations already definite, are presented only on a sector equal to 90° for the symmetry reasons. In these conditions, the blade is defined by an angular position equal to θ=45°. It appears that the flow is strongly dominated by the tangential component. Far from the region swept by the turbine, the rotating movement is no longer transmitted to the fluid, which remains quasi motionless.

4.1.3.2. Flow Patterns in r-z Plane

Figures 34, 35 and 36 show the secondary flow in three different r-z planes. These planes are defined by the angular coordinates equal to θ=40°, θ=45° and θ=85° respectively. In these conditions, the second presentation plan is confounded with the Rushton blade. However, the first and the last are situated respectively in the upstream and the downstream of the blade plane. In the case of the one impeller, the distribution of the field velocity shows the presence of a radial jet on the level of the turbine which changes against the walls of the tank with two axial flows thus forming two zones of recirculation on the two sides of the turbine. Also, we noted a slowing of the flow far from the Rushton turbine. However, with two Rushton impellers there is generation of four zones of symmetrical recirculation. Slowing is less elevated than the precedent configuration. The three impeller Rushton turbines are characterized by generation of six symmetrical recirculation zones. For this system, the field of velocity is more active than the two other cases; the weak flow zones are less frequent. Theses results confirmed that the multi-impellers systems appear necessary to decrease the weak flow zones in the stirred tanks. In these conditions, two Rushton impellers are obviously enough to improve the mechanical agitation operation.

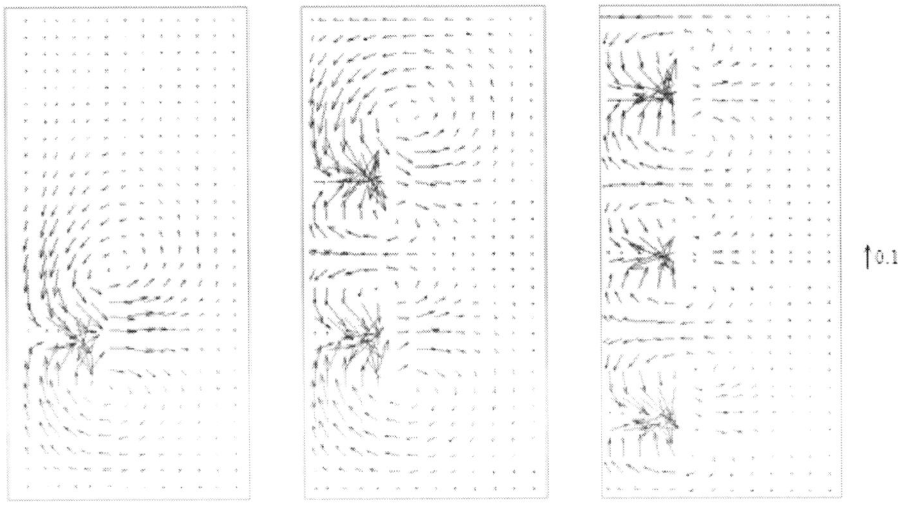

(a) One impeller (b) Two impellers (c) Three impellers

Figure 34. Flows patterns induced in r-z plane defined by θ=40°.

4.1.3.3. Viscous Dissipation Rate

Figures 37, 38, 39, 40 and 41 show the viscous dissipation rate in the r-θ planes defined respectively by the axial coordinates equal to z=0.33, z=0.66, z=1, z=1.33 and z=1.66. Figure 42 shows the viscous dissipation rate in the r-z plane defined by the angular coordinate equal to θ=45°. Globally, it's noted a maximal value in the blades tip. In fact, with one impeller the most important rate is reached at the plane defined by the axial coordinates equal to z=0.66 (Figures 38). With two impellers, the axial coordinates are defined by z=0.66 and z=1.33 (Figure 38 and 40). However, with three impellers the axial coordinates are defined by

z=0.33, z=1 and z=1.66 (Figure 37, 39 and 41). Out of the domain swept by impellers, the viscous dissipation rate becomes rapidely very weak. The same fact is observed in the r-z plane (Figures 42).

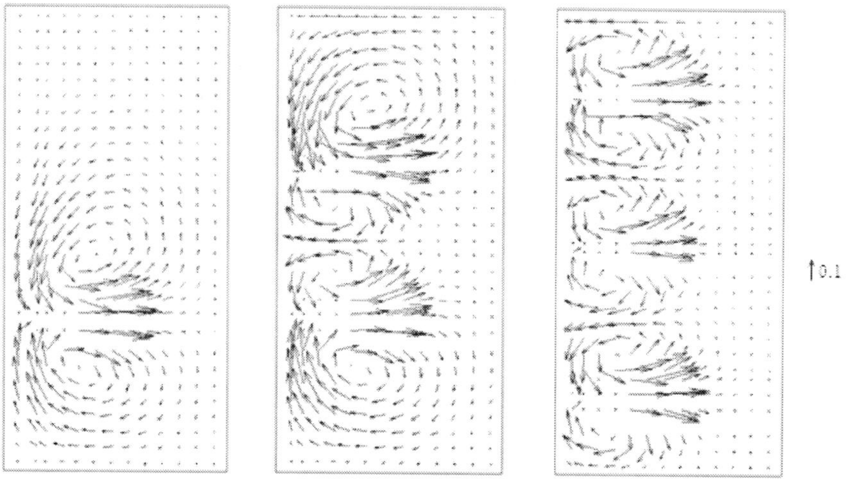

(a) One impeller (b) Two impellers (c) Three impellers

Figure 35. Flows patterns induced in r-z plane defined by θ=45°.

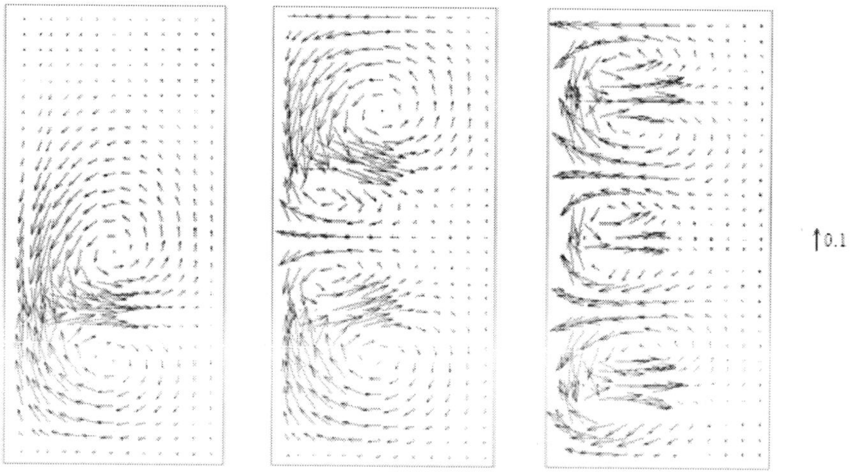

(a) One impeller (b) Two impellers (c) Three impellers

Figure 36. Flows patterns induced in r-z plane defined by θ=85°.

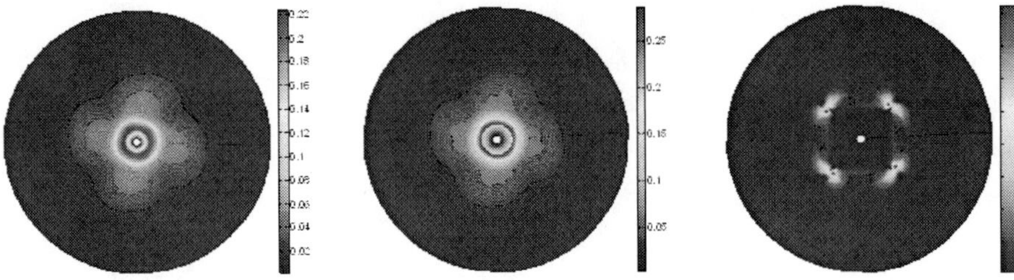

(a) One impeller(b) Two impellers (c)

Three impellers

Figure 37. Viscous dissipation rate in r-θ plane defined by z=0.33.

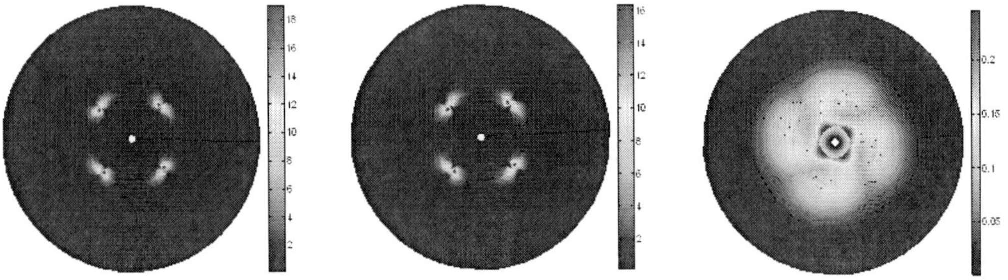

(a) One impeller(b) Two impellers (c)

Three impellers

Figure 38. Viscous dissipation rate in r-θ plane defined by z=0.66.

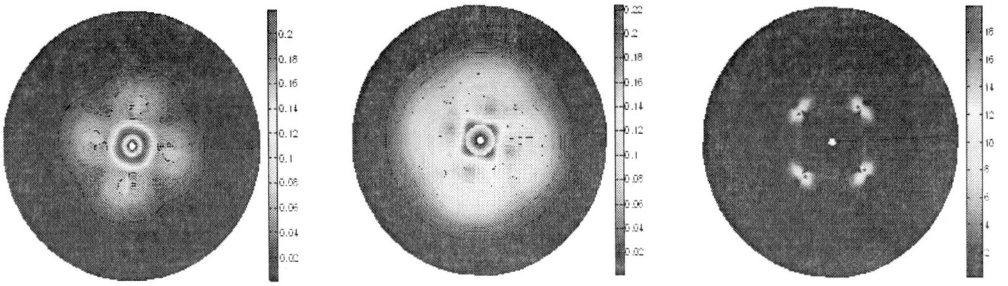

(a) One impeller(b) Two impellers (c)

Three impellers

Figure 39. Viscous dissipation rate in r-θ plane defined by z=1.

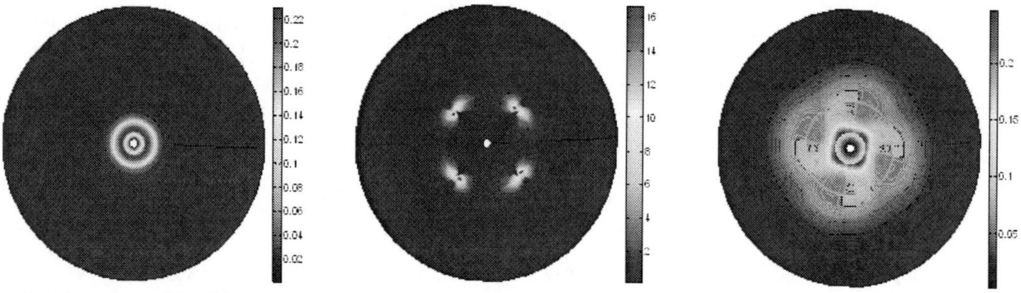

(a) One impeller(b)

Two

impellers

(c)

Three impellers

Figure 40. Viscous dissipation rate in r-θ plane defined by z=1.33.

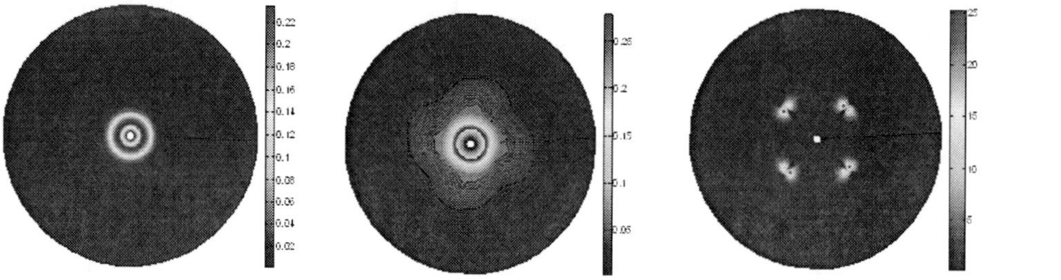

(a) One impeller(b)

Two

impellers

(c)

Three impellers

Figure 41. Viscous dissipation rate in r-θ plane defined by z=1.66.

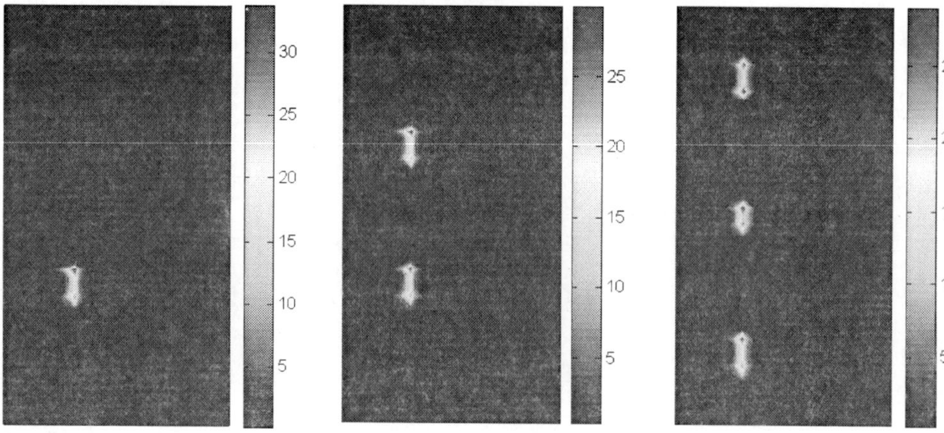

(a) One impeller(b)

Two

impellers

(c)

Three impellers

Figure 42. Viscous dissipation rate in r-z plane defined by θ=45°.

4.1.4. Double Helical Ribbons (DHR) and Double Helical Ribbons Screw (DHRS)

Viscous dispersive mixing is an important unit operation in polymerization, food and other industrial processes. To respond to the needs of industrial processes, various impellers have been developed. Among the different impellers available, the helical ribbon and the screw impellers are considered to be more efficient for the agitation of highly viscous liquids. In the present study, an attempt was made to compare the laminar flow results of the double helical ribbons with the double helical ribbons screw in agitator vessel. To evaluate the accuracy of the numerical method, the calculated results were compared with those obtained by the Nagata application [60] in the same geometry and size. The detailed of the stirred tanks equipped by double helical ribbons and double helical ribbons screw are presented in figure 43. The first impeller (Figure 43.a) has the same characteristics defined by the Nagata application [60]. The second is an association of a double helical ribbons and a screw impeller (Figure 43.b). These impellers dimensions are defined by b=0.1 d and h=d. The tip to tip impeller diameter ratio d/D is a 0.95. The shaft is placed concentrically with a diameter ratio s/D of 0.04. In this arrangement, D is the diameter of the cylindrical stirred tank having a plated bottom with a height-to-diameter ratio H/D of 1. The flow conditions are represented by the Reynolds number equal to Re=4 and the Froude number equal to Fr=0.19.

4.1.4.1. Flow Patterns in r-θ Plane

Figure 44 shows a velocity vector plot of the primary flow (U,V) presented in r-θ plane defined by the axial coordinate equal to z=1. It appeared that the flow was strongly dominated by the tangential component. Far from the region swept by the double helical ribbons, it has been noted a progressive slowing of the flow. Also, it has been noted that the velocity vector was directly affected by the proximity impellers type. Indeed, within the double helical screw ribbons it has been observed that the velocity field is very active in the two opposite sides that are confounded with the impeller tip.

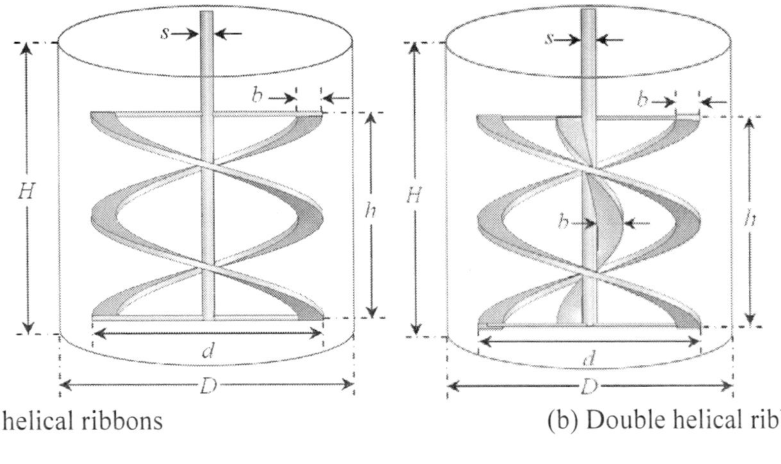

(a) Double helical ribbons (b) Double helical ribbons screw

Figure 43. Stirred vessel configurations.

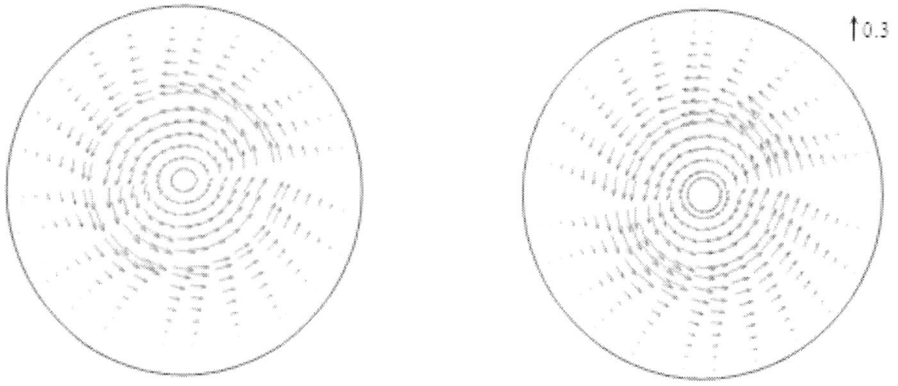

(a) Double helical ribbons (b) Double helical ribbons screw

Figure 44. Flows patterns induced in r-θ plane defined by z=1.

4.1.4.2. Flow Patterns in r-z Planes

Figures 45, 46 and 47 show a velocity vector plot of the secondary flow (U, W) presented in r-z planes. These presentation planes have been defined respectively by the angular coordinate equal to θ=122°, θ=136° and θ=316°. These positions have been chosen in order to show the velocity field evolution. With a double helical ribbons, it's noted that the flow have a centrifugal radial movement in the middle of the tank in the first plane (Figure 45.a). This movement decrease in the second plane (Figure 46.a). However, it becomes a centripetal type in the third plane (Figure 47.a). In the swept domain localised in the top of the tank, the flow have a descending axial movement. However in the bottom, this movement reverses and becomes an ascending and oblique character. While approaching to the lateral surface of the tank, the flow becomes weak. Within a double helical screw ribbons, the same observations are noted. But, it's clear that the velocity field is more active than the double helical ribbons.

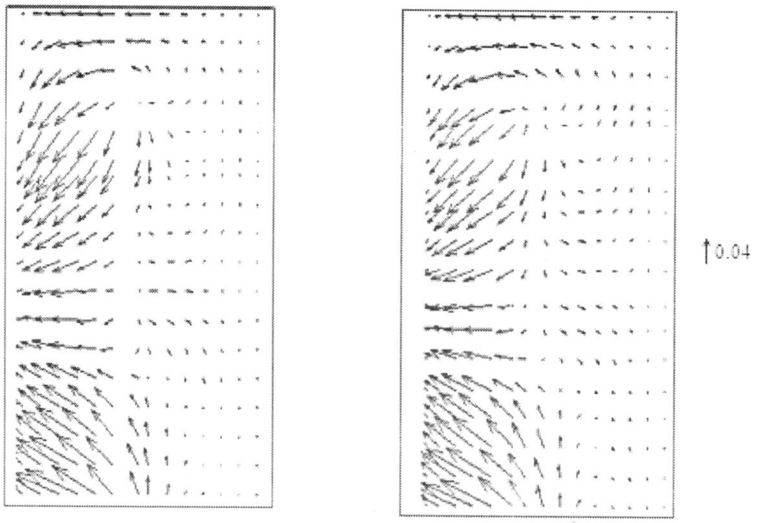

(a) Double helical ribbons (b) Double helical ribbons screw

Figure 45. Flows patterns induced in r-z plane defined by θ=122°.

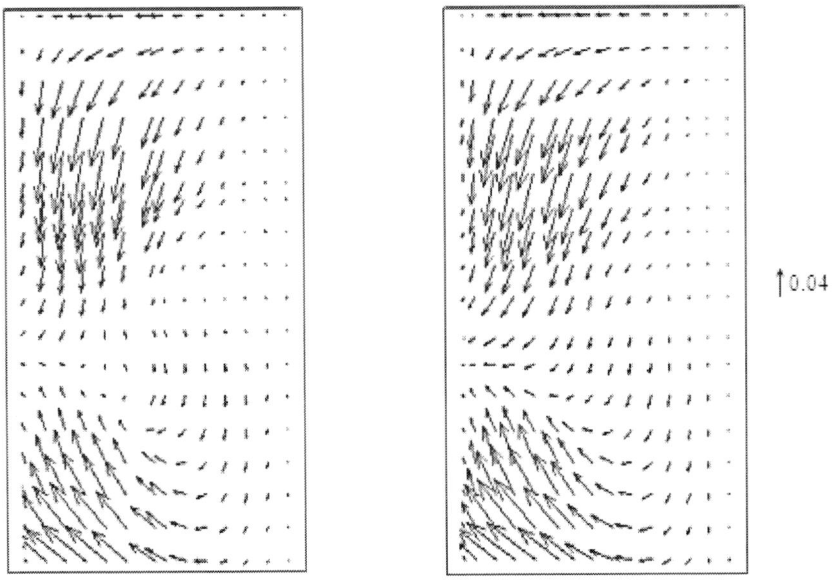

(a) Double helical ribbons (b) Double helical ribbons screw

Figure 46. Flows patterns induced in r-z plane defined by θ=136°.

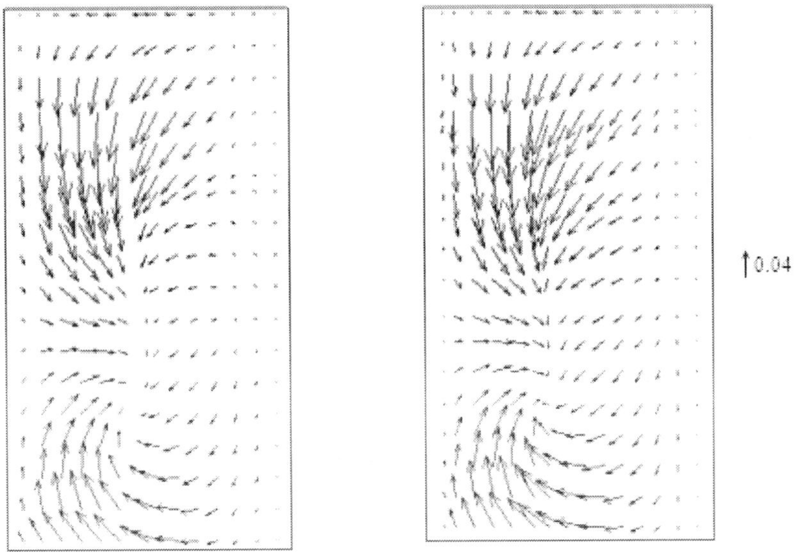

(a) Double helical ribbons (b) Double helical ribbons screw

Figure 47. Flows patterns induced in r-z plane defined by θ=316°.

4.1.4.3. Radial Profiles of the Dimensionless Velocity Components

Figure 48, 49 and 50 illustrate the predicted radial profiles of the dimensionless radial $U(r)$, tangential $V(r)$ and axial $W(r)$ velocity components of the double helical ribbons and the double helical screw ribbons. These profiles are presented in two different r-θ planes defined by the dimensionless axial coordinates equal to z=0.45 and z=0.8. These figures adequately portray the swirling radial, tangential and axial jets character. In these figures, the numerical results of the two impellers were superposed to compare the local characteristics.

Figure 48.a show that the radial velocity component *U(r)* reaches its maximal value equal to *U*=0.036 in the horizontal plane situated in the tank bottom. The radial position corresponding to this maximal value is defined by r=0.33. This result is also observed for tangential V(r) and axial W(r) velocity components. In fact, the tangential velocity component V(r) reaches its maximal value V=0.3 in the radial position equal to r=0.5 (Figure 49.a). Whereas, the axial velocity component W(r) reaches its maximal value W=0.03 in the radial position equal to r=0.58 (Figure 50.a). The minimal value W=-0.012 is reached in the radial position equal to r=0.16 (Figure 50.b). The negative value of the axial component corresponds to a downward movement toward the tank bottom.

In the tank top, it's noted a progressive reduction of the flow movements. In fact, the radial movement decreases and the radial velocity component *U(r)* reaches very weak values. The tangential velocity component *V(r)* shows a parabolic pace. It reaches very low values in the neighborhood of the axis and the tank walls. For the axial velocity component *W(r)*, it's noted a more intense downward movement. Globally, it's clear that the Double helical ribbons screw present a more active velocity field than the double helical ribbons.

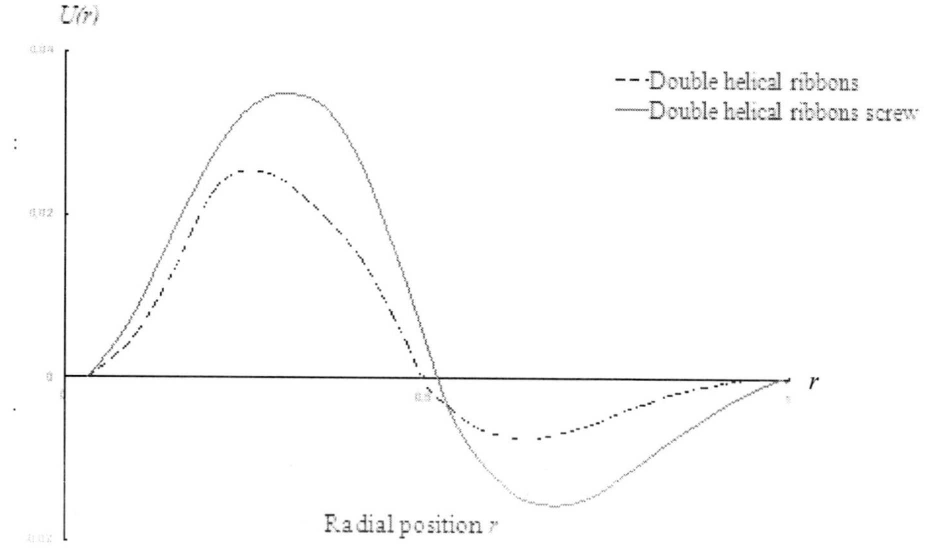

(a) Axial position of the presentation plane equal to *z*=0.45

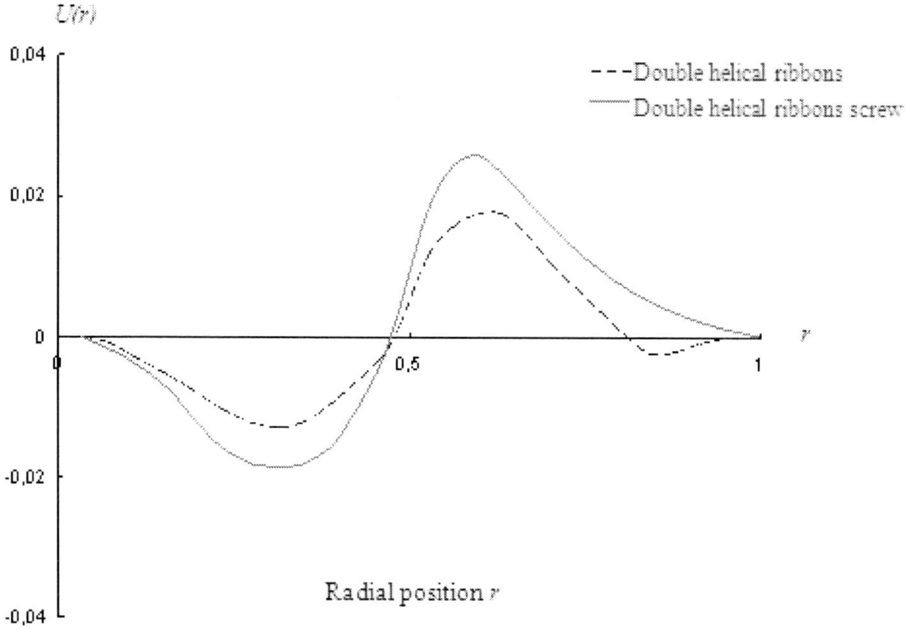

(b) Axial position of the presentation plane equal to z=0.8

Figure 48. Radial profiles of the radial velocity $U(r)$.

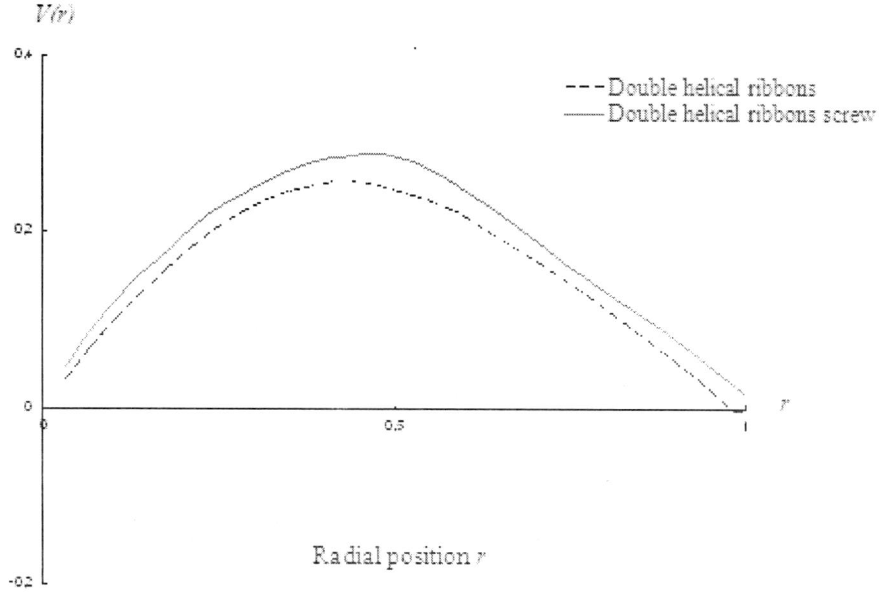

(a) Axial position of the presentation plane equal to z=0.45

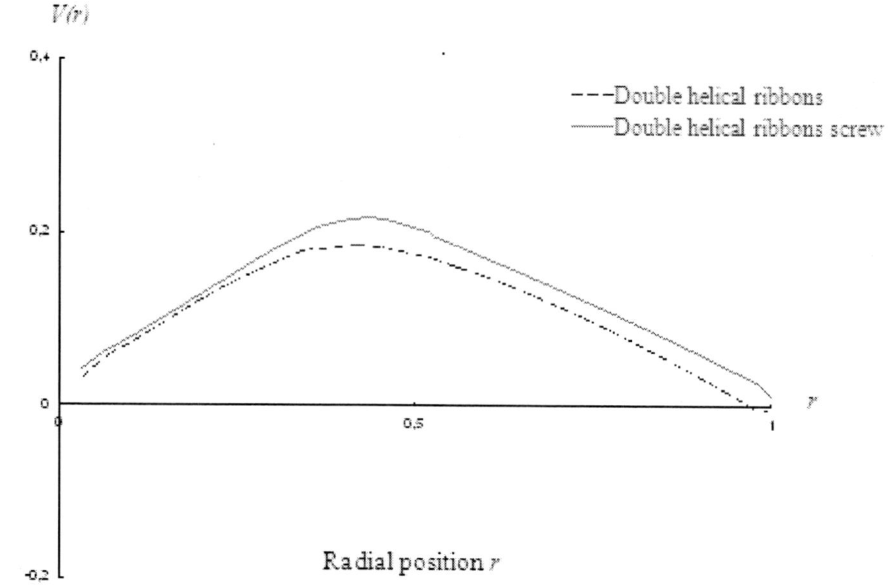

(b) Axial position of the presentation plane equal to z=0.8

Figure 49. Radial profiles of the tangential velocity *V(r)*.

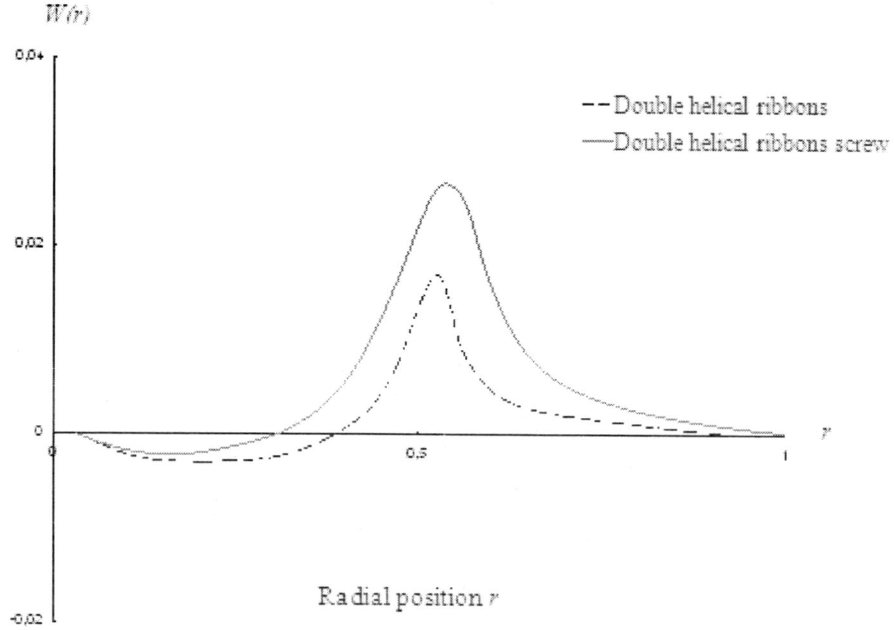

(a) Axial position of the presentation plane equal to z=0.45

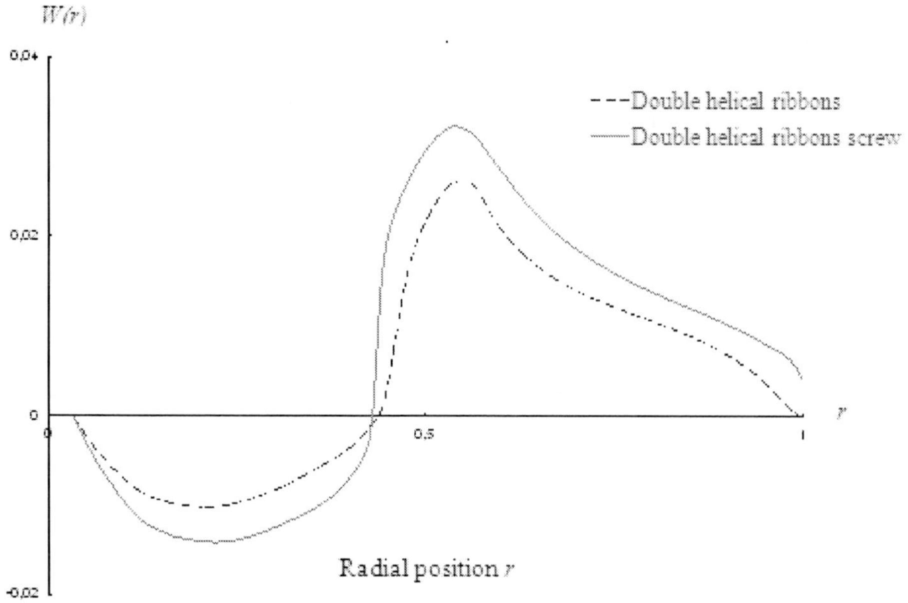

(b) Axial position of the presentation plane equal to z=0.8

Figure 50. Radial profiles of the axial velocity W(r).

4.1.4.4. Viscous Dissipation Rate

Figures 51, 52 and 53 show respectively the viscous dissipation rate in different r-θ planes defined respectively by the axial positions equal to z=0.1, z=0.95 and z=1.15. Figure 54 shows the viscous dissipation rate in r-z plane defined by the angular position equal to θ=108°. These presentations planes have been chosen in order to show the viscous dissipation rate evolution in the stirred tank. Seen the symmetry of the problem, two symmetry wake shape are observed in these r-θ planes. Globally, the maximal value of the viscous dissipation rate is reached in the meeting of the double helical ribbons with these presentation planes. Out of this domain, the viscous dissipation rate becomes rapidly very weak. All of these observations are available for the two impellers. But, it's noted that the viscous dissipation rate for the double helical screw ribbons is more important than the double helical ribbons.

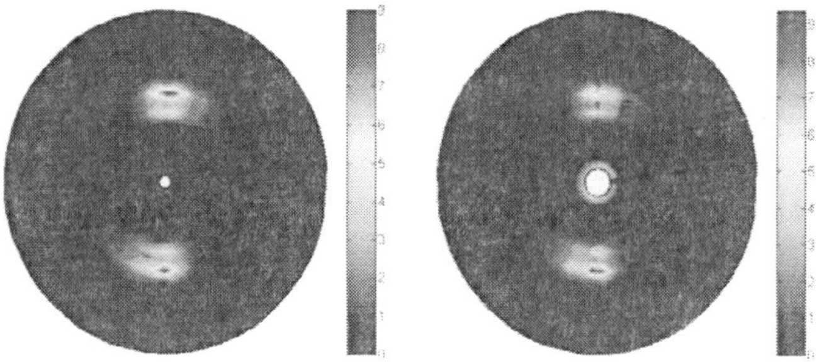

(a) Double helical ribbons (b) Double helical ribbons screw

Figure 51. Dissipation rate induced in r-θ plane defined by z=0.1.

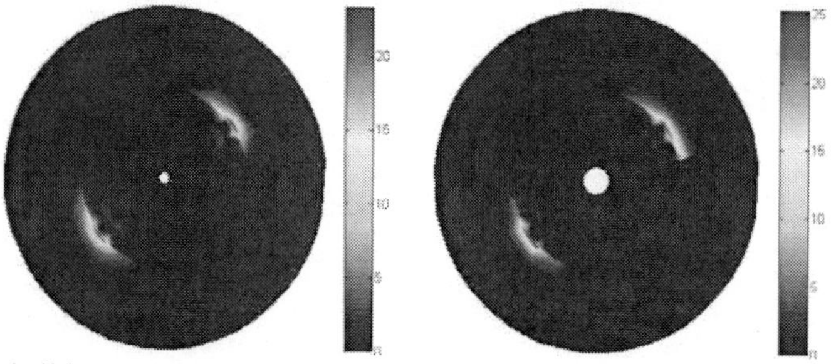

(a) Double helical ribbons (b) Double helical ribbons screw

Figure 52. Dissipation rate induced in r-θ plane defined by z=0.95.

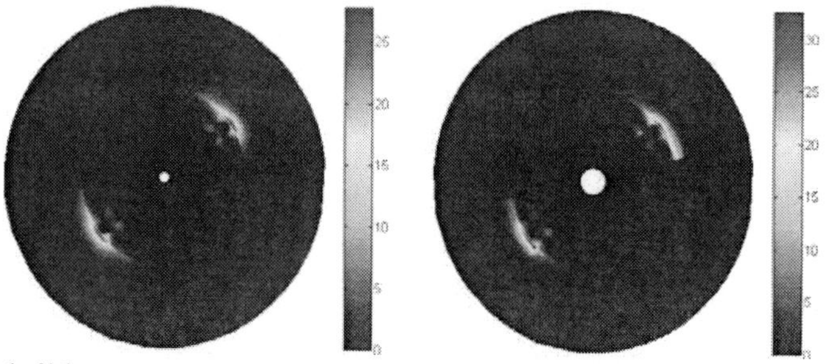

(a) Double helical ribbons (b) Double helical ribbons screw

Figure 53. Dissipation rate induced in r-θ plane defined by z=1.15.

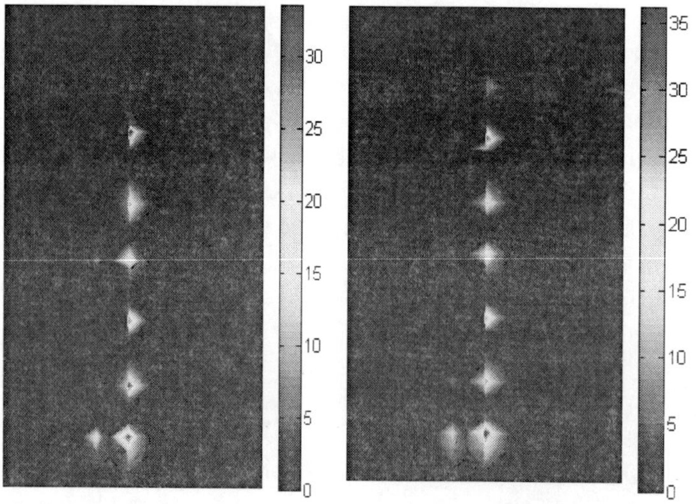

(a) Double helical ribbons (b) Double helical ribbons screw

Figure 54. Dissipation rate induced in r-z plane defined by θ=108°.

4.1.4.5. Global Characteristics

To compare the global characteristics of the double helical ribbons in closed stirred tanks to the double helical ribbons screw, the pumping flow number NQ_p, the pumping efficiency number E_p and the energy efficiency E_e are calculated from the CFD code. The dependence of these characteristics on Reynolds number Re was presented respectively in figures 55, 56 and

57. Globally, these parameters were proportional to the Reynolds number *Re* in the laminar flow regime. At the same Reynolds number, it's clear that the double helical ribbons screw characteristics values are superior to the double helical ribbons.

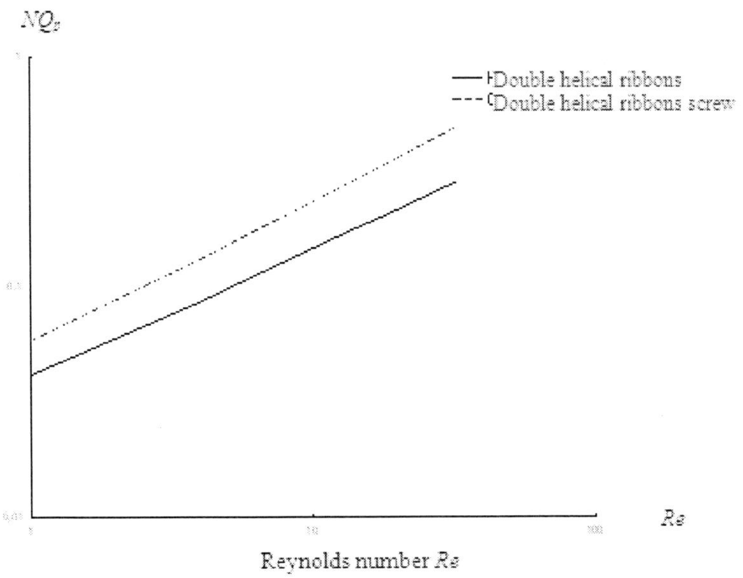

Figure 55. Pumping flow number variation N_{Qp}=f(*Re*).

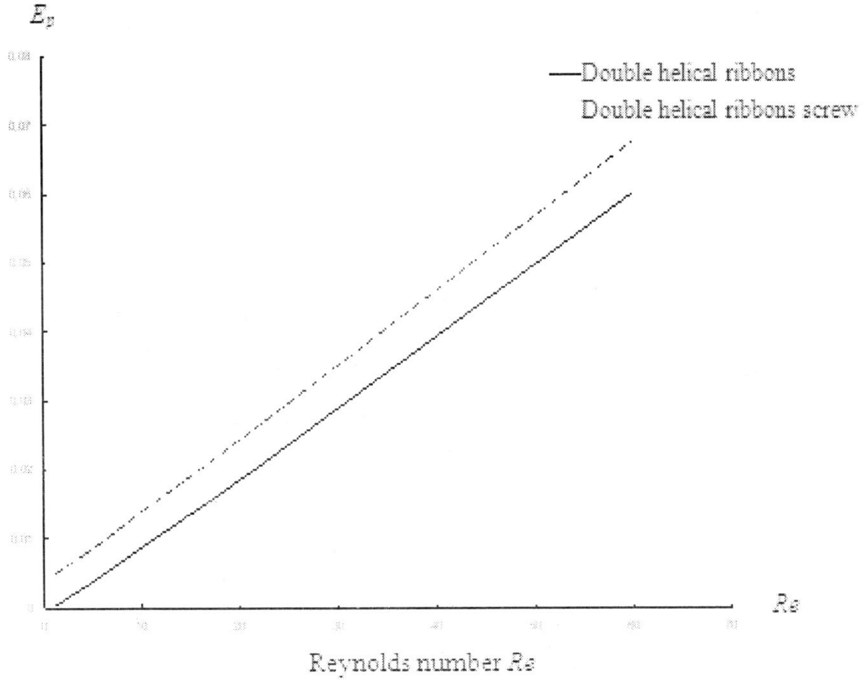

Figure 56. Pumping efficiency number variation E_p=f(*Re*).

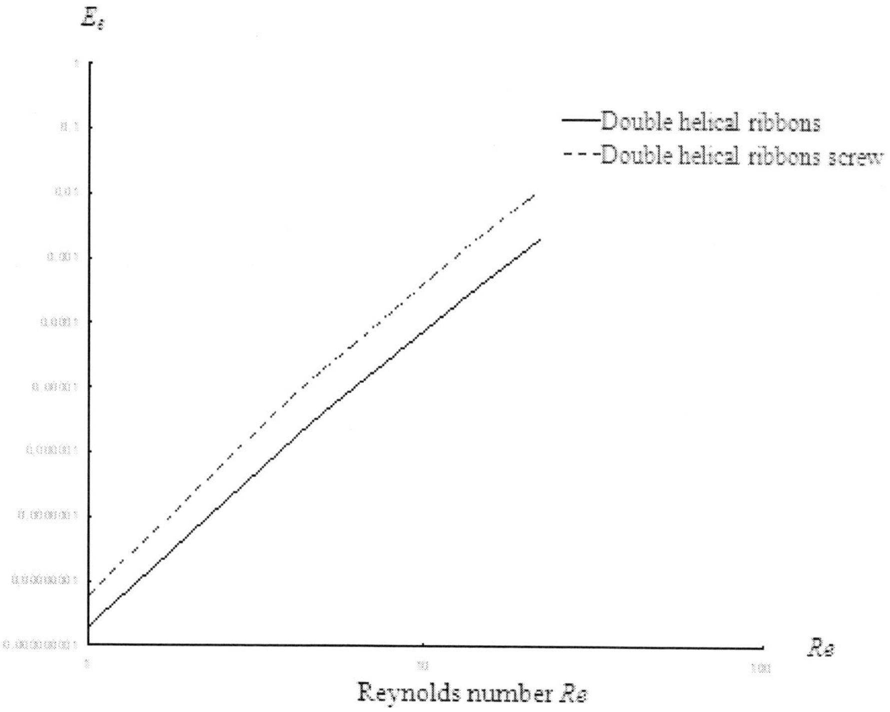

Figure 57. Energy efficiency variation E_e= f(Re) .

4.1.4.6. Comparison with Experimental Results

Figure 58 show the power number Np variation on Reynolds number Re in the laminar flow range of the double helical ribbons and the double helical ribbons screw impellers. In these conditions, the power number was inversely proportional to the Reynolds number and the value of the product $K_p = N_p\ Re$ remained constant. According to these results, it's noted that these curves present the same linear variation. Moreover, it's noted that the power number of the double helical ribbons screw is superior to the double helical ribbons at the same Reynolds. To verify our computer results, the power number calculated from the CFD code were compared with the experimental results found in the literature in the case of the double helical ribbons. In these conditions, the numerical value of the power number Np is slightly inferior to the experimental value when compared at the same Reynolds number Re. These experimental results founded by Nagata [60] were superposed over an average error of 10%. The good agreement between the experimental results and the numerical results quietly confirmed the analysis method.

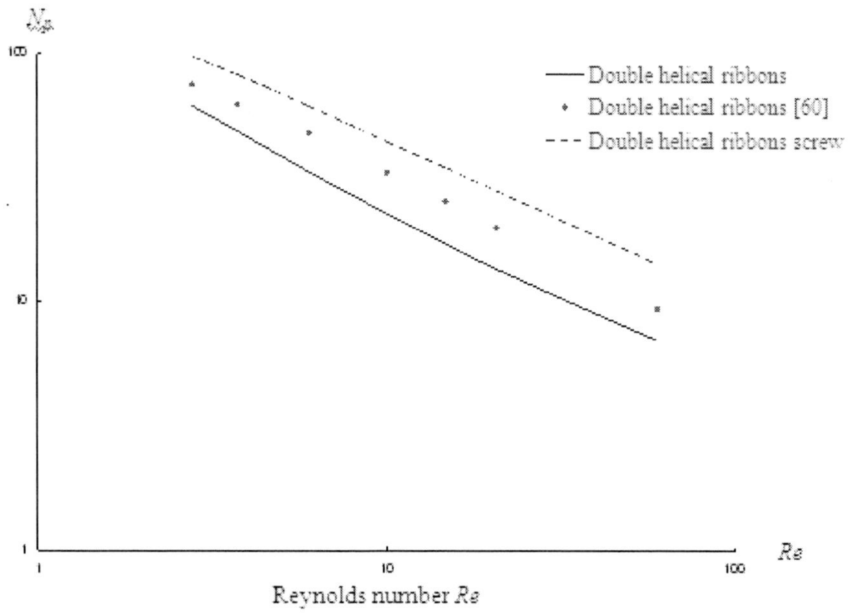

Figure 58. Power number variation N_p=f(Re).

4.2. Simulation of Turbulent Flow

Simulation of turbulent flow of the stired tanks is covered in many papers. For example, the main aims of the Aubin et al. [75] are to characterize the single phase turbulent flow in a tank stirred by a 45° PBT using CFD. The effect of the modeling approach, discretization scheme and turbulence model on mean velocities, turbulent kinetic energy and global quantities, such as the power and circulation numbers, has been investigated. The stationary and time-dependent modeling approaches were found to have little effect on the turbulent flow, however the choice of the numerical scheme was found to be important, especially for the predicted turbulent kinetic energy. Hollander et al. [76] presented a numerical study on the scale-up behaviour of orthokinetic agglomeration in stirred tanks. Large Eddy flow simulations were performed to obtain an accurate description of the turbulent flow encountered in stirred vessels, equipped with either a Rushton turbine or a 45° PBT. It was found that impeller shape, vessel size and Reynolds number have a profound effect on reactor performance. Dakshinamoorthy et al. [77] studied shortstopping results in vessels agitated with jets and impellers using the commercial CFD code Fluent. The simulated shortstopping results with the jet mixer are then compared with those obtained with Rushton turbine and 45° PBT. These results identify the conditions for effective shortstopping. The distribution of excess inhibitor is shown to be an important and essential design criterion for effective shortstopping when using impeller stirred vessels. Kumaresan and Joshi [78] presented the comparison of the flow pattern on the basis of equal power consumption to characterize the flow generated by different impeller geometries. Comparisons of laser doppler anemometry (LDA) measurements and CFD predictions have been presented. The good comparison indicates the validity of the CFD model. Deglon and Meyer [79] demonstrated that the multiple reference frames impeller rotation model and the standard k-ε turbulence model, as commonly used in engineering CFD simulations of stirred tanks, can accurately model turbulent fluid flow. In this study, the CFD software Fluent 6 is used to simulate flow in a small tank. Simulations are conducted on four grids of significantly different resolution using the upwind, central and Quick discretization schemes. CFD model results are evaluated in terms of the predicted flow field, power number, mean velocity components and turbulent kinetic energy using published experimental data. Armenante et al. [80] presented the velocity profiles and the turbulent kinetic energy distribution of a flow generated by a 6-blade, 45° PBT in an unbaffled, flat-bottom, cylindrical tank provided with a lid, and completely filled with water. The mean and fluctuating velocities in all three directions were experimentally measured with a laser doppler velocimeter (LDV) at five different heights and twenty radial positions within the vessel. Li et al. [81] have established a theoretical and numerical model to explore the fundamental problem of liquid-solid impact. Chang et al. [62] presented a dynamic analysis of a rub-impact rotor supported by two turbulent model journal bearings with non-linear suspension. Whitfield et al. [83] have developed a performance prediction procedure for radial and mixed flow turbines. Schobeiri [84] presented the exact solutions of the Navier-Stokes and the temperature differential equations for a viscous flow through a two-dimensional curved channel. Chen et al. [85] investigated the possibility of correlating the aerodynamic loading of radial and mixed flow turbines in some systematic manner. Zhao et al. [86] have established a general finite volume fluid in cell method for solving the Navier-Stokes equations. In stirred vessels, the quality of flow generated by the impeller mainly depends upon the impeller design. Although, many experimental studies on effect of impeller design on the flow pattern in cylindrical vessels have been published. For example, Suzukawa et al. [87] studied the effect of the blade attack angle on the roll and trailing vortex structures in a stirred vessel via laser doppler velocimetry (LDV). Biswas et al. [88] have established the suspension condition of silica sand water slurry using various design variables. Chapple et al. [89] investigated the effect of impeller and tank geometry on power number for pitched blade turbine (PBT). Dan Taca et Paunescu [90] have theoretically established the calculation relation for the impeller power number in the case of fully filled spherical vessels. Kuncewicz et Pietrzykowski [91] presented a 2D hydrodynamic model of a mixing vessel for PBT operating in laminar range of motion. To verify the model, the power consumption and axial forces acting on the bottom and wall of the mixing vessel were measured experimentally. Derksen et al. [92] studied the impeller region of a turbulently stirred tank via LDV. Karcz et Major [93] investigated the effect of the baffle length on power consumption in agitated vessel. The measurements were carried out for the Rushton and Smith turbines, the three 45° PBT and the propeller. Power characteristics for different geometrical parameter of the baffles were obtained within the turbulent regime of the fluid flow in the agitated vessel. Kumaresan et al. [94] examined the effect of impeller design such as blade angle, number of blades, blade width, blade twist, blade thickness and pumping direction on the flow pattern and mixing time for a set of axial flow impellers.

On the basis of the previous studies, we can confirm that the 45° PBT is frequently used in many applications. As a consequence, it appears important to study different configurations and to make a comparison between them through a CFD simulation. For this purpose, a computer method has been developed to study the pitched blade turbines design effect on the stirred tank flow characteristics. The present work aims the characterization of the turbulent flow of three types of PBT6 configurations defined by the inclined angles β equals to 45°, 60° and 75° [1-7]. Also, we have interested to study the turbulent flow of the PBT3 and two types of the Modified pitched blade turbines (MPBT3) [2].

4.2.1. Pitched Blade Turbines (PBT6)

The system under investigation in the current paper is a stirred tank equipped with a six-blade PBT. The tank is a vertical cylindrical vessel with a height-to-diameter ratio H/D of 1. The shaft is placed concentrically with a diameter ratio s/D of 0.04. The tip to tip turbine diameter ratio d/D is a 0.33 and the turbine position ratio y/D is a 0.25 (Figure 59). Three inclined angles β equals to 45°, 60° and 75° were considered (Figure 60). The first geometry system is similar to Armenante application system [80].

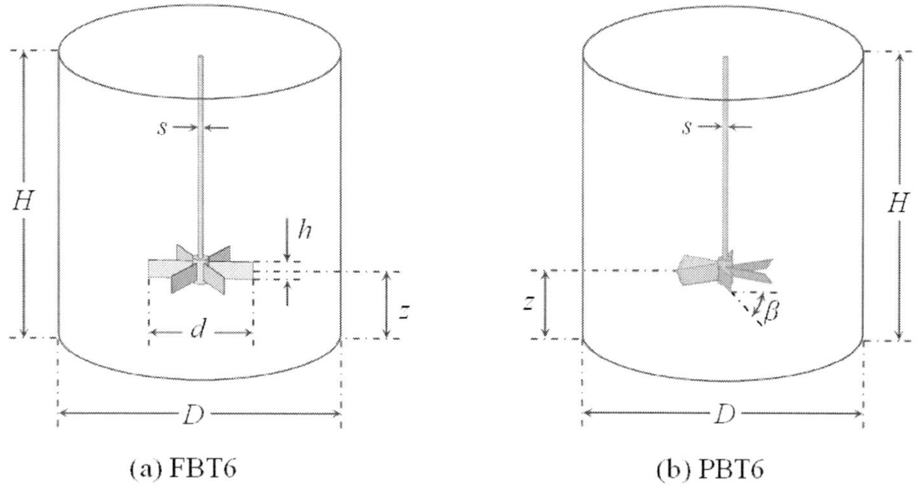

Figure 59. Stirred vessel configurations.

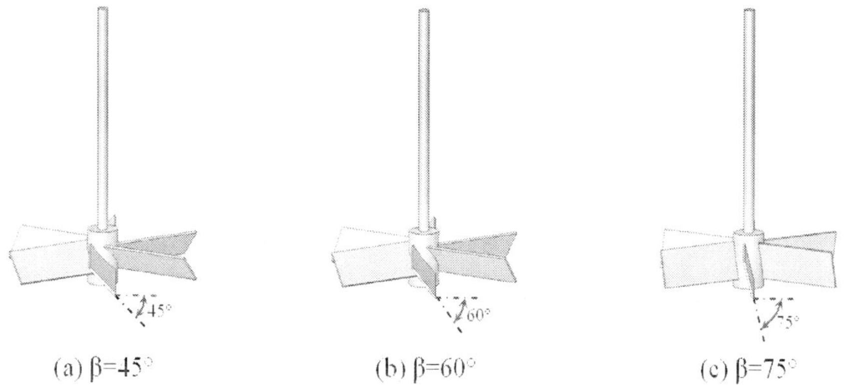

Figure 60. Pitched blade turbines (PBT6).

4.2.1.1. Flow Patterns

Figures 61 and 62 show a velocity vector plot of the primary flow (U,V) in r-θ plane and the secondary flow (U,W) in r-z plane. The r-θ plane has been defined by the axial coordinate equal to z=0.3. It appeared that the flow was strongly dominated by the tangential component. While comparing these three geometric configurations, it has been noted that the radial component of the velocity vector was weak in the case of the 45° PBT (Figure 61.a). For the two other configurations, this component increased (Figures 61.b and 61.c). The r-z plane has been defined by the angular coordinate equal to θ=52°. This presentation plane has been situated in the downstream of the blade plane. These figures showed the turbine stream impinging on the outer wall and being deflected in both directions in a wall jet motion producing the upper and lower vortices. Also, it has been clearly noticed that the velocity vector was directly affected by the turbine type. In fact, within the 75° PBT the radial flow was very important. This flow became more and more weak with the reduction of the inclined angle (Figure 62). Within the 45° PBT, the axial flow was very important near the wall and reaches the top of the liquid surface. This result has been explained by the turbine type allowing a high axial flow. However, within the 60° PBT and the 75° PBT the axial flow has been less intense in top of the stirred tank, which appears a zone of a very weak field. Also, the axial movement generated two zones of recirculation on all sides of these turbines. The zone shapes has been depending from the inclined angle used; the 75° PBT zones showed up a clear symmetry. However, with the feeble angle we noticed that the lower zone expanded mainly in the 45° PBT. Also, the recirculation center zone extended to the inner tank wall while the inclined angle increased, within r=0.6 for the 45° PBT and r=0.7 for the 75° PBT.

4.2.1.2. Distribution of Turbulent Kinetic Energy

Figure 63 and 64 present the evolution of the distribution of the turbulent kinetic energy in the case of different inclined angles equals to β=45°, β=60° and β=75°. The horizontal presentation planes, defined by the axial coordinate equal to z=0.5, were situated at the mid-height of the turbine. However, the vertical presentation planes were situated in the downstream of the blade plane. They were defined by the angular coordinate equal to θ=52°. Globally, it has been noted a maximal value in the region localized in the blades tip. In the domain swept by the turbine blades, the turbulent kinetic energy remains enough elevated. Out of this domain, it became quickly very weak. The horizontal presentation planes showed a wake localized in the domain swept by the turbine. This wake has been characterized by the elevated values of the turbulent kinetic energy. While moving away of this domain, these values decreased quickly. The turbine type was then responsible in the characterization of the wake shape. Indeed, according to these results the wake was developed with the increase of the inclined angle β. The wake took the oval shape within a 75° PBT. It shrunk further with the reduction of the inclined angle β. In the vertical presentation planes, the area sites of the maximum values were located in the wake which developed on the upstream of blades.

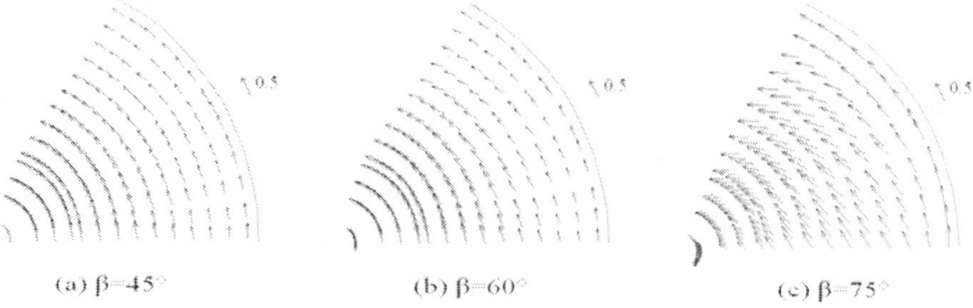

Figure 61. Flows patterns induced in r-θ plane define by the axial coordinate z=0.3.

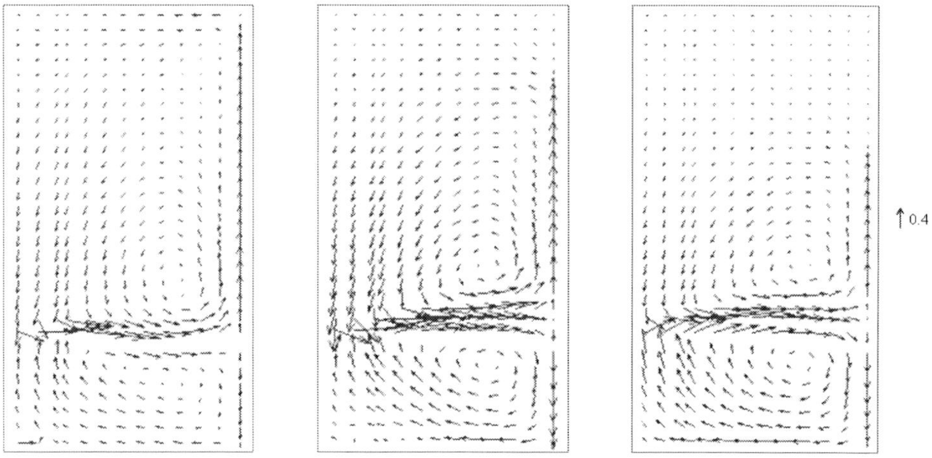

(a) β=45° (b) β=60° (c)
β=75°

Figure 62. Flows patterns induced in r-z plane define by the angular coordinate θ=52°.

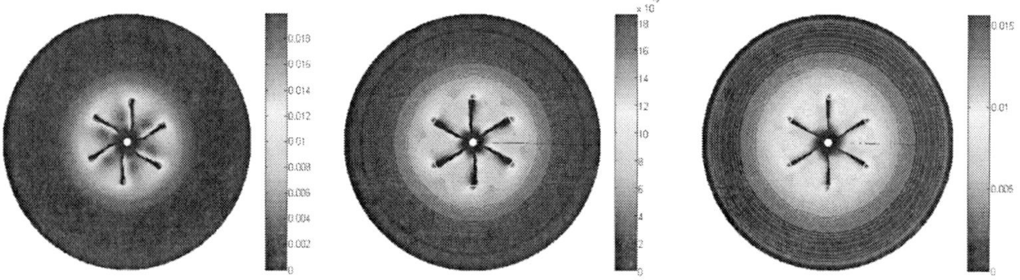

(a) β=45° (b) β=60° (c) β=75°

Figure 63. Distribution of the turbulent kinetic energy in r-θ plane define by z=0.5.

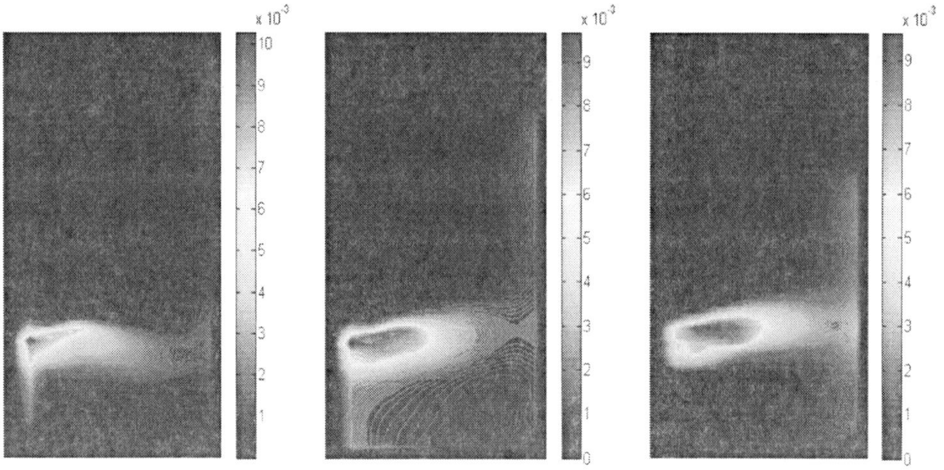

(a) β=45° (b) β=60° (c) β=75°

Figure 64. Distribution of the turbulent kinetic energy k in r-z plane define by θ=52°.

4.2.1.3. Distribution of Dissipation Rate of the Turbulent Kinetic Energy

Figure 65 and 66 present the distribution of dissipation rate of the turbulent kinetic energy ε. For the presentation of its features, the same planes adopted in the previous paragraph to study the turbulent kinetic energy were chosen. Therefore, results had shown a huge similarity of the dissipation rate with the turbulent kinetic energy. Indeed, it has been noted that the range of the maximum values of the dissipation rate of the turbulent kinetic energy ε were located in the wake which develops nearly to the blades ends. Mean while, the wake extended more and more with the augmentation of the inclined angle. This result was presented at the figure 69, which the power number (integration of the dissipation rate of the turbulent kinetic energy ε) decreased with the decrease of the inclined angle β.

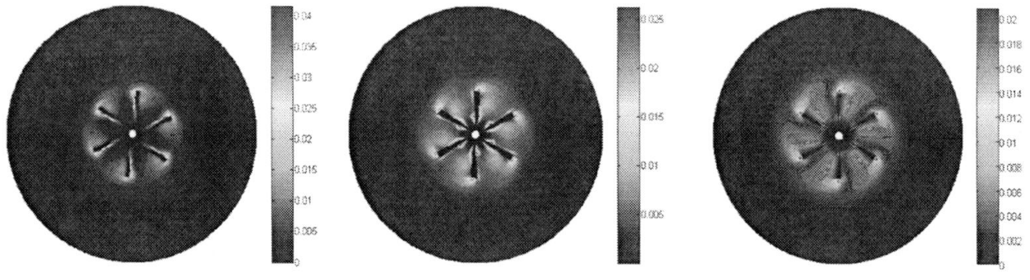

(a) β=45° (b) β=60° (c) β=75°

Figure 65. Distribution of dissipation rate of the turbulent kinetic energy ε in r-θ plane define by z=0.5.

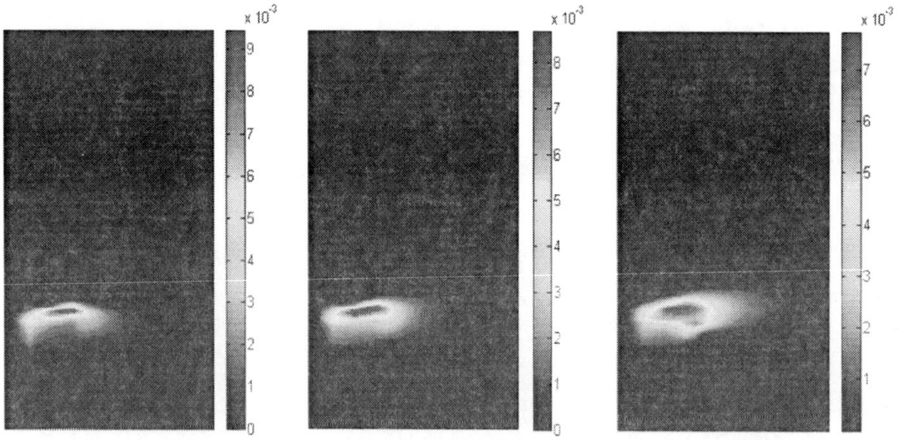

(a) β=45° (b) β=60° (c) β=75°

Figure 66. Distribution of dissipation rate of the turbulent kinetic energy ε in r-z plane define by θ=52°.

4.2.1.4. Distribution of Turbulent Viscosity

Figure 67 and 68 present the evolution of the turbulent viscosity distribution in the same planes adopted in the previous paragraphs. The turbulent viscosity has remained rather high in the field swept by blades of the turbine. Very near to the walls and around the turbine, the turbulent viscosity has made a very fast fall. By comparing the three configurations between them, it has been noted that the maximal values of the turbulent viscosity increased with the inclined angle. Indeed, within a 75° PBT the turbulent viscosity has been equal to υ_t =2000.

Than, for 60° and 45° PBT the turbulent viscosity has been equal to υ_t =1500 and υ_t =1400 respectively. The horizontal presentation plane showed two wakes localized on all sides of the turbine. These wakes were characterized by the elevated values of the turbulent viscosity. While moving away of this domain, these values decreased progressively until reaching some very weak values near the tank wall. The turbine type has been interfered in the characterization of the wake shape. Within a 75° PBT, the wake form has been developed in the turbine's bottom. However, the wake localized in the turbine's top saved the same form obtained with the 45° PBT.

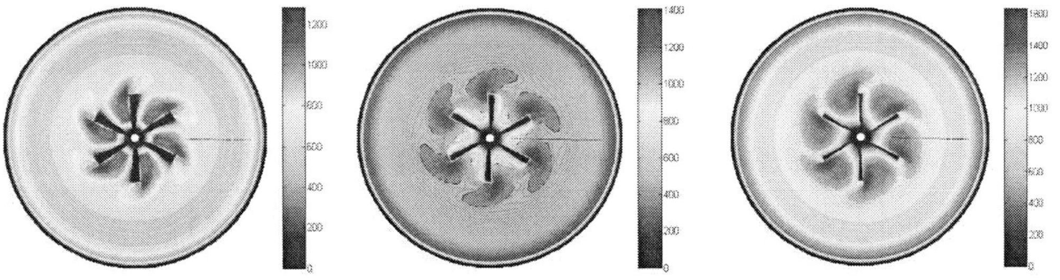

(a) β=45° (b) β=60° (c) β=75°

Figure 67. Distribution of turbulent viscosity υ_t in *r*-*θ* plane define by *z*=0.5.

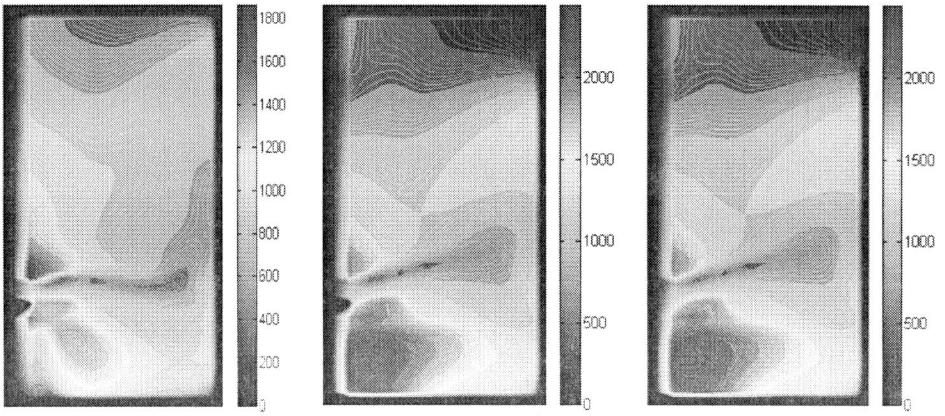

(a) β=45° (b) β=60° (c) β=75°

Figure 68. Distribution of turbulent viscosity υ_t in r-z plane define by θ=52°.

4.2.1.5. Comparison with Experimental Results

To verify our computer results, the power number *Np* were measured and compared with the number calculated from the CFD code as shown in Figure 69. These results made a strong proof of the inclined angle effects on the power number *Np*. It appears that the value of the *Np* decreases as an inclined angle *β* decreases when compared at the same Reynolds number *Re*. The downiest value equal to *Np*=0.55 has been obtained with the 45° PBT. Globally, *Np* remained constant in the turbulent flow range. But, in the laminar flow range, it is the product *Np Re* which remains constant.

In the other hand, the flow patterns have been compared with ones found by other experimental results [80]. Figures 74, 75, 76 and 77 illustrated the predicted radial profiles of the dimensionless radial velocity component *U(r)*, tangential velocity component *V(r)*, axial velocity component *W(r)* and turbulent kinetic energy *k(r)* within a 45° PBT.

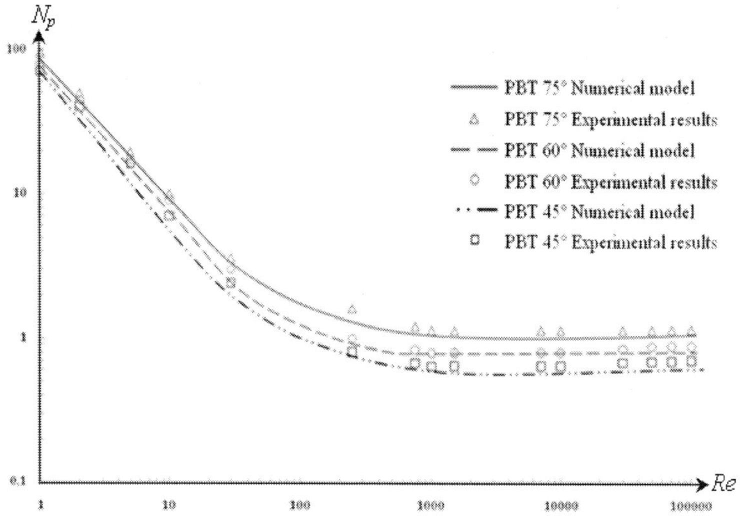

Figure 69. Power number N_p against Reynolds number *Re* of pitched-blade turbines.

These profiles have been defined in the downstream of the blade. The dimensionless axial coordinates has been equal to *z*=0.36, *z*=0.48, *z*=0.6, *z*=1.09 and *z*=1.59. The presentation planes defined by *z*=0.48 and *z*=0.6 were situated respectively in the bottom and in the top of the turbine. This figures adequately portrayed the swirling jets character. In these comparisons, the average error between the experimental results and the numerical results is equal to 7%. The good agreement confirmed the validity of the computer method.

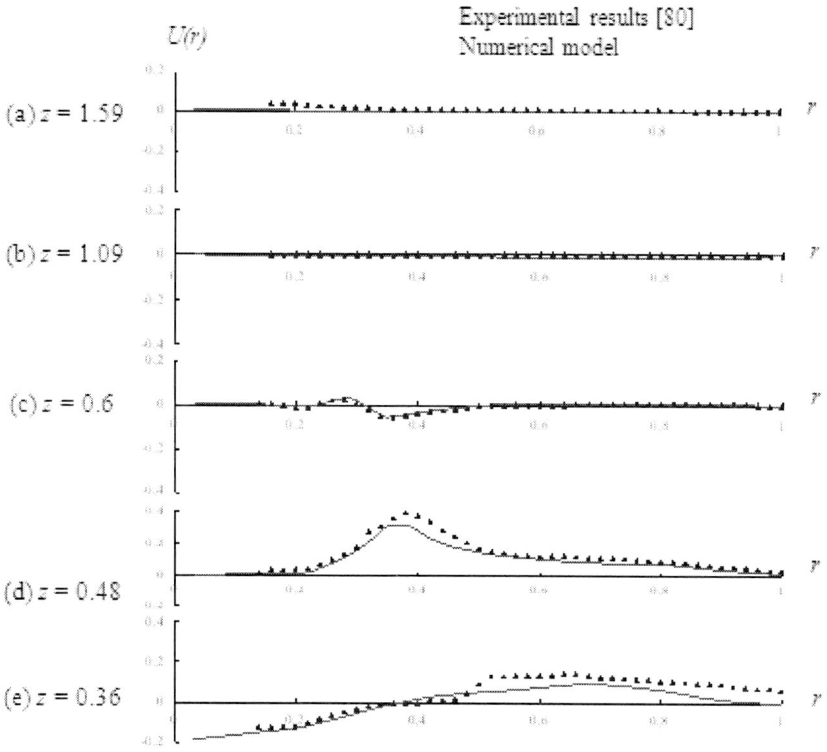

Figure 70. Radials profiles of the radial velocity for different axial positions U=f(r).

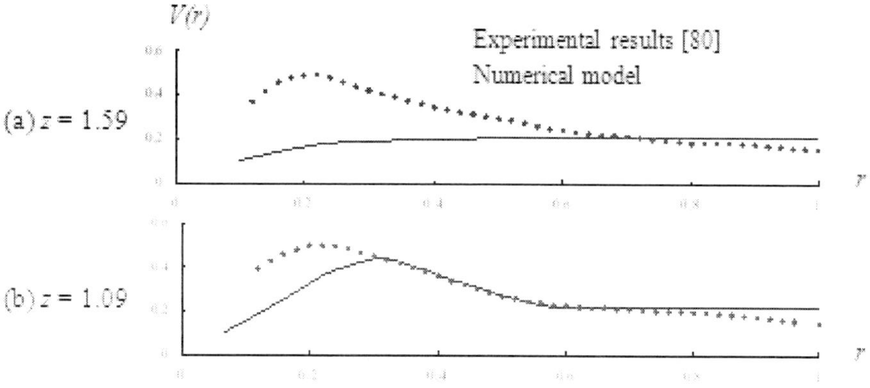

Figure 71. Continued on next page.

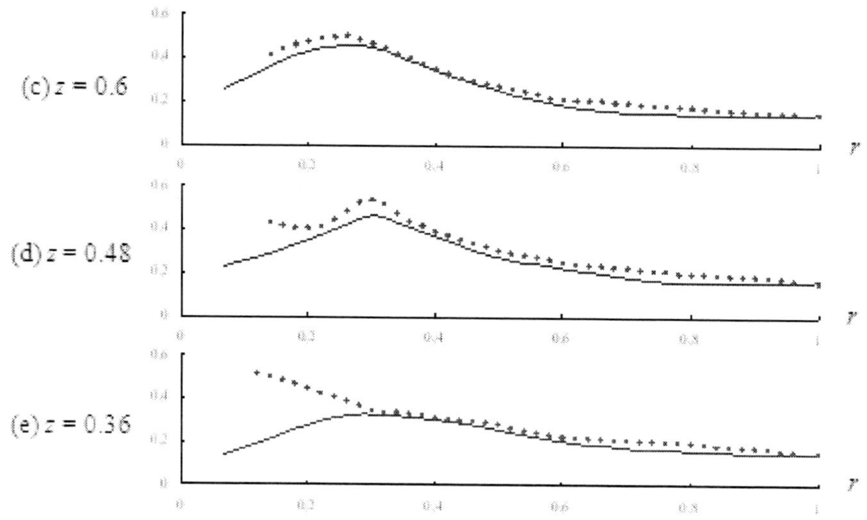

Figure 71. Radials profiles of the tangential velocity for different axial positions V=f(r).

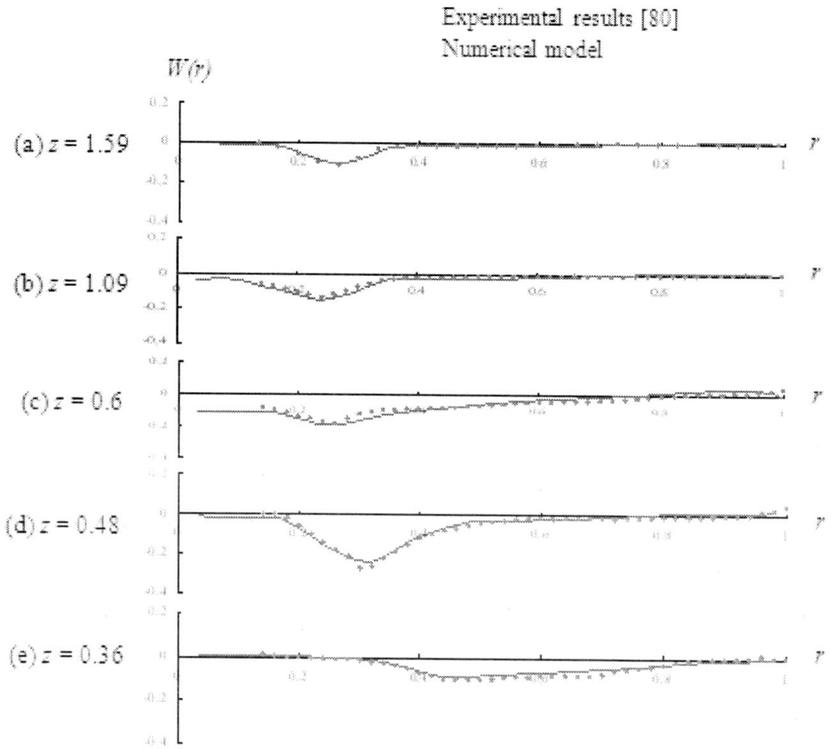

Figure 72. Radials profiles of the axial velocity for different axial positions W=f(r).

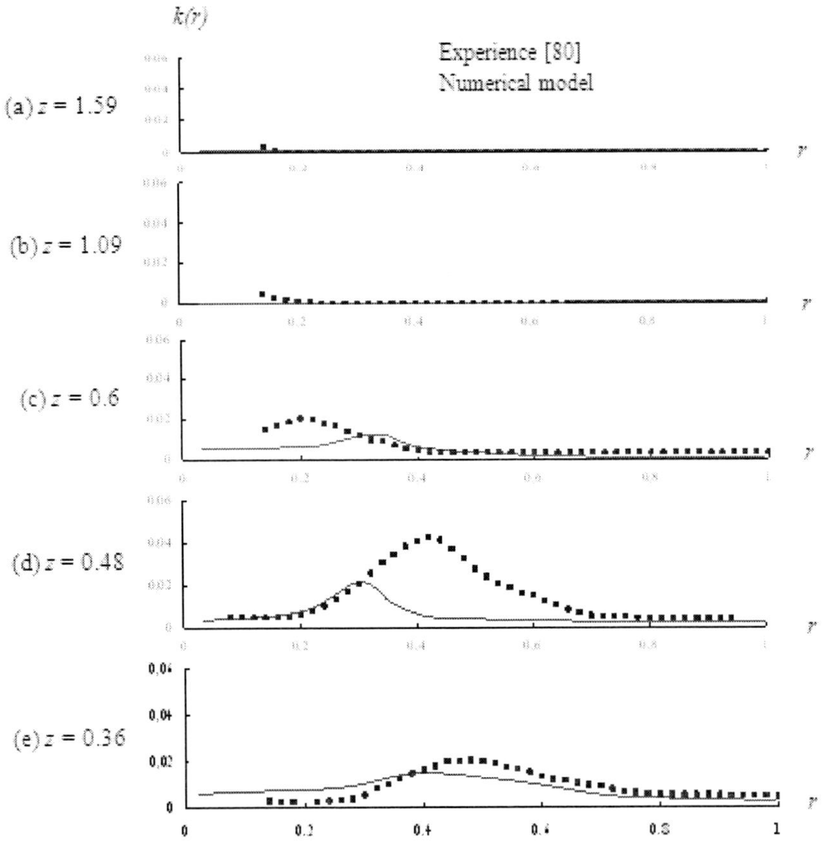

Figure 73. Radial profiles of the turbulent kinetic energy for different axial positions k=f(r).

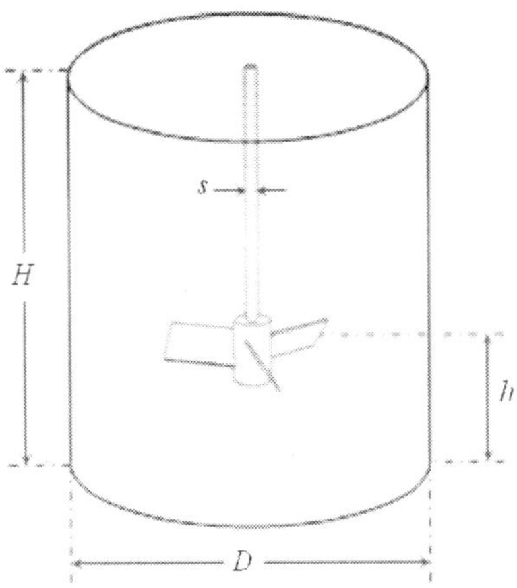

Figure 74. Stirred vessel configurations.

4.2.2. Modified Pitched Blade Turbines (MPBT3)

Figure 74 shows the geometry of the mixing vessel. The tank was a cylindrical, flat-bottomed, with a height diameter ratio H/D of 1 and agitated by a three blades turbine. More particularly, three different turbines were considered (Figure 74). The first turbine corresponds exactly to the geometry configuration already adopted by Karcz and Major [93] in their experimental work (figure 75.a). The diameter is equal to d=0.33 D. The clearance between the bottom of the agitated vessel and the middle of the turbine blades was h=0.33 H. Inclined angle β=45°, number of blades n_p, diameter of turbine d, and position of the blades h were maintained constant for all the configurations studied. However, the two other configurations were defined by an angle of modification along the blade equal to α=150° (MAPBT) and an angle of modification at the end of the blade φ=150° (MEPBT). These turbines are presented respectively in the figures 75.b and 75.c. The flow conditions are represented by the Reynolds number Re=10^5 and the Froude number Fr=0.19.

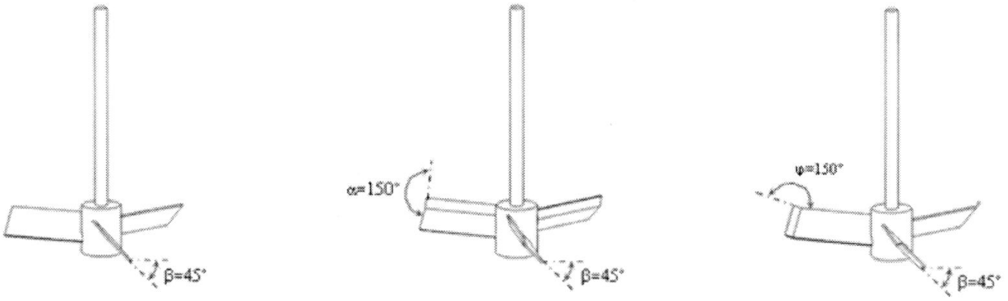

(a) PBT (b) MAPBT (c) MEPBT

Figure 75. Modified pitched blade turbines (MPBT3).

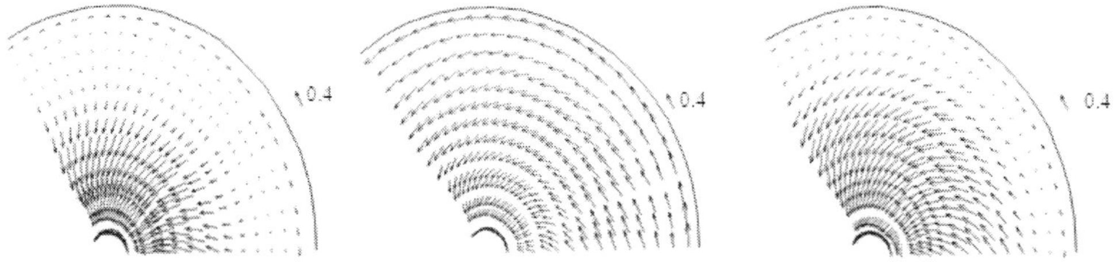

(a) PBT (b) MAPBT (c) MEPBT

Figure 76. Velocity fields induced in r-θ plane define by the axial coordinate z=0.3.

4.2.2.1. Velocity Fields

Simulated velocity fields produced by the modified pitched blade turbines are presented in figures 76 and 77. Figure 76 show the velocity vector plot (U, V) in r-θ plane situated at the axial coordinate equal to z=0.5. Due to symmetry, flow simulations were carried out only for 1/3 of the vessel. It appears that the flow is dominated by the tangential component. While comparing the three geometric configurations, it has been noted that the radial component of the velocity vector is more important in the case of a PBT. Figure 77 shows the flow (U, W) in r-z plane defined by the angular coordinate equal to θ=70°. This plane corresponds to a plane just downstream of the blade. Generally, this figures show the similar flow patterns with a radial jet on the level of the turbine which changes against the walls of the tank with two axial flows thus forming two zones of recirculation on the two sides of the turbine. However, some differences can be noticed. In fact, within a MAPBT, the radial and axial flow is very important. In addition, the position of the center of recirculation was directly affected by the blade type. Indeed, we observed that this center approaches the wall of the tank in the case of a MAPBT. For this same configuration, it has been noticed the apparition of a zone localized in the stirred tank bottom defined by a very weak field.

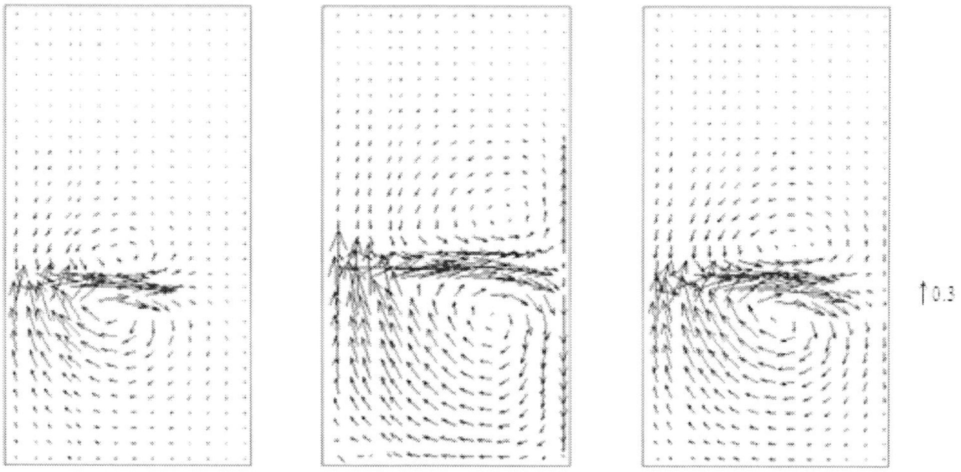

Figure 77. Velocity fields induced in r-z plane define by the angular coordinate θ=70°.

4.2.2.2. Distribution of the Turbulent Kinetic Energy

Figure 78 and 79 show the distribution of the turbulent kinetic energy k in the case of different modified inclined angles of blade β=45°, α=150° and φ=150°. The horizontal presentation planes was situated at the axial coordinate equal to z=0.66. However, the vertical presentation planes were situated in the downstream of the blade plane. They were defined by the angular coordinate equal to θ=70°. Globally, it has been noted that the maximal value is in the region localized in the blades tip. In the domain swept by the turbine blades, the turbulent kinetic energy remains enough elevated. Out of this domain, k became quickly very weak. While modifying the blade modified attack angle, it has been noted that the biggest turbulent kinetic energy is reached within a MEPBT (figure 78.c). The vertical presentation planes show a wake localized in the domain swept by the turbine. This wake is characterized by the elevated values of the turbulent kinetic energy. While moving away of this domain, these values decreased quickly. The modified attack angle intervenes in the characterization of the wake shape. Indeed, it has been remarqued that the wake has been developed more in the case of MEPBT (figure 79.c).

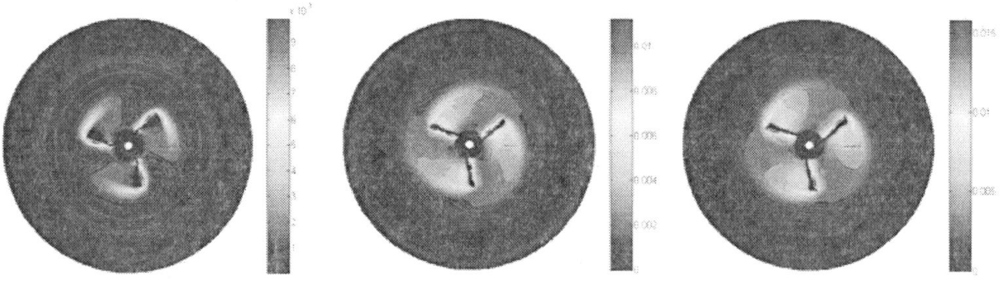

(a) PBT (b) MAPBT (c) MEPBT

Figure 78. Turbulent kinetic energy in r-θ plane define by the axial coordinate z=0.66.

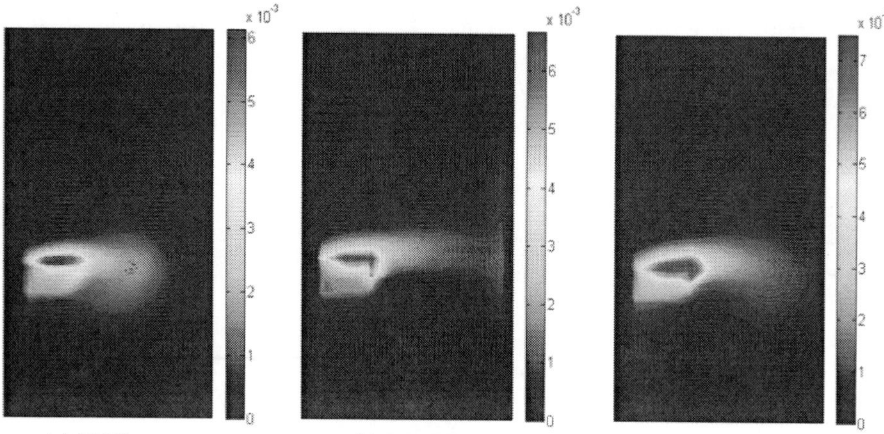

(a) PBT (b) MAPBT (c) MEPBT

Figure 79. Turbulent kinetic energy in *r-z* plane define by the angular coordinate θ=70°.

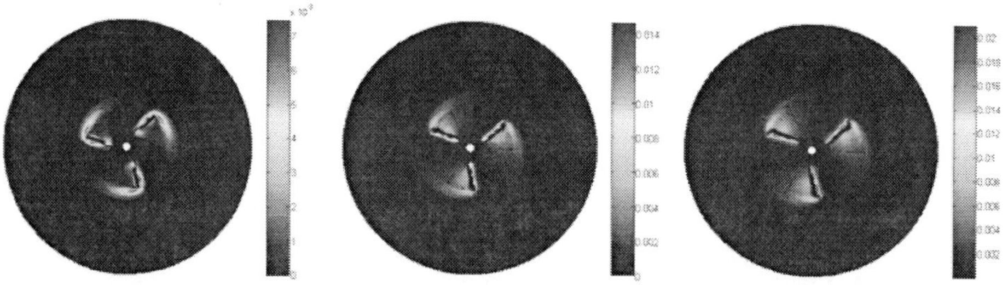

(a) PBT (b) MAPBT (c)
MEPBT

Figure 80. Dissipation rate of the turbulent kinetic energy in *r-θ* plane define by *z*=0.66.

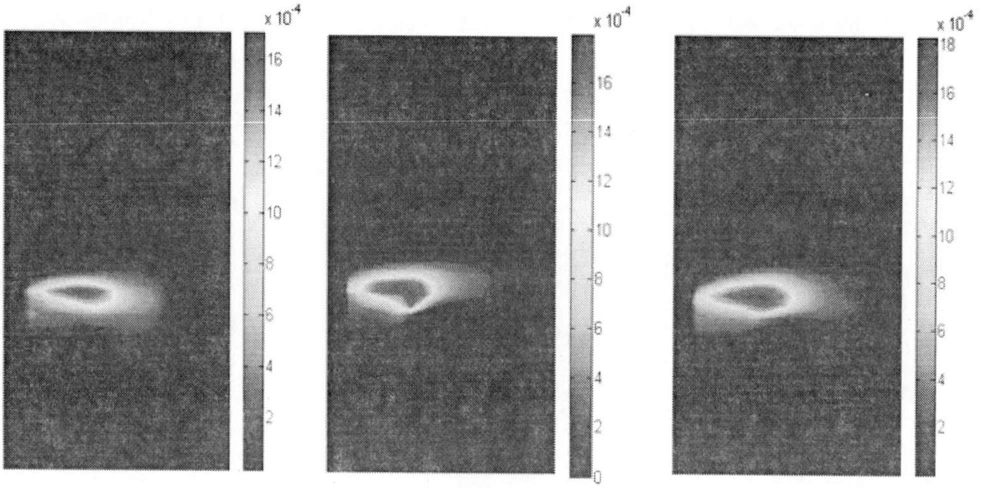

(a) PBT (b) MAPBT (c) MEPBT

Figure 81. Dissipation rate of the turbulent kinetic energy in *r-z* plane define by θ=70°.

4.2.2.3. Distribution of the Dissipation rate of the Turbulent Kinetic Energy

Figure 80 and 81 show the distribution of dissipation rate of the turbulent kinetic energy ε in the same planes adopted in the previous paragraphs. Globally, it has been observed that the distribution is similar to that already obtained with the turbulent kinetic energy. Indeed, it is noted that the area sits of the maximum values are located in the wake which develops nearly to the blades end. However, the dissipation rate becomes very weak outside the field swept by the blades. The greatest rate was reached within MEPBT.

4.2.2.4. Distribution of Turbulent Viscosity

Figure 82 and 83 show the evolution of the turbulent viscosity distribution υ_t. For the presentation of these features, we chose the same planes adopted in the previous paragraph to study the turbulent kinetic energy. In the field swept by blades of the turbine, the turbulent viscosity remained rather high. Very near to the walls and around the axis of the turbine, the turbulent viscosity undergoes a very fast fall. This is due to the deceleration of the flow. By comparing the three configurations between them, the maximal values of the turbulent viscosity increased with the modified inclined angle of blade. Indeed, within MEPBT, the turbulent viscosity is equal to υ_t=2500 (figure 82.c). In the two other cases corresponding to MAPBT and PBT the turbulent viscosity worth υ_t=1700 and υ_t=1100 respectively (figure 82.b and 82.a).

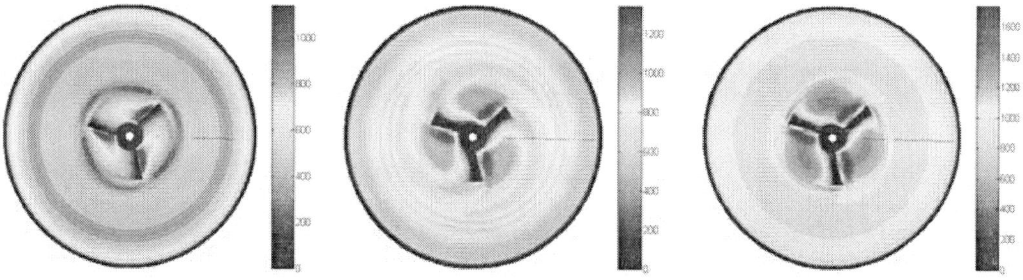

(a) PBT (b) MAPBT (c) MEPBT

Figure 82. Turbulent viscosity in r-θ plane define by the axial coordinate z=0.66.

4.2.2.5. Global Characteristics

On the basis of some theoretical considerations, has been deduced the general relation for calculating the impeller power number N_p, pumping flow number NQ_p, pumping efficiency number ηe and the energy efficiency number E_p in the case of fully filled cylindrical vessel, under laminar and turbulent regime of flow. The effect of blade modified attack angle, number of blades and the clearance between the bottom of the mixing vessel and the middle of the impeller blades was investigated.

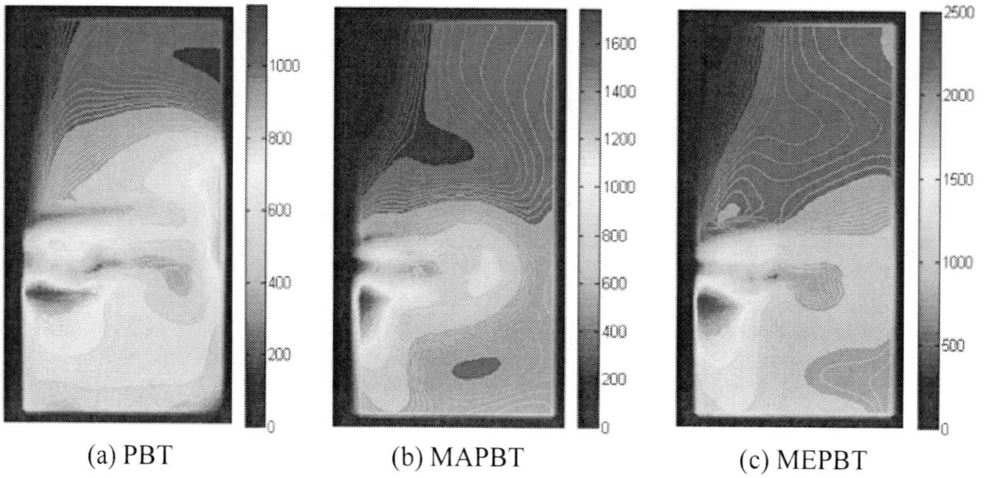

Figure 83. Turbulent viscosity in r-z plane define by the angular coordinate θ=70°.

4.2.2.5.1. Effect of the Blade Modified Attack Angle

In table 3, the global characteristics in the case of different modified pitched blade turbines β=90°, α=150° and φ=150° for a Reynolds number $Re=10^5$ are reported. Globally, the power number N_p increased by modifying the shape of blade attack angle. However, the pumping flow number was maximum for a PBT and reached NQ_p=0.44. While modifying the blade attack angle, it's noted that the energy efficiency and the pumping efficiency number decrease. The numerical results obtained from our CFD code are compared with the experimental results. For a PBT, the power number determined from our CFD code was equal to N_p=0.50. This value is close to that obtained by Karcz et Major [93] and reached N_p=0.57. The good agreement confirmed the validity of the computer method.

Table 3. Effect of blade with modified attack angle for Re=10^5

	Turbines geometry	N_p	NQ_P	η_e	E_p
Numerical model	45° PBT	0.50	0.44	0.88	0.17
	150° MAPBT	0.56	0.37	0.55	0.05
	150° MEPBT	0.65	0.40	0.62	0.10
Experimental results [93]	45° PBT	0.57			

4.2.2.5.2. Effect of Number of Blades

Figure 84 shows global characteristics for the 45° PBT with a different blades number n_p for a Reynolds number $Re=10^5$. With the increase in number of blades from three to six, both power number N_p and pumping flow number NQ_p increased. In addition, it has been noted that the biggest energy efficiency E_e was reached within a six blades turbine and is worth E_p=0.242 (figure 84). In addition, the effect of number of blades was investigated on the PBT in the case of the laminar flow Re=5. Figure 85 shows the effect of number of blades on product $K_p=N_p\ Re$. It was found that both K_p increase with the increase in number of blades.

For a PBT, the product $K_p=N_p\ Re$ determined from our CFD code is equal to K_p=79. This value is close to that obtained by Kuncewicz et Pietrzykowski [91] and is worth K_p=82.7.

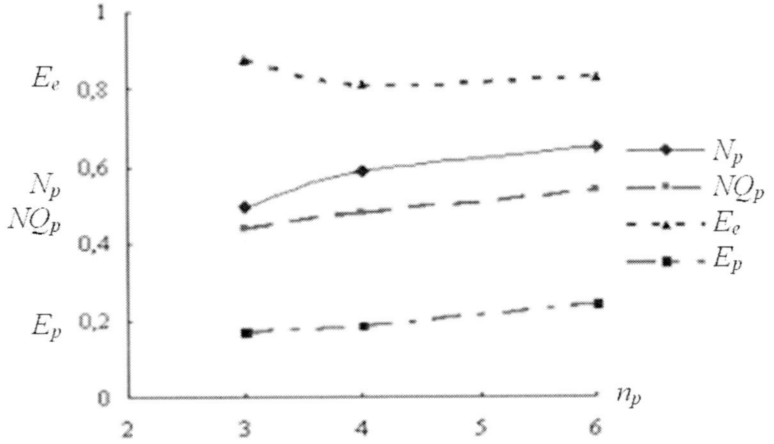

Figure 84. Dependence of global characteristics on number of turbine blades for $Re=10^5$.

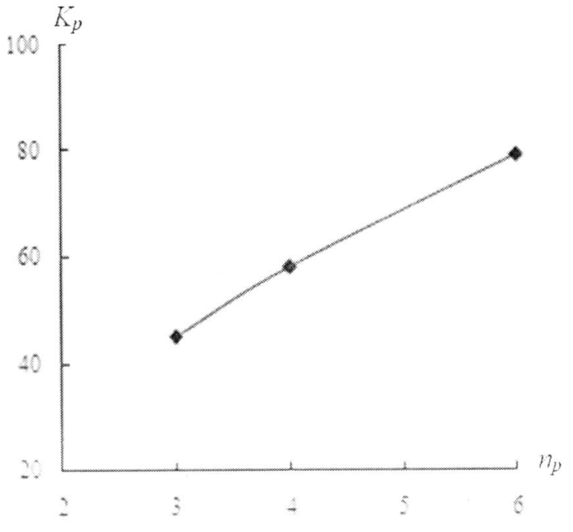

Figure 85. Dependence of product K_p on number of turbine blades for Re=5.

4.2.2.5.3. Effect of Impeller Clearance

Figure 86 shows the effect of the clearance between the bottom of the mixing vessel and the middle of the impeller blades on the global characteristics for a Reynolds number Re=105. Simulations were carried out for the three PBT with clearances of the impeller from the vessel bottom of h/H=0.33, h/H=0.5 and h/H=0.66. It appears that the value of the Np and NQp increased with the decrease in impeller clearance. In addition, it has been noted that the biggest energy efficiency was reached within a PBT defined by h/H=0.33.

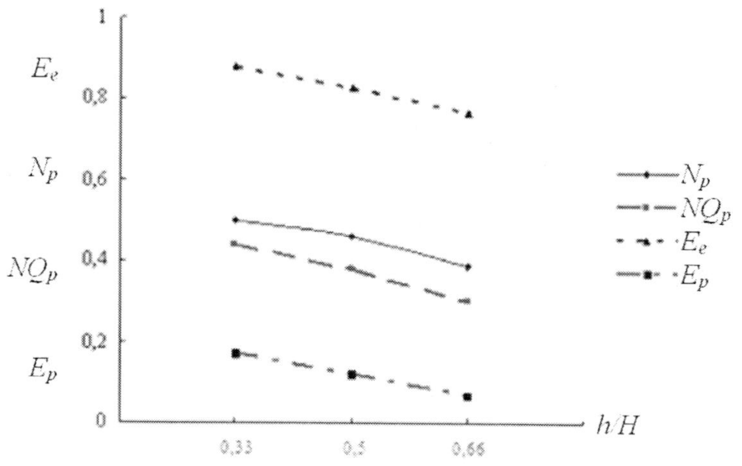

Figure 86. Dependence of global characteristics on impeller clearance for $Re=10^5$.

4.2.2.5.4. Dependence of Global Characteristics on Reynolds Number

Figure 87 and 88 show the dependence of power number N_p and pumping flow number NQ_p on Reynolds number in the case of different modified attack angles of blade. In the laminar flow range, the product $K_p=N_p$ Re remained constant. In the laminar flow range, the power number was inversely proportional to the Reynolds number and the value of the product K_p defined by $K_p=N_p$ Re depended on the impeller. Indeed, in the case of a MEPBT, K_p is equal to 54. However, for a MAPBT and a PBT, this product decreased and was equal to 50 and 45 respectively. In the transition flow range, the power number continuous to decrease when the Reynolds number increases. In the turbulent regime of flow, N_p remained constant. Figure 15 shows the dependence of power number NQ_p on Reynolds number *Re*. In the turbulent regime of flow, NQ_p remained constant. In the transition and laminar flow regime, the pumping flow number increased as a Reynolds number Re increased. According to these numerical results obtained from our CFD code, the following relations have been established:

In the case of a PBT:

$$N_p = 0.668\ Re^{-0.025} \tag{121}$$

In the case of a MAPBT:

$$N_p = 0.656\ Re^{-0.015} \tag{122}$$

In the case of a MEPBT:

$$N_p = 0.721\ Re^{-0.01} \tag{123}$$

In addition, the impeller power number and pumping flow number under turbulent regime of flow, in cylindrical vessels, can be calculated with the following relation:

$$N_p = 10.186\ Re^{-0.2}\ \alpha^{-0.622} \tag{124}$$

$$NQ_p = 1.483\ Re^{-0.2}\ \alpha^{0.951} \tag{125}$$

The angle α is expressed in rad.

We must point out that the relation (124) and (125) are valid only for the three blades turbine with h/H=0.33, used in a turbulent regime of flow. These relations can be exploited in future works for the design of the vessel tank.

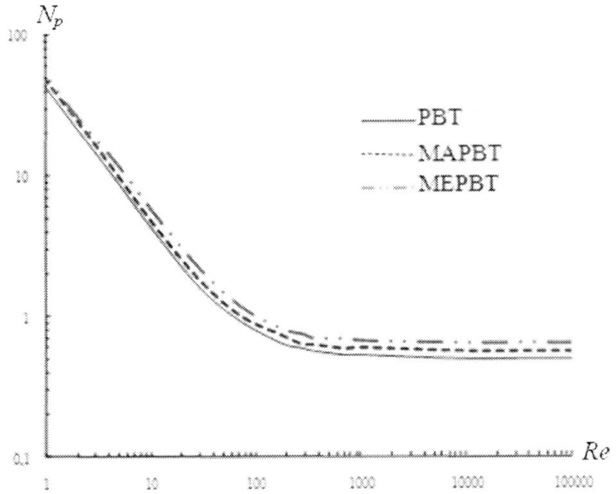

Figure 87. Relation between power number and Reynolds number.

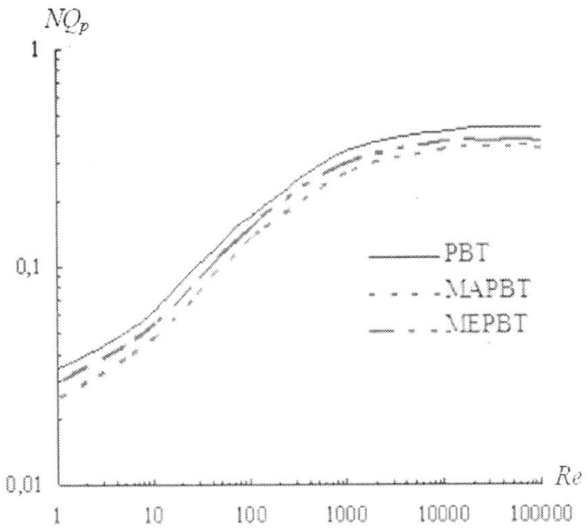

Figure 88. Relation between pumping flow number and Reynolds number.

Conclusion

In this chapter, a specific CFD code has been developed to study the laminar and the turbulent flow in stirred tanks equipped by various turbines impellers type. Particularly, we are interested to the Flat blade turbine (FBT8), the Retreated blade turbine (RBT8), the Ancre impeller (AI), the Maxblend impeller (MI), the Rushton turbines (RT), the Double helical ribbons (DHR), the Double helical ribbons screw (DHRS), the pitched blade turbines (PBT) and the modified pitched blade turbines (MPBT). The three-dimensional flow of a fluid is numerically analyzed using the Navier-Stokes equations in conjunction with the standard k-ε model. The basic equations are solved simultaneously with any additional equations implemented in a particular model to obtain the flow patterns, pressure, characteristics of turbulence and any derived quantities.

The computer method developed permits the numerical analyses of turbines with complex geometries. In fact, after defining the list of nodes belonging to the interface separating the turbine shape and the flow domain, the meshes in the flow domain are automatically generated on a finite volume discretization. The effect of the geometry shape of the impellers has been clearly observed on the local and global hydrodynamics results. In fact, the influence of the geometric parameters, like inclined and incurved angle, on the radial, tangential and axial flow has been shown. The effect of the impellers configurations on the value of the turbulent characteristics has also been demonstrated. Furthermore, the evolution of the pumping flow number NQ_p, the pumping efficiency number E_p, the energy efficiency E_e and the power number N_p have been presented in function of Reynolds number Re. The numerical results such as flow patterns profile and power numbers variation on Reynolds number were compared to our experimental results and those obtained by other authors. The comparison proved a good agreement. Our main objective has been to study and compare the shape effects on the field flow and turbulence characteristics to choose the most effective agitation system for the mixing applications.

In the future, we propose to develop our CFD code and introduce other models to investigate simulations in complex geometrical configurations.

Nomenclature

A	Van-Driest constant
$C_{1\varepsilon}$	constant in the standard k-ε model
$C_{2\varepsilon}$	constant in the standard k-ε model
C_{μ}	constant in the standard k-ε model
d	turbine diameter, m
D	internal diameter of the vessel tank, m
E_e	energy efficiency, dimensionless
E_p	pumping efficiency number, dimensionless

Fr	Froude number, dimensionless, $Fr = \dfrac{(2\pi N)^2 d}{g}$
G	turbulent kinetic energy production, dimensionless
g	gravity acceleration, $m^2.s^{-1}$
h	turbine position, m
H	vessel tank height, m
I	computational cell in radial direction, dimensionless
J	computational cell in tangential direction, dimensionless
K	computational cell in axial direction, dimensionless
k	turbulent kinetic energy, dimensionless
l_m	mixing length, dimensionless
N	velocity of the turbine, $rad.s^{-1}$
N_p	power number, dimensionless, $N_p = \dfrac{P}{\rho N^3 d^5}$
NQ_p	pumping flow number, dimensionless
NR	radial nodes number, dimensionless
NZ	axial nodes number, dimensionless
P	power, W
p	pressure, dimensionless
Q_p	pumping flow, $m^3.s^{-1}$
Re	Reynolds number, dimensionless, $Re = \dfrac{\rho N d^2}{\mu}$
r	radial coordinate, dimensionless
S_Φ	sink term, dimensionless
s	shaft diameter, m
t	time, s
U	radial velocity components, dimensionless
V	angular velocity components, dimensionless
V_0	rubbing velocity on the wall, $m.s^{-1}$
V_R	velocity neighbour to the wall, dimensionless
W	axial velocity components, dimensionless
y	distance of the adjacent node to the solid wall, m
y^+	local Reynolds number, dimensionless
z	axial coordinate, dimensionless
$\vec{V}$	velocity vector
$\vec{J}_\Phi$	flux term vector
$\bar{\bar{I}}$	identity tensor

Greek Symbols

β	inclined angle, degrees

ε	dissipation rate of the turbulent kinetic energy, dimensionless
θ	angular coordinate, rad
κ	Von Karman constant
μ	fluid viscosity, Pa.s
υ_e	effective viscosity, dimensionless
υ_t	turbulent viscosity, dimensionless
ρ	fluid density, $kg.m^{-3}$
σ_k	constant in the standard k-ε model
σ_ε	constant in the standard k-ε model
Γ_Φ	diffusion coefficient, dimensionless
ΔR	radial length between two nodes, dimensionless
ΔZ	axial length between two nodes, dimensionless
Φ	general transport parameter, dimensionless
Φ_v	viscous dissipation rate, dimensionless
$\overline{\overline{\tau}}$	stress tensor
$\overline{\overline{\tau}}_R$	Reynolds tensor

Abbreviations

CFD	computational fluid dynamics
CAD	computer aided design
CSD	computational structure dynamics
PBT	pitched blade turbine
LDA	Laser doppler anemometry
l	laminar
t	turbulent

References

[1] Z. Driss, G. Bouzgarrou, W. Chtourou, H. Kchaou, M.S. Abid. Computational studies of the pitched blade turbines design effect on the stirred tank flow characteristics. *European Journal of Mechanics B/Fluids*, **29**, pp. 236-245, 2010.

[2] G. Bouzgarrou, Z. Driss, M. S. Abid. CFD simulation of mechanically agitated vessel generated by modified pitched blade turbines. *International Journal of Engineering Simulation (IJES)*, Volume 10, Number 2, pp. 11-18, 2009.

[3] Z. Driss, G. Bouzgarrou, W. Chtourou, H. Kchaou, M. S. Abid. Simulation numérique de l'écoulement turbulent généré par des turbines à différents nombre de pales inclines. *Récents Progrès en Génie des Procédés*, Numéro 98, pp. (108)1-6, 2009.

[4] M. Ammar, W. Chtourou, Z. Driss, M.S. Abid. Numerical investigation of turbulent flow generated in baffled stirred vessels equipped with three different turbines in one and two-stage system. *Energy*, Vol. 36, Issue 8, pp. 5081-5093, 2011.

[5] W. Chtourou, M. Ammar, Z. Driss, M. S. Abid. Etude de l'influence de la géométrie de la cuve sur l'hydrodynamique des écoulements générés par une turbine à six pales inclinées. *Récents Progrès en Génie des Procédés*, Numéro 98, pp. 1-6 (344), 2009.

[6] G. Bouzgarrou, Z. Driss, W. Chtourou, M. S. Abid. Caractérisation expérimentale des turbines à pales inclinées dans une cuve mécaniquement agitée. *3ème Congrès International Conception et Modélisation des Systèmes Mécaniques (CMSM'09)*, pp. 1-8, 2009.

[7] H. Kchaou, Z. Driss, G. Bouzgarrou, W. Chtourou, M. S. Abid. Numerical Investigation of Internal Turbulent Flow Generated By A Flat-Blade Turbine and A Pitched-Blade Turbine in A Vessel Tank. *International Review of Mechanical Engineering (I.RE.M.E.)*, Volume 2, Number 3, pp. 427-434, 2008.

[8] Z. Driss, S. Karray, H. Kchaou, M. S. Abid. Computer Simulations of Fluid-Structure Interaction Generated by a Flat-Blade Paddle in a Vessel Tank. *International Review of Mechanical Engineering (I.RE.M.E.)*, Volume 1, Number 6, pp. 608-617, 2007.

[9] H. Kchaou, M. Baccar, M. Mseddi, M.S. Abid. Modélisation numérique de l'écoulement interne turbulent généré par une turbine radiale en cuve agitée. *Revue Française de Mécanique*, **1**, pp. 73-79, 2000.

[10] Z. Driss, S. Karray, G. Bouzgarrou, H. Kchaou, M.S. Abid. Couplage fluide structure d'une turbine à quatre pales flexibles. *Premier Colloque International Innovations en Mécaniques Passives et Actives (IMPACT 2010)*, pp. 1-8, 2010.

[11] Z. Driss, H. Kchaou, M. Mseddi, M. Baccar, M. S. Abid. Computer simulations of laminar mixing within a radial turbine vessel. *International Conference on Advances in Mechanical Engineering (ICAME'04)*, T6-T7, pp. 1-6, 2004.

[12] Z. Driss, H. Kchaou, M. Baccar, M. S. Abid. Numerical investigation of internal laminar flow generated by a retreated-blade paddle and a flat-blade paddle in a vessel tank. *International Journal of Engineering Simulation (IJES)*, pp. 10-16, Volume 6, Number 3, 2005.

[13] Z. Driss, H. Kchaou, M. Baccar, M. S. Abid. Simulation numérique de l'écoulement laminaire dans une cuve agitée par une turbine à pales incurvées. *Récents Progrès en Génie des Procédés*, Numéro 92, R-17, pp. 1-8, 2005.

[14] M. Baccar, M. S. Abid. Simulation numérique des comportements hydrodynamiques et thermiques des Echangeurs racleurs opérant en régime turbulent. *Int. J. Therm. Sci.*, Vol. 38, pp. 634-644, 1999.

[15] M. Baccar, M. S. Abid. Numerical analysis of three-dimensional flow and thermal behaviour in a scraped-surface heat exchanger. *Rev. Gén. Therm.*, Vol. 36, pp. 782-790, 1997.

[16] C. Xuereb, M. S. Abid, J. Bertrand. Chapter 16: Modeling of the Hydrodynamics behavior of highly viscous fluids in stirred tanks equipped with two blade impellers. pp. 455-485, Advances in Engineering Fluid Mechanics: Multiphase Reactor and Polymerization System Hydrodynamics, Edited by Nicholas P. Cheremisinoff, *Advances in Engineering Fluid Mechanics series*, Elsevier, 1996.

[17] M. S. Abid, C. Xuereb, J. Bertrand. Modeling of the 3D Hydrodynamics of 2-blade impellers in stirred tanks filled with a highly viscous fluid. *Can. J. Chem. Eng.*, Vol. 72, pp. 184-193, 1994.

[18] M. S. Abid, C. Xuereb, J. Bertrand. Hydrodynamics of 2-blade impellers in vessels stirred with anchors and gate agitators: Necessity of a 3-D Modeling. *Trans. IChemE.*, Vol. 70, pp. 377-384, july 1992.

[19] M. Baccar, M. S. Abid. Numerical modelling of the hydrodynamic and thermal behaviours of Newtonian and non-Newtonian pseudoplastic fluids in a stirred vessel by anchors and gate agitators. *Entropie*, Vol. 36, 227, pp. 22-29, 2000.

[20] M. Baccar, M. S. Abid. Caractérisation de l'écoulement turbulent et du transfert thermique générés par des mobiles à ancre et barrière dans une cuve agitée. *Int. J. Therm. Sci.*, Vol. 38, pp. 892-903, 1999.

[21] S. Karray, Z. Driss, H. Kchaou, M.S. Abid. Numerical simulation of fluid-structure interaction in a stirred vessel equipped with an anchor impeller. *Journal of Mechanical Science and Technology (JMST)*, 25, pp. 1749-1760, 2011.

[22] S. Karray, Z. Driss, H. Kchaou, M.S. Abid. Hydromechanics characterization of the turbulent flow generated by anchors impellers. *Engineering Applications of Computational Fluid Mechanics (EACFM)*, **5**, pp. 315-328, 2011.

[23] Z. Driss, S. Karray, H. Kchaou, M. S. Abid. *Computer simulations of laminar flow generated by an anchor blade and a Maxblend impellers*. Science Academy Transactions on Renewable Energy Systems Engineering and Technology (SATRESET), Vol. 1, N. 3, pp. 68-76, 2011.

[24] Z. Driss, H. Kchaou, M. Baccar, M. S. Abid. Etude hydrodynamique d'une turbine à pale trouée pour le mélange des revêtements en cuve agitée. *Journées Internationales Scientifiques & Pédagogiques de Mécanique & d'Energétique (JISPME 2005)*, Tome 2, pp. 133-140, 2005.

[25] W. Chtourou, Z. Driss, M. Ammar, M.S. Abid. The Effect of Multiple Rushton Turbine Use on Turbulent Flow in Stirred Tanks. *International Journal of Engineering Simulation (IJES)*, Vol. 11, Number 3, pp. 1-8, 2010.

[26] M. Baccar, H. Kchaou, M. Mseddi, M.S. Abid. Modelling of turbulent flow in standard vessel geometry stirred by Rushton turbine: Pseudo-stationary modelling approach. *Mécanique & Industries*, **1**, pp. 73-79, 2000.

[27] G. Bouzgarrou, Z. Driss, A. Kaffel, M. S. Abid. Caractérisation hydrodynamique des turbines de Rushton dans une cuve agitée en régime turbulent. *19ème Congrès Français de Mécanique (CFM'2009)*, pp. 1-6, 2009.

[28] Z. Driss, G. Bouzgarrou, W. Chtourou, H. Kchaou, M.S. Abid. CFD Modelling of Stirred Tanks Agitated With Multiple Rushton Impellers. *The Fouth International Conference on Advances in Mechanical Engineering and Mechanics (ICAMEM'08)*, pp. 1-6, 2008.

[29] Z. Driss, H. Kchaou, M. Mseddi, M. Baccar, M. S. Abid. Simulation numérique de l'écoulement laminaire dans une cuve agitée par une turbine de Rushton. *Colloque Franco-Tunisien sur les Méthodes Numériques Appliquées aux Ecoulements & aux Transferts (CFT'2004)*, pp. 159-164, 2004.

[30] Z. Driss, S. Karray, H. Kchaou, M.S. Abid. CFD simulation of the laminar flow in stirred tanks generated by double helical ribbons. *Central European Journal of Engineering (CEJE)*, **1**(4), pp. 413-422, 2011.

[31] Z. Driss, G. Bouzgarrou, H. Kchaou, M. S. Abid. Computer simulation of the laminar flow in stirred tanks generated by the proximity impellers of a mono and double screws

type with simple and modified profiles. *Mechanics & Industries (M&I)*, **12**, pp. 109-121, 2011.

[32] G. Bouzgarrou, Z. Driss, M.S. Abid. Etude de la structure hydrodynamique dans une cuve agitée générée par des helices. *Récents Progrès en Génie des Procédés*, Numéro 98, pp. (109)1-6, 2009.

[34] M. Mseddi, Z. Driss, M. Baccar, M.S. Abid. Etude d'une soupape de décharge de turbocompresseur de suralimentation. *Mécanique & Industries*, Numéro 5, pp. 539-551, 2005.

[35] Z. Driss, M. Mseddi, M. Baccar, M.S. Abid. Numerical and experimental study of a turbocharger's wastegate. International *Conference on Advances in Mechanical Engineering (ICAME'04)*, T3-T4, pp. 1-6, 2004.

[36] Z. Driss, M. Mseddi, M. Baccar, M.S. Abid. Modélisation 3D de l'écoulement turbulent à travers une soupape de décharge de turbocompresseur de suralimentation. *4èmes Journées Tunisiennes sur les Ecoulements et les Transferts (JTET'2002)*, Vol.1, pp. II8-II14, 2002.

[37] J. Blazek. *Computational Fluid Dynamics: Principles and Applications*. ELSEVIER, British Library Cataloguing in Publication Data, pp. 1-440, 2001.

[38] D.A. Anderson, J.C. Tannchill, R.M. Pletcher. Computational Fluid Mechanics and heat Transfer. Series in computational methods in mechanics and thermal sciences, *Hemisphere Publishing Corporation*, pp 1-599, 1984.

[39] E.L. Paul, V.A. Atiemo-Obeng, S.M. Kresta. *HandBook of Industrial Mixing: Science and Practice*, John Wiley & Sons, Inc., Hoboken, New Jersey, pp 1-1377, 2004.

[40] V.V. Ranade, *Computational Flow Modeling for Chemical Reactor Engineering*. Academic Press: A Harcourt Science and Technology Company, pp 1-452, 2002.

[41] C.R. Doering, J.D. Gibbon. *Applied analysis of the Navier-Stokes equations*. Cambridge University Press, pp 1-217, 2004.

[42] M. Ciofalo, CFD Prediction of Turbulent Flow and Mixing in Stirred Vessels: Single- and Two-Phase Flow. *Current Trends in Chemical Engineering*, pp 1-165, 2001.

[43] F. Durst. *Fluid Mechanics: An Introduction to the Theory of Fluid Flows*. Springer-Verlag Berlin Heidelberg, pp 1-723, 2008.

[44] J. Tu, G. H. Yeoh, C. Liu. *Computational Fluid Dynamics: A Practical Approach*. ELSEVIER, British Library Cataloguing in Publication Data, pp 1-459, 2008.

[45] M. Lesieur, Turbulence in Fluids: Third revised and Enlarged Edition. *Fluid Mechanics and its applications*, Kluwer Academic Publishers, pp 1-515, 1995.

[46] A.K. Coker, *Modeling of Chemical Kinetics and Reactor Design*. Gulf Publishing Company, Houston, Texas, pp 1-1095, 2001.

[47] O. C. Zienkiewicz,R. L. Taylor, The finite Element Method, Volume 3: *Fluid Dynamics*, Butterworth-Heinemann, pp 1-334, 2000.

[48] I. Tosun. Modeling in Transport Phenomena: *A Conceptual Approach*, ELSEVIER, The Netherlands, pp 1-590, 2002.

[49] M.M. Hafez. *Numerical Simulations of Incompressible Flows*, World Scientific, British Library Cataloguing-in-Publication Data, pp 1-680, 2003.

[50] P. Sagaut, S. Deck, M. Terracol, *Multiscale and Multiresolution Approaches in Turbulence*, Imperial College Press, pp 1-340, 2006.

[51] A. Bakker, H. Haidari, E. Marshall. Numerical modeling of mixing processes. What can LES offer? *AIChE Annual Meeting*, Miami Beach, FL, Paper 238j, pp. 15-20, 1998.

[52] J. Revstedt, L. Fuchs, C. Tragardh. Large eddy simulations of the turbulent flow in a stirred reactor, *Chem. Eng. Sci.*, **53**, pp. 4041-4053, 1998.

[53] J. Derksen, H.E.A. Van Den Akker. Large eddy simulations of the flow driven by a Rushton turbine. *AIChE J.*, **45**, pp. 209-221, 1999.

[54] V.T. Roussinova, S.M. Kresta, R. Weetman. Low frequency macroinstabilities in a stirred tank: scale-up and prediction based on large eddy simulations, *Chem Eng. Sci.* **58**, pp. 2297-2311, 2003.

[55] S. B. Pope. *Turbulent Flows*, Cambridge University Press, New York, 2000.

[56] B. E. Launder. Modeling the formation and dispersal of streamwise vortices in turbulent flow. Aeronaut. J., *35th Lanchester Lecture*, **99** (990), pp. 419-431, 1995.

[57] S.V. Patankar. Numerical heat transfer and fluid flow. *Series in Computational Methods in Mechanics and Thermal Sciences*, Mc Graw Hill, New York, 1980.

[58] J. Douglas, J. E. Gunn. A general formulation of alternating direction implicit methods, *Num. Math.*, **6**, pp. 428-453, 1964.

[59] C. Xuereb, J. Bertrand, 3-D Hydrodynamics in a tank stirred by a double-propeller system and filled with a liquid having evolving rheological properties. *Chemical Engineering Science*, **51**, pp. 1725-1734, 1996.

[60] S. Nagata. Mixing: principles and applications. John Wiley & Sons: Halstead press, Japan, 1975.

[61] J.M. Zalc, E.S. Szalai, M.M. Alvarez, F.J. Muzzio. Using CFD To Understand Chaotic Mixing in Laminar Stirred Tanks. *AIChE Journal*, Vol. 48, No. 10, pp. 2124-2134, 2002.

[62] M. M. Alvarez, J. M. Zalc, T. Shinbrot, P. E. Arratia, F. J. Muzzio. Mechanisms of Mixing and Creation of Structure in Laminar Stirred Tanks. *AIChE Journal*, Vol. 48, No. 10, pp. 2135-2148, 2002.

[63] M.M. Wang, A. Dorward, D. Vlaev, R. Mann. Measurements of Gas-Liquid Mixing in a Stirred Vessel Using Electrical Resistance Tomography (ERT). *1st World Congress on Industrial Process Tomography*, Buxton, Greater Manchester, April 14-17, pp. 78-83, 1999.

[64] A. Bakker, L.M. Oshinowo, E.M. Marshall. The Use of Large Eddy Simulation to Eddy Simulation to Study Stirred Vessel Hydrodynamics. *Proceedings of the 10th European Conference on Mixing, Delft*, The Netherlands, pp. 247-254, 2000.

[65] E. H. Stitt. Alternative multiphase reactors for fine chemicals A world beyond stirred tanks. *Chemical Engineering Journal*, **4025**, pp. 1-14, 2002.

[66] F. Guillard, C. Trägardh. Mixing in industrial Rushton turbine agitated reactors under aerated conditions. *Chemical Engineering and Processing*, pp. 1-13, 2002.

[67] Y. Kaneko, T. Shiojima, M. Horio. Numerical analysis of particle mixing characteristics in a single helical ribbon agitator using DEM simulation. *Powder Technology*, **108**, pp. 55-64, 2000.

[68] T. Espinosa-Solares, E. Britto-De La Fuente, A. Tecante, P. A. Tanguy. Power consumption of a dual turbine-helical ribbon impeller mixer in ungassed conditions. *Chemical Engineering Journal*, **67**, pp. 215-219, 1997.

[69] W. Yao, M. Mishima, K. Takahashi. Numerical investigation on dispersive mixing characteristics of MAXBLEND and double helical ribbons. *Chemical Engineering Journal*, **84**, pp. 565-571, 2001.

[70] P.A. Tanguy, F. Thibault, E.B. La Fuente, T. Espinosa-Solares, A. Tecante, Mixing performance induced by coaxial flat blade-helical ribbon impellers rotating at different speeds, *Chemical Engineering Science*, Vol. 52, No. 11, pp. 1733-1741, 1997.

[71] F. Bertrand, P.A. Tanguy, E. B. de la Fuente, P. Carreau. Numerical modeling of the mixing flow of second-order fluids with helical ribbon impellers. *Comput. Methods Appl. Mech. Engrg.*, **180**, pp. 267-280, 1999.

[72] A. Niedzielska, Cz. Kuncewicz. Effect of impeller geometry on the power consumption for helical ribbon impellers. *Tatranské Matliare*, Slovak Republic, **138**, 26-30, 2003.

[73] J.J. Wang, L.F. Feng, X.P. Gu, K. Wang, C.H. Hu. Power consumption of inner-outer helical ribbon impellers in viscous Newtonian and non-Newtonian fluids. *Chemical Engineering Science*, **55**, pp. 2339-2342, 2000.

[74] S.M.C.P. Pedrosa, J.R. Nunhez. The behaviour of stirred vessels with anchor type impellers. *Computers and Chemical Engineering*, **24**, pp. 1745-1751, 2000.

[75] J. Aubin, D.F. Fletcher, C. Xuereb. Modeling turbulent flow in stirred tanks with CFD: the influence of the modeling approach, turbulence model and numerical scheme. *Experimental Thermal and Fluid Science*, **28**, pp. 431-445, 2004.

[76] E.D. Hollander, J.J. Derksen, H.M.J. Kramer, G.M. Van Rosmalen, H.E.A. Van den Akker. A numerical study on orthokinetic agglomeration in stirred tanks. *Powder Technology*, **130**, pp. 169-173, 2003.

[77] D. Dakshinamoorthy, A.R. Khopkar, J.F. Louvar, V.V. Ranade. CFD simulation of shortstopping runaway reactions in vessels agitated with impellers and jets. *Journal of Loss Prevention in the Process Industries*, **19**, pp. 570-581, 2006.

[78] T. Kumaresan, J.B. Joshi. Effect of impeller design on the flow pattern and mixing in stirred tanks. *Chemical Engineering Journal*, **115**, pp. 173-193, 2006.

[79] D.A. Deglon, C.J. Meyer. CFD modelling of stirred tanks: Numerical considerations. *Minerals Engineering*, **19**, pp. 1059-1068, 2006.

[80] P. M. Armenante, L. Changgen, C. Chou, I. Fort, J. Medek. Velocity profiles in a closed, unbaflled vessel: comparison between experimental LDV data and numerical CFD predictions. *Chemical Engineering Science* **52** (1997) 3483-3492.

[81] N. Li, Q. Zhou, X. Chen, AT. Xu, S. Hui, D. Zhang. Liquid drop impact on solid surface with application to water drop erosion on turbine blades, Part I: Nonlinear wave model and solution of one-dimensional impact. *International Journal of Mechanical Sciences*, **50**, pp. 1526-1542, 2008.

[82] C.W. Chang-Jian, C.K. Chen. Non-linear dynamic analysis of rub-impact rotor supported by turbulent journal bearings with non-linear suspension. *International Journal of Mechanical Sciences*, **50**, pp. 1090-1113, 2008.

[83] A. Whitfield, N.C. Baines. A general computer solution for radial and mixed flow turbomachine performance prediction. *International Journal of Mechanical Sciences*, **18**, pp. 179-184, 1976.

[84] M.T. Schobeiri. The influence of curvature and pressure gradient on the flow temperature and velocity distribution. *International Journal of Mechanical Sciences*, **32**, pp. 851-861, 1990.

[85] H. Chen, N.C. Baines. The aerodynamic loading of radial and mixed flow turbines. *International Journal of Mechanical Sciences*, **36**, pp. 63-79, 1994.

[86] Y. Zhao, D.E. Winterbone. The finite volume flic method and its stability analysis. *International Journal of Mechanical Sciences*, **37**, pp. 1147-1160, 1995.

[87] K. Suzukawa, S. Mochizuki, H. Osaka. Effect of the attack angle on the roll and trailing vortex structures in an agitated vessel with a paddle impeller. *Chemical Engineering Journal*, **61**, pp. 2791-2798, 2006.

[88] P.K. Biswas, S.C. Dev, K.M. Godiwalla, C.S. Sivaramakrishnan. Effect of some design parameters on the suspension characteristics of a mechanically agitated sand-water slurry system. *Materials and Design*, **20**, pp. 253-265, 1999.

[89] D. Chapple, S.M. Kresta, A. Wall, A. Afacan. The effect of impeller and tank geometry on power number for a pitched blade turbine. *Institution of chemical Engineers*, **80**, Part A, 2002.

[90] C. Dan Taca, M. Paunescu. Power input in closed stirred vessels. *Chemical Engineering Science*, **56**, pp. 4445-4450, 2001.

[91] C. Kuncewicz, M. Pietrzykowski. Hydrodynamic model of a mixing vessel with pitched-blade turbines. *Chemical Engineering Science*, **56**, pp. 4659-4672, 2001.

[92] J.J. Derksen, M.S. Doelman, H.E.A. Van den Akker. Three-dimensional LDA measurements in the impeller region of a turbulently stirred tank. *Experiments in Fluids*, **27**, pp. 522-532, 1999.

[93] J. Karcz, M. Major, An Effect of a baffle Length on the power consumption in an agitated vessel. *Chemical Engineering Science*, **37**, pp. 249-256, 1998.

[94] T. Kumaresan, J. B. Joshi. Effect of impeller design on the flow pattern and mixing in stirred tanks. *Chemical Engineering Journal*, **115**, pp. 173-193, 2006.

In: Navier-Stokes Equations
Editor: R. Younsi

ISBN: 978-1-61324-590-3

Chapter 4

POSTINSTABILITY EXTENSION OF NAVIER-STOKES EQUATIONS

Michail Zak
Jet Propulsion Laboratory California Institute of Technology
Reasoning, Modeling, and Simulation Group
Pasadena, CA, U. S.

ABSTRACT

This chapter is devoted to investigation of the region of instability of the Navier-Stokes equations. This region is of critical impotence for modeling turbulent flows. Stochastic approach to turbulence based upon the Stabilization Principle is introduced. Onset of turbulence is interpreted as loss of stability of solutions to the Navier-Stokes equations *in the class of differentiable functions*. A nonlinear version of the Liouville- Fokker-Planck equation with negative diffusion is proposed for describing postinstability motions of the Navier-Stokes equations. As a result, the flow velocities are represented by a set of stochastic invariants found from the Stabilization Principle. The theory is illustrated by examples.

1. INTRODUCTION

a. Unsolved problems in Newtonian mechanics. The Navier–Stokes equations describe motions of viscous fluid. Their derivation is based on the assumption that the fluid, at the scale of interest, is a continuum, while all the fields of pressure, velocity, density, and temperature are differentiable, weakly at least. This "innocent" assumption that works well in other continua such as elastic bodies, polymers, etc, created the problem of turbulence - one of the most fundamental problems in theoretical physics that is still unsolved. Although applicability of the Navier-Stokes equations as a model for fluid mechanics is not in question, the instability of their solution for flows with supercritical Reynolds numbers raises a more general question: is Newtonian mechanics complete?

During several centuries, Newtonian mechanics enjoyed an unprecedented success being considered as the most general, the most accurate, and the most elegant branch of science. It became consensual that its structure is perfect and complete as far as the non-relativistic macroworld is concerned. However the problem of turbulence (stressed later by the discovery of chaos) demonstrated that the Newton's world is far more complex than those represented by classical models. It appears that the Lagrangian or Hamiltonian formulations do not suggest any tools for treating postinstability motions, and this is a major flaw of the classical approach to Newtonian mechanics. The explanation of that limitation was proposed in Zak, M., 1994, 1997: the classical formalism based upon the Newton's laws exploits *additional mathematical restrictions*, (such as space-time differentiability, the Lipchitz conditions etc.) that are not required by the Newton's laws. The only purpose for these restrictions is the application of a powerful technique of classical mathematical analysis. However in many cases such restrictions are incompatible with physical reality. This statement can be illustrated by a trivial example presented below, Fig. 1.

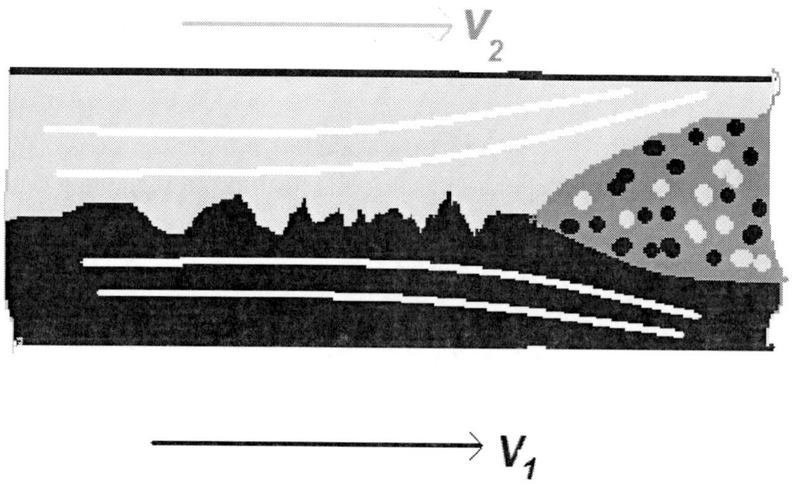

Figure 1. Loss of differentiability in fluid mechanics.

Consider a surface of a tangential jump of velocity $V_2 - V_1$ in a horizontal unidirectional flow of an inviscid incompressible fluid. Applying the principle of virtual work to a small volume V containing both flows of the fluid as well as the surface of the tangential jump of velocities separating the flows, one obtains

$$\int_V (\rho a_1 \bullet \delta U_1 + \rho a_2 \bullet \delta U_2) dV = 0 \tag{1}$$

where ρ, U_1, U_2 , a_1,a_2 are density, displacements, and accelerations of the fluid. The displacements U_1 and U_2 are mutually independent in the region that does not contain a separating surface, but they are dependent at the surface due to its impenetrability

$$U_1 \bullet n = U_2 \bullet n = U, \tag{2}$$

Hence, as follows from Eq. (1), at the surface the following equality holds:

$$(a_1 + a_2) \bullet n = 0, \qquad i.e. \qquad [(\frac{\partial}{\partial t} + V_1 \frac{\partial}{\partial S})^2 + (\frac{\partial}{\partial t} + V_2 \frac{\partial}{\partial S})^2]U + \alpha = 0 \tag{3}$$

where α is the term that does not contain the second order derivatives of U, and therefore, it does not effect the characteristic equation

$$(\lambda - V_1)^2 + (\lambda - V_2)^2 = 0 \tag{4}$$

and its characteristic roots

$$\lambda = \frac{1}{2}[(V_2 + V_1) \pm i(V_2 - V_1)] \tag{5}$$

Let us start with studying propagation of high frequency oscillations of the transverse displacements U. Recall that Eq. (3) being linear with respect to the second order time-space derivatives, strictly speaking, is nonlinear with respect to U and its first time-space derivatives that are contained in the term α. For small amplitudes and their first derivatives, this term can be linearized:

$$\alpha = \alpha_1 \frac{\partial U}{\partial t} + \alpha_2 \frac{\partial U}{\partial S} + \alpha_3 U + \alpha_4 \tag{6}$$

For further simplifications, all the coefficients in Eq.(3) can be linearized with respect to an arbitrarily selected point S_0 and instant of time $t_0 = 0$. Then Eq. (3) takes form of a linear elliptic PDE with constant coefficients

$$[(\frac{\partial}{\partial t} + V_1^0 \frac{\partial}{\partial S})^2 + (\frac{\partial}{\partial t} + V_2^0 \frac{\partial}{\partial S})^2 + \alpha^0{}_1 \frac{\partial}{\partial t} + \alpha^0{}_2 \frac{\partial}{\partial S} + \alpha^0{}_3]U + \alpha^0{}_4 = 0 \tag{7}$$

Obviously this equation is valid only for small amplitudes with small first derivatives, the small area around the above selected point S_0, and within a small period of time Δt.

Let us derive the solution to Eq. (7) subject to the following initial conditions

$$U^*_0 = \frac{1}{\lambda_0} e^{-\lambda_0 S i}, \qquad at \qquad t = 0 \tag{8}$$

assuming that λ_0 can be made as large as desired, i.e. $\lambda_0 > N \to \infty$. Consequently, the initial disturbances can be made as small as desired, i.e. $U^*_0 < N^{-1} \to 0$. The corresponding solution is written in the form

$$U^* = C_1 e^{-\lambda_0 (\lambda_1 t - S) i} + C_2 e^{-\lambda_0 t - S) i} \tag{9}$$

where λ_1, λ_2 are the roots of the characteristic equation (4) (see Eq. (5). Since the characteristic roots are complex, the solution (9) will contain the term

$$\frac{1}{\lambda_0} e^{|\mathrm{Im}\,\lambda_{1,2}|\Delta t} \sin \lambda_0 S, \qquad \lambda_0 \to \infty \tag{10}$$

that leads to infinity within an arbitrarily short period of time Δt and within an infinitesimal area around the point S_0.

Hence one arrived at the following situation: $|U^*| \to \infty$ in spite the fact that $|U_0| \to 0$. To obtain a geometrical interpretation of the above described instability, let's note that if the second derivatives in Eq. (7) are of order λ_0, then the first derivatives are of order of 1, and U is of order of $1/\lambda_0$. Hence, the period of time Δt can be selected in such a way that the second derivatives will be as large as desired, but U and its first derivatives are still sufficiently small. Taking into account that the original governing equation (3) is quasi-linear with respect to the second derivatives, and therefore, the linearization does not impose any restrictions on their values, one concludes that the linearized equation (7) is valid for the solution during the above mentioned period of time Δt. Turning to the term (10) of the solution (9), one can now interpret it as being represented by a function having an infinitesimal amplitude and changing its sign with an infinite frequency ($\lambda_0 \to \infty$). The first derivatives of this function can be small and change their signs by finite jumps (with the same infinite frequency), so that the second derivatives at the points of such jumps are infinite. From the mathematical viewpoint, this kind of function is considered as continuous, but *non-differentiable.*

The result formulated above was obtained under specially selected initial conditions (8), but it can be generalized to include any initial conditions. Indeed, let the initial conditions be defined as

$$U|_{t=0} = U^{**}_0(X) \tag{11}$$

and the corresponding solution to Eq. (3) is

$$U = U^{**}(X, t) \tag{12}$$

Then, by altering the initial conditions to

$$U|_{t=0} = U_0{}^*(X) + U_0{}^{**}(X) \tag{13}$$

where U^*_0 is defined by Eq. (8), one observes the preceding argument by superposition that vanishingly small change in the initial condition (13) leads to unboundedly large change in the solution (12) that occurs during an infinitesimal period of time.

That makes the solution (12) *non-differentiable*, but still continuous. However, the Euler's equation that governs the flow discussed above was derived under condition *of space-time differentiability* of the velocity field. This discrepancy manifests inapplicability of the Euler's equation for description of *postinstability* motion of an inviscid fluid, and therefore, for a developed turbulence. But this does not imply the incompleteness of Newtonian mechanics: it only means that the mathematical formalism that expresses the Newton's laws should include non-differentiable components of the velocity field.

Incorporation of thermodynamics into Newtonian mechanics does not eliminate the loss of differentiability: the Navier-Stokes equations appear to have even more sophisticated patterns of instability than its particular case – the Euler equations if the Reynolds number is supercritical, and from the point of view of mathematical formalism, it leads to the phenomenon described above. As a matter of fact, the Navier-Stokes equations impose even stronger mathematical restriction on the velocity field requiring it's twice differentiability with respect to space coordinates. Therefore, the Navier-Stokes equations require a modification that would allow one to include non-differentiable velocity field similar to those in the Euler's equations.

It should be noticed that the same kind of phenomena occurs in other branches of continuum mechanics, and in particular, in theory of flexible bodies. Indeed, consider an ideal filament in the gravity field stretched in the vertical direction with the additional tension T_0 as shown in Fig.2. Let us cut it at the middle point and observe a behavior of its lower part. As shown in Zak, M., 1994, the lower part has imaginary characteristic speeds

$$\lambda = \pm i\sqrt{\frac{S}{g}} \tag{14}$$

where S is the current length (see Fig. 2), and the solution to the corresponding governing equation contains the same destabilizing term (10) as those in Eq. (9).

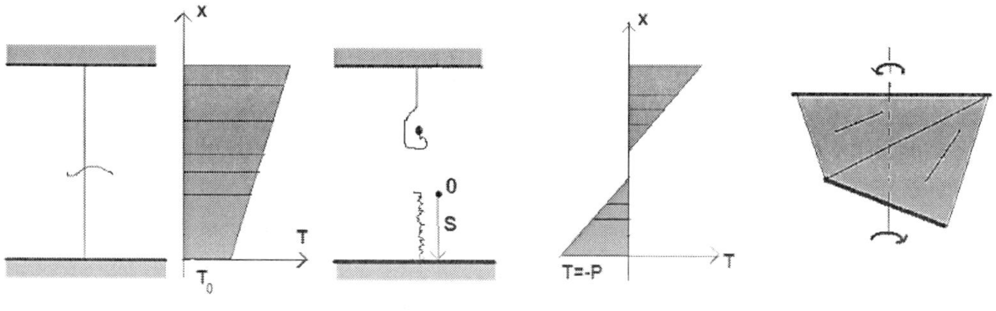

Figure 2. Loss of differentiability in flexible bodies.

As shown by Zak, M., 1994, the same loss of differentiability occurs in two- and three-dimensional flexible bodies (wrinkles in films, buckling in soft shells, and fractures in composite materials). In mathematics, all these phenomena are associated with the Hadamard's instability.

Remark. Returning to Fig. 2, it should be noticed that the motion of the upper part of the stretched and cut-at-the-middle string is also non-trivial. As shown in Zak, M., 1970, 1994, the configuration of the string remains differentiable within the open interval that excludes the free end; however the "price" for such a mathematical convenience is a loss of the snap-of-a-whip effect well known from experiments. Only inclusion of the free end allows one to observe this effect, but that leads to violation of another *mathematical* restriction: the Lipschitz condition. Therefore, two *mathematical* restrictions: space-time differentiability and the Lipschitz condition that do not follow from the Newton's laws are imposed on the formalism of Newtonian mechanics, and that leads to its apparent incompleteness. A removal both of these restrictions will be discussed below.

b. Reynolds approach. Over 100 years ago, Reynolds proposed a modification of the Navier-Stokes equations by decomposing the fluid velocity into the mean (time-averaged) component $\bar{v}$ and fluctuating component v', Reynolds, O., 1895. The new form of the Reynolds-averaged Navier-Stokes (RANS) equations for an incompressible fluid is the following

$$\frac{\partial \bar{v}_i}{\partial x_i} = 0 \tag{15}$$

$$\rho \frac{\partial \bar{v}_i}{\partial t} + \rho \frac{\partial \bar{v}_i \bar{v}_j}{\partial x_j} = \rho f_i + \frac{\partial}{\partial x_j}[-\bar{p}\delta_{ij} + \mu(\frac{\partial v_i}{\partial x_j} + \frac{\partial v_j}{\partial x_i}) - \rho \overline{v_i v_j})] \tag{16}$$

Here $\rho, \bar{p}, \mu$ are density, mean pressure, and coefficient of viscosity, while Eqs. (15) and (16) represent conservation of mass and momentum, respectively.

The last term in Eq. (16) (generally referred as Reynolds stresses) represents the contribution of the inertial forces of fluctuations into momentum an energy of a turbulent flow. However, along with the qualitative advantages, this term brings a fundamental quantitative disadvantage: the system Eqs. (15) and (16) becomes open since the number of unknowns now is larger than the number of equations; that creates the so called "closure" problem that is to find *additional* relationships between fluctuations and mean velocities. A typical closure was proposed by Prandtl

$$-\overline{v_i v_j} = l^2 \left| \frac{\partial \bar{v}_i}{\partial x_j} \right| \frac{\partial \bar{v}_i}{\partial x_j} \tag{17}$$

where a *non-local* parameter l (called mixing length) is supposed to be found from experiments. However the problem with this and other closures is that they are effective only for a narrow class of boundary conditions that are close to those for which the corresponding experiment was performed, while they fail otherwise. This failure could be expected. Indeed, Reynolds did not change the *physics* of the viscous fluid: he changed only a mathematical representation of this physics. Namely, he enlarged the class of functions in which turbulence should be described by moving from differentiable to maltivalued functions. Therefore, a closure should be searched not in new physical laws, but rather in relationships between fluctuations and the rate of instability that causes the failure of differentiability. These relationships follow from the Stabilization Principle introduced by Zak, M., 1986, 1994. It will be discussed and illustrated in the next section.

c. ***Stabilization Principle.*** A physical quantity is invariant if it has the same magnitude in all inertial reference frames. This definition disqualifies the concept of *stability* from being a physical invariant. Indeed, it can be shown that stability depends upon the frame of reference, upon the class of functions in which the motion is described, and upon the metrics of configuration space, and in particular, upon the way in which the distance between the basic and perturbed solution is defined.

The dependence upon the frame of reference can be illustrated by the following example, Arnold, V., 1988: consider an inviscid stationary flow with a *smooth* velocity field

$$v_x = A\sin z + C\cos y, \qquad v_y = B\sin x + A\cos z, \qquad v_z = C\sin y + b\cos x \tag{18}$$

Surprisingly, the trajectories of this flow are unstable (Lagrangian turbulence). It means that this flow is stable in the Eulerian coordinates, but it is unstable in the Lagrangian coordinates.

Several examples of dependence of the stability of a motion upon the metric of the configuration space are demonstrated by Synge, J., 1926.

Finally, the dependence of the stability of a motion upon the class of functions is illustrated by Zak, M., 1970, 1994.

Thus, the concept of stability is an *attribute of a mathematical formalism* rather than physics, and consequently, unsolved problems in Newtonian mechanics do not imply an incompleteness of the Newton's laws: they manifest inadequacy of the corresponding mathematical formalism, i.e. inconsistency between idealized models and reality.

d. Stabilization Principle for closure problem. Based upon comments made above, we will now formulate the stabilization principle for closure of the Reynolds-averaged equations. Consider a dynamical model that in some domain of its parameters loses its space-time differentiability, i.e. it becomes unstable in the class of differentiable functions. As noticed earlier, this means that the corresponding physical phenomenon cannot be adequately described without an appropriate modification of the mathematical formalism representing the model. The modification should be based upon an enlarging the original class of functions in such a way that the instability is eliminated. The mathematical formulation of this principle can be expressed in the following *symbolic* form:

$$A_R \otimes X = A_0 \otimes X + \sigma_R \tag{19}$$

If the original dynamical model

$$A_0 \otimes X = 0 \tag{20}$$

is unstable, but the Reynolds-averaged model

$$A_0 \otimes X + \sigma_R = 0 \tag{21}$$

is stable, obviously the stabilization is performed by the Reynolds stresses σ_R that represent contribution of all the non-smooth components into the modified model. Indeed, driven by the mechanism of instability of the original model, they grow until the instability is suppressed down to a neutral stability. From that viewpoint, the Prandil's closure Eq. (17) can be considered as a feedback that stabilizes an originally unstable laminar flow. Turning, for instance, to a plane Poiseuille flow with a parabolic velocity profile, one arrives at its instability if the Reynolds number is larger than $R \cong 5772$. Experiments show that the new steady turbulent profile is no longer parabolic: it is very flat near the center and is very steep near the walls. The same profile follows from the Prandtl solution based upon the closure (17). But since this profile can be experimentally observed, it must be stable, and this stabilization is carried out by the feedback (17). Mathematical justification of the *neutral* stability is presented in Zak, M., 1994. Experimental verification of neutral stability of free turbulent jets was reported in Lessen and Poillet, 1976. The meaningfulness of the Stabilization Principle formulated above and its application to the closure problem has been illustrated in Zak, M., 1994. However, a major limitation of this approach is a necessity to find a rate of instability of the original laminar flow prior to application of the Stabilization Principle, and this pre-condition is very complex and laborious.

Before introducing a general theory of postinstability, we illustrate the application of the stabilization principle to turbulence in shear flows.

2. Example: Analysis of Turbulence in Shear Flows

a. Background. If we confine our study to flows whose instability can be found from linear analysis, (plane Poiseuille flow, boundary layers), then the closure problem can be formulated as follows : let the original laminar flow described by the Navier-Stokes equations be unstable, i.e some of the eigenvalues for the corresponding Orr-Sommerfeld equation have positive imaginary parts. Then the closure is found from the condition that all these positive imaginary parts vanish, and therefore, the solution possesses a neutral stability. However the closure can be written in the explicit form only if the criteria for the onset of instability are formulated explicitly. Since such a situation is an exception rather than a rule, one can apply a step-by-step strategy .This strategy is based upon the fact that the Reynolds stress disturbances grow much faster than the mean motion disturbances. Hence one can assume that these stresses will be large enough to stabilize the mean flow which is still sufficiently close to its original unperturbed state .But the Reynolds stresses being substituted in the Reynolds equations will change the mean velocity profile, and consequently, the conditions of instability. These new conditions, in turn, will change the Reynolds stresses, etc. By choosing the iteration steps to be sufficiently small, one can obtain acceptable accuracy. In this example, the first step approximation will be applied to a plane Poiseuille flow.

b. Formulation of the Problem. Let us consider a plane shear flow with a dimensionless velocity profile:

$$\overline{U} = \overline{U}(y), \qquad 0 \le y \le 1 \tag{22}$$

with boundaries

$$y_1 = 0, \; y_2 = 1 \tag{23}$$

and the x coordinate being along the axis of symmetry. The stream function representing a single oscillation of the disturbance is assumed to be of the form

$$\psi(x,y,t) = \varphi(y)e^{i(\alpha x-\beta t)} \tag{24}$$

The function $\varphi(y)$ must satisfy the Orr-Sommerfeld equation

$$(U-C)(D^2-\alpha^2)\varphi - U''\varphi = (i\alpha R)^{-1}(D^2-\alpha^2)^2\varphi \tag{25}$$

in which α and β are constants, R is the Reynolds number, and

$$C = \frac{\beta}{\alpha}, \qquad D\varphi = \frac{d\varphi}{dy} = \varphi' \tag{26}$$

Equation (25) should be solved subject to the boundary conditions, which in case of a symmetric flow between rigid walls are

$$\varphi = D\varphi = 0 \quad at \quad y = y_2, \qquad D\varphi = D^3\varphi = 0 \quad at \quad y = y_1 \tag{27}$$

We will start with the velocity profile characterized by the critical Reynolds number

$$R = R_{cr} \tag{28}$$

Any increase in velocity when

$$R^* > R_{cr} \tag{29}$$

leads to instability of the laminar flow and to transition to a new turbulent flow.

We will concentrate our attention on the situation when the increase in the Reynolds number is sufficiently small

$$\frac{R^* - R_{cr}}{R_{cr}} << 1 \tag{30}$$

In this case we will be able to formulate a linearized version of the closure (21) explicitly based upon the conditions of instability of the Orr-Sommerfeld equation written for $R = R_{cr}$ and to obtain the mean velocity profile and Reynolds stress for the corresponding turbulent flow.

c. Generalized Orr-Sommerfeld Equation. In order to apply the stabilization principle and formulate the closure problem we have to incorporate the Reynolds stresses into the Orr-

Sommerfeld equation. For this purpose let us start with the Reynolds equations for a plane shear flow expressed in terms of small perturbations:

$$\frac{\partial \widetilde{U}}{\partial t}+U\frac{\partial \widetilde{U}}{\partial x}+\widetilde{V}\frac{dU}{dy}+\frac{1}{\rho}\frac{\partial \widetilde{P}}{\partial x}=\nu\nabla^2\widetilde{U}+\frac{\partial \widetilde{\tau}}{\partial y} \tag{31}$$

$$\frac{\partial \widetilde{V}}{\partial t}+U\frac{\partial \widetilde{V}}{\partial x}+\frac{1}{\rho}\frac{\partial \widetilde{P}}{\partial y}=\nu\nabla^2\widetilde{V}+\frac{\partial \widetilde{\tau}}{\partial x} \tag{32}$$

$$\frac{\partial \widetilde{U}}{\partial x}+\frac{\partial \widetilde{V}}{\partial y}=0 \tag{33}$$

using the bolngary layer approximation. Here *U(y)* is the mean velocity profile, $\widetilde{U},\widetilde{V},\widetilde{P}$ are small velocity and pressure perturbations, ν is the kinematic viscosity and $\widetilde{\tau}$ is the shear Reynolds stress which is sought in the form

$$\widetilde{\tau}=\hat{\tau}(y)e^{i(\alpha x-\beta t)} \tag{34}$$

Substituting Eq. (24) and Eq. (34) into Eqs.(31, 32, 33) we obtain after the elimination of pressure, the generalized Orr-Sommerfeld equation in dimensionless form:

$$(\overline{U}-C)(D^2-\alpha^2)\varphi-\overline{U}''\varphi-(iaR)^{-1}(D^2-\alpha^2)^2\varphi=-\frac{1}{\alpha}(D^2+\alpha^2)\tau \tag{35}$$

in which

$$\tau=\frac{\hat{\tau}}{\rho U_{\max}^{\ 2}} \tag{36}$$

Eq. (35) contains an additional term on the right hand side: the Reynolds stress disturbance, as yet unknown.

d. The Closure Problem. Returning to our problem, let us apply Eq. (35) to the case when

$$R=R^*,U=U(y) \tag{37}$$

Substituting the equalities (37) into Eq. (35), one obtains

$$(\overline{U}-C)(D^2-\alpha^2)\varphi-\overline{U}''\varphi-(iaR^*)^{-1}(D^2-\alpha^2)^2\varphi=-\frac{1}{\alpha}(D^2+\alpha^2)\tau \tag{38}$$

With zero Reynolds stress $(\tau = 0)$, Eq. (38) would have eigenvalues with positive imaginary parts since $R^* > R_{cr}$. These positive imaginary parts of the eigenvalues would vanish if R^* is replaced by R_{cr}. Hence, according to the stabilization principle, the Reynolds stresses should be selected such that Eq. (38) is converted to Eq. (25) at $R = R_{cr}$, i.e.

$$(iaR^*)^{-1}(D^2 - \alpha^2)^2 \varphi - \frac{1}{\alpha}(D^2 + \alpha^2)\tau = (i\alpha R_{cr})^{-1}(D^2 - \alpha^2)^2 \varphi \tag{39}$$

or

$$(D^2 + \alpha^2)\tau = (\frac{1}{R_{cr}} - \frac{1}{R^*})(D^2 - \alpha^2)^2 \varphi \tag{40}$$

Eq. (40) relates the disturbance of the mean flow velocity and the Reynolds stress τ. With reference to Eqs. (24, 34, 36), Eq. (40) allows us to formulate a linearized version of closure of the Reynolds equations (31), (32) and (33)

$$\bar{\tau}'' + \alpha^2 \bar{\tau} = (\frac{1}{R_{cr}} - \frac{1}{R^*})(\bar{\psi}'''' - 2\alpha^2 \bar{\psi}'' + \alpha^4 \bar{\psi}) \tag{41}$$

in which $\bar{\psi}$ and $\bar{\tau}$ are dimensionless Reynolds stress and the stream function characterizing the unperturbed flow (for instance, $\psi = -\partial U / \partial x$). Indeed, after perturbing Eq. (41) and substituting equations (24,34, 36), one returns to Eq. (40).

It is important to emphasize that Eq. (41) is not a universal closure: it contains two parameters (R_{cr} and α) which characterize a particular laminar flow. Here R_{cr} is the smallest value of the Reynolds number below which all initially imparted disturbances decay, whereas above this value those disturbances which are characterized by α (see Eq. (24) and (34)) are amplified. Both of these numbers can be found from

Eq. (25) as a result of classical analysis of hydrodynamics stability performed for a particular laminar flow. One should recall that the closure (41) implies a small increment of the Reynolds number over its critical value (see Eq. (30)). For large increments the procedure must be performed by steps: for each new mean velocity profile (which is sufficiently close to the previous one) the new R'_{cr} and α' are supposed to be found from the solution of the eigenvalue problem for the Orr-Sommerferd equation. Substituting R'_{cr} and α' into the closure Eq. (41) and solving it together with the corresponding Reynolds equation, one finds the mean velocity profile and the Reynolds stress for the next increase of the Reynolds number, etc.

e. Plane Poiseuille Flow. In this sub-section we will apply the approach developed above to a plane Poiseuille flow with the velocity profile

$$\overline{U}^0(y) = 1 - y^2 \tag{42}$$

and

$$R_{cr} = 5772.2, \quad \alpha = 1.021 \tag{43}$$

As a new (supercritical) Reynolds number we will take

$$R^* = 6000 \tag{44}$$

The closure (41) should be considered together with the governing equation for the unidirectional mean flow

$$\nu\overline{U}'' + \overline{\tau}' = C = const \tag{45}$$

or

$$\nu\overline{U}' + \overline{\tau} = \overline{C}_1 y + C_2 = const \tag{46}$$

The constants $\overline{C}_1$ and $\overline{C}_2$ can be found from the condition

$$\overline{\tau} = 0 \quad at \qquad y = 1 \quad and \qquad y = 0 \tag{47}$$

expressing the fact that the Reynolds stress vanishes at the rigid wall and at the middle of the flow. Hence

$$\overline{C}_2 = 0 \tag{48}$$

since $\overline{U}' = 0 \ at \quad y = 0$ and

$$\overline{C}_1 = \nu\overline{U}_1 \qquad (\overline{U}_1 = \overline{U} \qquad at \qquad y = 1) \tag{49}$$

Thus

$$\overline{\tau} = \nu(U_1' y - U') \tag{50}$$

or in dimensionless form,

$$\overline{\tau} = \frac{1}{R^*}(\overline{U}_1' y - \overline{U}') \tag{51}$$

Substituting Eq. (51) into the closure Eq. (41), one obtains the governing equation for the mean velocity profile in terms of the stream function $\overline{\psi}$ while $\overline{U} = \partial\overline{\psi} / \partial y$

$$\frac{1}{R_{cr}}\psi'''' - \alpha^2\left(\frac{2}{R_{cr}} - \frac{1}{R^*}\right)\overline{\psi}'' - \alpha^4\left(\frac{1}{R_{cr}} - \frac{1}{R^*}\right)\overline{\psi} = \frac{\alpha^2}{R^*}\overline{\psi}_1'' y \tag{52}$$

in which $\overline{\psi}_1'' = \overline{\psi}''$ *at* $y = y_1$

Without loss of generality it can be set

$$\overline{\psi}\,|_{y=0} = 0 \tag{53}$$

Since at the rigid wall $\overline{U} = 0$, one obtains

$$\overline{\psi}_1' = 0 \tag{54}$$

In the middle of the flow due to symmetry

$$\overline{U}_0' = 0,\ i.e. \qquad \overline{\psi}_0'' = 0 \tag{55}$$

Finally, the flux of the turbulent flow should be the same as the flux of the original (unperturbed) laminar flow

$$\overline{\psi}_1 = \int_0^1 (1 - y^2)dy = \frac{2}{3} \tag{56}$$

These four non-homogeneous boundaries conditions (53) – (56) allow one to find four arbitrary constants appearing as a result of integration of Eq. (52). After substituting the numerical values (43) and (44), one arrives at the following linear differential equation of the fourth order with respect to the dimensionless stream function

$$\overline{\psi}'''' - 1.08202\overline{\psi}'' - 0.04124\overline{\psi} = 1.044\overline{\psi}_1'' \tag{57}$$

whence

$$\begin{aligned} \overline{\psi} &= C_1 \sin 0.19199y + C_2 \cos 0.19199y + C_3 \sinh 0.19199y \\ &C_4 \cosh 0.19199y - 25.3152\overline{\psi}_1'' y \end{aligned} \tag{58}$$

Applying the conditions (53) and (55), one finds that

$$C_2 = C_4 = 0 \tag{59}$$

Taking into account that

$$\overline{\psi}_1'' = -0.00703C_1 + 0.00712C_3 \tag{60}$$

one obtains

$$\overline{\psi} = C_1 \sin 0.19199y + C_3 \sinh 0.19199y + (0.17797C_1 - 0.18924C_3)y \tag{61}$$

Now applying the conditions (54), (55) and (56) one arrives at the final form of thesolution

$$\overline{\psi} = 11.278\sin 0.19199y - 270.11\sinh 0.19199y + 50.692 \tag{62}$$

and therefore,

$$\overline{U} = 2.1653\cos 0.19199y - 51.8584\cosh 0.19199y + 50.692 \tag{63}$$

Substituting the solution (62) into Eq. (51), one obtains the Reynolds stress profile

$$R^* \tau = 0.41572\sin 0.19199y + 9.9563\sinh 0.19199y - 2.00259y \tag{64}$$

f. Analysis of the solution. We will start with the comparison of the original laminar velocity profile (42) and the mean velocity profile (63). Both of them envelop the same area, i.e., the fluxes of the original laminar and post-instability turbulent flows are the same. However, the maximum turbulent mean velocity is smaller than the maximum velocity of the original laminar flow:

$$\overline{U}^T{}_{\max} = 0{,}9989 < \overline{U}^L{}_{\max} = 1 \tag{65}$$

Also

$$|\overline{U}_0''|^T = 1.99132 < |\overline{U}_0''|^L = 2 \tag{66}$$

At the same time,

$$|\overline{U}_1'|^T = 1.99132 > |\overline{U}_1'|^L \tag{67}$$

Hence, the turbulent mean velocity profile is more flat at the centre and it is more steep at the walls in comparison with the corresponding laminar flow. This property is typical for turbulent flows.

Turning to the Reynolds stress profile (64), one finds that the maximum of the stress module $|\tau|$ is shifted toward the wall:

$$y^* = 0.58 \tag{68}$$

that expresses the well-known wall effect.

Finally, the pressure gradient

$$\frac{\partial \bar{p}}{\partial x} = \frac{1}{R^*}\bar{U}_0'' + \bar{\tau}_0' \quad (69)$$

for the new turbulent flow is greater than for the original laminar flow:

$$|\frac{\partial \bar{p}}{\partial x}|^T = \frac{2.002586}{R^*} > \frac{2}{R^*} = |\frac{\partial p}{\partial x}|^L \quad (70)$$

Therefore, despite the fact that the Reynolds number R^* slightly exceeds the critical value R_{cr}, all the typical features of turbulent flows are clearly pronounced in the solution obtained above.

Thus, it has been demonstrated again that the closure in turbulence theory is based upon the principle of stabilization of the original laminar flow by fluctuation velocities. We will stress again the mathematical meaning of this procedure. It is well known that the concept of stability is related to a certain class of functions: a solution which is unstable in a class of smooth functions can be stable in an enlarged (non-smooth) class of functions. Reynolds enlarged the class of smooth functions by introducing the field of fluctuation velocities which generated additional (Reynolds) stresses in the Navier-Stokes equations. Now it is reasonable to extend this procedure by choosing these Reynolds stresses such that they eliminate the original instability, i.e. by applying the stabilization principle.

Obviously one cannot expect that the solution of the type Eqs. (63, 64) will describe all the peculiarities of turbulent motion; it will rather extract the most essential properties of the motion, i.e., such properties which are reproducible, and therefore, have certain physical meaning. Description of finer details of turbulent motions will require different approach that will be introduced below.

3. Extension of the Liouville Formalism to Postinstability Dynamics

a. Background. Newtonian dynamics describes processes in which future can be derived from past, and past can be traced from future by time inversion. Because of such determinism and reversibility, classical dynamics becomes fully predictable, and therefore, it cannot explain the emergence of new dynamical patterns in nature. Dissipative version of Newtonian dynamics that includes elements of thermodynamics, is not reversible, but is still fully deterministic and predictable. The only phenomenon which breaks down determinism and creates "intrinsic stochasticity" is instability of the mathematical models that may lead to chaos and turbulence. The major limitation of Newtonian dynamics is that its mathematical formalism does not discriminate between stable and unstable motions, and therefore, an additional stability analysis is required for that. However, such an analysis is not constructive: in case of instability, it does not suggest any model modifications to efficiently describe postinstability motions. This flaw in classical dynamics has attracted the attention of many outstanding scientists (Gibbs, Planck, and Prigogine, among others). From the mathematical viewpoint, the instability of solutions manifests incompleteness of the corresponding model. For instance, the derivation of the Navier-Stocks equation (as well as other models of continua) is based upon the assumption that the velocity field is differentiable. Obviously, this assumption is violated when, driven by instability, a laminar flow is turning into a turbulent one. However, chaotic instability is the attribute of a mathematical model rather than a physical phenomenon, i.e. it is associated with a certain class of function in which the solution is sought. Indeed, a stochastic description of the same turbulent flow does not suffer any instability, and this gives a hint: enlarge the class of function to include stochasticity into the original mathematical model. In order to do that, we will use another hint: turning to quantum mechanics and representing the Schrödinger equation in the Madelung form, we observe the feedback from the Louville equation to the equation of motion in the form of the quantum potential, Landau, L.D., 1989. Preserving the same topology, we will replace the quantum potential by other type of feedback to introduce ***intrinsic*** stochasticity in the governing equation of Newtonian dynamics, and thereby, to enlarge the class of functions in which the Newton's laws are to be implemented.

b. Destabilizing effect of Liouville feedback. We will start with derivation of an auxiliary result that illuminates departure from classical models. For mathematical clarity, we will consider here a one-dimensional motion of a unit mass under action of a force f depending upon the *velocity* v and time t

$$\dot{v} = f(v,t), \tag{71}$$

If initial conditions are not deterministic, and their probability density is given in the form

$$\rho_0 = \rho_0(V), \quad where \;\; \rho \geq 0, \;\; and \quad \int_{-\infty}^{\infty} \rho dV = 1 \tag{72}$$

while ρ is a *single- valued* function, then the evolution of this density is expressed by the corresponding Liouville equation

$$\frac{\partial \rho}{\partial t} + \frac{\partial}{\partial V}(\rho f) = 0 \tag{73}$$

Remark. Here and below we make distinction between the random variable v(t) and its value V in probability space.

The solution of this equation subject to initial conditions and normalization constraints (72) determines probability density as a function of V and t: $\rho = \rho(V,t)$.

In order to deal with the constraint (72), let us integrate Eq. (73) over the whole space assuming that $\rho \to 0$ at $|V| \to \infty$ and $|f| < \infty$. Then

$$\frac{\partial}{\partial t}\int_{-\infty}^{\infty}\rho dV = 0, \qquad \int_{-\infty}^{\infty}\rho dV = const, \tag{74}$$

Hence, the constraint (74) is satisfied for $t > 0$ if it is satisfied for $t = 0$.

Let us now specify the force f as a feedback from the Liouville equation

$$f(v,t) = \varphi[\rho(v,t)] \tag{75}$$

and analyze the motion after substituting the force (75) into Eq.(71)

$$\dot{v} = \varphi[\rho(v,t)], \tag{76}$$

This is a fundamental step in our approach. Although the theory of ODE does not impose any restrictions upon the force as a function of space coordinates, the Newtonian formalism does: equations of motion are never coupled with the corresponding Liouville equation. Moreover, it can be shown that such a coupling leads to new properties of the underlying model. Indeed, substituting the force f from Eq. (75) into Eq. (73), one arrives at the *nonlinear* equation for evolution of the probability density

$$\frac{\partial\rho}{\partial t} + \frac{\partial}{\partial V}\{\rho\varphi[\rho(V,t)]\} = 0 \tag{77}$$

Let us now demonstrate the destabilizing effect of the feedback (75). For that purpose, it should be noted that the derivative $\partial\rho/\partial v$ must change its sign, at least once, within the interval $-\infty < v < \infty$, in order to satisfy the normalization constraint (72).

But since

$$Sign\frac{\partial\dot{v}}{\partial v} = Sign\frac{d\varphi}{d\rho}Sign\frac{\partial\rho}{\partial v} \tag{78}$$

there will be regions of v where the motion is unstable, and this instability generates randomness with the probability distribution guided by the Liouville equation (77). It should be noticed that the condition (78) may lead to exponential or polynomial growth of v (in the last case the motion is called neutrally stable, however, as will be shown below, it causes the emergence of randomness as well if prior to the polynomial growth, the Lipschitz condition is violated).

c. Emergence of randomness. In order to illustrate mathematical aspects of the concepts of Liouville feedback, as well as associated with it instability and randomness, let us take the feedback (75) in the form

$$f = -\sigma^2 \frac{\partial}{\partial v} \ln \rho \tag{79}$$

to obtain the following equation of motion

$$\dot{v} = -\sigma^2 \frac{\partial}{\partial v} \ln \rho, \tag{80}$$

This equation should be complemented by the corresponding Liouville equation (in this particular case, the Liouville equation takes the form of the Fokker-Planck equation)

$$\frac{\partial \rho}{\partial t} = \sigma^2 \frac{\partial^2 \rho}{\partial V^2} \tag{81}$$

Here v stands for the particle velocity, and σ^2 is the constant diffusion coefficient.

The solution of Eq. (81) subject to the sharp initial condition is

$$\rho = \frac{1}{2\sigma\sqrt{\pi t}} \exp(-\frac{V^2}{4\sigma^2 t}) \tag{82}$$

Substituting this solution into Eq. (81) at $V=v$ one arrives at the differential equation with respect to v *(t)*

$$\dot{v} = \frac{v}{2t} \tag{83}$$

and therefore,

$$v = C\sqrt{t} \tag{84}$$

Here C is an arbitrary constant. Since $v=0$ at $t=0$ for any value of C, the solution (84) is consistent with the sharp initial condition for the solution (82) of the corresponding Liouvile equation (71). The solution (84) describes the simplest irreversible motion: it is characterized by the "beginning of time" where all the trajectories intersect (that results from the violation of Lipcsitz condition at $t=0$, Fig.3), while the backward motion obtained by replacement of t with $(-t)$ leads to imaginary values of velocities. One can notice that the probability density (82) possesses the same properties.

For a fixed C, the solution (84) is *unstable* since

$$\frac{d\dot{v}}{dv} = \frac{1}{2t} > 0 \tag{85}$$

and therefore, an initial error always grows generating *randomness*. Initially, at $t=0$, this growth is of infinite rate since the Lipschitz condition at this point is violated

$$\frac{d\dot{v}}{dv} \to \infty \qquad at \qquad t \to 0 \tag{86}$$

This type of instability has been introduced and analyzed by (Zak, M., 1989, 1992).

Considering first Eq. (84) at fixed C as a sample of the underlying stochastic process (82), and then varying C, one arrives at the whole ensemble characterizing that process, (see Fig. 3). One can verify that, as follows from Eq. (82), Risken, 1989, the expectation and the variance of this process are, respectively

$$MV = 0, \qquad DV = 2\sigma \tag{87}$$

The same results follow from the ensemble (84) at $-\infty \le C \le \infty$. Indeed, the first equality in (87) results from symmetry of the ensemble with respect to $v=0$; the second one follows from the fact that

$$DV \propto v^2 \propto t \tag{88}$$

It is interesting to notice that the stochastic process (84) is an alternative to the following Langevin equation, Risken, 1989

$$\dot{v} = \Gamma(t), \qquad M\Gamma = 0, \qquad D\Gamma = \sigma \tag{89}$$

that corresponds to the *same* Fokker-Planck equation (81). Here $\Gamma(t)$ is the Langevin (random) force with zero mean and constant variance σ.

d. Negative diffusion. Let us consider the Langeven equation (89) and couple it with the corresponding Liouville equation in the same fashion as in Eq. (80) and present this equation in a dimensionless form assuming that the parameter α absorbs the velocity and the time scale factors

$$\dot{v} = \Gamma(t) + \alpha \frac{\partial}{\partial v} \ln \rho, \quad \Gamma(t)\Gamma(t') = q\delta(t-t') \tag{90}$$

If one chooses $\alpha = q^2$, then the corresponding Liouville equation (that takes the form of the Fokker-Planck equation) will change from Eq. (81) to the following

$$\frac{\partial \rho}{\partial t} = q^2 \frac{\partial^2 \rho}{\partial V^2} - \frac{\partial}{\partial V}[\rho \frac{q^2}{\rho} \frac{\partial \rho}{\partial V}] = 0, \qquad \rho = \rho_0(V) = const. \tag{91}$$

Thus, the Liouville feedback stops the diffusion. However, the feedback force can be even more effective: it can reverse the diffusion process and push the probability density back to the sharp value in *finite* time. Indeed, suppose that in the Liouville feedback.

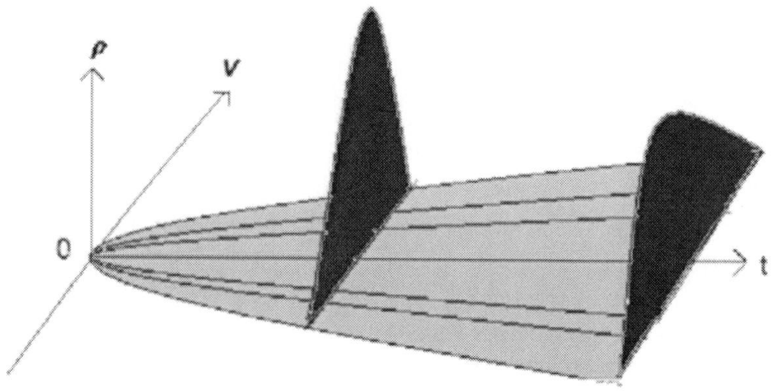

Figure 3. Stochastic process and probability density.

$$a = q^2 \exp\sqrt{D}, where \quad D(t) = \int_{-\infty}^{\infty} \rho V^2 dV \tag{92}$$

Then the Fokker-Planck equation takes the form

$$\frac{\partial \rho}{\partial t} = [q^2(1 - \exp\sqrt{D}]\frac{\partial^2 \rho}{\partial V^2} \tag{93}$$

Multiplying Eq.(44) by V^2, then integrating it with respect to V over the whole space, one arrives at ODE for the variance *D(t)*

$$\dot{D} = 2q^2(1 - \exp\sqrt{D}), \ i.e. \qquad \dot{D} \le 0 \qquad if \qquad D \ge 0 \tag{94}$$

Thus, as a result of *negative* diffusion, the variance D monotonously vanishes regardless of the initial value *D (0)*. It is interesting to note that the time T of approaching $D = 0$ is finite

$$T = \frac{1}{q^2}\int_{D(0)}^{0} \frac{dD}{1 - \exp\sqrt{D}} \le \frac{1}{q^2}\int_{0}^{\infty} \frac{dD}{\exp\sqrt{D} - 1} = \frac{\pi}{3q^2} \tag{95}$$

This terminal effect is due to violation of the Lipchitz condition, (Zak, M.,1992) at $D = 0$
Let us turn to a linear version of Eq. (93)

$$\frac{\partial \rho}{\partial t} = -q^2\frac{\partial^2 \rho}{\partial V^2}. \tag{96}$$

and discuss a negative diffusion in more details. As follows from the linear equivalent of Eq. (94)

$$\frac{d\dot{D}}{dD} = -q^2, i.e. \quad D = D_0 - q^2 t < 0 \qquad at \qquad t > D_0 / q^2 \tag{97}$$

Thus, eventually the variance becomes negative, and that disqualifies Eq. (96) from being meaningful. It has been shown (Zak, M., 2005) that the initial value problem for this equation is ill-posed: its solution is not differentiable at any point. (Such an ill-posedness expresses the Hadamard instability similar to those studied in Zak, M., 1994, see Appendix 1)). Therefore, a *negative diffusion must be nonlinear* in order to protect the variance from becoming negative, (see Fig. 6.). One of possible realization of this condition is placing a terminal attractor (Zak, M., 1992) at $D=0$ as it was done in Eq. (93).

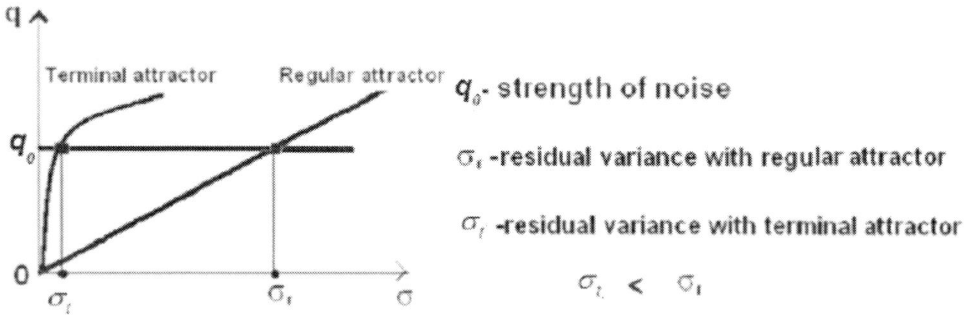

Figure 4. Effect of terminal attractor.

It should be emphasized that negative diffusion represents a major departure from Newtonian formalism.

4. Modified Newtonian Formalism

a. General model. In this section, based upon the stabilizing effect of negative diffusion considered above, we will introduce a general approach to modeling postinstability dynamics. The modified Newtonian formalism is based upon coupling the classical governing equations with the corresponding Liouville equation and suppression exponential divergence of trajectories by the effect of negative diffusion introduced above. The idea of the proposed approach is in introduction of such a Liouville feedback that as a fictitious force it acts only upon the erratic component of a trajectory without affecting its "expected", or mean value. For that purpose, introduce a system of n first order ODE with n unknowns

$$\dot{v}_i = f_i[\{v(t)\}t], \qquad \{v\} = v_1, ...v_n, \quad i = 1,2,...n \tag{98}$$

subject to initial conditions

$$v_i(0) = v_i^0 \tag{99}$$

Due to finite precision, the values (99) are not known exactly, and we assume that the error possesses some joint distribution

$$Err(V_i^0) = \rho(V_1^0, ...V_n^0) = \rho_0 \tag{100}$$

It is reasonable to assume that the initial conditions (99) coincide with the initial expectations i.e. that ρ has a maximum at $V_i^0 = v_i^0$, $i = 1,2,...n$. This means that

$$\frac{\partial \rho_0}{\partial V_0} = 0, \qquad \frac{\partial^2 \rho_0}{\partial V_i \partial V_j} < 0, \qquad i = 1,2,...n. \tag{101}$$

This is true for any symmetric initial density (for instance, the normal distribution) when the expected values have the highest probability to occur. The Liouville equation describing the evolution of the joint density ρ is

Its formal solution

$$\frac{\partial \rho}{\partial t} + \nabla \bullet (\rho f) = 0, \qquad f = f_1 ... f_n, \qquad f_i = f_i(\{V\}, t), \quad \rho = \rho(\{V\}, t). \tag{102}$$

$$P = P_0 \exp(-\int_0^t \nabla \bullet f d\tau) \tag{103}$$

suggests that the flattening of the error distribution is caused by the divergence of the trajectories of the governing equations (98) from the target trajectory that starts with the prescribed initial conditions (99), Fig.5.

Let us introduce the following Liouville feedback

$$F_i = \alpha_i \frac{\partial}{\partial v_i} \ln \rho, \qquad \alpha_i > 0 \tag{104}$$

Then the system (98) in modified to the following one

$$\dot{v}_i = f_i + \alpha_i \frac{\partial}{\partial v_i} \ln \rho, \tag{105}$$

that should be complemented by the corresponding Liouville equation

$$\frac{\partial \rho}{\partial t} + \sum_i \alpha_i \frac{\partial}{\partial V_i} (\rho f_i + \frac{\partial \rho}{\partial V_i}) = 0 \tag{106}$$

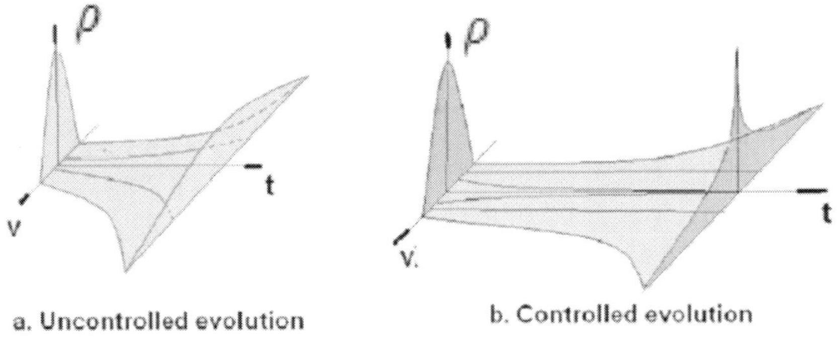

Figure 5. Divergence of trajectories.

The coupled ODE-PDE equations of this type have been studied by Zak, M., 2004, 2005, 2007a, 2007b, in connection with model of Livings. Here we will summarize only mathematical aspects of these systems.

Firstly, the force F_i makes the Liouville equation nonlinear, while ODE becomes dependent upon PDE. Secondly, this force introduces to PDE a negative diffusion that changes the type of the PDE from the hyperbolic to the parabolic one. At the same time, the behavior of the solution to Eq. (106) is fundamentally different from its Fokker-Planck analog.

Thirdly, as follows from Eq. (104), the force F_i does not affect the motion along that trajectory $v_i = v_i*$ which has the maximum probability of occurrence since

$$\frac{\partial \rho}{\partial v_i}(v_i = v_i*) = 0 \tag{107}$$

and that property makes this force fictitious.

Before formulating the proposed model in the final form, we will consider a trivial, but instructive example.

b. Example. Let us consider an unstable linear ODE

$$\dot{v} = \varepsilon v, \;\; \varepsilon << 1 \tag{108}$$

In this particular case, the expected trajectory is known in advance:

$$\bar{v} = 0 \tag{109}$$

However, any small error in initial conditions leads to a different trajectory that diverge exponentially from those in Eq.(108):

$$v = v_0 \varepsilon \exp t \tag{110}$$

Similar result follows from the corresponding Liouville equation:

$$\frac{\partial \rho}{\partial t} = -\varepsilon \frac{\partial}{\partial V}(\rho V) \tag{111}$$

$$V = V_0 \varepsilon \exp t \tag{112}$$

Let us introduce now the fictitious force as

$$F = \sqrt{D} \frac{\partial}{\partial v} \ln \rho, \tag{113}$$

where D is the variance

$$D(t) = \int_{-\infty}^{\infty} V^2 \rho(V, t) dV \tag{114}$$

and obtain the following modifies version of Eq. (108)

$$\dot{v} = \varepsilon v + \sqrt{D} \frac{\partial}{\partial v} \ln \rho$$

Due to this Liouville feedback, Eq. (111) is modified to the following Fokker-Planck equation

$$\frac{\partial \rho}{\partial t} = -\varepsilon \frac{\partial}{\partial V}(\rho V) - \sqrt{D} \frac{\partial^2 \rho}{\partial V^2} \tag{116}$$

Multiplying Eq.(116) by V, then using partial integration, one obtains for expectations the same Eq. (109) and its solution (112). Similarly one obtains for variances

$$\dot{D} = -2\sqrt{D} - \varepsilon D \approx 2\sqrt{D} \tag{117}$$

For the initial condition

$$D = D_0 \; at \qquad t = 0 \tag{118}$$

the solution to Eq.(117) is

$$D = (\sqrt{D_0} - t)^2 \qquad for \quad t < \sqrt{D_0}, \qquad and \quad D \equiv 0 \;\; for \quad t \geq \sqrt{D_0} \tag{119}$$

It is easily verifiable that the Lipschitz condition at $D=0$ is violated since

$$\frac{\partial \dot{D}}{\partial D} = -\frac{1}{\sqrt{D}} \to \infty \qquad at \qquad D \to 0 \tag{120}$$

As will be shown later, this property of the solution is of critical importance for multi-dimensional case.

Now the solution to the nonlinear version of the Fokker-Planck equation in Eq. (116) can be approximated by the first term in the Gram-Charlier series represented by the normal distribution with the variance D. For the case close to a sharp initial value at $V=0$

$$\rho = \frac{1}{D\sqrt{2\pi}} \exp(-\frac{V^2}{2D^2}), \tag{121}$$

Substituting Eq.(121) (with reference to the solution (119)) into Eq. (115) one obtains

$$\dot{v} = v[1 - \frac{1}{\sqrt{D_0 - t}}] \tag{122}$$

whence for $v = v_0$ at $t=0$ the solution is

$$v = \frac{v_0}{D_0} e^t (\sqrt{D_0} - t)^2 \quad 0 \le t \le D_0, \qquad v \equiv 0 \quad t > D_0. \tag{123}$$

For sufficiently small variance of initial error distribution $D_0 << 1$, an exponential growth of initial error v_0 is totally eliminated after $t > \sqrt{D_0}$, Fig.6.

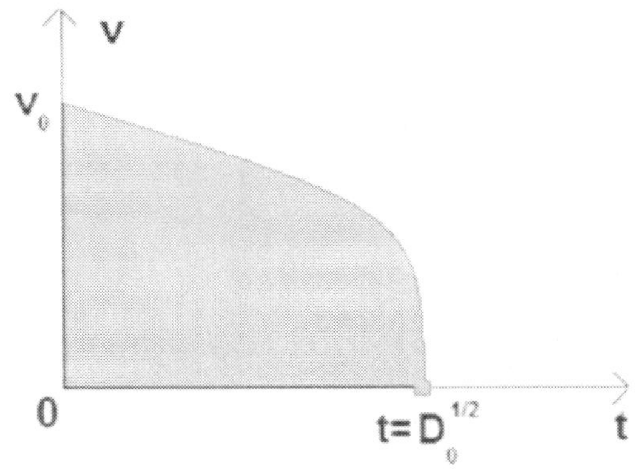

Figure 6. Suppression of instability.

It should be noticed that a finite time of approaching equilibrium are special properties of the terminal attractors discussed in Zak, M., 1970, 1989, 1992. One has to recall again that although the example we just considered is trivial, the stabilization mechanism performed by the negative-diffusion-based Liouville-feedback forces is the same. It is also important to learn from this example that the true expected solution is given by Eq. (123) rather than by Eq. (112) despite the fact that Eq. (112) directly follows from the Liouville equation (116). Indeed, the solution (112) is identical to the original solution (110), and any initial error will grow exponentially. This means that both of these solutions are unstable *in the class of differentiable functions.* But the same physical phenomenon described by Eq. (123) is stable in the *enlarged class of functions that includes stochastic components.* Obviously the stochastic components are found from the Stabilization Principle discussed in the previous section.

c. Final form of modified Newtonian formalism. Based upon the example considered above, we can now specify the coefficients α_i in Eqs. (105), and (106)

$$\dot{v}_i = f_i + \sqrt{D_{ii}}\frac{\partial}{\partial v_i}\ln\rho, \tag{124}$$

$$\frac{\partial\rho}{\partial t} + \frac{\partial}{\partial V_i}(\rho f_i + \sum_i \sqrt{D_{ii}}\frac{\partial\rho}{\partial V_i}) = 0 \tag{125}$$

where D_{ii} are principal variances

$$D_{ii} = \int_{-\infty}^{\infty}\dots\int_{-\infty}^{\infty}(V_i - \bar{V}_i)^2\rho(dV)^n \tag{126}$$

In order to verify the stabilizing effect of negative diffusion for n-dimensional case, let us linearize Eqs. (124) with respect to the initial state $v_i = 0$. Then the linearized versions of Eqs. (124) and (125) will be, respectively

$$\dot{v}_i = a_{ij}v_j + \sqrt{D_{ii}}\frac{\partial}{\partial v_i}\ln\rho, \qquad a_{ij} = (\frac{\partial f_i}{\partial v_j})_{v_j=0}\ , \tag{127}$$

$$\frac{\partial\rho}{\partial t} + \frac{\partial}{\partial V_i}(\rho a_{ij}v_j + \sum_i \sqrt{D_{ii}}\frac{\partial\rho}{\partial V_i}) = 0 \tag{128}$$

An n-dimensional analog of Eq. (117) can be obtained by multiplying Eq.(128) by V_i and then using partial integration

$$\dot{D}_{ij} = -a_{il}D_{lj} - a_{jl}D_{li} - 2\sqrt{D_{ij}} \tag{129}$$

Let us first analyze the effect of terminal attractor and, turning to Eq.(129), start with the matrix $[\partial\dot{D}_{ij}/\partial D_{lk}]$. Its diagonal elements become infinitely negative when the variances vanish

$$\frac{\partial \dot{D}_{ij}}{\partial D_{ij}} = (-2a_{ij} - \frac{1}{\sqrt{D_{ij}}}) \to -\infty \quad at \qquad D_{ii} \to 0 \tag{130}$$

while the off-diagonal elements are bounded. Therefore, due to the effect of terminal attractor (130), the system Eqs. (129) has infinitely negative characteristic roots, i.e. it is infinitely stable with respect to small errors regardless of the parameters a_{ij} of the original dynamical system (127). In addition to that, the terminal attractor (as well as any attractor) guarantees "impenetrability" of the state $D_{ii} = 0$, i.e. if the principle variances initially were non-negative, they will never become negative, and that prevent ill-possedness of the problem for the PDE (125).

Thus, all the properties of the modified model discovered in one-dimensional case are preserved in the n-dimensional case, namely: a simultaneous solution of the coupled ODE-PDE system (124) and (125) describes a stable "expected" motion regardless of the original instability.

d. Representation of higher moments. Although expected (mean) values of the state variables play an important role in description of postinstability motions, they do not expose the full dynamical picture. Indeed, measured velocities of turbulent flows look truly random, and therefore, the behavior of the higher moments is required within the framework of stochastic formalism. For that purpose, let us turn to Eqs. (98) and introduce new variables

$$v_{ij} = v_i v_j \tag{131}$$

After trivial transformations, the system (98) can be rewritten in an equivalent form being expressed via new variables

$$\dot{v}_{ij} = f_{ij}(v_{11}, \ldots v_{nn}) \tag{132}$$

in which

$$f_{ij} = \sqrt{v_{jj}}\, f_i(\sqrt{v_{11}}, \ldots \sqrt{v_{nn}}\} + \sqrt{v_{ii}}\, f_j(\sqrt{v_{11}}, \ldots \sqrt{v_{nn}}) \tag{133}$$

Let us now augment Eqs.(133) with the Liouville feedback similar to those in Eq.(75)

$$\dot{v}_{ij} = f_{ij} + \sqrt{D^*_{ijij}} \frac{\partial}{\partial v_{ij}} \ln \rho^*, \tag{134}$$

Then the corresponding Liouville equation will be similar to Eq. (76)

$$\frac{\partial \rho^*}{\partial t} + \frac{\partial}{\partial V_{ij}}(\rho f_{ij} + \sum_{ij} \sqrt{D^*_{ijij}} \frac{\partial \rho^*}{\partial V_i}) = 0 \tag{135}$$

where ρ^* and D^*_{ijij} are probability density and principal variances for the new variables.

Solving Eqs.(134) and (135) simultaneously, one obtains the evolution of the expectations of the new state variables that are equivalent to the second moments of the old variables (see Eqs.(131))

$$\overline{v}_{ij} = \overline{v_i v_j} \tag{136}$$

It should be noticed that ρ and ρ^* are different: for instance, if initially ρ is normally distributed, ρ^* must be recalculated by applying the rules for the change of variables (131); that is why the expectations and the second moments must be found from different equations.

The higher moments can be found in a similar way by introducing new variables v_{ijk}, v_{ijkl}, etc. Based upon the expectation and higher moments, one can reconstruct the joint probability distribution of state variables, and therefore, to obtain a complete information about dynamics of the underlying physical process in a *stable* form.

e. Application to postinstability models. The proposed approach can be stated as follows: in order to find the solution to a dynamical system (98) subject to the initial conditions (99) and avoid possible computational errors due to exponential divergence of the neighboring trajectories, it is sufficient to modify Eqs. (98) by applying a fictitious stabilizing force in the form of the Liouville feedback (104) and to solve the system (124) subject to the same initial conditions (99), simultaneously with the modified Liouville equation (125).

Actually, since Eq. (125) does not depend upon Eq. (124), the latter can be solved in advance, so that the Liouville feedback becomes known. Eq. (125) is to be solved subject to the normalization constraint and an initial density that can be established based upon the accuracy of the computations.

Obviously if stability of the system (98) in known in advance, the modification (124) and (125) is not necessary, although application of Eqs. (124) and (125) will only reaffirm this stability. However, in general, there is no analytical criterion that would predict chaos based only upon the system (98) without actual numerical runs. In view of that, the proposed approach seems quite universal.

f. Application to turbulence. In the case of turbulence, prior to application of the proposed methodology, the Navier-Stocks equations must be approximated by a system of ODE. Such an approximation can be performed using finite differences, finite elements, or the Galerkin method. The relevance of the finite-dimensional approximation to solutions of the fluid dynamics has been successfully demonstrated by Lorenz, E.N., 1963, who applied the Galerkin method to the Rayleigh- Benard convection model keeping only three Fourier components; as a result, he arrived at a strange attractor that now bears his name.

Let us illustrate the application of the Galerkin method to the Navier-Stokes equations. We will start with the following vector form

$$\hat{\rho}(\frac{\partial \vec{v}}{\partial t} + \vec{v} \bullet \nabla \vec{v}) = -\nabla p + \nabla \bullet T + \vec{f}), \quad \frac{\partial \hat{\rho}}{\partial t} + \nabla \bullet (\hat{\rho} \vec{v}) = 0 \tag{137}$$

where $\vec{v}$ is the flow velocity, $\hat{\rho}$ is the fluid density, p is the pressure, T is the (deviatoric) stress tensor, and $\vec{f}$ represents body forces (per unit volume) acting on the fluid. ∇

Let us represent Eqs. (137) in the following compressed form

$$\frac{\partial\vec{\Phi}(\vec{x},t)}{\partial t} = L(\vec{x})\vec{\Phi}(\vec{x},t) \tag{138}$$

where Φ is m-dimensional vector that specifies the state of the fluid, $\vec{x}$ is the configuration space (with components x, y, and z,) and $L(\vec{x})$ is a time-independent, nonlinear differential operator. Applying Fourier decomposition into mode amplitudes to the vector Φ, one obtains

$$\vec{\Phi}(\vec{x},t) = \sum_k \vec{\varphi}_k(t)e^{i\vec{k}\bullet\vec{x}} \tag{139}$$

where

$$\vec{\varphi}_k(t) = \frac{1}{(2\pi)^3}\int d^3x\vec{\Phi}(\vec{x},t)e^{-i\vec{k}\bullet\vec{x}} \tag{140}$$

Inserting (139) into (138) and using the orthogonality of the exponential functions in (139), one obtains for each mode an equation of the form

$$\dot{\vec{\varphi}}_k = \vec{\theta}_k(\vec{\varphi}_1,...\vec{\varphi}_k) \tag{141}$$

If only the first n modes are kept in the sum (139), then (141) represents a set of *mn* first order ODE describing the evolution in time of the mode amplitude components.

Eqs. (141) are of the same form as Eqs (98), and their solution can be performed using the approach introduced above. The effectiveness of the Galerkin approach in *discretization* procedure of the Navier-Stokes equations follows from the fact that despite a sharp truncation of the Galerkin expansion, there are obvious *qualitative* similarities between the original PDE model and its tree-dimensional approximation representing the Lorenz attractor.

A computational strategy for solution coupled ODE-PDE equations is discussed in Appendix 3.

CONCLUSION

Most dynamical processes are so complex that their universal theory, capturing all the details during all the time periods, is unimaginable. This is why the purpose of mathematical modeling is to extract only fundamental aspects of the process, and to neglect insignificant features, without losing core information. But "insignificant features" is not a simple concept. In many cases even vanishingly small forces can cause large changes in dynamical system parameters, and such situations are intuitively associated with the concept of instability. Obviously destabilizing forces cannot be considered as "insignificant features," and therefore, they cannot be ignored. But since they may be humanly indistinguishable, in the very beginning, there is no way to incorporate them into the model in advance. This simply means that a model is not adequate for quantitative description of the corresponding dynamical process: it must be changed or modified. However, instability delivers important qualitative information: it manifests the boundaries of applicability of the original model.

We will distinguish short and long-term instabilities. Short-term instabilities occur when a system has alternative stable states. For dissipative systems such states can be represented by static or periodic attractors. A typical example of short-term instability in conservative systems is an inverted pendulum in a gravity field. In the very beginning of the post-instability transition period, the unstable motion cannot be traced quantitatively, but it becomes more and more deterministic as it approaches the attractor or an alternative equilibrium position. Hence, a short-term instability does not necessarily require a model modification. Usually this type of instability is associated with bounded deviation of position coordinates whose changes affect the energy of the system. Indeed, if the growth of a position coordinate persists, the energy of the system would become unbounded. This type of instability represents a physical invariant of the dynamical system since it cannot be removed by change of mathematical description. Long term instability occurs when a system does not have an alternative stable state. Such instabilities can be associated only with ignorable coordinates since these coordinates do not affect the energy of the system. Long term instability is the main subjects of this chapter. A typical example of the long-term instability is the orbital instability. The first fundamental study of orbital instability was performed by Synge, 1926. Later Arnold, 1988, has shown that eventually this kind of instability leads to chaos. The simplest case of orbital instability is an inertial motion of a particle on a smooth pseudosphere. The same mathematical origin has the Lagrangian turbulence described by Eq.(18) while here the "orbit" takes place in the configuration space. Recently Zak, 1994, discovered that Hadamard's instability also leads to chaos and turbulence. The discovery of chaotic motions has "shaken-up" the scientific community and the number of publications in the area of chaos is still growing. However, notwithstanding some successes of applications of chaos to modeling complex phenomena, many open problems remain. Indeed, so far relationships between the parameters of chaotic dynamical systems and probabilistic structures of the solutions are not available. Moreover, based upon Gödel's incompleteness theorem, da Costa and Doria recently presented a rigorous proof of the algorithmic impossibility of deciding whether a given equation has chaotic domains or not in the class of "elementary" functions. Hence, even if a chaotic dynamical system describing a given nondeterministic process is found, its usefulness for prediction of the motion depends upon the values of positive Liapunov exponents. To complete this line of argumentation, one should recall that the Navier-Stokes equations are known for more than a century, but it does not help much in predictions of turbulent motions for even reasonably short time intervals. Several fundamental questions concerning the origin of chaos and turbulence are still unanswered, and one of them is the following: how a fully deterministic dynamical equation, whose solution subject to prescribed initial conditions is unique, can generate randomness without external random inputs. In order to elucidate the situation, let us consider a steady laminar flow whose instability is characterized by an exponential multiplier:

$$\tilde{v} = v_0 e^{\mu t}, \qquad 0 < \mu < \infty \tag{142}$$

Obviously, the solution with infinitely close initial condition

$$\tilde{v} = v_0 + \varepsilon, \qquad \varepsilon \to 0 \tag{143}$$

will remain infinitely close to the original one:

$$|\tilde{v} - v| = \varepsilon e^{\mu t} \to 0 \qquad if \qquad \varepsilon \to 0, t \leq N < \infty \tag{144}$$

during all the bounded time intervals. This means that random solutions can result only from random initial conditions when ε small, but finite rather than infinitesimal. The same arguments can be applied to discrete chaotic systems if divergence of actual trajectories is replaced by divergence of the trajectories in configuration space. Thus, as in stochastic differential equations, the changes in initial conditions for chaotic equations must be finite, although they may be humanly indistinguishable. However, unlike stochastic equations, the phenomenon of unpredictability in chaotic systems has a different origin: it is caused by exponential amplifications of the initial changes due to the mechanism of instability. Indeed, if two trajectories initially are "very close," and then they diverge exponentially, the same initial conditions can be applied to either of them, and therefore, the motion cannot be traced. From a mathematical viewpoint, such a multivaluedness can be interpreted as a failure of *differentiability* of the solution which means that the instability represents a boundary of applicability of the original model with the framework of differentiable dynamics. This means that in order to eliminate chaos one has to enlarge the class of smooth functions by introducing non-differentiable functions, and this leads us to the second question: is chaos an invariant of motion or is it an attribute of a mathematical model? From the mathematical viewpoint the concept of stability is related to a certain class of function, or a type of space, and therefore, the same solution can be stable in one space and unstable in another depending upon the distance between two solutions. Hence, the occurrence of chaos in description of mechanical motions means only that these motions cannot be properly described by smooth functions if the scale of observations is limited. Indeed, classical dynamics, in addition to Newton's laws, is based upon certain assumptions of a pure mathematical nature. They restrict the class of functions that describes the motions, to functions of sufficient smoothness. Such artificial limitations which do not follow from axioms of mechanics may become inconsistent with the physical nature of motions. As shown by Zak (1994) these inconsistencies lead to instabilities (in the class of smooth functions) of equations which govern turbulent and chaotic motions. The first step toward the enlarging the class of functions for modeling turbulence was made by Reynolds (1895) who decomposed the velocity field into the mean and pulsating components, and actually introduced a multivalued velocity field. However this decomposition brought new unknowns and that created the closure problem. As shown by Zak, (1986), the Reynolds equations can be obtained by referring the Navier-Stokes equations to a rapidly oscillating frame of reference while the Reynolds stresses represent contribution of inertia forces. From this viewpoint the closure has the same status as proof of Euclid parallel postulate since the motion of the frame of reference can be chosen arbitrarily. In other words, the closure of the Reynolds equations represents a case of undecidability in classical mechanics. However, based on the interpretation of the Reynolds stresses as inertia forces, it is reasonable to choose the motion of the frame of reference such that the inertia forces eliminate the original instability. In other words, the enlarged class of functions should be selected such that, the solution to the original problem in this class of functions will not possess an exponential sensitivity to changes in initial conditions. This stabilization principle has been formulated and illustrated by an example in this chapter.

A general approach to representation of postinstability motions in dynamics with application to the Navier-Stokes equations is the main objective of this chapter. The approach is based upon introduction of stabilizing forces that couple equations of motion and the evolution of the probability density of errors in initial conditions. These stabilizing forces create a powerful terminal attractor in the probability space that corresponds to occurrence of the target trajectory with the probability one. In configuration space, this effect suppresses exponential divergence of the close neighboring trajectories without affecting the target trajectory. As a result, the postinstability motion is represented by a set of functions describing the evolution of the statistical invariants such as expectations and higher moments, while this representation is stable. General analytical proof has been introduced. Since the proposed approach is not restricted by any special assumptions about the original dynamical system, it can be applied to both conservative and dissipative systems. The main applications for conservative systems are in celestial mechanics (for instance, many-body problems, see Appendix 2). The broad class of dissipative systems to which the proposed approach can be applied includes chaotic attractors and turbulence. Computational aspects of the propose approach are discussed in Appendix 3.

It should be noticed that the proposed approach combines several departures from the classical methods. Firstly, it introduces a nonlinear version of the Liouville equation that is coupled with the equation of motion (in Newtonian dynamics they are uncoupled). Secondly, it introduces terminal attractors characterized by violation of the Lipschitz conditions (in Newtonian dynamics as well as in theory of differential equations these conditions are preserved). Finally, the idea of a forced stabilization of unstable equations follows from the Stabilization Principle formulated in Zak, M., 1994. This is the most fundamental conceptual departure from the classical approach to mathematical modeling.

APPENDIX 1

Negative Diffusion and Hadamard Instability

Since a parabolic PDE with negative diffusion coefficients is of fundamental importance for the proposed model, we will take a closer look at its properties associated with the so called Hadamard's instability, or the ill-posedness of the initial value problem. Without loss of generality, the analysis will be focused on the one-dimensional case.

Consider a parabolic PDE

$$\frac{\partial \rho}{\partial t} = -q^2 \frac{\partial^2 \rho}{\partial X^2} \tag{A1}$$

subject to the following initial conditions

$$\rho^{00} = \rho|_{t=0} = \{\frac{\frac{1}{\lambda_0^2}\sin\lambda_0 X \quad if\ |X| \le X_0}{0 \quad if \quad |X| > X_0}\} \tag{A2}$$

with the parameter λ_0 being made as large as necessary,i.e

$$\lambda_0 \to \infty \tag{A3}$$

The region of the initial disturbance can be arbitrarily shrunk, i.e.

$$|X_0| \to 0 \tag{A4}$$

The solution to Eq. (A1) can be sought in the form

$$\rho = \frac{1}{\lambda_0^2} e^{\gamma \Delta t} \sin\lambda_0 X. \tag{A5}$$

Substituting this solution into Eq. (A1), one obtains

$$\lambda = q^2 \lambda_0^2 \to \infty \ at \qquad \lambda_0 \to \infty. \tag{A6}$$

Thus, the solution to Eq. (A1) subject to the initial conditions (A2) is

$$\rho = \frac{1}{\lambda_0^2} e^{q^2 \lambda_0^2 \Delta t} \sin\lambda_0 X \tag{A7}$$

This solution has very interesting properties: its modulus tends to infinity if

$$\lambda_0 \to \infty \tag{A8}$$

within an arbitrarily short period of time Δt_0 and within an infinitesimal length around the point $X = X_0$. In other words, vanishingly small changes in initial conditions lead to unboundedly large changes in the solution during infinitesimal period of time.

The result formulated above was obtained under specially selected initial conditions (A2), but it can be generalized to include any initial conditions. Indeed, let the initial conditions be defined as

$$\rho|_{t=0} = \rho^*(X) \tag{A9}$$

and the corresponding solution to Eq. (A1) is

$$\rho = \rho^{**}(X,t) \tag{A10}$$

Then, by altering the initial conditions to

$$\rho|_{t=0} = \rho^{*}(X) + \rho^{00}(X) \tag{A11}$$

where $\rho^{00}(x)$ is defined by Eq. (A2), one observes the preceding argument by superposition that vanishingly small change in the initial condition (A9) leads to unboundedly large change in the solution (A10) that occurs during an infinitesimal period of time. Such an unattractive property of the solution (that represents so called Hadamard's instability) repelled scientists from using Eq. (A1) as a model for physical phenomena. However, the situation becomes different if the variable ρ in Eq. (A1) cannot be negative, i.e. when Eq. (A1) is complemented by the constraint

$$\rho \geq 0 \tag{A12}$$

This constraint is imposed, for instance, when ρ stands for the probability density, or for the absolute temperature. It is easily verifiable that the proof of the Hadamard's instability presented above fails if the constraint (A12) is imposed, since negative values of ρ is essential for that proof. Thus, if the models of negative diffusion have attractors separating positive and negative areas of the solutions, they are free of the Hadamard's instability.

APPENDIX 2

Application to n-Body Problem

Unlike the Navier-Stokes equations that describe dissipative motions, the famous n-body problem describes conservative motions. But these two unsolved problems have a fundamental property in common: driven by supersensitivity to initial condition, they develop chaotic motions, and it turns out that the proposed methodology can be applied to both of them. The n-body problem is a classic astronomical and physical problem which naturally follows from the two-body problem first solved by Newton in his *Principia* in 1687. The efforts of many famous mathematicians have been devoted to this difficult problem, including Euler and Lagrange (1772), Jacobi (1836), Hill (1878), Poincaré (1899), Levi-Civita (1905), and Birkhoff (1915). However, despite centuries of exploration, there is no clear structure of the solution to the general n- or even three-body problem as there are no coordinate transformations that can simplify the problem, and there are more and more evidences that, in general, the solutions of n-body problems are chaotic. Failure to find a general analytical structure of the solution shifted the effort towards numerical methods. Many ODE solvers offer a variety of advance numerical methods for the solution. However due to the sensitivity of the solution to initial errors, different runs produces different results that is typical for chaotic phenomena. The governing equations of n-body problem can be written in the form of 2n ODE of the first order

$$\dot{\vec{v}}_i = -G\sum_{j=1,j\neq i}^{n} m_i \frac{\vec{r}_{ij}}{r_{ij}^3}, \qquad \dot{\vec{r}}_i = \vec{v}_i,\ \text{i=1,2,...n} \tag{A13}$$

Here r, v and m are positions, velocities and masses of the bodies centers, and r_{ij} is the distance between these centers. The sensitivity of solutions to these ODE's leading to chaos is measured by the Lyapunov exponents. Quantitatively, two trajectories in phase space with initial separation $\delta Z_0(t)$ diverge as

$$|\delta Z| \approx e^{\lambda t} |\delta Z_0| \tag{A14}$$

where λ is the Lyapunov exponent. A typical value of the positive Lyapunov exponent in a three body problem is $\lambda \approx 0.5$. Although the divergence that is associated with the Lyapunov instability is weaker that the Hadamard instability, it still leads to chaos. Here we will apply a new approach to solution of n-body problem proposed in this chapter. First of all, the system (3) should be presented in the form equivalent to Eqs. (124), (125)

$$\dot{v}_i^k = -G\sum_{j=1,j\neq i}^{n} m_i \frac{r_{ij}^k}{r_{ij}^3} + \sqrt{D_{ii}^k}\frac{\partial}{\partial v_i^k}\ln\rho, \quad i = 1,2,...n; \qquad k = 1,2,3. \tag{A15}$$

$$\dot{r}_i^k = v_i^k + \sqrt{D_{ii}^k}\frac{\partial}{\partial r_i^k}\ln\rho, \qquad i = 1,2...n; \qquad k = 1,2,3. \tag{A16}$$

$$\frac{\partial\rho}{\partial t} + \sum_{k=1}^{3}\sum_{i=1}^{n}\{\frac{\partial}{\partial V_i^k}[\rho(-G\sum_{j=1,j\neq i}^{n} m_i \frac{R_{ij}^k}{R_{ij}^3}) + \frac{\partial(\rho V_i^k)}{\partial R_i^k}] +$$

$$+\frac{\partial}{\partial V_i^k}(\sqrt{D_{ii}^k}\frac{\partial\rho}{\partial V_i^k}) + \frac{\partial}{\partial R_i^k}\sqrt{D_{ii}^k}\frac{\partial\rho}{\partial R_i^k})\} = 0 \tag{A17}$$

where r_i^k is the kth Cartesian projection of the radius-vector of the ith body

v_i^k is the kth Cartesian projection of the velocity vector of the ith body

r_{ij}^k is the kth Cartesian projection of the radius-vector of the distance between the ith and the jth bodies, and

$$D_{ii}^k = \int_{-\infty}^{\infty}\ldots\int_{-\infty}^{\infty}(V_i^k - \bar{V}_i^k)^2\rho(dXdYdZ) \tag{A18}$$

The last terms in Eqs. (A15) and (A16), and the second line in Eq. (A17) represent the effect of the fictitious forces in the form of the Liouville feedback.

Now the proposed methodology can be directly applied to solution of n-body problem described in Section 4, c,d.

APPENDIX 3

Computational Complexity of the Proposed Approach

Since the proposed approach, in general, can be implemented only by numerical simulations, its computational complexity becomes critical for a practical use. It should be emphasized that although our primary objective is to *stabilize the governing equations,* nevertheless our secondary objective is to make them practically tractable. As follows from the governing equations (124), and (125), one has to deal with $2n$ first order ODE and one second order parabolic PDE, and this PDE has exponential complexity in a sense that with a linear growth of the interacting bodies, the computational resources grow exponentially, and the problem becomes intractable. Indeed, Eq. (125) is a Fokker Planck equation for which the number of the independent variables is equal to the number of interacting bodies.

Here we will introduce a draft of the computational strategy for circumventing this obstacle by replacing simulation of Eq. (125) with direct collection statistics of the random trajectories. Assuming that the initial value of the joint probability density is a delta function, one can find all the state variables of the system (124), and (125) during a small time interval Δt. Repeating the same computations many times, one may obtain different results due to chaotic instability. These results can be used for collecting statistics and finding the joint probability density for the next small time interval, etc. It should be emphasized that the solution is obtained *without* exploiting the original PDE (125). The last property is very important since the complexity of the computing is coming from numerical solution of n-dimensional PDE. The price of this advantage is collection of statistics at each time step. Nevertheless, this procedure leads only to polynomial complexity, while computing or simulating an n-dimensional PDE has exponential complexity. Indeed, adopting Monte-Carlo approach applied for computation of multi-dimensional integrals, one can compute probability density by counting frequency of getting trajectories into preset areas, preset volumes, etc. while complexity of these simulations do not depend upon the problem dimensionality that is typical for Monte-Carlo methods. In particular, the computational complexity of integrating PDE is on the order of $(1/\varepsilon^2)^n$ — that is, the reciprocal of the error threshold rose to a power equal to the number of variables that is exponential in n. In contradistinction to that, the resources for simulations by Monte-Carlo method is on the order of $\left(1/\varepsilon^2\right)$, i.e., they do not depend upon the dimensionality of the problem. Therefore, the complexity of the whole approach is polynomial, and that is enormous advantage over standard approach to computing multi-dimensional Fokker-Planck equation. It should be noticed that the proposed approach is free of some limitations of the Monte-Carlo methods since success of the latter depends upon efficient implementations of multi-dimensional stochastic processes with prescribed density distribution, and that necessitates a fast and effective way to generate random numbers uniformly distributed on the interval [0,1]. It should be noticed that often-used computer-generated numbers are not really random, since computers are deterministic. In particular, if a random number seed is used more than once, one will get identical random numbers every time. Therefore, for multiple trials, different random number seeds must be applied. The proposed simulations approach does not need random number generator since *randomness is generated by the dynamical system itself via chaotic instability.* There is another advantage of proposed simulations: suppose that we are interested in behavior of the solution in a local region of the variables $\{x\}$; then, in case of computing, one has to find the global solution first, and only after that the local solution can be extracted, while the last procedure requires some additional integrations in order to enforce the normalization constraints. On the other hand, in our case, one can project all the simulations onto a desired sub-space $j_\alpha \otimes j_\beta$ of the total space $j_1 \otimes \dots j_\ell$ and directly obtain the local solution just disregarding the rest of the space.

ACKNOWLEDGMENT

Copyright 2008 California Institute of Technology. Government sponsorship acknowledged.

The research described in this paper was performed at Jet Propulsion Laboratory California Institute of Technology under contract with National Aeronautics and Space Administration.

REFERENCES

Arnold, V., 1988, *Mathematical methods of classical mechanics*, Springer, New York.

Lessen,M., and Poilllet, 1976, *Physics of Fluids*, **19**(7).

Lorenz,E.N., 1963, *J.Atmos. Sci.* **20**, 130.

Reynolds,O., 1895, On the dynamical theory of incompressible viscous fluid and the determination of the criterion. Phil. Trans. R. *Soc.Lond., Ser.A* **186**, 123-164.

Risken,H., 1989, *The Fokker-Planck Equation*, Springer, N.Y.

Synge, J., 1926, On the geometry of dynamics, *Phil. Trans.r. Soc. Lond., Ser.A* **226**, 31-106.

Zak, M., (1970), Uniqueness and Stability of the Solution of The Small Perturbation Problem of a Flexible Filament with a Free End, *Journal of Applied Mathematics and Mechanics*, Vol. 34, No. 6, Moscow, pp. 1048-1052, Transl. Engl. pp. 988-992.

Zak, M., 1986, Closure in turbulence theory using stabilization principle. *Phys. Letters A* **118**, 139-143.

Zak, M., 1989, "Terminal Attractors for Associative Memory in Neural Networks," *Physics Letters A*, Vol. 133, No. 1-2, pp. 18-22.

Zak,M.,1992, Terminal Model of Newtonian dynamics, *Int. J. of Theor. Phys.* **32**, 159-190.

Zak, M., 1994, Postinstability models in Dynamics, *Int. J. of Theor. Phys.*vol. 33, No. 11, 2215-2280.

Zak, M., Zbilut, J., & Meyers, R., 1997. *From instability to intelligence*, Springer, N.Y.

Zak,M.,2004, Self-supervised dynamical systems, *Chaos ,Solitons &fractals*, **19**, 645-666,

Zak, M. 2005, From Reversible Thermodynamics to Life. *Chaos, Solitons &Fractals*, 1019-1033.

Zak, M., 2007a, From quantum entanglement to mirror neuron, *Chaos, Solitons & Fractals*, **34**, 344-359.

Zak, M., 2007b, Physics of Life from First Principles, *EJTP* **4**, No. 16(II) (2007) 11–96

In: Navier-Stokes Equations
Editor: R. Younsi

ISBN: 978-1-61324-590-3

Chapter 5

3D SIMULATION OF TURBULENT TWO-PHASE FLOWS USING A STABILIZED FINITE ELEMENT METHOD

E. Hachem,* G. François† and T. Coupez‡
Center for Material Forming, MINES ParisTech
1, rue Claude Daunesse, 06 904 Sophia Antipolis, France§

Abstract

A recently developed stabilized finite element method is generalised for the incompressible Navier Stokes equations for high Reynolds numbers and free surface flows problems. The proposed method starts by the use of a finite element variational multiscale (VMS) method, which consists in here of decomposition for both the velocity and the pressure fields into coarse/resolved scales and fine/unresolved scales. This choice of decomposition is shown to be favorable for simulating flows at high Reynolds number. However, the main challenge remains in solving turbulent two-phase flows which occurs in a wide range of real life problems, industrial processes including molten metal flow, sloshing in tanks, wave mechanics, flows around structure. A successful approach to deal with such flows, especially in the presence of turbulent behaviour is the use of a local convected level set method coupled to a Large Eddy Simulation method. We assess the behaviour and accuracy of the proposed formulation in the simulation of three test cases. Results are compared with the literature and the experimental data and show that the present implementation is able to exhibit good stability for high Reynolds number flows using unstructured meshes.

Keywords: Stabilized finite element, high Reynolds number, level set, two-phase flow, large eddy simulation, parallel computation.

*E-mail address: elie.hachem@mines-paristech.fr
†E-mail address: guillaume.francois@mines-paristech.fr
‡E-mail address: thierry.coupez@mines-paristech.fr
§www.cemef.mines-paristech.fr

1. Introduction

The analysis of multiphase flows with high-Reynolds number are used to model a wide range of important physical phenomen. Significant emphasis has been placed in the literature for studying industrial and environmental applications such as sea waves, mold filling, casting and many others [1, 2, 3, 4, 5, 6]. The computation of free surface flows requires usually the development of advanced formulations capable at the same time to accurately solve time-dependent three-dimensional flow problems and to represent the interfaces separating different fluids. The main challenge remains dealing with high ratio of physical properties like density and viscosity of air and water.

In the literature, different numerical methods have been proposed. Ranging from laminar to turbulent flows, there is always an exigent need of robust high quality approaches to overcome the challenges related to updating the interface intrinsically. The great majority of these methods have adopted the Reynolds Averged Navier-Stokes (RANS) modeling in which only averaged quantities are computed [7] while others have used the Large Eddy Simulation (LES) [8], known to be more accurate for simulating two-phase flows [9].

In this chapter, we present the numerical simulation of 3D two-phase (air-water) turbulent flows and we test the use of finite element methods with unstructured meshes. The free surface flow is solved using a convected Level Set method introduced and detailed in [10, 11]. It is shown that some improvements are obtained if first we restrict the convection resolution to the neighbourhood of the interface and secondly if we replace the reinitialisation steps by an advective reinitialisation. Such modifications yield an efficient method to capture the evolving interfaces with a restraint computational cost [12].

Recall that the classical finite element approximation for the flow problem may fail because of two reasons: the compatibility condition known as the inf-sup condition or "Brezzi-Babuska" condition which requires an appropriate pair of the function spaces for the velocity and the pressure [13, 14, 15, 16, 17]; and when the convection dominates [18].

Therefore, a recently developed stabilised finite element method which draws upon features of both mixed [19, 20, 21] and stabilized finite element methods [22, 23] is used to solve the incompressible Navier Stokes equations for high Reynolds numbers [23]. The proposed method starts with the use of a finite element variational multiscale (VMS) method. It consists in here of decomposition for both the velocity and the pressure fields into coarse/resolved scales and fine/unresolved scales. This choice of decomposition is shown to be favorable for simulating flows at high Reynolds number.

For higher Reynolds number, an additional Large Eddy Simulation (LES) model is used [24, 25, 26, 27]. Two approaches are revisited; the Smagorinsky approximation [28] which can be seen as a convenient and simple way to model the subgrid scale behavior, and the dynamic procedure [8] was used to avoid the use of arbitrary and highly dissipative term yielding more accurate results. An emphasis on the LES-Level Set coupling method is also highlighted [29, 9].

The outline of this chapter is as follow: we first present the governing equations for both the free surface and the flow equations in Section 2. Section 3 presents the stabilized finite element method for solving these equations. A detailed description on the coupled LES-Level Set method is given in section 4. In section 5, the numerical performance of the presented method is demonstrated by means of 3D test cases. Comparisons with the experimental results are presented. Finally, conclusions and perspectives are outlined.

2. Governing Equations

Let $\Omega \subset \mathbb{R}^n$ be the spatial domain at time $t \in [0, T]$, where n is the number of space dimensions. Let Γ denote the boundary of Ω. We consider the following velocity-pressure formulation of the Navier-Stokes equations governing unsteady incompressible flows:

$$\rho(\partial_t \mathbf{u} + \mathbf{u} \cdot \nabla \mathbf{u}) - \nabla \cdot \boldsymbol{\sigma} = \mathbf{f} \text{ in } \Omega \times [0, T] \tag{2.1}$$

$$\nabla \cdot \mathbf{u} = 0 \text{ in } \Omega \times [0, T] \tag{2.2}$$

where ρ and $\mathbf{u}$ are the density and the velocity, $\mathbf{f}$ the body force vector per unity density and $\boldsymbol{\sigma}$ the stress tensor which reads:

$$\boldsymbol{\sigma} = 2\mu\, \varepsilon(\mathbf{u}) - p\, \mathbf{I_d} \tag{2.3}$$

with p and μ the pressure and the dynamic viscosity, $\mathbf{I_d}$ the identity tensor and ε the strain-rate tensor defined as

$$\varepsilon(\mathbf{u}) = \frac{1}{2}(\nabla \mathbf{u} + {}^t\nabla \mathbf{u}) \tag{2.4}$$

Essential and natural boundary conditions for equation (2.1) are:

$$\mathbf{u} = \mathbf{g} \quad \text{on } \Gamma_g \times [0, T] \tag{2.5}$$

$$\mathbf{n} \cdot \boldsymbol{\sigma} = \mathbf{h} \quad \text{on } \Gamma_h \times [0, T] \tag{2.6}$$

Γ_g and Γ_h are complementary subsets of the domain boundary Γ. Functions $\mathbf{g}$ and $\mathbf{h}$ are given and $\mathbf{n}$ is the unit outward normal vector of Γ. As initial condition, a divergence-free velocity field $\mathbf{u}_0(\mathbf{x})$ is specified over the domain Ω_t at $t = 0$:

$$\mathbf{u}(\mathbf{x}, 0) = \mathbf{u}_0(\mathbf{x}) \tag{2.7}$$

2.1. Interface Tracking

The interface between the air and liquid is tracked by solving a convected Level-Set function as introduced in [10, 12, 11]. The obtained results show that mass conservation is ensured and computation is simpler, efficient and robust, even in three dimensions. The basic idea of this method is to use both the physical time and the convective time derivative in the classical Hamilton-Jaobi reinitialisation equation. Consequently, a modified level-set function is given first as follows:

$$\alpha = \begin{cases} 2E/\pi & \text{for } \phi > E \\ \dfrac{2E}{\pi} \sin\left(\dfrac{\pi}{2E}\phi\right) & \text{for } |\phi| < E \\ -2E/\pi & \text{for } \phi < -E \end{cases} \tag{2.8}$$

where ϕ stands for the standard distance function, and E is the truncation thickness. The level-set evolution equation is then given by

$$\left\{ \begin{array}{l} \dfrac{\partial \alpha}{\partial t} + \boldsymbol{u}.\nabla\alpha + \lambda s \left(|\nabla\alpha| - \sqrt{1 - \left(\frac{\pi}{2E}\alpha\right)^2} \right) = 0 \\ \alpha(t=0,x) = \alpha_0(x) \end{array} \right. \tag{2.9}$$

where λ is a coupling constant depending on time discretisation and spatial discretisation, typically $\lambda \simeq h/\Delta t$ and $\boldsymbol{u}$ is the convection velocity. Finally, the authors in [11] show that by setting $\boldsymbol{v} = s\dfrac{\nabla\alpha}{|\nabla\alpha|}$ and $g(\alpha) = \sqrt{1 - \left(\frac{\pi}{2E}\alpha\right)^2}$, a rearranged form of (2.9) leads to the following simple convection equation:

$$\left\{ \begin{array}{l} \dfrac{\partial \alpha}{\partial t} + (\boldsymbol{u} + \lambda\boldsymbol{v}) \cdot \nabla\alpha = \lambda \cdot s \cdot g(\alpha) \\ \alpha(t=0,x) = \alpha_0(x) \end{array} \right. \tag{2.10}$$

The finite element formulation for the level set method is based on the use of the classical SUPG (Streamline upwind Petrov-Galerkin) method. It controls the suprious oscillations in the advection dominated regime. More details is given in [29, 11]. In brief, the finite element formulation of equation (2.8) can be written as follows: find $\alpha_h \in V_h$, such that, $\forall w_h \in W_h$

$$\begin{aligned} &\int w_h \left(\frac{\partial \alpha_h}{\partial t} + (u_h + \lambda v_h) \cdot \nabla\alpha_h \right) d\Omega - \int w_h \lambda.s.g(\alpha) d\Omega \\ &+ \sum_{e=1}^{n_{el}} \int_{\Omega^e} \tau_{SUPG} u_h \cdot \nabla w_h \left(\frac{\partial \alpha_h}{\partial t} + (u_h + \lambda v_h) \cdot \nabla\alpha_h - \lambda.s.g(\alpha) \right) d\Omega^e \quad = 0 \end{aligned} \tag{2.11}$$

where V_h and W_h are standard test and weight finite element spaces. The classical Galerkin terms are represented by the first two integrals whereas the element-wise summation, tuned by the stablization parameter τ_{SUPG}, represents the SUPG term needed to control the convection in the streamline direction. More details about the use of stabilized finite element methods for the convection equation and the evaluation of this parameter can be found in [29].

2.2. Physical Properties

Once the level-set function is defined, it can be used to easily seperate both phases. At the interface, the sharp discontinuity of fluid properties is smoothed over a transition thickness using the following expressions:

$$\rho = H(\alpha)\rho_1 + (1 - H(\alpha))\rho_2 \tag{2.12}$$
$$\mu_a = H(\alpha)\mu_1 + (1 - H(\alpha))\mu_2 \tag{2.13}$$

where H is a smoothed Heaviside function given by:

$$H(\alpha) = \left\{ \begin{array}{ll} 1 & \text{if } \alpha > \varepsilon \\ \dfrac{1}{2}\left(1 + \dfrac{\alpha}{\varepsilon} + \dfrac{1}{\pi}\sin\left(\dfrac{\pi\alpha}{\varepsilon}\right)\right) & \text{if } |\alpha| \le \varepsilon \\ 0 & \text{if } \alpha < -\varepsilon \end{array} \right. \tag{2.14}$$

Here ε is a small parameter such that $\varepsilon = O(h_I)$, known as the interface thickness, and h_I is the mesh size in the normal direction to the interface.

The use of linear interpolation for the dynamic viscosity is advised only for an interface orthogonal shear. However this would lead to inaccurate results for an interface parallel shear. In order to handle the abrupt changes at the interface, we use the following harmonic mean formulation instead:

$$\frac{1}{\mu_h} = \frac{H(\alpha)}{\mu_1} + \frac{1 - H(\alpha)}{\mu_2} \tag{2.15}$$

2.3. Surface Tension

In many fluid flow problems, surface tension may play a significant role by inducing microscopic localized surface forces that exerts itself in both normal and tangential directions. Again, by using the signed distance function, such interfacial force is expressed in function of the unit normal vector, $\boldsymbol{n}$, and added to the momentum equation by the following expression:

$$\boldsymbol{F}^{CSF} = \sigma\kappa\boldsymbol{n} \tag{2.16}$$

where σ is a physical constant depending on fluids and κ is the interface curvature given by [30]:

$$\kappa = -\nabla \cdot \boldsymbol{n} \tag{2.17}$$

The Continuum Surface Force (CSF) model introduced in [31], enables to consider the interfacial force, non coincident with the mesh, as a volumetric term. The method consists in multiplying (2.16) by a smoothed dirac approximation $\delta_\Gamma^E(\alpha)$ as follows:

$$f^{ST}(\alpha) = \sigma\nabla \cdot \left(\frac{\nabla\alpha}{|\nabla\alpha|}\right)\frac{\nabla\alpha}{|\nabla\alpha|}.\delta_\Gamma^E(\alpha) \tag{2.18}$$

where E depends on mesh size, typically $E \simeq 2h$ and

$$\delta_\Gamma^E(\alpha) = \begin{cases} 0 & \text{if } |\alpha| > E \\ \frac{1}{2E}\left[1 + \cos\left(\frac{\pi\alpha}{2E}\right)\right] & \text{if } |\alpha| < E \end{cases} \tag{2.19}$$

This additional force term once added to the Navier-Stokes equations represents the surface tension as a continuous three dimensional effect across a given interface. However, in order to take into account the density distribution on each side of the interface, we propose the following modified weighted formulation:

$$f^{ST}(\alpha) = \frac{2\rho}{\rho_1 + \rho_2}\sigma\nabla \cdot \left(\frac{\nabla\alpha}{|\nabla\alpha|}\right)\frac{\nabla\alpha}{|\nabla\alpha|}.\delta_\Gamma^E(\alpha) \tag{2.20}$$

3. Navier-Stokes Discretisation

3.1. Weak Formulation of the Incompressible Navier-Stokes Equations

The function spaces for the velocity, the weighting function space and the scalar function space for the pressure are respectively defined by:

$$V = \left\{\mathbf{u}(\mathbf{x}, t) \mid \mathbf{u}(\mathbf{x}, t) \in H^1(\Omega)^n,\ \mathbf{u} = \mathbf{g} \text{ on } \Gamma_g\right\} \tag{3.1}$$

$$Q = \left\{p(\mathbf{x}, t) \mid p(\mathbf{x}, t) \in L^2(\Omega),\ \int_\Omega p\, \mathrm{d}\Omega = 0\right\} \tag{3.2}$$

The weak form of the Navier-Strokes equations reads: find $(\mathbf{u}, p) \in V \times Q$ such that:

$$\begin{cases} \rho\,(\partial_t \mathbf{u}, \mathbf{w})_\Omega + \rho\,(\mathbf{u}\cdot\nabla\mathbf{u}, \mathbf{w})_\Omega + (\boldsymbol{\sigma}(p,\mathbf{u}), \varepsilon(\mathbf{w}))_\Omega \\ \qquad = (\mathbf{f}, \mathbf{w})_\Omega + (\mathbf{h}, \mathbf{w})_{\Gamma_h} \quad \forall w \in V_0 \\ (\nabla\cdot\mathbf{u}, q)_\Omega = 0 \quad \forall q \in Q_0 \end{cases} \tag{3.3}$$

where $(\varphi, \psi)_\Omega = \int_\Omega \varphi\psi \mathrm{d}\Omega$ is the standard scalar product in $L^2(\Omega)$. The standard Galerkin approximation consists in decomposing the domain Ω into N_{el} elements K such that they cover the domain and there are either disjoint or share a complete edge (or face in 3D). Using this partition $\mathcal{T}_h$, the above-defined functional spaces (3.1) and (3.2) are approached by finite dimensional spaces spanned by continuous piecewise polynomials such that:

$$V_h = \left\{\mathbf{u}_h \mid \mathbf{u}_h \in C^0(\Omega)^n,\ \mathbf{u}_{h|K} \in P^1(K)^n,\ \forall K \in \mathcal{T}_h\right\} \tag{3.4}$$

$$Q_h = \left\{p_h \mid p_h \in C^0(\Omega),\ p_{h|K} \in P^1(K),\ \forall K \in \mathcal{T}_h\right\} \tag{3.5}$$

The Galerkin discrete problem consists therefore in solving the following mixed problem:

Find a pair $(\mathbf{u}_h, p_h) \in V_h \times Q_h$, such that: $\forall\ (\mathbf{w}_h, q_h) \in V_{h,0} \times Q_h$

$$\begin{cases} \rho\,(\partial_t \mathbf{u}_h, \mathbf{w}_h)_\Omega + \rho\,(\mathbf{u}_h\cdot\nabla\mathbf{u}_h, \mathbf{w}_h)_\Omega \\ \qquad + (2\mu\varepsilon(\mathbf{u}_h) : \varepsilon(\mathbf{w}_h))_\Omega - (p_h, \nabla\cdot\mathbf{w}_h)_\Omega = (\mathbf{f}, \mathbf{w}_h)_\Omega + (\mathbf{h}, \mathbf{w}_h)_{\Gamma_h} \\ (\nabla\cdot\mathbf{u}_h, q_h)_\Omega = 0 \end{cases} \tag{3.6}$$

It is known that the Galerkin approximation of the Navier-Stokes equations may fail because of two reasons. Firstly, in convection dominated flows, for which it appears layers where the solution and its gradient exhibit rapid variation, the classical Galerkin approach leads to oscillations of the solution that can spread quickly and pollute the entire solution domain. Secondly, the use of inappropriate combinations of interpolation functions to represent the velocity and pressure fields [32, 33, 34, 35, 36, 37, 27] yields instable schemes. These instabilities associated are circumvented by addition of stabilization terms, e.g. residual terms weighted by tuned parameters.

3.2. Variational MultiScale Method

The stabilizing schemes from a variational multiscale point of view are described and presented in this section. The velocity and the pressure spaces are enriched by a space of bubbles that cures the spurious oscillations in the convection-dominated regime as well as the pressure instability.

Following the lines in [33, 36, 23, 27], we consider an overlapping sum decomposition of the velocity and the pressure fields into resolvable coarse-scale and unresolved fine-scale $\boldsymbol{u} = \boldsymbol{u}_h + \boldsymbol{u}'$ and $p = p_h + p'$. Likewise, we regard the same decomposition for the weighting functions $\boldsymbol{w} = \boldsymbol{w}_h + \boldsymbol{w}'$ and $q = q_h + q'$. The unresolved fine-scales are usually modelled using residual based terms that are derived consistently. The static condensation consists in substituting the fine-scale solution into the large-scale problem providing additional terms, tuned by a local time-dependent stabilizing parameter, that enhance the stability and accuracy of the standard Galerkin formulation for the transient non-linear Navier-Stokes equations. Thus, by separating the two scales, and by integrating by parts within each element, we split equation (3.6) into two sub-problems and we obtain the so-called coarse-scale problem:

$$\begin{cases} \rho\left(\partial_t(\boldsymbol{u}_h + \boldsymbol{u}'), \boldsymbol{w}_h\right)_\Omega + \rho\left((\boldsymbol{u}_h + \boldsymbol{u}') \cdot \boldsymbol{\nabla}(\boldsymbol{u}_h + \boldsymbol{u}'), \boldsymbol{w}_h\right)_\Omega \\ \qquad + \left(2\eta\, \underline{\underline{\dot{\epsilon}}}(\boldsymbol{u}_h) : \underline{\underline{\dot{\epsilon}}}(\boldsymbol{w}_h)\right)_\Omega - \left((p_h + p'), \boldsymbol{\nabla} \cdot \boldsymbol{w}_h\right)_\Omega \\ \qquad = (\boldsymbol{f}, \boldsymbol{w}_h)_\Omega \qquad \forall \boldsymbol{w}_h \in V_{h,0} \\ \left(\boldsymbol{\nabla} \cdot (\boldsymbol{u}_h + \boldsymbol{u}'), q_h\right)_\Omega = 0 \qquad \forall q_h \in Q_{h,0} \end{cases} \tag{3.7}$$

and the fine-scale problem:

$$\begin{cases} \rho\left(\partial_t(\boldsymbol{u}_h + \boldsymbol{u}'), \boldsymbol{w}'\right)_K + \rho\left((\boldsymbol{u}_h + \boldsymbol{u}') \cdot \boldsymbol{\nabla}(\boldsymbol{u}_h + \boldsymbol{u}'), \boldsymbol{w}'\right)_K \\ \qquad + \left(2\eta\, \underline{\underline{\dot{\epsilon}}}(\boldsymbol{u}') : \underline{\underline{\dot{\epsilon}}}(\boldsymbol{w}')\right)_K - \left((p_h + p'), \boldsymbol{\nabla} \cdot \boldsymbol{w}'\right)_\Omega \\ \qquad = (\boldsymbol{f}, \boldsymbol{w}')_\Omega \qquad \forall \boldsymbol{w}' \in V_0' \\ \left(\boldsymbol{\nabla} \cdot (\boldsymbol{u}_h + \boldsymbol{u}'), q'\right)_\Omega = 0 \quad \forall q' \in Q_0' \end{cases} \tag{3.8}$$

To derive the stabilized formulation, we first solve the fine scale problem, defined on the sum of element interiors and written in terms of the time-dependant large-scale variables [38, 39, 40]. Then we substitute the fine-scale solution back into the coarse problem (3.7), thereby *eliminating the explicit appearance of the fine-scale while still modelling their effects* [41, 42, 43]. At this stage, four important remarks and assumptions have to be made in order to deal with the time-dependency and the non-linearity of the momentum equation of the subscale system:

i) when using linear interpolation functions, the second derivatives vanish as well as all terms involving integrals over the element interior boundaries;

ii) as the fine-scale space is assumed to be orthogonal to the finite element space, the crossed viscous terms vanish in (3.7) and (3.8) [44];

iii) the subscales are not tracked in time, therefore, quasi-static subscales are considered here (see [45] for a justification of this choice); however, the subscale equation remains quasi time-dependent since it is driven by the large-scale time-dependent residual;

iv) the convective velocity of the non-linear term may be approximated using only large-scale part so that $(\boldsymbol{u}_h + \boldsymbol{u}') \cdot \boldsymbol{\nabla}(\boldsymbol{u}_h + \boldsymbol{u}') \approx \boldsymbol{u}_h \cdot \boldsymbol{\nabla}(\boldsymbol{u}_h + \boldsymbol{u}')$.

Consequently, by rearranging the terms in the fine-scale problem, it reduces to the following:

$$\begin{cases} \rho\left(\boldsymbol{u}_h \cdot \boldsymbol{\nabla}\boldsymbol{u}', \boldsymbol{w}'\right)_\Omega + \left(2\eta\, \dot{\underline{\underline{\epsilon}}}(\boldsymbol{u}') : \dot{\underline{\underline{\epsilon}}}(\boldsymbol{w}')\right)_\Omega + \left(\boldsymbol{\nabla}p', \boldsymbol{w}'\right)_\Omega \\ \qquad = \left(\boldsymbol{\mathcal{R}}_{\mathrm{M}}, \boldsymbol{w}'\right)_\Omega \qquad \forall \boldsymbol{w}' \in V_0' \\ \left(\boldsymbol{\nabla} \cdot \boldsymbol{u}', q'\right)_\Omega = \left(\mathcal{R}_{\mathrm{C}}, q'\right)_\Omega \quad \forall q' \in Q_0' \end{cases} \tag{3.9}$$

with $\boldsymbol{\mathcal{R}}_{\mathrm{M}}$ and $\mathcal{R}_{\mathrm{C}}$ the momentum and continuity residuals, respectively.

It is known, from the works of Wall *et al.* [46], Tezduyar and Osawa [47], that considering the small-scale pressure as an additional variable enables to complete the continuity condition on the small-scale level. It provides additional stability especially when increasing Reynolds number. However, solving the small-scale equation for both the velocity and the pressure is somewhat complicated. Franca and co-workers [35] proposed a separation technique of the small-scale unknowns. They replaced the small-scale continuity equation by the small-scale pressure Poisson equation (PPE). Since only the effect of the small-scale pressure Poisson equation on the large-scale equation must be retained, Franca and Oliveira [38] showed that rather than solving this equation it could be approximated by way of an additional term in the fashion of a stabilizing term as follows:

$$p' \approx \tau_{\mathrm{C}} \mathcal{R}_C \tag{3.10}$$

For the numerical implementation, we adopt the definition proposed by Codina in [41] for the stabilizing coefficient:

$$\tau_{\mathrm{C}} = \left(\left(\frac{\mu}{\rho}\right)^2 + \left(\frac{c_2}{c_1}\frac{\|\boldsymbol{u}\|_K}{h}\right)^2\right)^{1/2} \tag{3.11}$$

where c_1 and c_2 are two constants independent from h (h being the characteristic length of the element). Once this stabilizing coefficient τ_{C} has been defined, expression (3.10) can be inserted into the large scale equation (3.7). Then, it remains to deal with the small scale momentum equation. Codina has shown in [41] that the small scale velocity is exclusively driven by the residual of the large scale momentum equation and not by the residual of the continuity equation. Without loss of generality and by following the work of Masud and Khurram [39] and assuming that the large scale momentum residual $\boldsymbol{\mathcal{R}}_{\mathrm{M}}$ is constant, the fine scale velocity can read:

$$\boldsymbol{u}' = \tau_K \boldsymbol{\mathcal{R}}_{\mathrm{M}} \tag{3.12}$$

where τ_K is the stabilization parameter which has been naturally obtained after the resolution of the fine scale sub-problem. The effect of the bubble is now condensed in this elemental parameter. Obviously, the choice of the bubble functions affects the value of the stability parameter. More details about the choice of the bubbles is given in [48, 23, 39]. Applying integration by parts to the third terms in the first equation of (3.7) and to the second term in the second equation, then substituting the expressions of both the fine-scale pressure (3.10) and the fine-scale velocity (3.12), we get:

$$\left\{\begin{array}{l} \rho\,(\partial_t \boldsymbol{u}_h, \boldsymbol{w}_h)_\Omega + (\rho \boldsymbol{u}_h \cdot \boldsymbol{\nabla} \boldsymbol{u}_h, \boldsymbol{w}_h)_\Omega \\ \quad - \sum\limits_{K\in\Omega_h} (\tau_K \boldsymbol{\mathcal{R}}_{\mathrm{M}}, \rho \boldsymbol{u}_h \boldsymbol{\nabla} \boldsymbol{w}_h)_K + \left(2\eta\, \underline{\underline{\dot{\epsilon}}}(\boldsymbol{u}_h) : \underline{\underline{\dot{\epsilon}}}(\boldsymbol{w}_h)\right)_\Omega \\ \quad - (p_h, \boldsymbol{\nabla} \cdot \boldsymbol{w}_h)_\Omega - \sum\limits_{K\in\Omega_h} (\tau_{\mathrm{C}} \mathcal{R}_{\mathrm{C}}, \boldsymbol{\nabla} \cdot \boldsymbol{w}_h)_K \\ \qquad = (\boldsymbol{f}, \boldsymbol{w}_h)_\Omega \;\; \forall \boldsymbol{w}_h \in V_{h,0} \\ (\boldsymbol{\nabla} \cdot \boldsymbol{u}_h, q_h)_\Omega - \sum\limits_{K\in\Omega_h} (\tau_K \boldsymbol{\mathcal{R}}_{\mathrm{M}}, \boldsymbol{\nabla} q_h)_K = 0 \quad \forall q_h \in Q_{h,0} \end{array}\right. \tag{3.13}$$

When compared with the Galerkin method (3.6), the proposed stable formulation involves additional integrals that are evaluated element wise. These additional terms, obtained by replacing the approximated $\boldsymbol{u}\,'$ and p' into the large-scale equation, represent the effects of the sub-grid scales and they are introduced in a consistent way to the Galerkin formulation. All of these terms enable to overcome the instability of the classical formulation arising in convection dominated flows and to satisfy the inf-sup condition for the velocity and pressure interpolations. Moreover, the last term in equation (3.13) provides additional stability at high Reynolds number [40, 23].

Note also that a lot of estimations of stabilizing parameters can be found in the litterature. For illustration, the most common used definition for the transient Navier-Stokes problems and linear elements comes from references [47, 49, 43, 27]

$$\tau_K = \left(\left(\frac{2}{\Delta t}\right)^2 + \left(\frac{4\eta}{h^2}\right)^2 + \left(\frac{4|u_k|}{h}\right)^2\right)^{-1/2} \tag{3.14}$$

3.3. Coupled Large-Eddy-Simulations/Level Set Method

In the LES approach applied here, the Navier-Stokes equations are spatially filtered such that the large scale, energy carrying eddies are resolved and the small-scale, dissipative eddies are modeled by a subgrid-scale model. The basic idea of the filtering consists in decomposing a variable Φ into two scales $\Phi = \bar{\Phi} + \Phi'$:

- the resolved scale term $\bar{\Phi}$ obtained by a spatial averaging on the domain ;
- the subgrid scale term Φ' that has to be modelled to approach the problem and to obtain $\bar{\Phi}$.

Taking into account multi-phase flow and heterogeneous densities, a new filtering operator can also be introduced :

$$\widetilde{\Phi} = \frac{\overline{\rho\Phi}}{\bar{\rho}} \tag{3.15}$$

Following the lines in [9, 50], the authors showed that a new filtered Navier-Stokes equations are obtained having an additional subgrid term τ_t. The Smagorinsky approach [28] consists in modelling τ_t by the contribution of a viscosity stress tensor, such as $\boldsymbol{\tau}_t = 2\nabla \cdot (\bar{\rho}\nu_t\bar{\varepsilon})$. Here, the eddy viscosity ν_t can be written as follows:

$$\nu_t = (C_S\Delta)^2 \, |\bar{\varepsilon}| \tag{3.16}$$

where $|\varepsilon| = \sqrt{2\varepsilon_{ij}\varepsilon_{ij}}$, Δ is the space filter length and C_S is the Smagorinsky constant.

It is usually assumed that the spatial filter application domain is the element volume, therefore the filter length can be set as $\Delta = V^{1/d}_{element}$. This assumption can be used as well for either isotropic or anisotropic meshes. The value of C_S is not fully determined by physics and may vary from one case to another, its value is generally set between 0.1 and 0.2.

The Smagorinsky model is widely used in literature for mono-fluid and multi-fluid computations. However, rather than taking C_S as a constant arbitrary value, one can use the Germano dynamic procedure that enables to determine it depending on the flow physics. This procedure is based on the use of a second filter, whose length is larger than the previous filter [51, 52, 53] . The correlation between computed data and their filtered value allows a variation of C_S in space and time. Eddy viscosity computation is then more accurate and dissipation is limited.

4. Numerical Examples

The implemented finite element solvers for unsteady incompressible Navier-Stokes two-phase flows with high Reynolds number coupled to the convected Level Set method are validated first against two test cases: the 3D driven cavity and the 3D dam breaking benchmarks. A new constructed scale model of water filling in two-communicating tanks is also simulated and added at the end of this section.

All the numerical implementations were carried out by using the CimLib parallel finite element library [54]. The CimLib library [55] is fully parallel involving the use of SPMD (Single Program, Multiple Data) modules and the MPI (Message Passing Interface) library standard [56]. All the steps are parallelized including the assembly of algebric problems through PETSc as well as the partitioner and the meshing [57]. The 3D computations have been obtained using 32 2.4 Ghz Opteron cores in parallel (linked by an Infiniband network). The algebric problems resulting from the finite element formulation are assembled and solved using the conjugate residual method associated to the incomplete LU preconditioner from the PETSc (Portable Extensive Toolkit for Scientific Computation) library.

4.1. Three-dimensional Lid-driven Cavity Flow

In many industrial and environmental flows, the 3D lid-driven cavity flow can be seen as an interesting re-circulating flow as well as a challenging benchmark to validate the implemented flow solver. The flow is confined in a cubic domain with the upper wall moving at a constant speed of 1m/s (see figure 1 for details). Although the geometry is simple, complex physical phenomena occurs inside the cubic cavity.

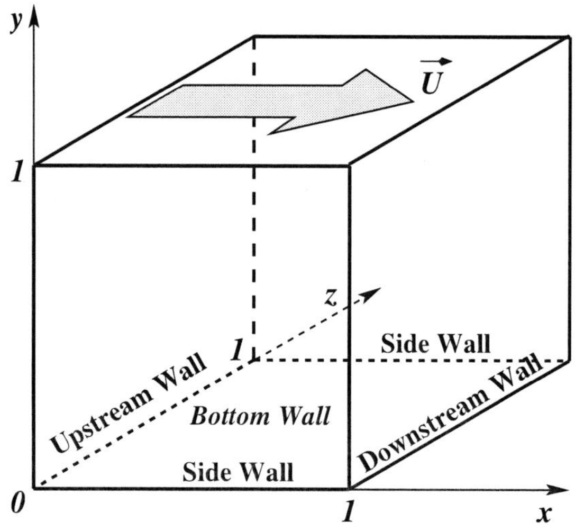

Figure 1. The 3D lid driven cavity problem.

A literature review on this 3D problem shows that in fact by examining a plane parallel to the downstream wall, corner eddies were caused at the juncture of the side-walls and the ground while downstream secondary vortices appeared. Moreover, due to centrifugal forces along the downstream, eddy separation surface were found along the span. These vortices are often known as Taylor-Görter-like (TGL) vortices in reference to their

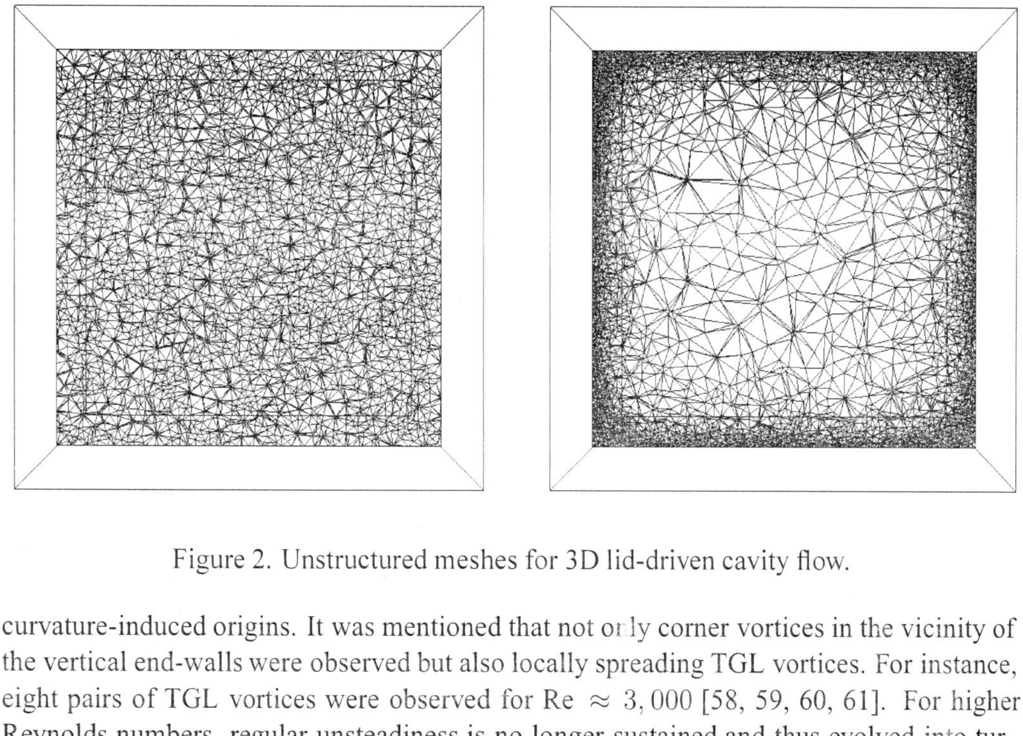

Figure 2. Unstructured meshes for 3D lid-driven cavity flow.

curvature-induced origins. It was mentioned that not only corner vortices in the vicinity of the vertical end-walls were observed but also locally spreading TGL vortices. For instance, eight pairs of TGL vortices were observed for Re $\approx 3,000$ [58, 59, 60, 61]. For higher Reynolds numbers, regular unsteadiness is no longer sustained and thus evolved into turbulence. The need and the use of LES modelling is then well justified. Two unstructured tetrahedral grids were employed to simulate the lid-driven cavity flow: the first refered as the coarse mesh, consists of $36,282$ nodes and $192,080$ linear tetrahedral elements, and the second refereed as the fine mesh consists of $238,580$ nodes and $1,229,089$ linear tetrahedral elements with grid clustering near the six cavity walls. These grids are displayed in figure 2. Aiming at comparing our results with the given reference, the two different meshes employed here are formed by approximately the same number of elements (see [65] for the coarse mesh and [27] for the fine mesh). The initial velocity in the flow is set to zero everywhere except on the lid surface. The viscosity is adjusted to obtain the desired Reynolds number. The computational results for Re $= 1,000$, Re $= 3,200$ and Re $= 12,000$ are compared to the results from Tang *et al.* [62], Zang *et al.* [63] and Prasad and Koseff [64] respectively. The mean velocity profiles in the mid-plane are plotted and compared with the reference solutions. All the velocity profiles are in good agreement with profiles reported by the given references. The differences with the experimental results is most probably due to the fact that the grid is not fine enough to simulate accurately such complex fluid phenomena. However, as a first implementation, the agreement between the present and the experimental results has been considered satisfactory.

4.2. 3D Dam Breaking

To assess the ability to simulate 3D free surface flows using the proposed numerical implementation, we consider the 3D dam breaking benchmark. We refer to the experimental and numerical data presented in [66] using a static Smagorinsky method.

The computational domain of a $0.42\ m$ wide square cavity extruded by a $0.114\ m$ in the third direction is discretized using 100 000 tetrahedral elements. A $0.228\ m$ high dam is released on the left corner. A fixed time step of $0.01\ s$ is used and an additional 2D simulation with 9 000 triangular elements is used for comparisons reasons. Both respected air/water densities and viscosities are employed. Contrary to the mentioned reference, no turbulent modeling is used, only the Navier-Stokes solver is applied to compute the velocity and the pressure.

Figure 4 shows snaphots of the water surface position that clearly indicates the expected flow pattern at different time step. The strong deformation of the interface and the unsteady character of the flow presented in here are essential for validating the proposed stabilized finite element method. The water column collapses and accelerates toward the air due to the pressure difference between the adjacent water and air. We see that from the time the water front reaches the maximum height on a side wall, it falls back and creates some propagares waves toward the opposite wall.

Figure 5 examines the evolution of the water height on the left side of the tank when using slippery or non-slippery conditions. Both 2D and 3D numerical results are in good agreement with the experimental data and the given reference [66]. This enables to validate the capability to simulate highly perturbed flows and reproduce realistic behavior for a simple case.

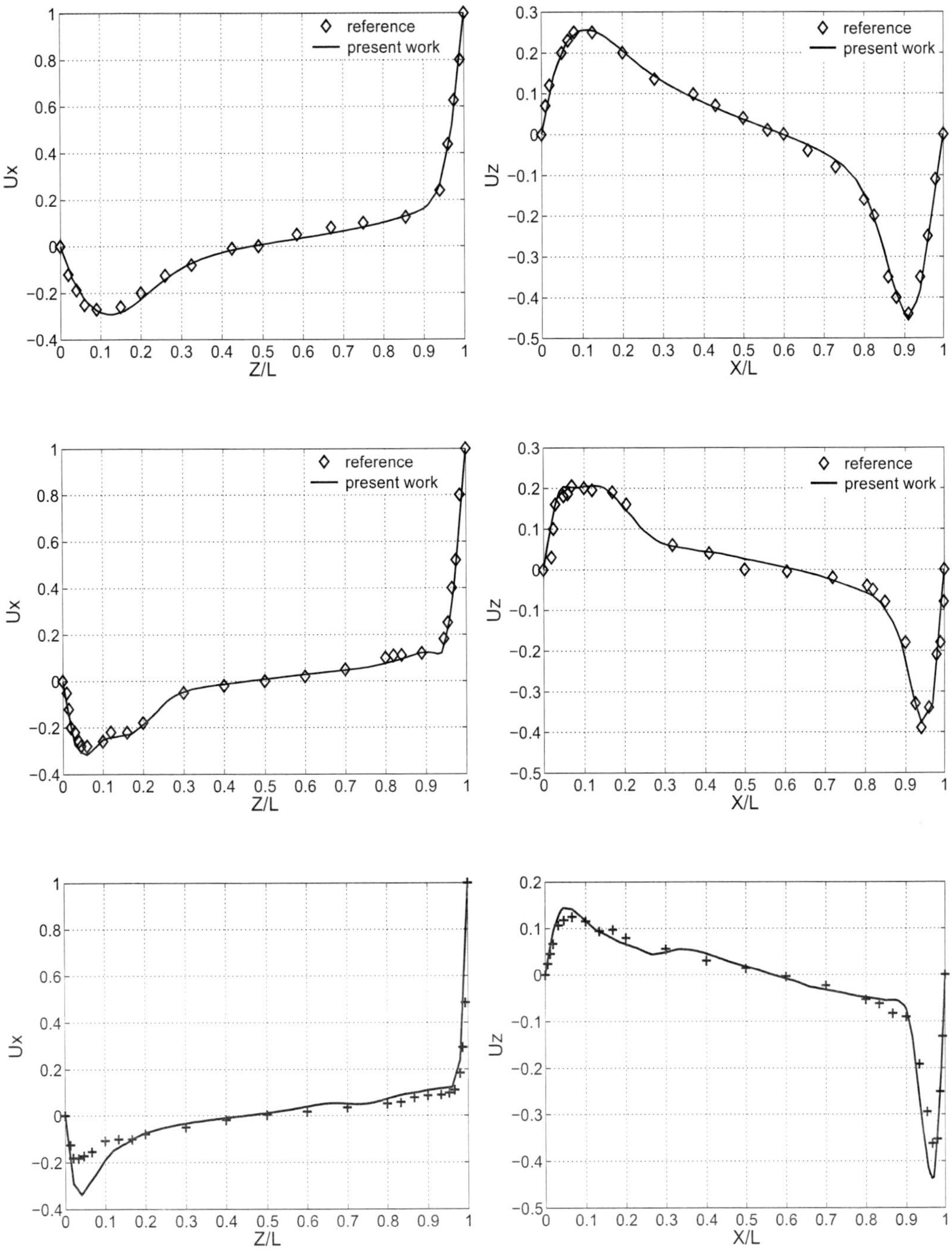

Figure 3. Comparison of velocity profiles in the mid-plane $y = 0.5$ with reference data (symbols) from [62] for Re $= 1,000$ (top), [63] for Re $= 3,200$ (middle) and [64] for Re $= 12,000$ (bottom). Left: mean value of velocity in the x-direction. Right: mean value of velocity in the z-direction.

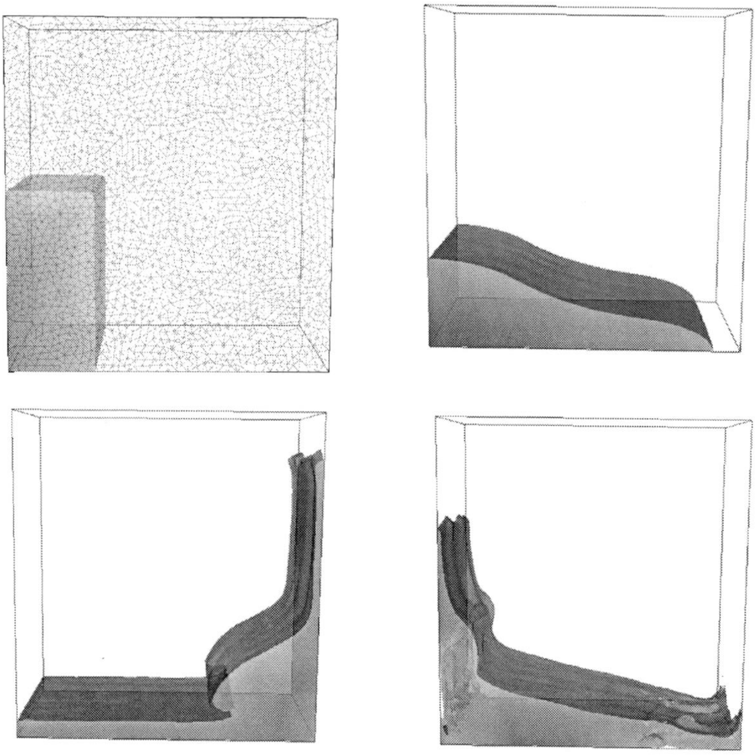

Figure 4. Interface positions at different time step: for $t = 0\ s$, $t = 0.2\ s$, $t = 0.624\ s$ and $t = 1.024\ s$.

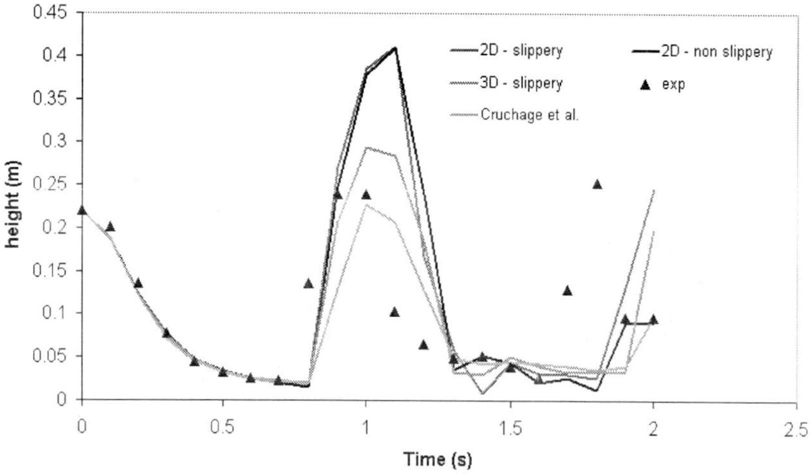

Figure 5. Evolution of the water height on left side of the tank.

4.3. 3D Filling

In order to characterize the behaviour of splashing water inside an initially air filled tank, we constructed a small scale model. It consists of a left water filled tank which starts to empty in the right side tank (see figure 6). Both two tanks have the following dimensions: 200x150x100 (LxHxW) and communicate by a certain joint tap. A water volume of 200x120x100 (LxHxW) intially rests and occupies the first tank.

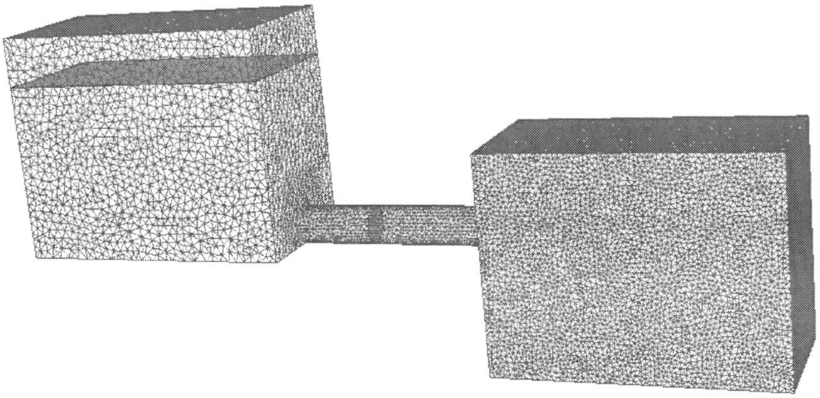

Figure 6. The computational domain and discretization - 1 000 000 tetrahedrical elements of the two-communicating tanks.

The finite element mesh consists of 925 000 tetrahedra elements and 166 000 nodes. The time step is set to half CFL condition [67] in order to improve computational cost and limit instabilities. Both static and dynamic Smagorinsky models have been used to simulate the flow with both actual air/water densities and viscosities. The slippery conditions appeared to be most realistic conditions, but a wall friction model is advised to be used for better boundary layer behavior. Different snaphots of the filling process were taken for comparisons.

Figure 7 shows some snaphots of the water surface position at different time step using both static and dynamic subgrid-scale models and compared with the experimental results. As expected, the Smagorinsky static model fails to reproduce the realistic flow description since the subgrid dissipation is quite overestimated, whereas the dynamic model provides an appropriate description of the flow pattern during the whole simulation.

The 3D unsteady stabilized Navier-Stokes two-phase flow solver using only the dynamic subgrid-scale model shows a very good ability to predict water impacting on both walls. The filling speed is also well respected until the end. This enables once more to validate the capability of the implemented methods to simulate highly perturbed two-phase flows and reproduce realistic behavior in a 3D complex configurations.

5. Conclusion

A turbulent two-phase flow solver using a stabilized finite element method is presented in this chapter. The unsteady three-dimensional Navier-Stokes equations were discretized using a Variational MultiScale method including surface tension and both classical Smagorinsky and Germano models. The decomposition of both unknowns, the velocity and the pressure fields, into coarse and fine-scales is considered. This choice of decomposition is shown to be favorable for simulating flows at high Reynolds number. By substituting the modelled fine-scale velocity and pressure fields in the coarse-scale problem, additional stabilization terms were added to the standard Galerkin formulation. These terms enhance the stability and accuracy of the classical formulation. Since the main focus of the chapter was to couple and prepare the developed flow solver to deal with multiphase flows, we used a convected Level Set method to obtain an accurate description of the flow and interface. A SUPG stabilized method was used to resolve the modified advection transport equation. Subgrid terms are obtained explicitly taking into account fluid heterogeneity.

Results obtained via this monolithic formulation have been compared and analyzed with existing data and are in very good agreement. In particular, both the 3D lid-driven cavity problem at Reynolds number up to 12,000 and the 3D Dam breaking case were considered. New results using this method were also presented on a three dimensional water filling experiment. The numerical simulations showed the limits of the static Smagorinsky model while the eddy viscosity overestimation is avoided using the dynamic procedure. Results show that the proposed scheme can produce the accurate numerical solutions to unsteady free surface flows. The reasonable nature of the results in the last test case increased our confidence in and demonstrated a good potential for formulations developed. Taking into account solid objects and turbulent wall functions is a possible extension of the method.

References

[1] M. J. Ketabdari, M. R. H. Nobari, M. M. Larmaei, Simulation Of Waves Group Propagation And Breaking In Coastal Zone Using A Navier-Stokes Solver With An Improved Vof Free Surface Treatment, *Applied Ocean Research* **30** (2) (2008) 130–143. doi:10.1016/J.Apor.2008.08.005.

[2] P. Hieu, T. Katsutohi, V. Ca, Numerical Simulation Of Breaking Waves Using A Two-Phase Flow Model, *Applied Mathematical Modelling* **28** (11) (2004) 983–1005. doi:10.1016/J.Apm.2004.03.003.

[3] P. Lubin, S. Vincent, S. Abadie, J.-P. Caltagirone, Three-Dimensional Large Eddy Simulation Of Air Entrainment Under Plunging Breaking Waves, *Coastal Engineering* **53** (8) (2006) 631–655. doi:10.1016/J.Coastaleng.2006.01.001.

[4] T. Li, P. Troch, J. De Rouck, Interactions Of Breaking Waves With A Current Over Cut Cells, *Journal Of Computational Physics* **223** (2) (2007) 865–897. doi:10.1016/J.Jcp.2006.10.003.

[5] F. Ilinca, J. Hetu, Finite Element Solution Of Three-Dimensional Turbulent Flows Applied To Mold-Filling Problems, *International Journal For Numerical Methods In Fluids* **34** (8) (2000) 729–750.

[6] H.-C. Chen, K. Yu, Cfd Simulations Of Wave-Current-Body Interactions Including Greenwater And Wet Deck Slamming, *Computers & Fluids* **38** (5) (2009) 970–980. doi:10.1016/J.Compfluid.2008.01.026.

[7] B. Launder, D. Spalding, Lectures in mathematical models of turbulence, Academic Press Inc. (London) Ltd.

[8] M. Germano, U. Piomelli, P. Moin, W. Cabot, A Dynamic Subgrid-Scale Eddy Viscosity Model, *Physics Of Fluids A-Fluid Dynamics* **3** (7) (1991) 1760–1765.

[9] S. Vincent, J. Larocque, D. Lacanette, A. Toutant, P. Lubin, P. Sagaut, Numerical Simulation Of Phase Separation And A Priori Two-Phase Les Filtering, *Computers & Fluids* **37** (7) (2008) 898–906. doi:10.1016/J.Compfluid.2007.02.017.

[10] T. Coupez, Réinitialisation convective et locale des fonctions level set pour le mouvement de surfaces et d'interfaces, *JournÉes ActivitÉs Universitaires De MÉcanique - La Rochelle*, 31 AoÛt Et 1Er Septembre 2006.

[11] L. Ville, L. Silva, T. Coupez, Convected level set method for the numerical simulation of fluid buckling, *International Journal For Numerical Methods In Fluids* .

[12] T. Coupez, Convection Of Local Level Set Function For Moving Surfaces And Interfaces In Forming Flow, in: *Numiform '07: Materials Processing And Design: Modeling, Simulation And Applications* , Vol. 908, 2007, pp. 61–66.

[13] F. Brezzi, J. Douglas, Stabilized mixed methods for the Stokes problem, *Numerische Mathematik* **53** (1988) 225–236.

[14] F. Brezzi, J. Pitkäranta, On the stabilization of finite element approximations of the Stokes problem, Efficient Solutions Of Elliptic Systems, *Notes On Numerical Fluid Mechanics* **10** (1984) 11–19.

[15] L. Franca, T. J. R. Hughes, Two classes of mixed finite element methods, *Computer Methods In Applied Mechanics And Engineering* **69** (1988) 89–129.

[16] R. Codina, J. M. GonzÁlez-Ondina, G. DÍaz-HernÁndez, J. Principe, *Finite element approximation of the modified boussinesq equations using a stabilized formulation* **57** (2008) 1305–1322.

[17] R. Codina, Comparison of some finite element methods for solving the diffusion-convection-reaction equation, *Computer Methods In Applied Mechanics And Engineering* **156** (1998) 185–210.

[18] A. Brooks, T. Hughes, *Streamline upwind/Petrov-Galerkin formulations for convection dominated flows with particular emphasis on the incompressible Navier-Stokes equations* **32** (1982) 199–259.

[19] D. Arnold, F. Brezzi, M. Fortin, A stable finite element for the Stokes equations, *Calcolo* **23** (4) (1984) 337–344.

[20] F. Brezzi, M. Bristeau, L. Franca, M. Mallet, G. Rogé, A relationship between stabilized finite element methods and the Galerkin method with bubble functions, *Computer Methods In Applied Mechanics And Engineering* **96** (1992) 117–129.

[21] R. Bank, B. Welfert, A comparison between the mini element and the Petrov-Galerkin formulations for the generalized Stokes problem, *Computer Methods In Applied Mechanics And Engineering* **83** (1990) 61–68.

[22] A. Masud, R. A. Khurram, A multiscale finite element method for the incompressible navier-stokes equations, *Computer Methods in Applied Mechanics and Engineering* **195** (2006) 1750–1777.

[23] E. Hachem, B. Rivaux, T. Kloczko, H. Digonnet, T. Coupez, Stabilized finite element method for incompressible flows with high reynolds number, *Journal of Computational Physics* **229** (23) (2010) 8643–8665.

[24] V. Gravemeier, M. W. Gee, M. Kronbichler, W. A. Wall, An Algebraic Variational Multiscale-Multigrid Method For Large Eddy Simulation Of Turbulent Flow, *Computer Methods In Applied Mechanics And Engineering* **199** (13-16, Sp. Iss. Si) (2010) 853–864. doi:10.1016/J.Cma.2009.05.017.

[25] V. Gravemeier, A Consistent Dynamic Localization Model For Large Eddy Simulation Of Turbulent Flows Based On A Variational Formulation, *Journal Of Computational Physics* **218** (2) (2006) 677–701. doi:10.1016/J.Jcp.2006.03.001.

[26] V. Gravemeier, Scale-Separating Operators For Variational Multiscale Large Eddy Simulation Of Turbulent Flows, *Journal Of Computational Physics* **212** (2) (2006) 400–435. doi:10.1016/J.Jcp.2005.07.007.

[27] E. F. Lins, R. N. Elias, G. M. Guerra, F. A. Rochinha, A. L. G. A. Coutinho, Edge-Based Finite Element Implementation Of The Residual-Based Variational Multiscale Method, *International Journal For Numerical Methods In Fluids* **61** (1) (2009) 1–22. doi:10.1002/Fld.1941.

[28] J. Smagorinsky, General circulation experiments with primitive equations, *Mon. Weather Rev.* **91** (1963) 99–164.

[29] R. N. Elias, A. L. G. A. Coutinho, Stabilized Edge-Based Finite Element Simulation Of Free-Surface Flows, *International Journal For Numerical Methods In Fluids* **54** (6-8) (2007) 965–993. doi:10.1002/Fld.1475.

[30] A. Beliveau, A. Fortin, Y. Demay, A Two-Dimensional Numerical Method For The Deformation Of Drops With Surface Tension, *International Journal Of Computational Fluid Dynamics* **10** (3) (1998) 225–240.

[31] J. Brackbill, D. Kothe, C. Zemach, A Continuum Method For Modeling Surface-Tension, *Journal Of Computational Physics* **100** (2) (1992) 335–354.

[32] A. Brooks, T. Hughes, Streamline upwind /petrov-galerkin formulations for convection dominated flows with particular emphasis on the incompressible navier-stokes equations, *Computer Methods In Applied Mechanics And Engineering* **32** (1982) 199–259.

[33] T. Hughes, G. Feijoo, L. Mazzei, J. Quincy, The Variational Multiscale Method - A Paradigm For Computational Mechanics, *Computer Methods In Applied Mechanics And Engineering* **166** (1-2) (1998) 3–24.

[34] T. Hughes, L. Franca, M. Balestra, A New Finite-Element Formulation For Computational Fluid-Dynamics .5. Circumventing The Babuska-Brezzi Condition - A Stable Petrov-Galerkin Formulation Of The Stokes Problem Accommodating Equal-Order Interpolations, *Computer Methods In Applied Mechanics And Engineering* **59** (1) (1986) 85–99.

[35] L. Franca, A. Nesliturk, M. Stynes, On The Stability Of Residual-Free Bubbles For Convection-Diffusion Problems And Their Approximation By A Two-Level Finite Element Method, *Computer Methods In Applied Mechanics And Engineering* **166** (1-2) (1998) 35–49.

[36] G. Scovazzi, A Discourse On Galilean Invariance, Supg Stabilization, And The Variational Multiscale Framework, *Computer Methods In Applied Mechanics And Engineering* **196** (4-6) (2007) 1108–1132. doi:10.1016/J.Cma.2006.08.012.

[37] A. Nesliturk, Approximating the incompressible Navier-Stokes equations using a two level finite element method, *Ph.D. thesis*, University Of Colorado (1999).

[38] L. Franca, S. Oliveira, Pressure Bubbles Stabilization Features In The Stokes Problem, *Computer Methods In Applied Mechanics And Engineering* **192** (16-18) (2003) 1929–1937. doi:10.1016/S0045-7825(02)00628-X.

[39] A. Masud, R. Khurram, A Multiscale/Stabilized Finite Element Method For The Advection-Diffusion Equation, *Computer Methods In Applied Mechanics And Engineering* **193** (21-22) (2004) 1997–2018. doi:10.1016/J.Cma.2003.12.047.

[40] M. Behr, L. Franca, T. Tezduyar, Stabilized Finite-Element Methods For The Velocity Pressure Stress Formulation Of Incompressible Flows, *Computer Methods In Applied Mechanics And Engineering* **104** (1) (1993) 31–48.

[41] R. Codina, Stabilization Of Incompressibility And Convection Through Orthogonal Sub-Scales In Finite Element Methods, *Computer Methods In Applied Mechanics And Engineering* **190** (13-14) (2000) 1579–1599.

[42] R. Codina, Stabilized finite element method for the transient Navier-Stokes equations based on a pressure gradient projection, *Computer Methods In Applied Mechanics And Engineering* **182** (3-4) (2000) 277–300.

[43] R. Codina, , J. Blasco, Analysis of a stabilized finite element approximation of the transient convection-diffusion-reaction equation using orthogonal subscales, *Comput. Visual. Sci.* **4** (3) (2002) 167–174.

[44] T. Coupez, Stable-stabilized finite element for 3D forming calculation, *Tech. rep.*, Cemef (1996).

[45] T. Dubois, F. Jauberteau, R. Temam, *Dynamic Multilevel Methods And The Numerical Simulation Of Turbulence*, Cambridge University Press, Cambridge, 1999.

[46] W. Wall, M. Bischoff, E. Ramm, A Deformation Dependent Stabilization Technique, Exemplified By Eas Elements At Large Strains, *Computer Methods In Applied Mechanics And Engineering* **188** (4) (2000) 859–871.

[47] T. Tezduyar, Y. Osawa, Finite Element Stabilization Parameters Computed From Element Matrices And Vectors, *Computer Methods In Applied Mechanics And Engineering* **190** (3-4, Sp. Iss. Si) (2000) 411–430.

[48] E. Hachem, Stabilized finite element method for heat transfer and turbulent flows inside industrial furnaces, *Ph.D. thesis*, Ecole Nationale Supérieure des Mines de Paris (2009).

[49] R. Codina, Stabilized Finite Element Approximation Of Transient Incompressible Flows Using Orthogonal Subscales, *Computer Methods In Applied Mechanics And Engineering* **191** (39-40) (2002) 4295–4321.

[50] E. Labourasse, D. Lacanette, A. Toutant, P. Lubin, S. Vincent, O. Lebaigue, J. P. Caltagirone, P. Sagaut, Towards Large Eddy Simulation Of Isothermal Two-Phase Flows: Governing Equations And A Priori Tests, *International Journal Of Multiphase Flow* **33** (1) (2007) 1–39. doi:10.1016/J.Ijmultiphaseflow.2006.05.010.

[51] D. Lilly, A Proposed Modification Of The Germano-Subgrid-Scale Closure Method, *Physics Of Fluids A-Fluid Dynamics* **4** (3) (1992) 633–635.

[52] S. Bhushan, A Proposed Modification To The Dynamic Approach, *International Journal For Numerical Methods In Fluids* **54** (9) (2007) 1075–1095. doi:10.1002/Fld.1415.

[53] C. Meneveau, T. Lund, The Dynamic Smagorinsky Model And Scale-Dependent Coefficients In The Viscous Range Of Turbulence, *Physics Of Fluids* **9** (12) (1997) 3932–3934.

[54] Y. Mesri, H. Digonnet, T. Coupez, Advanced parallel computing in material forming with cimlib, *European Journal Of Computational Mechanics* **18** (2009) 669–694.

[55] H. Digonnet, T. Coupez, Object-oriented programming for fast and easy development of parallel applications in forming processes simulation, in: *Computational Fluid And Solid Mechanics* 2003, 2003, pp. 1922–1924.

[56] H. Digonnet, S. Luisa, T. Coupez, Cimlib: A fully parallel application for numerical simulations based on components assembly, *Numiform*, 2007.

[57] T. Coupez, H. Digonnet, R. Ducloux, Parallel meshing and remeshing, *Applied Mathematical Modelling* **25** (2) (2000) 153–175.

[58] J. R. Koseff, R. L. Street, P. M. Gresho, C. D. Upson, J. A. C. Humphrey, J. W. To, A three-dimensional lid-driven cavity flow: Experiment and simulation, in: *Third International Conference On Numerical Methods In Laminar And Turbulent Flow* , Seattle, Wa, 1983, pp. 564–581.

[59] J. R. Koseff, R. L. Street, Visualization studies of a shear driven three-dimensional recirculating flow, *Journal Of Fluids Engineering* **106** (1984) 21—-29.

[60] C. J. Freitas, R. L. Street, A. N. Findikakis, J. R. Koseff, *Numerical simulation of three dimensional flow in a cavity* **5** (1985) 561—-575.

[61] C. J. Freitas, R. L. Street, *Non-linear transport phenomena in a complex recirculating flow: A numerical investigation* **8** (1988) 769—-802.

[62] L. Q. Tang, T. Cheng, T. T. H. Tsang, *Transient solutions for three-dimensional lid-driven cavity flows by a least-squares finite element method* **21** (1995) 413—-432.

[63] Y. Zang, R. L. Street, J. R. Koseff, A dynamic mixed subgrid-scale model and its application to turbulent recirculating flows, *Physics Of Fluids* **A5** (1993) 3186—-3196.

[64] A. K. Prasad, J. R. Koseff, Reynolds number and end-wall effects on a lid-driven cavity flow, *Physics Of Fluids* **A1** (1989) 208—-218.

[65] T. L. Popiolek, A. M. Awruch, P. R. F. Teixeira, Finite element analysis of laminar and turbulent flows using Les and subgrid-scale models, *Applied Mathematical Modelling* **30** (2) (2006) 177—-199.

[66] M. Cruchaga, D. J. Celentano, T. Tezduyar, Collapse of a liquid coumn : Numerical simulation and experimental validation, *Comput. Mech.* **39** (2007) 453–476.

[67] R. Courant, K. Friedrichs, H. Lewy, Uber die partiellen diefferenzengleichungen der mathematischen physik, *Mathematische Annalen* **100** (1) (1982) 32–74.

(a) $t = 0.5s$ (b) $t = 0.75s$ (c) $t = 1s$

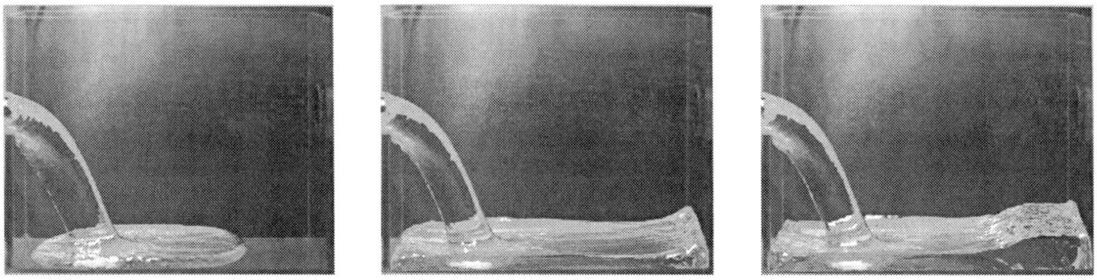

(d) $t = 0.5s$ (e) $t = 0.75s$ (f) $t = 1s$

(g) $t = 0.5s$ (h) $t = 0.75s$ (i) $t = 1s$

Figure 7. Time evolution of water-air interface: experimental (top), and numerical results computed with static model (middle) and with dynamic model (bottom).

In: Navier-Stokes Equations
Editor: R. Younsi

ISBN: 978-1-61324-590-3

Chapter 6

ON SOLUTIONS OF THE NAVIER-STOKES EQUATIONS

***K. Fakhar*[1,*] *and A.H. Kara*[2†]**
[1]Department of Mathematics, Faculty of Science,
Universiti Teknologi Malaysia, 81310 UTM Skudai, Johor, Malaysia
[2]School of Mathematics and Centre for Differential Equations,
Continuum Mechanics and Applications, University of the Witwatersrand,
Johannesburg, P Bag 3 Wits 2050, South Africa

Abstract

The equations which govern the motion of an incompressible viscous fluid are precisely the Navier-Stokes equations. It is a known fact that the Navier-Stokes equations in general are nonlinear partial differential equations and few analytic solutions of these equations are reported in the existing literature. In this chapter we have considered unsteady Navier-Stokes equations for viscous, incompressible three (two) dimensional flow in cylindrical polar coordinates (polar coordinates). The Lie point symmetries for these equations have been obtained by utilizing a direct approach. Using the symmetries obtained via this approach, similarity reductions in order have been performed. A successive similarity reduction has reduced the governing partial differential equations to a system of ordinary deferential equations. The resulting differential equations have been solved analytically and then by process of back substitutions, the solutions of the original system have been obtained. Some interesting discussions and conclusions have been presented at the end of the chapter.

1. Introduction

The Navier-Stokes equations are nonlinear partial differential equations describing the phenomena of fluid flow. These equations originated a century and a half ago and are still of wide interest and form an active area of research. These equations have varied applicability in the different kinds of flows ranging from thin films to large-scale atmospheric, even

*E-mail address: kamranfakhar@yahoo.com
†E-mail address: Abdul.Kara@wits.ac.za

cosmic, flows and also have numerous applications in different industries and various other fields. Specifically, it is the nonlinearity of these equations which have attracted researches for further exploitation around the globe.

Since the Navier-Stokes equations are nonlinear in nature, much of the attention so far has been given to the formulation of efficient numerical techniques and algorithms for dealing with the equations. Some simplifications, such as linearization, assumptions of weak nonlinearity, small fluctuations and discretization, inter alia, have been employed by many researchers in their work to obtain numerical solutions (see [1-7] for recent results). Although such numerical techniques have proven to be successful, many still carry the usual error associated with approximations. Further, to obtain accurate numerical simulations, the computational costs are high except for simple 'small scale' problems in engineering.

On the other hand, the perturbation technique which is widely used by engineers, physicists and computational mathematicians to deal with these types of nonlinear physical phenomena also have their own limitations since all perturbation techniques are based on small or large parameters for which at least one unknown must be expressed in a series of small parameters. Not every non-linear differential equation has this kind of structure. Even if there exists such a small parameter, the result given by perturbation methods are valid, in most cases, only for the small values of the parameter. Mostly, the simplified linear equations have different properties from the original non-linear differential equation and occassionally some initial or boundary conditions are superfluous for the simplified linear equations. As a result, the corresponding initial approximations are perhaps far from exact.

The above mentioned limitations highlight the need and importance of analytical solutions for these equations. To this end, inter alia, the group method or Lie group analysis approach has been used to tackle the underlying systems of differential equations [8-16]. However, in cylindrical and polar coordinates, only limited analytical work is available in the literature. Therefore, in this chapter, some similarity solutions for unsteady Navier-Stokes equations in cylindrical polar coordinates and also in polar coordinates have been provided using the Lie group analysis approach [17-20].

We will often resort to the theorem below to transform *vector fields* from one coordinate system to another. The vector fields would represent the *one-parameter Lie group of transformations* that leave invariant the system of differential equations under investigation.

Theorem 1 (Ibragimov [20]) *Consider the partial differential operator of the first order*

$$Q = \xi^1(\boldsymbol{x})\frac{\partial}{\partial x^1} + \cdots + \xi^n(\boldsymbol{x})\frac{\partial}{\partial x^n}, \tag{1.1}$$

where $\boldsymbol{x} = (x^1, \ldots, x^n)$. *Let* $x'^i = \varphi^i(\boldsymbol{x}), i = 1, \ldots, n$. *Then the operator (1.1) is written in the new variables in the form*

$$\overline{Q} = Q(\varphi^1)\frac{\partial}{\partial x'^1} + \cdots + Q(\varphi^n)\frac{\partial}{\partial x'^n},$$

where

$$Q(\varphi^i) = \xi^1(\boldsymbol{x})\frac{\partial \varphi^i}{\partial x^1} + \cdots + \xi^n(\boldsymbol{x})\frac{\partial \varphi^i}{\partial x^n}.$$

This would allow one to use, mainly for reduction of the differential equations, some basic transformations such as translation, rotation, scaling and their combinations to deal with more complicated ones. Once the resulting system of differential equations have been solved analytically, the process of back substitutions would lead to a solution of the original system.

At the end of the chapter we have presented some discussions/conclusions and comments for future work.

2. Three Dimensional Cylindrical Polar Navier-Stokes System

We first write the three dimensional unsteady Navier-Stokes equations for viscous, incompressible flow in cylindrical polar coordinates in component form as [21]

$$\begin{aligned}
&\frac{\partial u}{\partial t}+u\frac{\partial u}{\partial r}+\frac{v}{r}\frac{\partial u}{\partial \theta}-\frac{v^2}{r}+w\frac{\partial u}{\partial z}=-\frac{\partial p}{\partial r}+\mu\bigg(\frac{\partial^2 u}{\partial r^2}+\frac{1}{r}\frac{\partial u}{\partial r}-\frac{u}{r^2}\\
&+\frac{1}{r^2}\frac{\partial^2 u}{\partial \theta^2}-\frac{2}{r^2}\frac{\partial v}{\partial \theta}+\frac{\partial^2 u}{\partial z^2}\bigg),\\
&\frac{\partial v}{\partial t}+u\frac{\partial v}{\partial r}+\frac{v}{r}\frac{\partial v}{\partial \theta}+\frac{vu}{r}+w\frac{\partial v}{\partial z}=-\frac{1}{r}\frac{\partial p}{\partial \theta}+\mu\bigg(\frac{\partial^2 v}{\partial r^2}+\frac{1}{r}\frac{\partial v}{\partial r}-\frac{v}{r^2}\\
&+\frac{1}{r^2}\frac{\partial^2 v}{\partial \theta^2}+\frac{2}{r^2}\frac{\partial u}{\partial \theta}+\frac{\partial^2 v}{\partial z^2}\bigg),\\
&\frac{\partial w}{\partial t}+u\frac{\partial w}{\partial r}+\frac{v}{r}\frac{\partial w}{\partial \theta}+w\frac{\partial w}{\partial z}=-\frac{\partial p}{\partial z}+\mu\bigg(\frac{\partial^2 w}{\partial r^2}+\frac{1}{r}\frac{\partial w}{\partial r}+\frac{1}{r^2}\frac{\partial^2 w}{\partial \theta^2}+\frac{\partial^2 w}{\partial z^2}\bigg),\\
&\frac{1}{r}\frac{\partial}{\partial r}(ru)+\frac{1}{r}\frac{\partial}{\partial \theta}(v)+\frac{\partial}{\partial z}(w)=0.
\end{aligned}\tag{2.1}$$

Here, the last equation in system (2.1) is a continuity equation. Further $r \neq 0$, u is the radial, v is the circumferential, w is the axial velocities respectively, μ is (constant) viscosity and p is the generalized pressure with density $\rho = 1$.

2.1. The Infinitesimals for System (2.1)

We look for transformations of the independent variables t, r, θ and z and the dependent variables u, v, w and p of the form

$$\begin{aligned}
&\overline{t}=\overline{t}(t,r,\theta,z,u,v,w,p),\ \overline{r}=\overline{r}(\cdots),\ \overline{\theta}=\overline{\theta}(\cdots),\ \overline{z}=\overline{z}(\cdots)\\
&\overline{u}=\overline{u}(\cdots),\ \overline{v}=\overline{v}(\cdots),\ \overline{w}=\overline{w}(\cdots),\ \overline{p}=\overline{p}(\cdots),
\end{aligned}\tag{2.2}$$

which constitute a one-parameter Lie group of transformations leaving system (2.1) invariant. From Lie's theory, transformations in equation (2.2) are obtained in terms of the infinitesimal transformations are of the form

$$\begin{aligned}
&\overline{t}\simeq t+\epsilon T(t,r,\theta,z,u,v,w,p),\ \overline{r}\simeq r+\epsilon R(\cdots),\ \overline{\theta}\simeq\theta+\epsilon\Theta(\cdots),\ \overline{z}\simeq r+\epsilon Z(\cdots),\\
&\overline{u}\simeq u+\epsilon U(\cdots),\ \overline{v}\simeq v+\epsilon V(\cdots),\ \overline{w}\simeq w+\epsilon W(\cdots),\ \overline{p}\simeq p+\epsilon P(\cdots),
\end{aligned}\tag{2.3}$$

or in terms of the the infinitesimal operator

$$\mathbf{X} = T\frac{\partial}{\partial t} + R\frac{\partial}{\partial r} + \Theta\frac{\partial}{\partial \theta} + Z\frac{\partial}{\partial z} + U\frac{\partial}{\partial u} + V\frac{\partial}{\partial v} + W\frac{\partial}{\partial w} + P\frac{\partial}{\partial p}, \tag{2.4}$$

which is a generator of Lie point symmetry of system (2.1). Here, $T,\ R,\ \Theta,\ Z,\ U,\ V,\ W,\ P$ are the infinitesimals that need to be determined. As we mentioned earlier, instead of using classical way to determine the infinitesimal here we will make use of theorem 1. For this we first need the infinitesimal operator for three dimensional Navier-Stokes system in Cartesian coordinates. From [8], we have the following form of infinitesimal along with infinitesimal operator

$$Q = \eta_1\frac{\partial}{\partial \grave{t}} + \eta_2\frac{\partial}{\partial \grave{x}} + \eta_3\frac{\partial}{\partial \grave{y}} + \eta_4\frac{\partial}{\partial \grave{z}} + \eta_5\frac{\partial}{\partial \grave{u}} + \eta_6\frac{\partial}{\partial \grave{v}} + \eta_7\frac{\partial}{\partial \grave{w}} + \eta_8\frac{\partial}{\partial \grave{p}}, \tag{2.5}$$

where the infinitesimals $\eta_i = \eta_i(\grave{t},\ \grave{x},\ \grave{y},\ \grave{z},\ \grave{u},\ \grave{v},\ \grave{w},\ \grave{p}),\ i = 1, 2, \cdots 8$, are given as

$$\begin{aligned}
&\eta_1 = \alpha + 2\beta\grave{t}, \quad \eta_2 = \beta\grave{x} - \gamma\grave{y} - \lambda\grave{z} + f(\grave{t}), \quad \eta_3 = \beta\grave{y} + \gamma\grave{x} - \sigma\grave{z} + g(\grave{t}),\\
&\eta_4 = \beta + \lambda\grave{x} + \sigma\grave{y} + h(\grave{t}), \quad \eta_5 = -\beta\grave{u} - \gamma\grave{v} - \lambda\grave{w} + f'(\grave{t}),\\
&\eta_6 = -\beta\grave{v} + \gamma\grave{u} - \sigma\grave{w} + g'(\grave{t}), \quad \eta_7 = -\beta\grave{w} + \lambda\grave{u} + \sigma\grave{v} + h'(\grave{t}),\\
&\eta_8 = -2\beta\grave{p} + j(\grave{t}) - xf''(\grave{t}) - yg''(\grave{t}) - zh''(\grave{t}),
\end{aligned}$$

where $\grave{u}(\grave{t}, \grave{x}, \grave{y}, \grave{z}), \grave{v}(\grave{t}, \grave{x}, \grave{y}, \grave{z}), \grave{w}(\grave{t}, \grave{x}, \grave{y}, \grave{z})$ are the velocity components in Cartesian coordinates, $\alpha,\ \beta,\ \gamma,\ \lambda,\ \sigma$ are arbitrary parameters and $f(\grave{t}),\ g(\grave{t}),\ h(\grave{t}),\ j(\grave{t})$ are arbitrary sufficiently smooth functions of time $\grave{t}$. For cylindrical polar coordinates, we define the transformations as

$$\begin{aligned}
&t = \grave{t}, r = \sqrt{(}\grave{x}^2 + \grave{y}^2), \theta = \arctan(\frac{\grave{y}}{\grave{x}}), z = \grave{z},\\
&u = \grave{u}\cos\theta + \grave{v}\sin\theta, v = \grave{v}\cos\theta - \grave{u}\sin\theta, w = \grave{w}, p = \grave{p}.
\end{aligned}$$

The operator (2.5), after application of theorem 1, takes the following form

$$\begin{aligned}
\overline{Q} = {} & Q(t)\frac{\partial}{\partial t} + Q(r)\frac{\partial}{\partial r} + Q(\theta)\frac{\partial}{\partial \theta} + Q(z)\frac{\partial}{\partial z}\\
& + Q(u)\frac{\partial}{\partial u} + Q(v)\frac{\partial}{\partial v} + Q(w)\frac{\partial}{\partial w} + Q(p)\frac{\partial}{\partial p}.
\end{aligned} \tag{2.6}$$

Here $Q(t) = T, Q(r) = R,\ Q(\theta) = \Theta,\ Q(z) = Z,\ Q(u) = U,\ Q(v) = V,\ Q(w) = W$ and $Q(p) = P$ and $\mathbf{X} = \overline{Q}$ are our required infinitesimals and the associated operator respectively. The values of these infinitesimals can be easily found by using theorem 1. We omit the details and directly write the values of infinitesimals, viz.,

$$\begin{aligned}
&T = \alpha + 2\beta t, \quad R = \beta r - z(\lambda\cos\theta + \sigma\sin\theta) + f(t)\cos\theta + g(t)\sin\theta,\\
&\Theta = \frac{1}{r}\{\gamma r + z(\lambda\sin\theta - \sigma\cos\theta) - f(t)\sin\theta + g(t)\cos\theta\},\\
&Z = \beta z + r(\lambda\cos\theta + \sigma\sin\theta) + h(t),
\end{aligned}$$

$$
\begin{aligned}
U &= \frac{v}{r}\{z(\lambda\sin\theta - \sigma\cos\theta) - f(t)\sin\theta + g(t)\cos\theta\} - \beta u - w(\lambda\cos\theta + \sigma\sin\theta) \\
&\quad + f'(t)\cos\theta + g'(t)\sin\theta, \\
V &= \frac{u}{r}\{f(t)\sin\theta - z(\lambda\sin\theta - \sigma\cos\theta) - g(t)\cos\theta\} - \beta v + w(\lambda\sin\theta - \sigma\cos\theta) \\
&\quad - f'(t)\sin\theta + g'(t)\cos\theta, \\
W &= -\beta w + u(\lambda\cos\theta + \sigma\sin\theta) - v(\lambda\sin\theta - \sigma\cos\theta) + h'(t), \\
P &= -2\beta p + j(t) - r\cos\theta f''(t) - r\sin\theta g''(t) - zh''(t).
\end{aligned}
\tag{2.7}
$$

2.2. Three Dimensional Similarity Solution

The presence of arbitrary functions of time $f,\ g,\ h$ and j suggest that the Lie algebra associated with system (2.1) is an infinite dimensional. In order to reduce system (2.1) to a system of ordinary differential equations we apply the linear combination of translation symmetries in t and θ and the scaling symmetry in independent and dependent variables, i.e., $\frac{\partial}{\partial t} + \frac{\partial}{\partial \theta} + (2t\frac{\partial}{\partial t} + r\frac{\partial}{\partial r} + z\frac{\partial}{\partial z} - u\frac{\partial}{\partial u} - v\frac{\partial}{\partial v} - w\frac{\partial}{\partial w} - 2p\frac{\partial}{\partial p})$ which is a 3-d subalgebra of infinite dimensional algebra of system (2.1). For reduction, we first use translation in t variable and hence $\alpha = 1$, while $\beta = \gamma = \lambda = \sigma = 0$ and $f(t) = g(t) = h(t) \equiv 0$ is taken. The similarity variables and functions associated with the translation symmetry are

$$
\begin{aligned}
& r = \overline{r}, \theta = \overline{\theta},\ z = \overline{z},\ u = \overline{u}(\overline{r}, \overline{\theta}, \overline{z}),\ v = \overline{v}(\overline{r}, \overline{\theta}, \overline{z}),\ w = \overline{w}(\overline{r}, \overline{\theta}, \overline{z}), \\
& p = \overline{p}(\overline{r}, \overline{\theta}, \overline{z}) + I(t), \quad \texttt{where}\ I(t) = \int j(t)dt.
\end{aligned}
\tag{2.8}
$$

System (2.1), under (2.8), reduces to the following steady state system

$$
\begin{aligned}
& \overline{u}\frac{\partial \overline{u}}{\partial \overline{r}} + \frac{\overline{v}}{\overline{r}}\frac{\partial \overline{u}}{\partial \overline{\theta}} - \frac{\overline{v^2}}{\overline{r}} + \overline{w}\frac{\partial \overline{u}}{\partial \overline{z}} = -\frac{\partial \overline{p}}{\partial \overline{r}} + \mu\bigg(\frac{\partial^2 \overline{u}}{\partial \overline{r}^2} + \frac{1}{\overline{r}}\frac{\partial \overline{u}}{\partial \overline{r}} - \frac{\overline{u}}{\overline{r}^2} + \frac{1}{\overline{r^2}}\frac{\partial^2 \overline{u}}{\partial \overline{\theta^2}} \\
& - \frac{2}{\overline{r^2}}\frac{\partial \overline{v}}{\partial \overline{\theta}} + \frac{\partial^2 \overline{u}}{\partial \overline{z}^2}\bigg), \\
& \overline{u}\frac{\partial \overline{v}}{\partial \overline{r}} + \frac{\overline{v}}{\overline{r}}\frac{\partial \overline{v}}{\partial \overline{\theta}} + \frac{\overline{vu}}{\overline{r}} + \overline{w}\frac{\partial \overline{v}}{\partial \overline{z}} = -\frac{1}{\overline{r}}\frac{\partial \overline{p}}{\partial \overline{\theta}} + \mu\bigg(\frac{\partial^2 \overline{v}}{\partial \overline{r}^2} + \frac{1}{\overline{r}}\frac{\partial \overline{v}}{\partial \overline{r}} - \frac{\overline{v}}{\overline{r}^2} + \frac{1}{\overline{r^2}}\frac{\partial^2 \overline{v}}{\partial \theta^2} \\
& + \frac{2}{\overline{r^2}}\frac{\partial \overline{u}}{\partial \overline{\theta}} + \frac{\partial^2 \overline{v}}{\partial \overline{z}^2}\bigg), \\
& \overline{u}\frac{\partial \overline{w}}{\partial \overline{r}} + \frac{\overline{v}}{\overline{r}}\frac{\partial \overline{w}}{\partial \overline{\theta}} + \overline{w}\frac{\partial \overline{w}}{\partial \overline{z}} = -\frac{\partial \overline{p}}{\partial \overline{z}} + \mu\bigg(\frac{\partial^2 \overline{w}}{\partial \overline{r}^2} + \frac{1}{\overline{r}}\frac{\partial \overline{w}}{\partial \overline{r}} + \frac{1}{\overline{r^2}}\frac{\partial^2 \overline{w}}{\partial \overline{\theta^2}} + \frac{\partial^2 \overline{w}}{\partial \overline{z}^2}\bigg), \\
& \frac{1}{\overline{r}}\frac{\partial}{\partial \overline{r}}(\overline{ru}) + \frac{1}{\overline{r}}\frac{\partial \overline{v}}{\partial \overline{\theta}} + \frac{\partial}{\partial \overline{z}}(\overline{w}) = 0.
\end{aligned}
\tag{2.9}
$$

Now, any steady state solution of (2.9) can be transformed by means of (2.8) into a time-dependent solution involving an arbitrary function of the time variable. As the system (2.9) is still a partial differential equation system, we continue with the reductions. Again, adopting the same procedure as above, the infinitesimals corresponding to the reduced system (2.9) are

$$
\nu_1 = \beta\overline{r} - \overline{z}(\lambda\cos\overline{\theta} + \sigma\sin\overline{\theta}), \quad \nu_2 = \frac{1}{\overline{r}}\{\gamma\overline{r} + \overline{z}(\lambda\sin\overline{\theta} - \sigma\cos\overline{\theta})\},
$$

$$
\begin{aligned}
\nu_3 &= \beta\overline{z} + \overline{r}(\lambda\cos\overline{\theta} + \sigma\sin\overline{\theta}) + \zeta,\\
\nu_4 &= \frac{\overline{v}}{\overline{r}}\{\overline{z}(\lambda\sin\overline{\theta} - \sigma\cos\overline{\theta})\} - \beta\overline{u} - \overline{w}(\lambda\cos\overline{\theta} + \sigma\sin\overline{\theta}),\\
\nu_5 &= \frac{\overline{u}}{\overline{r}}\{-\overline{z}(\lambda\sin\overline{\theta} - \sigma\cos\overline{\theta})\} - \beta\overline{v} + \overline{w}(\lambda\sin\overline{\theta} - \sigma\cos\overline{\theta}),\\
\nu_6 &= -\beta\overline{w} + \overline{u}(\lambda\cos\overline{\theta} + \sigma\sin\overline{\theta}) - \overline{v}(\lambda\sin\overline{\theta} - \sigma\cos\overline{\theta}),\ \nu_7 = -2\beta\overline{p} + \Delta, \qquad (2.10)
\end{aligned}
$$

where ζ and Δ are new arbitrary parameters. We now use translation in the θ variable to reduce the system (2.9). For this case, $\gamma = 1$, while other parameters are zero. For this generator the similarity variables and functions are

$$
\overline{r} = \tilde{r},\ \overline{z} = \tilde{z},\ \overline{u} = \tilde{u}(\tilde{r}, \tilde{z}),\ \overline{v} = \tilde{v}(\tilde{r}, \tilde{z}),\ \overline{w} = \tilde{w}(\tilde{r}, \tilde{z}),\ \overline{p} = \tilde{p}(\tilde{r}, \tilde{z}), \qquad (2.11)
$$

By substituting the similarity variables and functions into (2.9) we obtain

$$
\begin{aligned}
&\tilde{u}\frac{\partial\tilde{u}}{\partial\tilde{r}} - \frac{\tilde{v}^2}{\tilde{r}} + \tilde{w}\frac{\partial\tilde{u}}{\partial\tilde{z}} = -\frac{\partial\tilde{p}}{\partial\tilde{r}} + \mu\left(\frac{\partial^2\tilde{u}}{\partial\tilde{r}^2} + \frac{1}{\tilde{r}}\frac{\partial\tilde{u}}{\partial\tilde{r}} - \frac{\tilde{u}}{\tilde{r}^2} + \frac{\partial^2\tilde{u}}{\partial\tilde{z}^2}\right),\\
&\tilde{u}\frac{\partial\tilde{v}}{\partial\tilde{r}} + \frac{\tilde{v}\tilde{u}}{\tilde{r}} + \tilde{w}\frac{\partial\tilde{v}}{\partial\tilde{z}} = \mu\left(\frac{\partial^2\tilde{v}}{\partial\tilde{r}^2} + \frac{1}{\tilde{r}}\frac{\partial\tilde{v}}{\partial\tilde{r}} - \frac{\tilde{v}}{\tilde{r}^2} + \frac{\partial^2\tilde{v}}{\partial\tilde{z}^2}\right),\\
&\tilde{u}\frac{\partial\tilde{w}}{\partial\tilde{r}} + \tilde{w}\frac{\partial\tilde{w}}{\partial\tilde{z}} = -\frac{\partial\tilde{p}}{\partial\tilde{z}} + \mu\left(\frac{\partial^2\tilde{w}}{\partial\tilde{r}^2} + \frac{1}{\tilde{r}}\frac{\partial\tilde{w}}{\partial\tilde{r}} + \frac{\partial^2\tilde{w}}{\partial\tilde{z}^2}\right),\\
&\frac{1}{\tilde{r}}\frac{\partial}{\partial\tilde{r}}(\tilde{r}\tilde{u}) + \frac{\partial}{\partial\tilde{z}}(\tilde{w}) = 0. \qquad (2.12)
\end{aligned}
$$

The infinitesimals of the reduced one-parameter Lie group of transformations leaving system (2.12) invariant are

$$
\tau_1 = \beta\tilde{r},\ \tau_2 = \beta\tilde{z} + \zeta,\ \tau_3 = -\beta\tilde{u}, \tau_4 = -\beta\tilde{v},\ \tau_5 = -\beta\tilde{v}_z,\ \tau_6 = -2\beta\tilde{p} + \Delta. \quad (2.13)
$$

Finally, we use the scaling symmetry ($\beta = 1$) to reduce (2.12). The characteristic equations for finding the similarity transformation would then be

$$
\frac{d\tilde{r}}{\tilde{r}} = \frac{d\tilde{z}}{\tilde{z}} = \frac{d\tilde{u}}{-\tilde{u}} = \frac{d\tilde{v}}{-\tilde{v}} = \frac{d\tilde{w}}{-\tilde{w}} = \frac{d\tilde{p}}{-2\tilde{p}}.
$$

The similarity variables and functions are

$$
\xi = \frac{\tilde{z}}{\tilde{r}},\ \tilde{u} = \frac{1}{\tilde{r}}F(\xi),\ \tilde{v} = \frac{1}{\tilde{r}}G(\xi),\ \tilde{w} = \frac{1}{\tilde{r}}H(\xi),\ \tilde{p} = \frac{1}{\tilde{r}^2}J(\xi). \qquad (2.14)
$$

Substituting the new variables into (2.12) yield the following ordinary differential equation system

$$
\begin{aligned}
&-F(F + \xi F') - G^2 + HF' = 2J + \xi J' + \mu\{3\xi F' + (1 + \xi^2)F''\},\\
&-\xi FG' + HG' = \mu\{3\xi G' + (1 + \xi^2)G''\},\\
&-F(H + \xi H') + HH' = -J' + \mu\{H + 3\xi H' + (1 + \xi^2)H''\},\\
&H' - \xi F' = 0. \qquad (2.15)
\end{aligned}
$$

The constant of viscosity, μ , can be eliminated from (2.15) by means of the substitutions

$$F_1 = \mu^{-1}F, G_1 = \mu^{-1}G, H_1 = \mu^{-1}H, J_1 = \mu^{-2}J,$$

to give us the following system of equations

$$F_1^2 + G_1^2 + \xi F_1 F_1' - H_1 F_1' + 3\xi F_1' + (1+\xi^2)F_1'' + 2J_1 + \xi J_1' = 0, \quad (2.16)$$
$$(\xi F_1 - H_1 + 3\xi)G_1' + (1+\xi^2)G_1'' = 0, \quad (2.17)$$
$$F_1 H_1 + H_1 + \xi F_1 H_1 - H_1 H_1' + 3\xi H_1' + (1+\xi^2)H_1'' - J_1' = 0, \quad (2.18)$$
$$H_1' - \xi F_1' = 0. \quad (2.19)$$

From equation (2.17),

$$G_1 = c\ (constant) \quad (2.20)$$

Eliminating J_1 between equations (2.16) and (2.18) and using equations (2.19) and (2.20), we obtain

$$(1+\xi^2)H_1F_1'' - \xi(1+\xi^2)F_1F_1'' - 3(1+2\xi^2)F_1F_1' + \xi(2+\xi)H_1F_1' - \xi^3 F_1'F_1'$$
$$+ H_1^2 - 3H_1F_1 = (1+\xi^2)^2 F_1''' + 10\xi(1+\xi^2)F_1'' + 3(2+7\xi^2)F_1' + 3H_1. \quad (2.21)$$

It is not difficult to check that the equation (2.21) together with equation (2.19) form an analytic system for any fixed point. Therefore, by using the existence theorem (8.1) of analytic solution [22, p.34], the solutions of system (2.19) and (2.21) can be expressed as a series in the form

$$F_1 = \sum_{k=0}^{\infty} a_k \xi^k, \quad H_1 = \sum_{k=0}^{\infty} b_k \xi^k. \quad (2.22)$$

Substituting the series (2.22) into (2.19), one obtains the recursion relation

$$b_{k+1} = \frac{k}{k+1} a_k, \ k = 0, 1, 2 \cdots .$$

Similarly, from (2.20), one obtains the sets of equations

$$2a_2b_0 - 3a_1a_0 + 3a_0b_0 + b_0^2 - 6a_3 - 6a_1 - 3b_0 = 0\ (k=0),$$
$$6a_3b_0 + 2a_2b_1 - 8a_2a_0 - 3a_1^2 - a_1b_0 - 3a_0b_1 + 2b_0b_1$$
$$- 24a_4 - 32a_2 - 3b_1 = 0\ (k=1),$$
$$\cdots$$
$$(k+1)(k+2)(k+3)a_{k+3} + 2(k+1)^2(k+3)a_{k+1}$$
$$+ (k-1)\{(k+3)(k+2)+1\}a_{k-1}$$
$$+ 3b_k + \sum_{n=2}^{k+2} b_{k+2-n}a_n n(n-1) + \sum_{n=2}^{k} b_{k-n}a_n n(n-1)$$
$$- \sum_{n=2}^{k-1} a_{k-1-n}a_n n(n-1) - \sum_{n=2}^{k+1} a_{k+1-n}a_n n(n-1) - 6\sum_{n=1}^{k-1} a_{k-1-n}a_n n$$

$$-3\sum_{n=1}^{k+1} a_{k+1-n}a_n n + \sum_{n=1}^{k-1} b_{k-1-n}a_n n + 2\sum_{n=1}^{k} b_{k-n}a_n n$$
$$-\sum_{n,m=1}^{k-2} a_n a_m nm + \sum_{n+m=k} b_n b_m - 3\sum_{n+m=k} a_n a_m = 0, \; (k = 2, 3, 4, \cdots). \tag{2.23}$$

The coefficients a_k and b_k can be determined by solving the above equations. F_1 and H_1 functions are

$$F_1(\xi) = a_0 + a_1\xi + a_2\xi^2 + \left[\left\{\frac{a_2}{3} + \frac{a_0}{2} + \frac{b_0}{6} - \frac{1}{2}\right\}b_0 - \left\{1 + \frac{a_0}{2}\right\}a_1\right]\xi^3 + \cdots,$$
$$H_1(\xi) = b_0 + \frac{a_1}{2}\xi^2 + \frac{2a_2}{3}\xi^3 \cdots,$$

$J_1(\xi)$ can be determined by using equation (2.18) and the result is

$$J_1(\xi) = c_0 + [(a_0+1)b_0 + a_1]\xi + 2a_2\xi^2 + \left[\left\{\frac{2a_2}{3} + \frac{3a_0}{2} + \frac{b_0}{2} - \frac{3}{2}\right\}b_0 - \left\{\frac{1}{6} + a_0\right\}a_1\right]\xi^3 + \cdots.$$

In terms of original variables the solution of the system (2.1) is given as

$$u(r,\theta,z,t) = a_0\frac{\mu}{r} + a_1\frac{\mu z}{r^2} + a_2\frac{\mu z^2}{r^3} + \left[\left\{\frac{a_2}{3} + \frac{a_0}{2} + \frac{b_0}{6} - \frac{1}{2}\right\}b_0 - \left\{1 + \frac{a_0}{2}\right\}a_1\right]\frac{\mu z^3}{r^4} + \cdots,$$
$$v(r,\theta,z,t) = \frac{\mu c}{r},$$
$$v_z(r,\theta,z,t) = b_0\frac{\mu}{r} + \frac{a_1}{2}\frac{\mu z^2}{r^3} + \frac{2a_2}{3}\frac{\mu z^3}{r^4} + \cdots,$$
$$p(r,\theta,z,t) = I(t) + c_0\frac{\mu^2}{r^2} + [(a_0+1)b_0 + a_1]\frac{\mu^2 z}{r^3} + 2a_2\frac{\mu^2 z^2}{r^4} + \left[\left\{\frac{2a_2}{3} + \frac{3a_0}{2} + \frac{b_0}{2} - \frac{3}{2}\right\}b_0 - \left\{\frac{1}{6} + a_0\right\}a_1\right]\frac{\mu^2 z^3}{r^5} + \cdots. \tag{2.24}$$

3. Two Dimensional Polar Navier-Stokes System

From system (2.1), the equations of motion for an incompressible fluids in polar coordinates can be written as

$$\frac{\partial u}{\partial t} + u\frac{\partial u}{\partial r} + \frac{v}{r}\left(\frac{\partial u}{\partial \theta}\right) - \frac{v^2}{r} = -\frac{\partial p}{\partial r} + \eta\left(\frac{\partial^2 u}{\partial r^2} + \frac{1}{r^2}\frac{\partial^2 u}{\partial \theta^2} + \frac{1}{r}\frac{\partial u}{\partial r} - \frac{2}{r^2}\frac{\partial v}{\partial \theta} - \frac{u}{r^2}\right),$$
$$\frac{\partial v}{\partial t} + u\frac{\partial v}{\partial r} + \frac{v}{r}\left(\frac{\partial v}{\partial \theta}\right) - v\frac{u}{r} = -\frac{1}{r}\frac{\partial p}{\partial \theta} + \eta\left(\frac{\partial^2 v}{\partial r^2} + \frac{1}{r^2}\frac{\partial^2 v}{\partial \theta^2} + \frac{1}{r}\frac{\partial v}{\partial r} + \frac{2}{r^2}\frac{\partial u}{\partial \theta} - \frac{v}{r^2}\right),$$
$$\frac{\partial u}{\partial r} + \frac{u}{r} + \frac{\partial v}{\partial \theta} = 0. \tag{3.1}$$

Again we require that the system (3.1) remain invariant under the infinitesimal transformations

$$\bar{t} \simeq t + \epsilon\psi_1(t, r, \theta, u, v, p),\ \bar{r} \simeq r + \epsilon\psi_2(\cdots),\ \bar{\theta} \simeq \theta + \psi_3(\cdots),$$
$$\bar{u} \simeq u + \epsilon\psi_4(\cdots),\ \bar{v} \simeq v + \epsilon\psi_5(\cdots),\ \bar{p} \simeq p + \epsilon\psi_6(\cdots), \tag{3.2}$$

here $\psi_j,\ j = 1, 2, \cdots 6$, are the infinitesimals. In order to calculate infinitesimals we again make use of theorem 1. The expression for the infinitesimals are given as

$$\begin{aligned}
&\psi_1 = \alpha + 2\beta t, \quad \psi_2 = \beta r + f(t)\cos\theta + g(t)\sin\theta,\\
&\psi_3 = \frac{1}{r}\{\gamma r - f(t)\sin\theta + g(t)\cos\theta\},\\
&\psi_4 = \left\{\frac{v}{r}g(t) + f'(t)\right\}\cos\theta - \left\{\frac{v}{r}f(t) - g'(t)\sin\theta\right\} - \beta u,\\
&\psi_5 = -\left\{\frac{u}{r}g(t) - g'(t)\right\}\cos\theta + \left\{\frac{u}{r}f(t) - f'(t)\sin\theta\right\} - \beta v,\\
&\psi_6 = -2\beta p + j(t) - r\cos\theta f''(t) - r\sin\theta g''(t).
\end{aligned} \tag{3.3}$$

4. Two-Dimensional Similarity Solutions

In this section we present the two similarity solutions.

4.1. First Solution

Here we consider a linear combination of $\frac{\partial}{\partial t}$, $\frac{\partial}{\partial \theta}$ and $\frac{\partial}{\partial j(t)}$. This linear combination is an infinite dimensional subalgebra of infinite dimensional algebra for system (3.1). The similarity variables and functions associated with these translation symmetries are

$$r = \tau,\ u = \hat{u}(\tau),\ v = \hat{v}(\tau),\ p = \hat{p}(\tau) + l(t), \tag{4.1}$$

where $l(t) = \int j(t)dt$. In these variables, system (3.1) reduces to the following system of ordinary differential equations

$$\begin{aligned}
&\hat{u}\frac{d\hat{u}}{d\tau} - \frac{\hat{v}^2}{\tau} = -\frac{d\hat{p}}{d\tau} + \eta\left(\frac{d^2\hat{u}}{d\tau^2} + \frac{1}{\tau}\frac{d\hat{u}}{d\tau} - \frac{\hat{u}}{\tau^2}\right),\\
&\hat{u}\frac{d\hat{v}}{d\tau} - \hat{v}\frac{\hat{u}}{\tau} = \eta\left(\frac{d^2\hat{v}}{d\tau^2} + \frac{1}{\tau}\frac{d\hat{v}}{d\tau} - \frac{\hat{v}}{\tau^2}\right),\\
&\frac{d\hat{u}}{d\tau} + \frac{\hat{u}}{\tau} = 0.
\end{aligned} \tag{4.2}$$

Solving (4.2) and then using (4.1), we get the solution of (3.1) in the form

$$\begin{aligned}
&u = \frac{c}{r},\\
&v = c_1 r^{\frac{(c-\eta)}{\eta}} + c_2 r,\\
&p = l(t) + \frac{c_2^2}{2}r^2 + \frac{\eta c_1^2}{2(c-\eta)}r^{\frac{2(c-\eta)}{\eta}} + \frac{2\eta c_1 c_2}{c}r^{\frac{c}{\eta}} - \frac{c^2}{2r^2} + c_3,
\end{aligned}$$

where c_1, c_2 and c_3 are arbitrary constants.

4.2. Second Solution

Now we consider a linear combination of the scaling symmetry and $\partial_{j(t)}$. The similarity variables and functions associated with this combination of symmetries are

$$\theta = \xi,\ u = \frac{1}{r}\breve{u}(\xi),\ v = \frac{1}{r}\breve{v}(\xi),\ p = \frac{1}{r^2}(\breve{p}(\xi) + l(t)). \tag{4.3}$$

Invoking the new variables above, (3.1) yields

$$\begin{aligned}
&-\breve{u}^2 + \breve{v}\frac{d\breve{u}}{d\xi} - \breve{v}^2 = 2\breve{p} + \eta\left(\frac{d^2\breve{u}}{d\xi^2} - 2\frac{d\breve{v}}{d\xi}\right),\\
&-2\breve{u}\breve{v} + \breve{u}\frac{d\breve{v}}{d\xi} = -\frac{d\breve{p}}{d\xi} + \eta\left(\frac{d^2\breve{v}}{d\xi^2} + 2\frac{d\breve{u}}{d\xi}\right),\\
&\frac{1}{r^2}\frac{d\breve{v}}{d\xi} = 0
\end{aligned} \tag{4.4}$$

which leads to the following equations

$$\breve{v} = b, \tag{4.5}$$

$$\eta\frac{d^2\breve{u}}{d\xi^2} - b\frac{d\breve{u}}{d\xi} + \breve{u}^2 + b^2 + 2\breve{p} = 0, \tag{4.6}$$

$$\frac{d\breve{p}}{d\xi} = 2b\breve{u} + 2\eta\frac{d\breve{u}}{d\xi}. \tag{4.7}$$

By using (4.6) in (4.7) with $k = \frac{1}{\eta}$ leads to the following Cauchy problem

$$\frac{d^3\breve{u}}{d\xi^3} - 4bk\frac{d^2\breve{u}}{d\xi^2} + 2(k\breve{u} + 2)\frac{d\breve{u}}{d\xi} + 4bk\breve{u} = 0, \tag{4.8}$$

$$\breve{u}(\xi = 0) = 1,\ \frac{d\breve{u}}{d\xi}(\xi = 0) = 0,\ \frac{d^2\breve{u}}{d\xi^2}(\xi = 0) = 0. \tag{4.9}$$

The second solutions for u, $v = b$ and p with $r = 1$ along with different values of viscosity are plotted in Figs. 1 and 2. These Figs. indicate monotonic behavior of u and p when η is varying.

Special Case
It is noted that for $b = 0$ the θ-component of velocity $\breve{v} = 0$ and therefore (4.8) reduces to

$$\frac{d^3\breve{u}}{d\xi^3} + 2k\breve{u}\frac{d\breve{u}}{d\xi} + 4\frac{d\breve{u}}{d\xi} = 0,$$

which is a type of generalized stationary KDV equation and hence integrable.

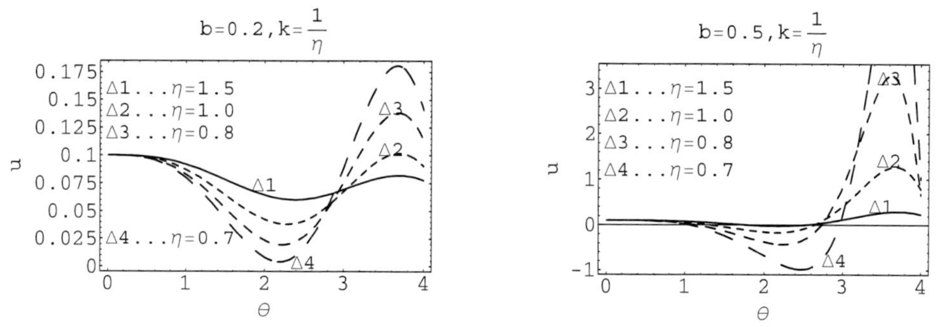

(i) (ii)

Figure 1. Plot of r-component of velocity u for $b = 0.2,\ 0.5$.

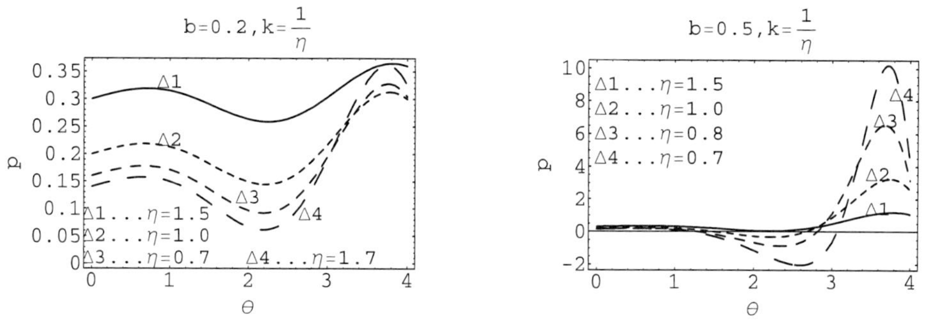

(i) (ii)

Figure 2. Plot of pressure gradient p for $b = 0.2,\ 0.5$.

5. Discussions and Comments on Future Work

In this chapter, we have investigated the unsteady Navier-Stokes in cylindrical polar and polar coordinates. Lie group theory is applied to the equations of motion. A direct approach has been utilized to obtain the symmetries of the groups. It is observed that the direct approach is more convenient, straightforward and involves less calculation as compared with the computer packages or other symbolic manipulation systems available for calculation of the symmetries of a system of differential equations. Using symmetries obtained through the direct approach, a similarity reduction of the governing nonlinear partial differential equations to ordinary differential equations was achieved. As the resulting ordinary differential equations are more involved, only a series type of approximate solution is constructed in the cylindrical polar case. From the solution (2.24) we observed that the velocity is independent of time whereas the generalized pressure depends upon time. Also the absence of θ in the solution indicates that the velocity is symmetric about the z-axis. Further the magnitude of the velocity $\rightarrow$ zero or ∞ as $r \rightarrow \infty$ or 0. Also the magnitude will increase with increase in z. We also noticed that when $r \rightarrow \infty$ the ratio of components v_r and v_θ , $\frac{v_r}{v_\theta} \rightarrow \frac{a_0}{c}$(constant) and $\frac{v_r}{v_\theta} \rightarrow \infty$ when $r \rightarrow 0$. The streams lines are a type of helix start from (or end to) origin with decreasing pitch and radius as $z \rightarrow +0$. Further when $r \gg 1$, the asymptotes of the lines are a circular cone helixes. For the polar case, we have obtained two similarity solutions. In the second solution, an initial value problem is reduced into a generalized stationary KDV and hence integrable.

At present, we only obtained the exact solution through basic transformations as our main aim was to highlight the effectiveness of a direct approach to obtain symmetries. The similarity reduction through other subalgebras (finite and infinite dimensions) is left for future research. Furthermore, it is worth mentioning that the present technique provides an alternate to the symmetry analysis performed by various researchers in the field. In this chapter, we put forward the analysis of the analytical solutions of the time-dependent Navier-Stokes equations. In view of the paucity of the analytic solutions this chapter seems a reasonable contribution.

References

[1] J. Yan, F. Thiele, L. Xue, A modified full multigrid algorithm for the NavierStokes equations, *Computers & Fluids*, **36**(2007), 445–454.

[2] F. Ghadi, V. Ruas, M. Wakrim, Numerical solution of the time-dependent incompressible NavierStokes equations by piecewise linear finite elements, *Journal of Computational and Applied Mathematics*, **215**(2008), 429–437.

[3] Y. He and A. Wang, A simplified two-level method for the steady NavierStokes equations, *Computer Methods in Applied Mechanics and Engineering*, **197**(2008), 1568–1576.

[4] J.C. Kalita, A.K. Dass and N. Nidhi, An efficient transient NavierStokes solver on compact nonuniform space grids, *Journal of Computational and Applied Mathematics*, **214**(2008), 148–162.

[5] J. Frochte, W. Heinrichs, A splitting technique of higher order for the Navier-Stokes equations, *Journal of Computational and Applied Mathematics*, **228**(2009), 373–390.

[6] S. Deparis, G. Rozza, Reduced basis method for multi-parameter-dependent steady NavierStokes equations: Applications to natural convection in a cavity, *Journal of Computational Physics*, **228**(2009), 4359–4378.

[7] K. Shahbazi, D.J. Mavriplis, N.K. Burgess, Multigrid algorithms for high-order discontinuous Galerkin discretizations of the compressible NavierStokes equations, *Journal of Computational Physics*, **228**(2009), 7917–7940.

[8] R.E. Boisvert, W.F. Ames, U.N. Srivastava, Group analysis and new solutions of Navier-Stokes equations, *Journal of Engineering Mathematics*, **17**(1983), 203–221.

[9] D.K. Ludlow, P.A. Clarkson and A.P. Bassom, Nonclassical symmetry reductions of the three-dimensional incompressible Navier-Stokes equations, *Journal of Physics A: Mathematical and General*, **31**(1998), 7965–7980.

[10] D.K. Ludlow, P.A. Clarkson and A.P. Bassom, Nonclassical symmetry reductions of the two-dimensional incompressible Navier-Stokes equations, *Studies in Applied Mathematics*, **103**(1999), 183–240.

[11] S.V. Meleshko, A particular class of partially invariant solutions of the Navier-Stokes equations, *Nonlinear Dynamics*, **36**(2004), 47–68.

[12] K. Thailert, One class of regular partially invariant solutions of the Navier-Stokes equations, *Nonlinear Dynamics*, **43**(2006), 343–364.

[13] K. Fakhar, Chen Zu-Chi, Ji Xiaoda, Cheng Yi, *Engineering Computations: International Journal for Computer-Aided Engineering and Software* , **23**(2006), 632–643.

[14] K. Fakhar, T. Hayat, Cheng Yi, Kun Zhao, Symmetry transformation of solutions for Navier-Stokes Equations, *Applied Mathematics & Computations*, **207**(2009), 213–224.

[15] K. Fakhar, T. Hayat, Cheng Yi, N. Amin, A note on the similarity solutions of the Navier-Stokes equations, *Communications in Theoretical Physics*, **53**(2010), 575–578.

[16] G. Nugroho, A.M.S. Ali, Z.A.A. Karim, A class of exact solutions to the three-dimensional incompressible Navier-Stokes equations, *Applied Mathematics Letters*, **23**(2010), 1388–1396.

[17] L.V. Ovsiannikov1982, *Group Analysis of Differential Equations* , Academic Press, New York.

[18] P.J. Olver 1986, *Application of Lie Groups to Differential Equations* , Springer, Berlin.

[19] G.W. Bluman, S. Kumei 1989, *Symmetries and Differential Equations* , Springer, New York.

[20] N.H. Ibragimov 1999, *Elementrary Lie Group Analysis and Ordinary Differential Equations*, John Wiley & Sons, New York.

[21] L.O. Landau and E.M. Lifshitz 1978, *Fluid Mechanics*, Pergamon Press, New York.

[22] E.A. Coddington and S. Levison 1954, *Theory of Ordinary Differential Equations* , Mc Graw-Hill, New York.

In: Navier-Stokes Equations
Editor: R. Younsi

ISBN: 978-1-61324-590-3

Chapter 7

HIGH ORDER SHOCK CAPTURING SCHEMES FOR NAVIER-STOKES EQUATIONS

Yiqing Shen * ***and Gecheng Zha*** †
Dept. of Mechanical and Aerospace Engineering
University of Miami
Coral Gables, Florida, U. S.

Abstract

High order low diffusion numerical schemes with the capability of shock and contact discontinuities capturing is essential to study high speed flows, turbulence, acoustics, fluid-structural interactions, combustions, etc. Developing these schemes is numerically challenging. This paper introduces several recent studies of high order accuracy shock capturing numerical algorithms for Navier-Stokes equations. It includes implicit time marching algorithms, high order weighted essentially non-oscillatory(WENO) schemes for the inviscid fluxes, high order conservative schemes for the viscous terms, and preconditioning methods for solving unified compressible and incompressible flows at all speeds.

This paper is organized as follows. Section 1 gives the full Navier-Stokes equations in the Cartesian coordinates and the generalized computational coordinates, including the form of Reynolds averaged Navier-Stokes equations and spatially filtered Navier-Stokes equations for large eddy simulation. Section 2 presents the study on implicit time marching algorithms, a comparison of three different methods including the unfactored implicit Gauss-Seidel relaxation scheme, the lower-upper symmetric Gauss-Seidel method, and a new hybrid method. Section 3 describes an improvement for the fifth order WENO schemes near shock points, a generalized finite compact scheme for shock capturing, an improved seventh order WENO scheme. Section 4 introduces the high order conservative central difference schemes for viscous terms and applications. Section 5 depicts the preconditioning methods(Roe-type method and E-CUSP scheme) for solving unified compressible and incompressible flows. Each section has its independent sub-section of introduction, results, and conclusions.

* E-mail address: yqshen@miami.edu
† E-mail address: gzha@miami.edu

1. Governing Equations

The normalized Navier-Stokes equations governing compressible viscous flows can be written in the Cartesian coordinates as:

$$\frac{\partial Q}{\partial t}+\frac{\partial E}{\partial x}+\frac{\partial F}{\partial y}+\frac{\partial G}{\partial z}=\frac{1}{Re}\left(\frac{\partial R}{\partial x}+\frac{\partial S}{\partial y}+\frac{\partial T}{\partial z}\right) \tag{1}$$

where

$$Q=\begin{bmatrix}\rho\\ \rho u\\ \rho v\\ \rho w\\ \rho e\end{bmatrix}, E=\begin{bmatrix}\rho u\\ \rho u^2+p\\ \rho uv\\ \rho uw\\ (\rho e+p)u\end{bmatrix}, F=\begin{bmatrix}\rho v\\ \rho uv\\ \rho v^2+p\\ \rho vw\\ (\rho e+p)v\end{bmatrix}, G=\begin{bmatrix}\rho w\\ \rho uw\\ \rho vw\\ \rho w^2+p\\ (\rho e+p)w\end{bmatrix},$$

$$R=\begin{bmatrix}0\\ \tau_{xx}\\ \tau_{xy}\\ \tau_{xz}\\ u_k\tau_{xk}-q_x\end{bmatrix}, S=\begin{bmatrix}0\\ \tau_{xy}\\ \tau_{yy}\\ \tau_{yz}\\ u_k\tau_{yk}-q_y\end{bmatrix}, T=\begin{bmatrix}0\\ \tau_{xz}\\ \tau_{yz}\\ \tau_{zz}\\ u_k\tau_{zk}-q_z\end{bmatrix},$$

The repeated index k stands for the Einstein summation over x, y and z. The stress τ and heat flux q are,

$$\tau_{ik}=\mu\left[\left(\frac{\partial u_i}{\partial x_k}+\frac{\partial u_k}{\partial x_i}\right)-\frac{2}{3}\delta_{ik}\frac{\partial u_j}{\partial x_j}\right]$$

$$q_j=\frac{-\mu}{(\gamma-1)M_\infty^2 Pr}\frac{\partial T}{\partial x_j}$$

The equation of state is

$$\rho e=\frac{p}{\gamma-1}+\frac{1}{2}\rho(u^2+v^2+w^2)$$

In the above equations, ρ is density, u, v, and w are the Cartesian velocity components in x, y and z directions, p is static pressure, and e is total energy per unit mass, μ is molecular viscosity, γ, Re, M_∞, Pr and Pr_t are the ratio of specific heat, Reynolds number, freestream Mach number, Prandtl number and turbulent Prandtl number, respectively.

The dimensionless flow variables in the governing equations are defined as the following,

$$x^*=\frac{x}{L}, y^*=\frac{y}{L}, z^*=\frac{z}{L},$$

$$u^*=\frac{u}{U_\infty}, v^*=\frac{v}{U_\infty}, w^*=\frac{w}{U_\infty},$$

$$\rho^*=\frac{\rho}{\rho_\infty}, \mu^*=\frac{\mu}{\mu_\infty}, t^*=\frac{t}{L/U_\infty},$$

$$T^*=\frac{T}{T_\infty}, p^*=\frac{p}{\rho_\infty U_\infty^2}, e^*=\frac{e}{U_\infty^2}, \mu^*=\frac{\mu}{\mu_\infty}$$

where L is the reference length, the free stream conditions are denoted by the subscript ∞. For simplicity, the subscript * is omitted in Eq.(1).

In the generalized computational coordinates, Eq.(1) can be written as:

$$\frac{\partial Q'}{\partial t} + \frac{\partial E'}{\partial \xi} + \frac{\partial F'}{\partial \eta} + \frac{\partial G'}{\partial \zeta} = \frac{1}{Re}\left(\frac{\partial R'}{\partial \xi} + \frac{\partial S'}{\partial \eta} + \frac{\partial T'}{\partial \zeta}\right) \tag{2}$$

where,

$$Q' = \frac{1}{J}Q,$$

$$E' = \frac{1}{J}(\xi_t Q + \xi_x E + \xi_y F + \xi_z G),$$

$$F' = \frac{1}{J}(\eta_t Q + \eta_x E + \eta_y F + \eta_z G),$$

$$G' = \frac{1}{J}(\zeta_t Q + \zeta_x E + \zeta_y F + \zeta_z G),$$

$$R' = \frac{1}{J}(\xi_x R + \xi_y S + \xi_z T),$$

$$S' = \frac{1}{J}(\eta_x R + \eta_y S + \eta_z T),$$

$$T' = \frac{1}{J}(\zeta_x R + \zeta_y S + \zeta_z T),$$

where J is the transformation Jacobian. The inviscid flux, for example E', can be written as

$$E'(\mathbf{n}) = \frac{1}{J}\left\{\begin{array}{l} \rho U \\ \rho u U + \xi_x p \\ \rho v U + \xi_y p \\ \rho w U + \xi_z p \\ (\rho e + p)U \end{array}\right\},$$

where $U = \xi_t + \xi_x + \xi_y v + \xi_z w$. For simplicity, the prime $'$ in Eq.(2) will be omitted in the rest of this paper.

1.1. Reynolds Averaged Navier-Stokes Equations [1]

The Reynolds averaged Navier-Stokes equations (RANS equations) are time-averaged equations of motion for fluid flow.

For an arbitrary function $f(x_i, t)$, the time-averaged variable $\bar{f}(x_i, t)$ is defined as

$$\bar{f}(x_i, t) = \frac{1}{T}\int_{-T/2}^{T/2} f(x_i, t+\tau)d\tau \tag{3}$$

where T is to be chosen large enough compared to the period of the random fluctuations associated with the turbulence, but small with respect to the time constant for any slow variations in the flow field associated with ordinary unsteady flows.

If the flow variables are replaced by time averages plus fluctuations, for example,

$$u = \bar{u} + u'$$

then, the time average of the fluctuating quantity is zero,

$$\overline{u'} = \frac{1}{T}\int_{-T/2}^{T/2} u' d\tau = 0 \tag{4}$$

For compressible flows, it is convenient to introduce a mass-weighted average(here, only for velocity components and thermal variables) as

$$\bar{u} = \frac{\overline{\rho u}}{\bar{\rho}} \tag{5}$$

To substitute into the Eqs. (1), the new fluctuating quantity is defined by

$$f = \bar{f} + f'' \tag{6}$$

Then the entire equation (1) is time averaged, yields the same form as (1) with

$$\mathbf{Q} = \begin{pmatrix} \bar{\rho} \\ \bar{\rho}\bar{u} \\ \bar{\rho}\bar{v} \\ \bar{\rho}\bar{w} \\ \bar{\rho}\bar{e} \end{pmatrix}, \mathbf{F} = \begin{pmatrix} \bar{\rho}\bar{u} \\ \bar{\rho}\bar{u}^2 + \bar{p} \\ \bar{\rho}\bar{u}\bar{v} \\ \bar{\rho}\bar{u}\bar{w} \\ (\bar{\rho}\bar{e} + \bar{p})\bar{u} \end{pmatrix}, \mathbf{G} = \begin{pmatrix} \bar{\rho}\bar{v} \\ \bar{\rho}\bar{v}\bar{u} \\ \bar{\rho}\bar{v}^2 + \bar{p} \\ \bar{\rho}\bar{v}\bar{w} \\ (\bar{\rho}\bar{e} + \bar{p})\bar{v} \end{pmatrix}, \mathbf{H} = \begin{pmatrix} \bar{\rho}\bar{w} \\ \bar{\rho}\bar{w}\bar{u} \\ \bar{\rho}\bar{w}\bar{v} \\ \bar{\rho}\bar{w}^2 + \bar{p} \\ (\bar{\rho}\bar{e} + \bar{p})\bar{w} \end{pmatrix}$$

$$\mathbf{R} = \begin{pmatrix} 0 \\ \bar{\tau}_{xx} + \sigma_{xx} \\ \bar{\tau}_{xy} + \sigma_{xy} \\ \bar{\tau}_{xz} + \sigma_{xz} \\ Q_x \end{pmatrix}, \mathbf{S} = \begin{pmatrix} 0 \\ \bar{\tau}_{yx} + \sigma_{yx} \\ \bar{\tau}_{yy} + \sigma_{yy} \\ \bar{\tau}_{yz} + \sigma_{yz} \\ Q_y \end{pmatrix}, \mathbf{T} = \begin{pmatrix} 0 \\ \bar{\tau}_{zx} + \sigma_{zx} \\ \bar{\tau}_{zy} + \sigma_{zy} \\ \bar{\tau}_{zz} + \sigma_{zz} \\ Q_z \end{pmatrix},$$

where, $\bar{\tau}_{ij}$ is the same as in Eq. (1) and has the same form in terms of the time-averaged velocity. σ_{ij} is the Reynolds stresses,

$$\sigma_{ij} = -\overline{\rho u_i'' u_j''} + \mu\left[\left(\frac{\partial \overline{u_i''}}{\partial x_j} + \frac{\partial \overline{u_j''}}{\partial x_i}\right) - \frac{2}{3}\delta_{ij}\frac{\partial \overline{u_k''}}{\partial x_k}\right]. \tag{7}$$

The energy flux Q is expressed as

$$Q_i = \bar{u}_j(\bar{\tau}_{ij} + \sigma_{ij}) - \bar{q}_i + \Phi_i \tag{8}$$

where

$$\Phi_i = -\bar{\rho}C_p\overline{T'u_i'} - C_p\overline{\rho'T'u_i'} - \bar{u}_iC_p\overline{\rho'T'} \tag{9}$$

After a closure equation, which provides the additional Reynolds stresses and heat flux quantities, is introduced, the RANS equations can be written as the same form of Eqs. (1), but the viscosity μ in the stress terms and the term (μ/Pr) in the heat conduction terms are modeled as

$$\begin{cases} \mu = \mu_l + \mu_t \\ \dfrac{\mu}{Pr} = \dfrac{\mu_l}{Pr_l} + \dfrac{\mu_t}{Pr_t} \end{cases} \tag{10}$$

where the subscripts l and t represent laminar and turbulent contributions, respectively. Pr_t is the turbulent Prandtl number. μ_t is the turbulent viscosity calculated by the closure turbulent model. There are various models for RANS equations, for example, Baldwin-Lomax(BL) model[2], Spalart-Allmaras(SA) model[3], two-equation models[4, 5, 6]. In our research, the BL and SA models are employed.

1.2. Governing Equations for Large Eddy Simulation

For the large eddy simulation, the Eqs. (1) are spatially filtered. The spatial filtering removes the small scale high frequency components of the fluid motion, while keeping the unsteadiness associated with the large scale turbulent motion.

For an arbitrary function $f(x_i, t)$, the filtered variable $\bar{f}(x_i, t)$ is defined as

$$\bar{f}(x_i, t) = \int_D G(x_i - \xi_i, \Delta) f(\xi_i, t) d\xi_i \tag{11}$$

where G is the filter function and Δ is the filter width and is associated with the mesh size.

Similar to the cases of RANS, for compressible flows, it is convenient to introduce the Farve-filtered variable $\tilde{u}(x_i, t)$ as

$$\tilde{u}(x_i, t) = \frac{\overline{\rho u}}{\bar{\rho}} \tag{12}$$

A variable can be thus decomposed into its Favre-filtered component and fluctuating component as

$$u(x_i, t) = \tilde{u}(x_i, t) + u''(x_i, t) \tag{13}$$

Applying these definitions and following the derivation of Knight et al.[7], the filtered compressible Navier-Stokes equations in Cartesian coordinates can be expressed as the same Eqs. (1) with

$$\mathbf{Q} = \begin{pmatrix} \bar{\rho} \\ \bar{\rho}\tilde{u} \\ \bar{\rho}\tilde{v} \\ \bar{\rho}\tilde{w} \\ \bar{\rho}\tilde{e} \end{pmatrix}, \mathbf{F} = \begin{pmatrix} \bar{\rho}\tilde{u} \\ \bar{\rho}\tilde{u}^2 + \bar{p} \\ \bar{\rho}\tilde{u}\tilde{v} \\ \bar{\rho}\tilde{u}\tilde{w} \\ (\bar{\rho}\tilde{e} + \bar{p})\tilde{u} \end{pmatrix}, \mathbf{G} = \begin{pmatrix} \bar{\rho}\tilde{v} \\ \bar{\rho}\tilde{v}\tilde{u} \\ \bar{\rho}\tilde{v}^2 + \bar{p} \\ \bar{\rho}\tilde{v}\tilde{w} \\ (\bar{\rho}\tilde{e} + \bar{p})\tilde{v} \end{pmatrix}, \mathbf{H} = \begin{pmatrix} \bar{\rho}\tilde{w} \\ \bar{\rho}\tilde{w}\tilde{u} \\ \bar{\rho}\tilde{w}\tilde{v} \\ \bar{\rho}\tilde{w}^2 + \bar{p} \\ (\bar{\rho}\tilde{e} + \bar{p})\tilde{w} \end{pmatrix}$$

$$\mathbf{R} = \begin{pmatrix} 0 \\ \bar{\tau}_{xx} + \sigma_{xx} \\ \bar{\tau}_{xy} + \sigma_{xy} \\ \bar{\tau}_{xz} + \sigma_{xz} \\ Q_x \end{pmatrix}, \mathbf{S} = \begin{pmatrix} 0 \\ \bar{\tau}_{yx} + \sigma_{yx} \\ \bar{\tau}_{yy} + \sigma_{yy} \\ \bar{\tau}_{yz} + \sigma_{yz} \\ Q_y \end{pmatrix}, \mathbf{T} = \begin{pmatrix} 0 \\ \bar{\tau}_{zx} + \sigma_{zx} \\ \bar{\tau}_{zy} + \sigma_{zy} \\ \bar{\tau}_{zz} + \sigma_{zz} \\ Q_z \end{pmatrix},$$

The overbar denotes a regular filtered variable as given in Eq. (11), and the tilde is used to denote the Favre filtered variable defined in Eq. (12).

The $\bar{\tau}$ is the molecular viscous stress tensor. The σ is the subgrid scale stress tensor due to the filtering process and is expressed as

$$\sigma_{ij} = -\bar{\rho}(\widetilde{u_i u_j} - \tilde{u}_i\tilde{u}_j) \tag{14}$$

The energy flux Q is expressed as

$$Q_i = \tilde{u}_j(\bar{\tau}_{ij} + \sigma_{ij}) - \bar{q}_i + \Phi_i \tag{15}$$

where Φ is the subscale heat flux

$$\Phi_i = -C_p\bar{\rho}(\widetilde{u_i T} - \tilde{u}_i\tilde{T}) \tag{16}$$

The $\bar{q}_i$ is the molecular heat flux

$$\bar{q}_i = -\frac{\tilde{\mu}}{(\gamma - 1)M^2 Pr}\frac{\partial \tilde{T}}{\partial x_i}$$

where M is the freestream Mach number.

$$\bar{\rho}\tilde{e} = \frac{\bar{p}}{(\gamma - 1)} + \frac{1}{2}\bar{\rho}(\tilde{u}^2 + \tilde{v}^2 + \tilde{w}^2) + \rho k \tag{17}$$

where γ is the ratio of specific heats, ρk is the subscale kinetic energy per unit volume.

$$\rho k = \frac{1}{2}\bar{\rho}(\widetilde{u_i u_i} - \tilde{u}_i\tilde{u}_i) = -\frac{1}{2}\sigma_{ii} \tag{18}$$

The closure of the filtered compressible Navier-Stokes equations requires a model for the subgrid scale stress σ_{ij} and heat flux Φ_i, for example, Smagorinsky SGS model[8], dynamic SGS model[9], implicit subgrid scale model[10, 11, 12].

2. Implicit Time Marching Algorithms

2.1. Introduction

The implicit methods for compressible flow calculation have been widely employed due to their less stiffness and faster convergence rate than the explicit schemes. In general, implicit methods require the inversion of a linearized system of equations. The direct inversion of the linear equations is usually preventively expensive. The implicit linear equations are therefore commonly inverted by iterative methods.

It is known that the approximately factored (AF) implicit schemes such as the Beam-Warming scheme [13] will introduce the factorization errors, which limit the size of the allowable time steps. For 3D linear wave equation, the AF scheme is also not unconditionally stable. The unfactored schemes with no factorization errors such as the line Gauss-Seidel iterations can have larger time steps with faster convergence rate than the AF methods[14, 15, 16, 17, 18, 19]. However, the unfactored schemes typically require more CPU time per iteration since the matrices are usually the full Jacobian matrices and can not be diagonalized.

The lower-upper symmetric Gauss-Seidel (LU-SGS) method suggested by Jameson and Yoon [20, 21] has been widely used due to their relatively easier implicit implementation[22, 23, 24]. The attractive feature of the LU-SGS is that the evaluation and storage of the Jacobian matrices can be eliminated by making some approximations to the implicit operator. Although the LU-SGS method could be more efficient than explicit schemes and is unconditionally stable for linear wave equation, the factorization is approximated and will necessarily introduce the factorization errors.

For the unfactored implicit Gauss-Seidel relaxation scheme used to solve the 2D incompressible Navier-Stokes equations, Rogers[25] compared the efficiency of point-Jacobi relaxation (PR), Gauss-Seidel line relaxation (GSLR), incomplete lower-upper decomposition, and the generalized minimum residual method preconditioned with each of the three other schemes. If a forward sweep plus a backward sweep is counted as one sweep, Rogers found that these methods can obtain different efficiency when the different number of the sweeps are used. For three-dimensional incompressible flows, Yuan[26] compared the efficiency of the point-Jacobi relaxation, line Gauss-Seidel relaxation, and diagonalized ADI schemes. Yuan[26] observed that the PR(2) (PR with two sweeps) is optimum in all PR(n), and GSLR(1) is optimum in all GSLR(n). For the line Gauss-Seidel relaxation methods, one can choose one or more of the coordinate directions as the sweep direction[15, 27]. For compressible flows, there is few study on how the sweep directions will affect the convergence rate and CPU time. There is also no study to compare the efficiency of the unfactored GSLR and the factored LU-SGS.

This section introduces a new implicit method, which combines the GSLR and LU-SGS and is hence named Lower Upper Gauss Seidel Line Relaxation (LU-GSLR) method. It uses the same unfactored GSLR formulation, however, all the matrices are replaced by those of LU-SGS. It is hoped that LU-GSLR can achieve the combined advantages of the high convergence rate of GSLR and the simplicity of LU-SGS.

In addition, this section applies the different implicit methods to calculate compressible flows to compare the convergence rate and CPU efficiency. This is very important to guide researchers to select the most efficient method.

Using the finite volume method, Eq.(2) is discretized into an implicit form as

$$\frac{\Delta V}{\Delta t}\Delta Q^{n+1} + (E^{n+1}_{i+\frac{1}{2}} - E^{n+1}_{i-\frac{1}{2}}) + (F^{n+1}_{j+\frac{1}{2}} - F^{n+1}_{j-\frac{1}{2}}) + (G^{n+1}_{k+\frac{1}{2}} - G^{n+1}_{k-\frac{1}{2}})$$
$$= \frac{1}{Re}[(R^{n+1}_{i+\frac{1}{2}} - R^{n+1}_{i-\frac{1}{2}}) + (S^{n+1}_{j+\frac{1}{2}} - S^{n+1}_{j-\frac{1}{2}}) + (T^{n+1}_{k+\frac{1}{2}} - T^{n+1}_{k-\frac{1}{2}})] \quad (19)$$

where Δt is the time interval, ΔV is the volume of control cell.

Even though this paper is focused on high order shock capturing schemes, this section describing the implicit schemes only uses the 2nd order scheme. The reason is that a common implicit solver is utilized in this numerical algorithm regardless the accuracy order of the right hand side of the discretized Navier-Stokes equations. That is the high order schemes described in the later sections use the same implicit solver given in this section.

2.2. Implicit Discretization

For different discretization scheme, the linearized matrices are different. In this section, the Roe scheme[28] is used. The interface flux is evaluated as:

$$E_{i+\frac{1}{2}} = \frac{1}{2}[E_L + E_R - \tilde{A}(Q_R - Q_L)]_{i+\frac{1}{2}} \quad (20)$$

The fluxes E_L (E_R is similar) on time level $n+1$ can be expressed as,

$$E_L^{n+1} = E_L^n + \frac{\partial E}{\partial Q}|_L^n \Delta Q_L^{n+1} = E_L + A_{i+\frac{1}{2}}^L \Delta Q_L^{n+1} \tag{21}$$

and

$$\tilde{A}(Q_R - Q_L)^{n+1} = \tilde{A}(Q_R - Q_L)^n + \tilde{A}^n(\Delta Q_R^{n+1} - \Delta Q_L^{n+1}) \tag{22}$$

To enhance the diagonal dominance, the first order scheme is used for the implicit terms of the inviscid fluxes. That is:

$$A_{i+1/2}^L \Delta Q_L^{n+1} = A_i^L \Delta Q_i^{n+1}, A_{i+1/2}^R \Delta Q_R^{n+1} = A_{i+1}^R \Delta Q_{i+1}^{n+1} \tag{23}$$

then,

$$\begin{aligned} E_{i+\frac{1}{2}}^{n+1} - E_{i-\frac{1}{2}}^{n+1} = (E_{i+\frac{1}{2}}^n - E_{i-\frac{1}{2}}^n) + \hat{A}_{i+\frac{1}{2}}^R \Delta Q_{i+1}^{n+1} + \hat{A}_{i+\frac{1}{2}}^L \Delta Q_i^{n+1} \\ - \hat{A}_{i-\frac{1}{2}}^R \Delta Q_i^{n+1} - \hat{A}_{i-\frac{1}{2}} \Delta Q_{i-1}^{n+1} \end{aligned} \tag{24}$$

where,

$$\hat{A}_{i\pm\frac{1}{2}}^R = \frac{1}{2}(A_{i\pm\frac{1}{2}}^R - \tilde{A}_{i\pm\frac{1}{2}}), \hat{A}_{i\pm\frac{1}{2}}^L = \frac{1}{2}(A_{i\pm\frac{1}{2}}^L + \tilde{A}_{i\pm\frac{1}{2}}) \tag{25}$$

Therefore, the final implicit form can be written as,

$$\begin{aligned} [I + A + B + C]\Delta Q_{i,j,k}^{n+1} + A^+ \Delta Q_{i+1,j,k}^{n+1} + A^- \Delta Q_{i-1,j,k}^{n+1} + B^+ \Delta Q_{i,j+1,k}^{n+1} \\ + B^- \Delta Q_{i,j-1,k}^{n+1} + C^+ \Delta Q_{i,j,k+1}^{n+1} + C^- \Delta Q_{i,j,k-1}^{n+1} = RHS^n \end{aligned} \tag{26}$$

where the coefficient matrices A, $A^{\pm}$, B, $B^{\pm}$ and C, $C^{\pm}$ are defined as

$$A^+ = \frac{\Delta t}{\Delta V}(\hat{A}_{i+\frac{1}{2}}^R - L_{i+\frac{1}{2}}^R) \tag{27}$$

$$A = \frac{\Delta t}{\Delta V}[(\hat{A}_{i+\frac{1}{2}}^L - L_{i+\frac{1}{2}}^L) - (\hat{A}_{i-\frac{1}{2}}^R - L_{i-\frac{1}{2}}^R)] \tag{28}$$

$$A^- = -\frac{\Delta t}{\Delta V}(\hat{A}_{i-\frac{1}{2}}^L - L_{i-\frac{1}{2}}^L) \tag{29}$$

$$B^+ = \frac{\Delta t}{\Delta V}(\hat{B}_{j+\frac{1}{2}}^R - M_{j+\frac{1}{2}}^R) \tag{30}$$

$$B = \frac{\Delta t}{\Delta V}[(\hat{B}_{j+\frac{1}{2}}^L - M_{j+\frac{1}{2}}^L) - (\hat{B}_{j-\frac{1}{2}}^R - M_{j-\frac{1}{2}}^R)] \tag{31}$$

$$B^- = -\frac{\Delta t}{\Delta V}(\hat{B}_{j-\frac{1}{2}}^L - M_{j-\frac{1}{2}}^L) \tag{32}$$

$$C^+ = \frac{\Delta t}{\Delta V}(\hat{C}_{k+\frac{1}{2}}^R - N_{k+\frac{1}{2}}^R) \tag{33}$$

$$C = \frac{\Delta t}{\Delta V}[(\hat{C}_{k+\frac{1}{2}}^L - N_{k+\frac{1}{2}}^L) - (\hat{C}_{k-\frac{1}{2}}^R - N_{k-\frac{1}{2}}^R)] \tag{34}$$

$$C^- = -\frac{\Delta t}{\Delta V}(\hat{C}^L_{k-\frac{1}{2}} - N^L_{k-\frac{1}{2}}) \tag{35}$$

where, similar to the definition of $A_{i\pm1/2}$, the matrices $B_{j\pm1/2}$ and $C_{k\pm1/2}$ are for the inviscid fluxes F and G. The matrices $L_{i\pm1/2}$, $M_{j\pm1/2}$ and $N_{k\pm1/2}$ are the Jacobian matrices on a cell interface for the viscous fluxes R, S and T, respectively, and are obtained by using 2nd order central differencing. Superscript R and L denote the right and left side of the interface.

The RHS is the net flux going through a control volume:

$$\begin{aligned} RHS^n = \frac{\Delta t}{\Delta V}\{\frac{1}{Re}[(R^n_{i+\frac{1}{2}} - R^n_{i-\frac{1}{2}}) + (S^n_{j+\frac{1}{2}} - S^n_{j-\frac{1}{2}}) + (T^n_{k+\frac{1}{2}} - T^n_{k-\frac{1}{2}})] \\ - [(E^n_{i+\frac{1}{2}} - E^n_{i-\frac{1}{2}}) + (F^n_{j+\frac{1}{2}} - F^n_{j-\frac{1}{2}}) + (G^n_{k+\frac{1}{2}} - G^n_{k-\frac{1}{2}})]\} \end{aligned} \tag{36}$$

Eq. (26) can be written as a general semi-discretized form as:

$$[\frac{1}{\Delta t}I - (\frac{\partial RHS}{\partial Q})^n]\Delta Q^{n+1} = RHS^n, \tag{37}$$

or

$$LHS^{n+1} = RHS^n \tag{38}$$

For steady state solutions, when a solution is converged, $\delta Q^{n+1} \approx 0$. The accuracy of the solution is hence determined by the RHS. In other words, the left hand side (LHS) of Eqs. (26), (37) and (38) does not affect the accuracy of the solution. The principle of designing the LHS is hence to enhance the numerical stability. In this research, the LHS always employs first order scheme to enhance the diagonal dominance[17] no matter what order schemes are used for the RHS.

2.3. Convergence Criterion

The solution convergence is measured by the maximum of L2-Norm residuals, which is defined as:

$$\max_{i,j,k} |RHS^n|_2 \leq \varepsilon \tag{39}$$

where, ε is a small non-negative number.

When the residual is reduced to machine zero, the solution is considered as converged. For engineering problems, it is usually sufficient for the residual to be reduced by 4 order of magnitude.

2.4. Gauss-Seidel Line Relaxation(GSLR)

The Gauss-Seidel line iteration with a certain sweep direction (in this paper, one sweep is defined as a forward and a backward sweep), for example, in ξ direction with the index from small to large, can be written as

$$B^-\Delta Q^{n+1}_{i,j-1,k} + \bar{B}\Delta Q^{n+1}_{i,j,k} + B^+\Delta Q^{n+1}_{i,j+1,k} = RHS' \tag{40}$$

where,

$$\bar{B} = I + A + B + C \tag{41}$$

$$RHS' = RHS^n - A^+ \Delta Q^n_{i+1,j,k} - A^- \Delta Q^{n+1}_{i-1,j,k} - C^+ \Delta Q^n_{i,j,k+1} - C^- \Delta Q^{n+1}_{i,j,k-1} \tag{42}$$

2.5. LU-SGS Method [20, 21]

When the LU-SGS method is used, the full Jacobians of Eqs. (26) and (37) are not used. Instead, the discretization is written as

$$[I + \delta_\xi^- A^+ + \delta_\xi^+ A^- + \delta_\eta^- B^+ + \delta_\eta^+ B^- + \delta_\zeta^- C^+ + \delta_\zeta^+ C^-]\Delta Q^{n+1} = RHS^n \tag{43}$$

and,

$$\begin{aligned} A^\pm &= \tfrac{1}{2}\tfrac{\Delta t}{\Delta V}[A \pm \tilde{\rho}(A)I], \\ B^\pm &= \tfrac{1}{2}\tfrac{\Delta t}{\Delta V}[B \pm \tilde{\rho}(B)I], \\ C^\pm &= \tfrac{1}{2}\tfrac{\Delta t}{\Delta V}[C \pm \tilde{\rho}(C)I]. \end{aligned} \tag{44}$$

where $\tilde{\rho}(A) = \max[|\lambda(A)|]$ represents a spectral radius of the Jacobian Matrix A with the eigenvalues $\lambda(A)$. δ^- and δ^+ represent backward and forward differencing, respectively. If first-order one side differences are used, Eq.(43) can be approximately factored into lower, diagonal and upper matrices as:

$$LD^{-1}U\Delta Q^{n+1} = RHS^n \tag{45}$$

where

$$\begin{aligned} L &= \tilde{\rho}I - A^+_{i-1,j,k} - B^+_{i,j-1,k} - C^+_{i,j,k-1} \\ D &= \tilde{\rho}I \\ U &= \tilde{\rho}I + A^-_{i+1,j,k} + B^-_{i,j+1,k} + C^-_{i,j,k+1} \end{aligned}$$

where

$$\tilde{\rho}I = \frac{\Delta t}{\Delta V}[\tilde{\rho}(A) + \tilde{\rho}(B) + \tilde{\rho}(C)]$$

Further rearrangement leads to

$$L\Delta Q^* = RHS^n$$

$$U\Delta Q^{n+1} = D\Delta Q^*$$

or

$$\begin{aligned} \Delta Q^* &= D^{-1}[RHS^n + A_{i-1}\Delta Q^*_{i-1} + B_{j-1}\Delta Q^*_{j-1} + C_{k-1}\Delta Q^*_{k-1}] \\ \Delta Q^{n+1} &= \Delta Q^* - D^{-1}[A_{i+1}\Delta Q^{n+1}_{i+1} + B_{j+1}\Delta Q^{n+1}_{j+1} + C_{k+1}\Delta Q^{n+1}_{k+1}] \end{aligned} \tag{46}$$

In the present study, for the computation of RHS^n, the Roe scheme with the 3rd order MUSCL(monotone upstream-centered schemes for conservation laws) differencing approach [29] is used for the inviscid terms, and the second-order central differencing scheme is used for the viscous terms.

2.6. GSLR with LU-SGS Matrices[30]

A modified GSLR method, namely LU-GSLR, is proposed to replace the matrices of the unfactored line Gauss-Seidel iteration by the LU-SGS matrices. The purpose is to make use of the advantages of the high convergence rate of the unfactored line Gauss-Seidel iteration and the simplicity of the LU-SGS matrices and to hope it may achieve high CPU efficiency.

For Eqs.(40)-(42), the matrices A^+, $\cdots$, C^- and $\bar{B}$ are replaced by the following matrices based on the LU-SGS[20],

$$A^- = \frac{1}{2}\frac{\Delta t}{\Delta V}[A + \tilde{\rho}(A)I]_{i-1}$$

$$A^+ = \frac{1}{2}\frac{\Delta t}{\Delta V}[A - \tilde{\rho}(A)I]_{i+1}$$

$$B^- = \frac{1}{2}\frac{\Delta t}{\Delta V}[B + \tilde{\rho}(B)I]_{j-1}$$

$$B^+ = \frac{1}{2}\frac{\Delta t}{\Delta V}[B - \tilde{\rho}(B)I]_{j+1}$$

$$C^- = \frac{1}{2}\frac{\Delta t}{\Delta V}[C + \tilde{\rho}(C)I]_{k-1}$$

$$C^+ = \frac{1}{2}\frac{\Delta t}{\Delta V}[C - \tilde{\rho}(C)I]_{k+1}$$

and

$$\bar{B} = I + \frac{\Delta t}{\Delta V}[\tilde{\rho}(A) + \tilde{\rho}(B) + \tilde{\rho}(C)]I$$

then, the LU-GSLR method is obtained by using the process of the unfactored Gauss-Seidel line iteration Eqs.(40)-(42).

2.7. Implicit Scheme for Unsteady Calculations

The previous sections describe the implicit procedure for the steady state calculations. For unsteady simulation, the implicit method can be extended straightforwardly by using the dual time stepping method[31] with a pseudo temporal derivative term $\frac{\partial Q}{\partial \tau}$ added in the Navier-Stokes equations.

The Navier-Stokes equations becomes

$$\frac{\partial Q}{\partial \tau} + \frac{\partial \mathbf{Q}}{\partial t} + \frac{\partial \mathbf{F}}{\partial x} + \frac{\partial \mathbf{G}}{\partial y} + \frac{\partial \mathbf{H}}{\partial z} = \frac{1}{Re}\left(\frac{\partial \mathbf{R}}{\partial x} + \frac{\partial \mathbf{S}}{\partial y} + \frac{\partial \mathbf{T}}{\partial z}\right) \tag{47}$$

The physical temporal term $\frac{\partial Q}{\partial t}$ is discretized implicitly using a 2nd-order three point, backward differencing as the following

$$\frac{\partial Q}{\partial t} = \frac{3Q^{n+1} - 4Q^n + Q^{n-1}}{2\Delta t},$$

and the pseudo temporal term is discretized with first-order Euler scheme to enhance diagonal dominance. The semi-discretized implicit equation can then be expressed as

$$\left[\left(\frac{1}{\Delta\tau}+\frac{1.5}{\Delta t}\right)I-\left(\frac{\partial RHS}{\partial Q}\right)^{n+1,m}\right]\Delta Q^{n+1,m+1}$$
$$=RHS^{n+1,m}-\frac{3Q^{n+1,m}-4Q^{n}+Q^{n-1}}{2\Delta t}, \quad (48)$$

where, n stands for the physical time step and m is the pseudo time step. Within each physical time step, the solution is iterated with the pseudo time step until the residual is reduced at least by 3 order of magnitude. All the previously described iteration methods such as the Gauss-Seidel iteration can be used.

This implicit method is used for large eddy simulation[32], fluid-structural interaction[33, 34, 35].

2.8. Numerical Results

For all cases, the RHS is calculated by using Roe scheme with the 3rd order MUSCL(monotone upstream-centered schemes for conservation laws) differencing approach [29] for the inviscid terms and the second-order central differencing scheme for the viscous terms. All the cases presented in this section are steady state.

2.8.1. Subsonic Flat Plate Turbulent Boundary Layer

The first case is a subsonic flat plate turbulent boundary layer. In this case, the Baldwin-Lomax turbulence model is used. The mesh size is 180×80. The non-dimensional distance y^+ of the first cell center to the wall is kept under 0.2. The inlet Mach number is 0.5, and the Reynolds number is 4×10^6 based on the plate length. The flow is subsonic at inlet and outlet.

Fig. 1 shows that the computed velocity profile agrees well with the law of the wall. Fig. 2 shows the comparison of convergence histories for the subsonic flat plate turbulent boundary layer. GSLR can obtain the converged solution at about 1/16 of the CPU time required by LU-GSLR, and 1/26 of the CPU time required by the standard LU-SGS method.

2.8.2. Transonic Converging-Diverging Nozzle

The transonic converging-diverging nozzle is calculated to study the behavior of the Gauss-Seidel iteration for internal flows. The nozzle was designed and tested at NASA and was named as Nozzle A1[36] and is symmetric about the centerline. Hence only the upper half of nozzle is computed. The mesh size is 175×50. The grid is clustered near the wall. The inlet Mach number is 0.22. This case is calculated as inviscid flow.

Fig. 2.9. is the computed Mach contour of this flow, which shows that two shocks emanating from the walls downstream of the throat, interacting each other and reflecting on the walls. Fig. 4 is the comparison of convergence histories of the transonic converging-diverging nozzle. GSLR behaves the same as for the previous case with the highest CPU efficiency and convergence rate. However, for this case, the LU-SGS outperforms the LU-GSLR with less CPU time to converge to machine zero.

2.8.3. Transonic RAE2822 Airfoil

The steady state solution of the transonic RAE2822 airfoil is calculated using the Reynolds averaged Navier-Stokes equation with the Baldwin-Lomax turbulent model. The computational region is divided into two blocks, and the mesh size of 128×55 is used for each block, the freestream Mach number M_∞ is 0.729, the Reynolds number based on chord length is 6.5×10^6, and the angle of attack is 2.31^o.

Fig. 5 is the Mach contours, which shows the shock wave captured well. Fig. 6 shows the comparison of convergence histories for the transonic RAE2822 airfoil. Again, GSLR is the most efficient method. Before the residual reaches the level of 10^{-10}, LU-GSLR is faster than LU-SGS. For the last 3 orders of magnitude to machine zero, the convergence rate of the LU-GSLR is decreased and the LU-SGS maintains a linear convergence rate.

2.9. Conclusions

Considering the CPU time used to calculate the Jacobian matrix for implicit method is different, hence, only the CPU times are compared. All the cases tested in this section demonstrate that the GSLR method is the most efficient one. On the other hand, since more CPU time is needed for the GSLR matrix, it is clear that the total iteration step needed by GSLR is much smaller than other two methods.

It needs to point out that the comparison of the convergence behavior of the GSLR, LU-SGS and LU-GSLR in this section is for the Roe scheme only. Hence, the conclusion that the GSLR is the most efficient method may not be general. For the GSLR, the Jacobian matrix based on the Roe scheme is used. For the LU-SGS or LU-GSLR, the general Jacobian matrix suggested by Jameson [20] is used. It is almost certain that there is a matching or compatibility issue. That is when the implicit matrix based on the Roe scheme on the LHS matches the Roe scheme on the RHS such as the GSLR method, high convergence rate can be obtained. When the implicit matrix on the LHS does not match the Roe scheme on the RHS such as the LU-SGS or LU-GSLR method, the convergence rate may be slowed down. In other words, it does not rule out that a high CPU efficiency may be obtained when the LU-SGS or LU-GSLR is used with other schemes on the RHS.

3. High Order WENO Schemes

3.1. Introduction

Since the application of computational fluid dynamics becomes more and more popular, the demand on high accuracy and high efficiency CFD solutions also becomes stronger to satisfy the needs of the broad engineering problems. So far, most of the engineering applications employ the 2nd order numerical accuracy. The high order schemes (higher than 3rd order) are mostly limited to the fundamental research such as high fidelity turbulence simulation (e.g. Large Eddy Simulations and Direct Numerical Simulation), fluid-structural interaction, and aeroacoustic calculation. The reason is that the high order schemes are generally not mature enough for robust engineering applications.

For aerospace engineering applications with shock waves or contact surfaces, the essentially non-oscillatory (ENO) or weighted essentially non-oscillatory (WENO) schemes

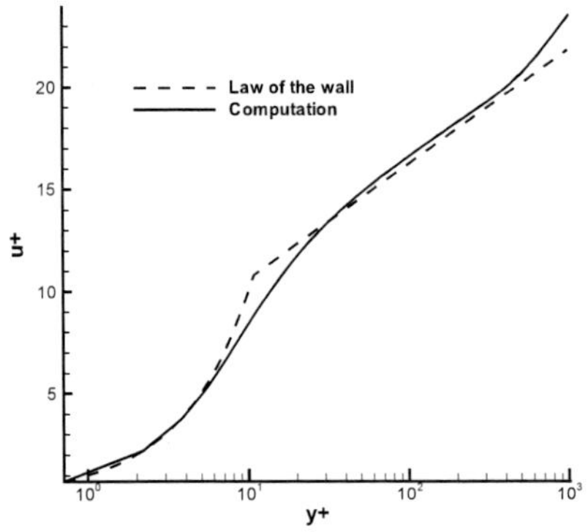

Figure 1. Computed velocity profile compared with the law of the wall for the subsonic turbulent boundary layer.

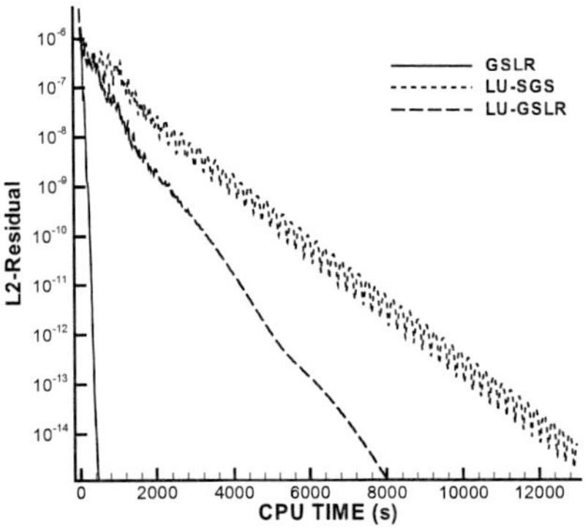

Figure 2. Comparison of L2-norm residual vs CPU time for the turbulent boundary layer.

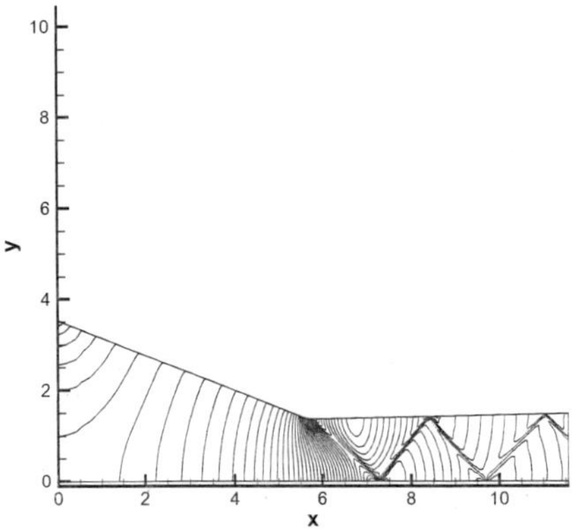

Figure 3. Mach contours of the transonic converging-diverging nozzle flow.

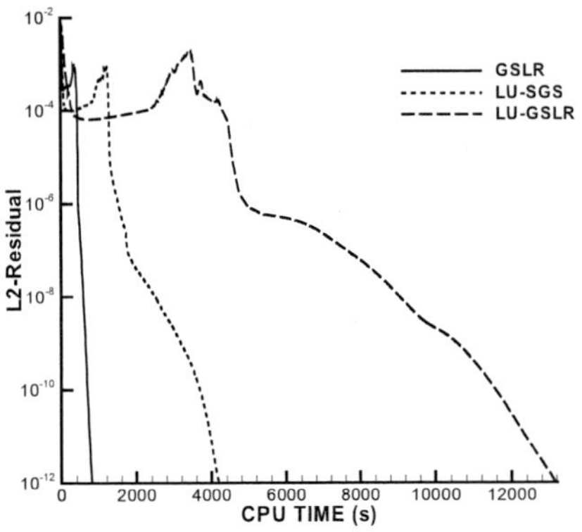

Figure 4. Comparison of L2-norm residual vs CPU time for the transonic converging-diverging nozzle flow.

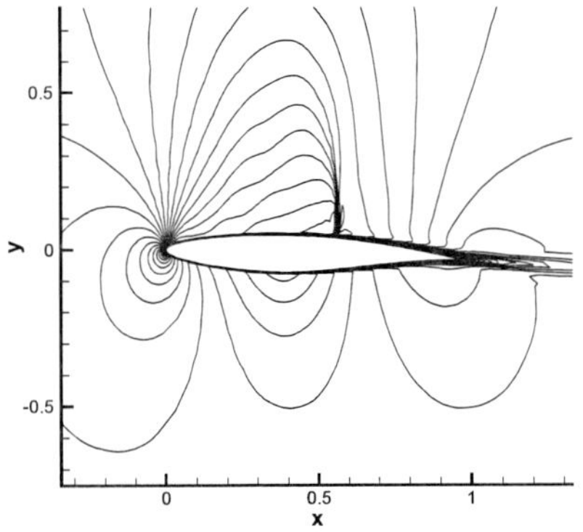

Figure 5. Mach contours of of the transonic RAE2822 airfoil.

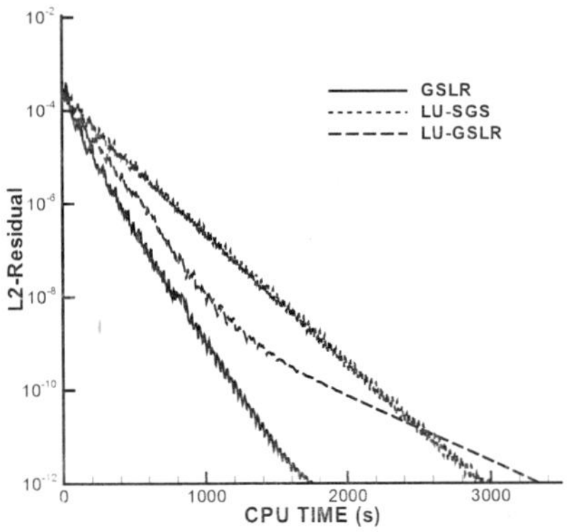

Figure 6. Comparison of L2-norm residual vs CPU time for the transonic flow of RAE2822 airfoil.

are attractive for their capability to treat the discontinuities and achieve the consistent high order accuracy in the smooth regions. By using a convex combination of all candidate stencils to replace the smoothest one in the ENO scheme, the WENO scheme has more advantages over its ENO counterpart. For example, it approaches certain optimal accuracy in smooth regions and has better convergence rate due to the smoother numerical flux used. From its appearance[37, 38] to present, the WENO schemes have been extensively applied to different flow problems in many areas.

The WENO scheme concept was first proposed by Liu et al[37] and then improved by Jiang and Shu[38]. WENO schemes are based on ENO (essentially non-oscillatory) schemes[39, 40], but use a convex combination of all candidate stencils instead of the smoothest one in the ENO schemes. The WENO schemes achieve high order accuracy in smooth regions with more compact stencil and have better convergence due to the smoother numerical flux used.

Jiang and Shu[38] analyze and modify the 5th order WENO scheme proposed by Liu et al[37] and suggest a new way of measuring the smoothness of a numerical solution. Thus a WENO scheme with the optimal $(2r-1)$th order accuracy rather than $(r+1)$th order is obtained. Henrick et al[41] pointed out that the original smoothness indicators of Jiang and Shu fail to improving the accuracy order of WENO scheme at critical points, where the first derivatives are zero. A mapping function is proposed by Henrick et al[41] to obtain the optimal order near critical points. Borges et al[42] devised a new set of WENO weights that satisfies the necessary and sufficient conditions for fifth-order convergence proposed by Henrick et al[41] and enhances the accuracy at critical points. A class of higher than 5th order weighted essentially non-oscillatory schemes are designed by Balsara and Shu in [43]. Wang and Chen [44] proposed optimized WENO schemes for linear waves with discontinuity. Martin et al[45] proposed a symmetric WENO method by means of a new candidate stencil, the new schemes are $2r$th-order accurate and symmetric, and less dissipative than Jiang and Shu's scheme.

The above mentioned WENO schemes are constructed to have $(2r-1)$th or $2r$th[45] order of accuracy in the smooth regions directly from rth order ENO schemes. For a solution containing discontinuities, these methods can not obtain the optimal accuracy near the discontinuity points. Shen et al[46] indicate that the smoothness indicator IS_k of Jiang and Shu's WENO scheme does not satisfy the condition $\beta_k = D(1+O(\Delta x^2))$ at a critical point ($f_i'=0$), and proposed a step-by-step reconstruction to avoid the strict condition.

In recent years, there are many efforts to make compact schemes possess shock-capturing capability. Cockburn and Shu[47] develop the nonlinear stable compact schemes using the TVDM(total variation diminishing in the means) property. The schemes require an implicit symmetric matrix and a reconstruction from the mean variable obtained by TVDM compact schemes. Ravichandran[48] improves this type of schemes with a class of the compact upwind schemes developed without the limitation of a symmetric matrix. Tu and Yuan[49] construct a fifth-order shock-capturing compact upwind scheme by using a characteristic-based flux splitting limited method.

Adams and Shariff[50] propose a hybrid compact-ENO scheme for shock-turbulence interaction problems. Following the same basic approach, Pirozzoli[51] derives a conservative hybrid compact-WENO scheme. Ren et al.[52] present a fifth-order conservative hybrid compact-WENO scheme for shock-capturing calculation, which is constructed through the weighted average of conservative compact scheme and WENO scheme. Zhou et al. [53] suggest a new family of high order compact upwind difference schemes, which are later made to have shock-capturing capability by combining them with WENO schemes.

Shen et al.[54] propose a finite compact scheme, which treats the discontinuity as the internal boundary and avoids the global dependence of the traditional compact schemes. Combined with the TVD of ENO limiters, a set of high resolution finite compact (FC) difference schemes with only bi-diagonal matrix inversion are constructed[54, 55].

Some of the above methods need to calculate the preliminary fluxes first by using a standard compact scheme[47, 48, 49, 50, 51]. This may result in contaminated (oscillatory) fluxes in the regions near discontinuities, and hence the compact fluxes will lose their high order accuracy in those regions. Some other methods using limiter functions will degrade the accuracy at extrema to first order[47, 48, 49, 54]. All the hybrid schemes[47, 50, 51, 52, 53, 54] introduce a free parameter to judge the flow gradient and to switch to the ENO/WENO schemes at discontinuities. Such parameters are usually problem dependent and hence lose their generality. The weighted compact nonlinear schemes proposed by Deng and Zhang[56] need to use more nodes than a standard compact scheme and hence lose the compactness. The same problem exists in the higher order extensions[57] of Deng and Zhang's method.

Artificial dissipation and compact filters are also introduced into compact schemes to help stabilize numerical solutions and reduce oscillations near discontinuities[58, 59, 60, 61, 62, 63, 64, 65]. Nonlinear characteristics-based(artificial compression method, ACM) filters is used to construct the low-dissipative high order shock-capturing scheme by Yee et al.[66] with a problem dependent parameter introduced. A WENO-type smoothness estimator is used in [67] as a sensor to switch between the high-order compact spatial filters and the ACM filters, for which a threshold parameter for the sensor is also needed.

This section will cover three areas of high order shock capturing schemes: 1) an improved fifth-order WENO scheme near the discontinuities, 2) a generalized finite compact scheme without any free parameter introduced, and 3) an improved seventh-order WENO (WENO-Z7) scheme. Numerical results are given to demonstrate the high accuracy and robustness of the schemes.

3.2. The WENO Scheme

For the conservation law,

$$\frac{\partial u}{\partial t} + \frac{\partial f}{\partial x} = 0 \tag{49}$$

The semi-discretized form of Eq. (49) by using conservative finite difference formula can be written as

$$\frac{du_i(t)}{dt} = \frac{1}{\Delta x}(h_{i+\frac{1}{2}} - h_{i-\frac{1}{2}}) \tag{50}$$

High-order WENO interpolation $\hat{f}_{i\pm\frac{1}{2}}$ approaching to $h_{i\pm\frac{1}{2}}$ are constructed through the convex combination of interpolated values $\hat{f}^k(x_i \pm \frac{1}{2})$, in which $\hat{f}^k(x)$ is the rth-degree

Table 1. Coefficients of c_{kj} and d_k

c_{kj}	j=1	j=2	j=3	j=4	d_k
k=0	-1/4	13/12	-23/12	25/12	1/35
k=1	1/12	-5/12	13/12	1/4	12/35
k=2	-1/12	7/12	7/12	-1/12	18/35
k=3	1/4	13/12	-5/12	1/12	4/35

polynomial, defined as[37, 38, 43, 42]:

$$\hat{f}_{i\pm\frac{1}{2}} = \sum_{k=0}^{r-1} \omega_k \hat{f}^k(x_{i\pm\frac{1}{2}}) \tag{51}$$

where

$$\hat{f}^k_{i+\frac{1}{2}} = \hat{f}^k(x_{i+\frac{1}{2}}) = \sum_{j=0}^{r-1} c_{kj} f_{i-k+j}, \quad i = 0, \ldots, N \tag{52}$$

The weights ω_k are defined as

$$\omega_k = \frac{\alpha_k}{\sum_{l=0}^{r-1} \alpha_l}, \quad \alpha_k = \frac{d_k}{(\beta_k + \varepsilon)^p} \tag{53}$$

The coefficient d_k are the optimal weights, which generate the $(2r-1)$th order upstream central scheme. The coefficients of c_{kj} and d_k of $r = 4$ for seventh-order WENO schemes[43] are listed in Table 1.

The smoothness indicators β_k suggested by Jiang and Shu[38] are given by

$$\beta_k = \sum_{l=1}^{r-1} \Delta x^{2l-1} \int_{x_{i-\frac{1}{2}}}^{x_{i+\frac{1}{2}}} \left(\frac{d^l}{dx^l} \hat{f}^k(x) \right)^2 dx \tag{54}$$

3.3. Conditions for (2r-1)th Order Accuracy

By applying Taylor expansions, it is easy to prove that the necessary and sufficient conditions for $(2r-1)$th order convergence are the following[41, 42, 68]

$$\sum_{k=0}^{r-1} A_k(\omega_k^+ - \omega_k^-) = O(\Delta x^r) \tag{55}$$

$$\omega_k^{\pm} - d_k = O(\Delta x^{r-1}) \tag{56}$$

where A_k is the coefficient of Taylor series of $\hat{f}^k_{i\pm1/2}$, it is independent of Δx. Superscripts $\pm$ corresponds to the $\pm$ in the $\hat{f}^k_{i\pm1/2}$.

First, it can be shown that the Taylor series expansion of (52) is

$$\hat{f}^k_{i\pm\frac{1}{2}} = h_{i\pm\frac{1}{2}} + A_k \Delta x^r + O(\Delta x^{r+1}) \tag{57}$$

Then, using (51), there is

$$\frac{\hat{f}_{i+\frac{1}{2}} - \hat{f}_{i-\frac{1}{2}}}{\Delta x} = \frac{h_{i+\frac{1}{2}} - h_{i-\frac{1}{2}}}{\Delta x} + O(\Delta x^{2r-1})$$
$$+ \frac{\sum_{k=0}^{r-1}(\omega^+ - d_k)\hat{f}^k_{i+\frac{1}{2}} - \sum_{k=0}^{r-1}(\omega^- - d_k)\hat{f}^k_{i-\frac{1}{2}}}{\Delta x}$$
$$= f'(x_i) + O(\Delta x^{2r-1}) + \Delta x^{r-1}\sum_{k=0}^{r-1} A_k(\omega^+ - \omega^-)$$
$$+ \sum_{k=0}^{r-1}\left[(\omega_k^+ - d_k)O(\Delta x^r)\right] - \sum_{k=0}^{r-1}\left[(\omega_k^- - d_k)O(\Delta x^r)\right] \tag{58}$$

Here, the condition $\sum_{k=0}^{r-1}(\omega_k^{\pm} - d_k) = 0$ or $\sum_{k=0}^{r-1}\omega_k^{\pm} = \sum_{k=0}^{r-1} d_k = 1$ is used. Eq. (58) shows that Eqs. (55) and (56) are the necessary and sufficient conditions to achieve $(2r-1)$th order convergence.

Similarly, a sufficient condition for $(2r-1)$th order of convergence is obtained as

$$\omega_k^{\pm} - d_k = O(\Delta x^r) \tag{59}$$

For the fifth-order WENO scheme, a mapping function $g_k(\omega)$ is proposed by Henrick et al[41] to make Eq. (59) be satisfied at critical points. However, if at a critical point the second derivative also vanishes, the WENO scheme of Henrick et al only has the same second order of convergence[42] as the classical WENO of Jiang and Shu[38]. Borges et al [42] used the whole 5-points stencil to devise a smoothness indicator, which has higher order than the classical smoothness indicators.

The smoothness indicators β_k^z defined by Borges et al [42] are

$$\beta_k^z = \frac{\beta_k + \varepsilon}{\beta_k + \tau_5 + \varepsilon}, \quad k = 0, 1, 2 \tag{60}$$

and the corresponding WENO weights ω_k^z are

$$\omega_k^z = \frac{\alpha_z^k}{\sum_{l=0}^{2}\alpha_l^z}, \quad \alpha_k^z = \frac{d_k}{\beta_k^z} = d_k\left[1 + \left(\frac{\tau_5}{\beta_k + \varepsilon}\right)^q\right], \quad k = 0, 1, 2 \tag{61}$$

where

$$\tau_5 = |\beta_0 - \beta_2| \tag{62}$$

The parameter ε is used to avoid the division by zero and q are chosen to increase the magnitude difference of distinct weights at non-smooth parts of the solution as p used in Eq. (53). For a smooth function, increasing the value of q makes the scheme closer to the optimal central upwind scheme. On the other hand, for a discontinuous solution, increasing q makes the scheme more dissipative[42].

The study of Borges et al [42] shows that their smoothness indicators β_k^z have higher order truncation terms than the original β_k of Jiang and Shu's[38]. The resulted WENO scheme (WENO-Z5) has higher order accuracy with less dissipation and higher resolution than Jiang and Shu's WENO scheme(WENO-JS).

From Eq.(54), there is

$$\tau_5 = \frac{13}{3}|f_i'' f_i'''|\Delta x^5 + O(\Delta x^6) \tag{63}$$

So, for the smooth solution,

$$\frac{\tau_5}{\beta_k} = O(\Delta x^3) \tag{64}$$

and from Eq.(61),

$$\omega_k^z = d_k + O(\Delta x^3) \tag{65}$$

At a critical point ($f_i' = 0$, but $f_i'' \neq 0$), the fifth order also can be achieved with $q = 2$[42].

3.4. Improvement of 5th-order WENO Scheme Near Shocks [69]

Fifth-order WENO schemes can capture shock wave and have fifth-order accuracy in smooth regions. However, because the WENO scheme is constructed directly from rth-order interpolation to achieve $(2r-1)$th-order, the accuracy is reduced at the transition point from smooth region to discontinuous point and vice versa. In order to demonstrate this problem, Fig. 7 is taken as an example.

At point $(i-1)$, the stencil $S^5_{(i-1)-1/2}$ is

$$S^5_{(i-1)-1/2} = \{x_{i-4}, x_{i-3}, x_{i-2}, x_{i-1}, x_i\} \tag{66}$$

and it is a smooth stencil, $h_{(i-1)-1/2}$ is obtained by the process of WENO-Z5 or WENO-JS as a fifth-order flux.

However, for

$$S^5_{(i-1)+1/2} = \{x_{i-3}, x_{i-2}, x_{i-1}, x_i, x_{i+1}\} \tag{67}$$

there is a discontinuity at stencil $S^3_2 = \{x_{i-1}, x_i, x_{i+1}\}$, so

$$\beta_2 >> \beta_0, \beta_1 \tag{68}$$

no matter whether WENO-Z5 or WENO-JS is used. To calculate the flux $h_{(i-1)+1/2}$ from either Eq. (53) or (61), it is easy to find

$$\omega_0 \to \frac{1}{7}, \quad \omega_1 \to \frac{6}{7}, \quad \omega_2 \to 0 \tag{69}$$

The situation at point $(i+3)$ is similar to at the point $i-1$. $S^5_{(i+3)-1/2}$ contains a discontinuity at stencil $S^3_0 = \{x_i, x_{i+1}, x_{i+2}\}$, whereas $S^5_{(i+3)+1/2} = \{x_{i+1}, x_{i+2}, x_{i+3}, x_{i+4}, x_{i+5}\}$ is a smooth stencil. For the flux $h_{(i+3)-1/2}$,

$$\omega_0 \to 0, \quad \omega_1 \to \frac{2}{3}, \quad \omega_2 \to \frac{1}{3} \tag{70}$$

Let us have a look at a numerical example of a discontinuous function[42]

$$u(0,x) = f(x) = \begin{cases} -\sin(\pi x) - \frac{1}{2}x^3, & -1 < x \le 0, \\ -\sin(\pi x) - \frac{1}{2}x^3 + 1, & 0 < x \le 1, \end{cases} \tag{71}$$

consisting of a piecewise sine function with a jump discontinuity at $x_i = 0$. The weights calculated by WENO-Z5 scheme (eq.(61)) is shown in Fig. 8, it demonstrates the accuracy degrading problem. For the flux $h_{(i-1)+1/2}$, $\omega_0 \approx \frac{1}{7}$ (point A), $\omega_1 \approx \frac{6}{7}$(point B). For $h_{(i+3)-1/2}$, $\omega_1 \approx \frac{2}{3}$ (point D), $\omega_1 \approx \frac{1}{3}$(point C).

Under the condition of $\Delta x \to 0$, there are

$$h_{(i-1)-\frac{1}{2}} = \frac{1}{30}f_{i-4} - \frac{13}{60}f_{i-3} + \frac{47}{60}f_{i-2} + \frac{9}{20}f_{i-1} - \frac{1}{20}f_i \tag{72}$$

and

$$h_{(i-1)+\frac{1}{2}} = \frac{1}{20}f_{i-3} - \frac{13}{42}f_{i-2} + \frac{41}{42}f_{i-1} + \frac{2}{7}f_i \tag{73}$$

Applying Taylor series expansion, we obtain

$$\frac{1}{\Delta x}\left(h_{(i-1)+\frac{1}{2}} - h_{(i-1)-\frac{1}{2}}\right) = f'_{i-1} + O(\Delta x^3) \tag{74}$$

The accuracy at the downstream point $(i+3)$ can be analyzed similarly.

That is, at the points (continuous points) immediately upstream or downstream of a discontinuity, all the current fifth-order WENO schemes only give third-order accuracy.

Hence, a new method is proposed to overcome the drawback of the 5th order WENO schemes mentioned above. The method combines the idea of the step-by-step construction of a higher order WENO scheme[46] and the properties of τ_5 introduced by Borges et al[42]. For completeness, two important properties of τ_5 are listed here:

(1) If the stencil S^5 does not contain discontinuities, then $\tau_5 << \beta_k$ for $k = 0, 1, 2$;

(2) if the solution is continuous at some of the stencil S_i^3, but discontinuous in the whole stencil S^5, then $\beta_i << \tau_5$.

The new method can be described using the sketch of Fig. 9.

First, the stencils S_0^4 and S_1^4 are defined as

$$\begin{cases} S_0^4 = S_0^3 \cup S_1^3 = \{x_{i-2}, x_{i-1}, x_i, x_{i+1}\}, \\ S_1^4 = S_1^3 \cup S_2^3 = \{x_{i-1}, x_i, x_{i+1}, x_{i+2}\} \end{cases} \tag{75}$$

and τ_4^0 and τ_4^1 are defined as,

$$\begin{cases} \tau_4^0 = |\beta_0 - \beta_1|, \\ \tau_4^1 = |\beta_1 - \beta_2| \end{cases} \tag{76}$$

Here, τ_4^0 and τ_4^1 have the same property(2) as τ_5, i.e., if the solution is continuous at some of the stencil S_{l+i}^3, but discontinuous in the whole S_l^4, then $\beta_{l+i} << \tau_4^l$.

Next, we will analyze the relationship between τ_4^l and β_k with the case of

(1) the solution is smooth on stencil S_0^4;

(2) the solution is discontinuous on S_1^4, i.e., the discontinuity is on (x_{i+1}, x_{i+2}).

With condition (1), there are Taylor expansions of β_0 and β_1 as following

$$\begin{cases} \beta_0 = f_i'^2\Delta x^2 + \left(\frac{13}{12}f_i''^2 - \frac{2}{3}f_i'f_i'''\right)\Delta x^4 - \left(\frac{13}{6}f_i''f_i''' - \frac{1}{2}f_i'f_i^{(4)}\right)\Delta x^5 \\ \qquad + \left(\frac{43}{36}f_i'''^2 + \frac{91}{72}f_i''f_i^{(4)} - \frac{7}{30}f_i'f_i^{(5)}\right)\Delta x^6 + O(\Delta x^7) \\ \beta_1 = f_i'^2\Delta x^2 + \left(\frac{13}{12}f_i''^2 + \frac{1}{3}f_i'f_i'''\right)\Delta x^4 \\ \qquad + \left(\frac{1}{36}f_i'''^2 + \frac{13}{72}f_i''f_i^{(4)} + \frac{1}{120}f_i'f_i^{(5)}\right)\Delta x^6 + O(\Delta x^8) \end{cases} \tag{77}$$

Hence, if $(f_i' \neq 0)$ or $(f_i' = 0$ and $f_i'' \neq 0)$, there is

$$\tau_4^0 = |\beta_0 - \beta_1| \leq \min(\beta_0, \beta_1) \tag{78}$$

With condition (2), there are

$$\beta_2 >> \beta_1, \quad \tau_1^4 = |\beta_1 - \beta_2| >> \beta_1$$

Hence, under the conditions (1) and (2), and if $(f_i' \neq 0)$ or $(f_i'' = 0$ and $f_i' \neq 0)$, there is

$$\begin{cases} \tau_4^0 \leq \min(\beta_0, \beta_1, \beta_2), \\ \tau_4^1 > \min(\beta_0, \beta_1, \beta_2) \end{cases} \tag{79}$$

The same conclusion can be drawn for the case with

(1) the solution is smooth on stencil S_1^4;

(2) the solution is discontinuous on S_0^4, i.e., the discontinuity is on (x_{i-2}, x_{i-1}).

Hence, the new method is constructed as

$$h_{i+\frac{1}{2}} = \begin{cases} h_0^4, \text{ if } \tau_4^0 \leq \min(\beta_0, \beta_1, \beta_2) \text{ and } \tau_4^1 > \min(\beta_0, \beta_1, \beta_2), \\ h_1^4, \text{ if } \tau_4^0 > \min(\beta_0, \beta_1, \beta_2) \text{ and } \tau_4^1 \leq \min(\beta_0, \beta_1, \beta_2), \\ h^{WENO-Z5}, \text{ otherwise} \end{cases} \tag{80}$$

where

$$h_0^4 = C_0^{4,0} \hat{f}_{i+1/2}^0 + C_1^{4,0} \hat{f}_{i+1/2}^1, \quad h_1^4 = C_0^{4,1} \hat{f}_{i+1/2}^1 + C_1^{4,1} \hat{f}_{i+1/2}^2 \tag{81}$$

and

$$C_0^{4,0} = \frac{1}{4}, \; C_1^{4,0} = \frac{3}{4}; \quad C_0^{4,1} = \frac{1}{2}, \; C_1^{4,1} = \frac{1}{2}$$

That is

$$\begin{cases} h_0^4 = \dfrac{1}{12} f_{i-2} - \dfrac{5}{12} f_{i-1} + \dfrac{13}{12} f_i + \dfrac{1}{4} f_{i+1} \\ h_1^4 = -\dfrac{1}{12} f_{i-1} + \dfrac{7}{12} f_i + \dfrac{7}{12} f_{i+1} - \dfrac{1}{12} f_{i+2} \end{cases}$$

For a smooth solution with three or more vanishing derivatives, the new scheme Eq. (80) switches to the 5th-order WENO-Z5 scheme.

Again, the point $(i - 1)$ in Fig. 7 is taken as an example, $S_0^4|_{i-1/2} = \{x_{i-3}, x_{i-2}, x_{i-1}, x_i\}$ is a smooth stencil, according to the properties of τ_4^l, there is

$$\tau_4^0 < \min(\beta_0, \beta_1, \beta_2) \quad \text{and} \quad \tau_4^1 >> \min(\beta_0, \beta_1, \beta_2)$$

so

$$h_{(i-1)+1/2} = h_0^4 = \frac{1}{12} f_{i-3} - \frac{5}{12} f_{i-2} + \frac{13}{12} f_{i-1} + \frac{1}{4} f_i$$

Meanwhile, $S_{i-3/2}^5$ is a smooth stencil, $h_{(i-1)-1/2}$ keeps the fifth-order flux $h_{(i-1)-1/2}^{WENO-Z5}$ (Eq.(72)). Hence, applying Taylor series expansion, there is

$$\frac{1}{\Delta x}\left(h_{(i-1)+\frac{1}{2}} - h_{(i-1)-\frac{1}{2}}\right) = f_{i-1}' + O(\Delta x^4) \tag{82}$$

Compared with the accuracy of the original WENO-Z5 or WENO-JS scheme (Eq. (74)), the new method improves one accuracy order at the point right next to the discontinuity $(i - 1)$.

3.5. The Finite Compact Differencing Scheme [70]

The general form of a compact difference scheme[71] for the derivative $f'_i = \frac{\partial f}{\partial x}|_i$ can be written as

$$\mathbf{A}f'_i = \frac{1}{\Delta x}\mathbf{B}f_i \tag{83}$$

or use the flux form:

$$\begin{cases} f'_i = \dfrac{1}{\Delta x}(\hat{h}_{i+1/2} - \hat{h}_{i-1/2}) \\ \mathbf{A}\hat{h}_{i+1/2} = \mathbf{C}f_i, \quad i = 1, 2, \ldots, N \end{cases} \tag{84}$$

where $\mathbf{A}$, $\mathbf{B}$ and $\mathbf{C}$ are the coefficient matrices depending on different requirements.

In [54, 55], the global compact difference scheme (84) is divided into multiple smooth zones by discontinuities. In those smooth zones, a bi-diagonal compact scheme avoiding flux contamination at discontinuities due to the standard compact scheme is suggested. The scheme is hence called finite compact(FC) difference scheme[54, 55].

In [70], a lemma of τ_5 is suggested and used to construct a generalized finite compact difference algorithm.

Lemma 1. *If $\tau_5 > \min(\beta_0, \beta_1, \beta_2)$, then S^5 is a non-smooth stencil, which includes the function discontinuities and derivative discontinuities up to order n provided all the derivatives lower than order n are zero.*

The procedure of the new finite compact (FC) scheme is as following:

(1) A zone containing a discontinuity is detected by using Lemma 1, and the fluxes within this zone including at the zone boundaries are calculated by a WENO scheme.

(2) those smooth point(s) between two zones containing discontinuities(or between boundary point and discontinuous interface)are defined as a compact stencil, and a compact scheme is used to calculate the numerical fluxes on the compact stencil.

For example, a tridiagonal compact scheme,

$$\alpha h_{i-1/2} + \gamma h_{i+1/2} + \beta h_{i+3/2} = d_{i+1/2}, \tag{85}$$

is solved for the flux $h_{i+1/2}$ in the compact stencil.

Note: in step (2), the fluxes on the discontinuity zones obtained from step (1) are automatically used as the internal boundary fluxes.

Fig. 10 is a sketch showing this finite compact scheme. M denotes the number of compact stencil divided by discontinuities.

As pointed in [70], for step1 and 2, any WENO and compact schemes can be used to construct the fluxes. Several results by combining the fifth-order WENO scheme (WENO-Z)[42] and the 6th-order Pade scheme[71] are given in this paper. Hence, in Eq. (85),

$$\alpha = \beta = \frac{1}{3}, \quad \gamma = 1 \tag{86}$$

and

$$d_{i+1/2} = \frac{1}{36}(f_{i+2} + 29f_{i+1} + 29f_i + f_{i-1}) \tag{87}$$

The finite compact scheme given in [54, 55] can only use the compact scheme with bi-diagonal matrix **A**(upwind-type), whereas the present generalized finite compact scheme can use any compact scheme, for example, the Pade scheme (85). Thus, with the same number of grid points, the present scheme can achieve higher order accuracy. In addition, since the lemma is used to judge the discontinuity zones, there is no free parameter introduced to switch the compact scheme and the WENO scheme.

3.6. The Seventh-order WENO Scheme

3.6.1. The Original 7th-order WENO Scheme of Balsara and Shu [43]

To obtain the 7th-order WENO scheme ($r = 4$), Balsara and Shu[43] suggest the smoothness indicators β_k:

$$\begin{aligned}
\beta_0 &= f_{i-3}(547f_{i-3} - 3882f_{i-2} + 4642f_{i-1} - 1854f_i) + f_{i-2}(7043f_{i-2} \\
&\quad - 17246f_{i-1} + 7042f_i) + f_{i-1}(11003f_{i-1} - 9402f_i) + 2107f_i^2 \\
\beta_1 &= f_{i-2}(267f_{i-2} - 1642f_{i-1} + 1602f_i - 494f_{i+1}) + f_{i-1}(2843f_{i-1} \\
&\quad - 5966f_i + 1922f_{i+1}) + f_i(3443f_i - 2522f_{i+1}) + 547f_{i+1}^2 \\
\beta_2 &= f_{i-1}(547f_{i-1} - 2522f_i + 1922f_{i+1} - 494f_{i+2}) + f_i(3443f_i - 5966f_{i+1} \\
&\quad + 1602f_{i+2}) + f_{i+1}(2843f_{i+1} - 1642f_{i+2}) + 267f_{i+2}^2 \\
\beta_3 &= f_i(2107f_i - 9402f_{i+1} + 7042f_{i+2} - 1854f_{i+3}) + f_{i+1}(11003f_{i+1} \\
&\quad - 17246f_{i+2} + 4642f_{i+3}) + f_{i+2}(7043f_{i+2} - 3882f_{i+3}) + 547f_{i+3}^2
\end{aligned} \tag{88}$$

and they can be rewritten as[72]:

$$\beta_k = \sum_{l=1}^{r-1} \gamma_l \left[\frac{1}{l!} f_i^{(l)}(i+k-r+1, \cdots, i+k)\Delta x^l \right]^2 \tag{89}$$

where

$$\gamma_1 = 240, \quad \gamma_2 = 1040, \quad \gamma_3 = 9372, \tag{90}$$

and $f_i^{(l)}(i+k-r+1, \cdots, i+k)$ denotes the differencing approximation of lth order derivative $f_i^{(l)}$ by using points $x_{i+k-r+1}, \cdots, x_{i+k}$. Because r points are used, the highest order approximation of $f_i^{(l)}$ is $(r-l)$th order interpolation.

Taylor expansion of β_k gives

$$\left\{\begin{array}{l}\beta_0 = \gamma_1[f_i'\Delta x - \frac{6}{4!}f_i^{(4)}\Delta x^4 + \frac{36}{5!}f_i^{(5)}\Delta x^5 - \frac{150}{6!}f_i^{(6)}\Delta x^6 + O(\Delta x^7)]^2 \\ \quad +\gamma_2[\frac{1}{2!}f_i''\Delta x^2 - \frac{11}{4!}f_i^{(4)}\Delta x^4 + \frac{60}{5!}f_i^{(5)}\Delta x^5 + O(\Delta x^6)]^2 \\ \quad +\gamma_3[\frac{1}{3!}f_i'''\Delta x^3 - \frac{6}{4!}f_i^{(4)}\Delta x^4 + \frac{25}{5!}f_i^{(5)}\Delta x^5 + O(\Delta x^6)]^2 \\ \beta_1 = \gamma_1[f_i'\Delta x + \frac{2}{4!}f_i^{(4)}\Delta x^4 - \frac{4}{5!}f_i^{(5)}\Delta x^5 + \frac{10}{6!}f_i^6\Delta x^6 + O(\Delta x^7)]^2 \\ \quad +\gamma_2[\frac{1}{2!}f_i''\Delta x^2 + \frac{1}{4!}f_i^{(4)}\Delta x^4 + O(\Delta x^6)]^2 \\ \quad +\gamma_3[\frac{1}{3!}f_i'''\Delta x^3 - \frac{2}{4!}f_i^{(4)}\Delta x^4 + \frac{5}{5!}f_i^{(5)}\Delta x^5 + O(\Delta x^6)]^2 \\ \beta_2 = \gamma_1[f_i'\Delta x - \frac{2}{4!}f_i^{(4)}\Delta x^4 - \frac{4}{5!}f_i^{(5)}\Delta x^5 - \frac{10}{6!}f_i^{(6)}\Delta x^6 + O(\Delta x^7)]^2 \\ \quad +\gamma_2[\frac{1}{2!}f_i''\Delta x^2 + \frac{1}{4!}f_i^{(4)}\Delta x^4 + O(\Delta x^6)]^2 \\ \quad +\gamma_3[\frac{1}{3!}f_i'''\Delta x^3 + \frac{2}{4!}f_i^{(4)}\Delta x^4 + \frac{5}{5!}f_i^{(5)}\Delta x^5 + O(\Delta x^6)]^2 \\ \beta_0 = \gamma_1[f_i'\Delta x + \frac{6}{4!}f_i^{(4)}\Delta x^4 + \frac{36}{5!}f_i^{(5)}\Delta x^5 + \frac{150}{6!}f_i^{(6)}\Delta x^6 + O(\Delta x^7)]^2 \\ \quad +\gamma_2[\frac{1}{2!}f_i''\Delta x^2 - \frac{11}{4!}f_i^{(4)}\Delta x^4 - \frac{60}{5!}f_i^{(5)}\Delta x^5 + O(\Delta x^7)]^2 \\ \quad +\gamma_3[\frac{1}{3!}f_i'''\Delta x^3 + \frac{6}{4!}f_i^{(4)}\Delta x^4 + \frac{25}{5!}f_i^{(5)}\Delta x^5 + O(\Delta x^6)]^2 \end{array}\right. \tag{91}$$

Hence, there is

$$\beta_k = D(1 + O(\Delta x^2)),\ if\ f_i' \neq 0\ or\ f_i'' \neq 0, \tag{92}$$

this results in

$$\omega_k = d_k + O(\Delta x^2),$$

It means that the original WENO scheme of Balsara and Shu[7] that intends to achieve 7th order accuracy does not satisfy the necessary and sufficient conditions (55) and (56) for seventh-order accuracy($r = 4$). In other words, the scheme is not expected to achieve 7th order accuracy. The accuracy deficiency is also shown by the numerical experiments given in this paper.

3.6.2. Improved Smoothness Indicators for 7th-order WENO Scheme [73, 74]

In [73], τ_7 is constructed by only using the first and last β_k to cover the whole stencil width. It hence achieves lower truncated order and requires higher q value, which may increase diffusion in the smooth region with high gradients. In[74], the revised version of [73] (submitted to J.Compu.Phys. in Feb. 2010, see also Castro et al.[75], J.Compu.Phys. 2011, vol. 230), a new parameter τ_7 is defined as the absolute value of a linear combination of β_k and is required to cover the whole stencil width and achieve maximum order of the truncated term,

$$\tau_7 = \left|\sum_{k=0}^{r-1} a_k\beta_k\right| \tag{93}$$

and

$$a_0 = 1,\ a_1 = 3,\ a_2 = -3,\ a_3 = -1$$

which makes τ_7 achieve the maximum truncated order of 7.

$$\tau_7 = \left|-\frac{2}{3}\gamma_1 f_i' f_i^{(6)} + \gamma_2 f_i'' f_i^{(5)} - \frac{2}{3}f_i''' f_i^{(4)}\right|\Delta x^7 + O(\Delta x^8) \tag{94}$$

And then the new 7th-order WENO weights ω_k^z is defined as

$$\omega_k^z = \frac{\alpha_k^z}{\sum_{l=0}^{3}\alpha_l^z}, \quad \alpha_k^z = d_k\left[1 + (\frac{\tau_7}{\beta_k + \varepsilon})^q\right] \tag{95}$$

3.7. The Role of ε in Practical Applications [76]

Shen et al[76] observed that, in practical applications, the mesh size Δx can not be infinite, and the flow fields do not vary uniformly. Hence, even for very smooth flow fields, there aways exist difference between the smoothness indicators β_k and disturbance in the calculation of β_k. In most of the situations, the difference or disturbance induces the weights oscillation, and results in convergence problem, and decreases the numerical accuracy. Shen et al[76] suggested an optimized ε value of 10^{-2} to remove the weights oscillation, improve convergence, and minimize the numerical dissipation.

In this paper, the studies of multi-dimensional flow applications find that the convergence stability problem of the seventh-order WENO schemes also exists. Different ε values of 10^{-6} and 10^{-2} are studied to compare the behavior of the seventh-order WENO schemes. The numerical results show that $\varepsilon = 10^{-2}$ does improve the convergence, accuracy and robustness of the WENO schemes.

However, the $\varepsilon = 10^{-2}$ is relative in scale to the normalized smoothness indicator defined as

$$\beta'_k = \beta_k/\gamma_1$$

where β_k is defined as Eq. (88) or (89), γ_1 is the coefficient of the first derivative in the Taylor expansion of smoothness indicator, for example, Eq. (89) or (91). Hence, in all calculations in this paper, β'_k is used instead of β_k.

3.8. Numerical Results

3.8.1. Accuracy Comparison of a Discontinuous Function

Table 2 gives the comparison of values and errors of WENO-Z5 scheme and the present method of first-order derivative of $f(x)$ given by Eq. (71) near the discontinuity points. For this case, $x_i = 0$ and the next point x_{i+1} are the discontinuity points. At points x_{i-1} and x_{i+3}, the present method is clearly more accurate than WENO-Z5 scheme.

3.8.2. Linear Transport Equation

The linear transport equation is used to test the accuracy of WENO schemes.

$$\frac{\partial u}{\partial t} + \frac{\partial u}{\partial x} = 0, \quad -1 < x < 1 \tag{96}$$

$$u(x,0) = u_0(x), \quad \text{periodic}$$

Three cases are calculated.

(1) First, a solution with an initial discontinuity[42] is calculated

$$u(0,x) = f(x) = \begin{cases} -\sin(\pi x) - \frac{1}{2}x^3, & -1 < x \le 0, \\ -\sin(\pi x) - \frac{1}{2}x^3 + 1, & 0 < x \le 1, \end{cases} \tag{97}$$

Table 2. Results and errors

Point	x_i	$f'(x_i)$	f_i'(WENO-Z5)	f_i'(present)	error(WENO-Z5)	error(present)
N=40	-0.2000	-.26016E+01	-.26016E+01	-.26016E+01	0.31953E-05	0.31953E-05
	-0.1500	-.28329E+01	-.28329E+01	-.28329E+01	0.25566E-05	0.25566E-05
	-0.1000	-.30028E+01	-.30028E+01	-.30028E+01	0.18554E-05	0.18554E-05
	-0.0500	-.31067E+01	-.31048E+01	-.31066E+01	**0.19045E-02**	**0.95956E-04**
	0.0000	-.31416E+01	-.31343E+01	-.31325E+01	0.72982E-02	0.91067E-02
	0.0500	-.31067E+01	0.16883E+02	0.16883E+02	0.19989E+02	0.19989E+02
	0.1000	-.30028E+01	-.29997E+01	-.30012E+01	0.31294E-02	0.15856E-02
	0.1500	-.28329E+01	-.28346E+01	-.28331E+01	**0.16706E-02**	**0.12686E-03**
	0.2000	-.26016E+01	-.26016E+01	-.26016E+01	0.26522E-05	0.26522E-05
N=80	-0.1000	-.30028E+01	-.30028E+01	-.30028E+01	0.53282E-07	0.53282E-07
	-0.0750	-.30632E+01	-.30632E+01	-.30632E+01	0.41587E-07	0.41587E-07
	-0.0500	-.31067E+01	-.31067E+01	-.31067E+01	0.29636E-07	0.29636E-07
	-0.0250	-.31328E+01	-.31323E+01	-.31328E+01	**0.52653E-03**	**0.59822E-05**
	0.0000	-.31416E+01	-.31436E+01	-.31431E+01	0.20075E-02	0.14870E-02
	0.0250	-.31328E+01	0.36868E+02	0.36868E+02	0.40000E+02	0.40000E+02
	0.0500	-.31067E+01	-.31052E+01	-.31057E+01	0.14201E-02	0.99143E-03
	0.0750	-.30632E+01	-.30637E+01	-.30632E+01	**0.43659E-03**	**0.79597E-05**
	0.1000	-.30028E+01	-.30028E+01	-.30028E+01	0.43276E-07	0.43276E-07
N=160	-0.0500	-.31067E+01	-.31067E+01	-.31067E+01	0.84590E-09	0.84590E-09
	-0.0375	-.31219E+01	-.31219E+01	-.31219E+01	0.65623E-09	0.65623E-09
	-0.0250	-.31328E+01	-.31328E+01	-.31328E+01	0.46551E-09	0.46551E-09
	-0.0125	-.31394E+01	-.31393E+01	-.31394E+01	**0.14246E-03**	**0.37364E-06**
	0.0000	-.31416E+01	-.31425E+01	-.31423E+01	0.89829E-03	0.75621E-03
	0.0125	-.31394E+01	0.76861E+02	0.76861E+02	0.80000E+02	0.80000E+02
	0.0250	-.31328E+01	-.31324E+01	-.31325E+01	0.42399E-03	0.31005E-03
	0.0375	-.31219E+01	-.31220E+01	-.31219E+01	**0.11444E-03**	**0.49793E-06**
	0.0500	-.31067E+01	-.31067E+01	-.31067E+01	0.68340E-09	0.68340E-09

Fig. 11 shows the numerical solutions at $t = 10$. It can be seen that, near the discontinuity, the present method obtains more accurate solution.

(2) And then a more complex initial solution is tested

$$u_0(x) = \begin{cases} -x\sin(3\pi x^2/2), & -1 \le x < -1/3 \\ |\sin(2\pi x)|, & -1/3 \le x \le 1/3 \\ 2x - 1 - \sin(3\pi x)/6, & \text{otherwise} \end{cases} \tag{98}$$

Fig. 12 shows the numerical solutions at $t = 6$. Again, it can be seen that the present method is more accurate.

(3) Finally, a more challenging test case that contains a smooth combination of Gaussians, a square wave, a sharp triangle wave, and a half ellipse is calculated.

$$u_0(x) = \begin{cases} \frac{1}{6}(G(x,\beta,z-\delta)+G(x,\beta,z+\delta)+4G(x,\beta,z)), & -0.8 \le x \le -0.6, \\ 1, & -0.4 \le x \le -0.2, \\ 1-|10(x-0.1)|, & 0 \le x \le 0.2, \\ \frac{1}{6}(F(x,\alpha,\alpha-\delta)+F(x,\alpha,\alpha+\delta)+4F(x,\alpha,a)), & 0.4 \le x \le 0.6, \\ 0, & \text{otherwise} \end{cases} \tag{99}$$

As in Ref.[38], the constants for this case are taken as $a = 0.5$, $z = -0.7$, $\delta = 0.005$, $\alpha = 10$, and $\beta = log2/36\delta^2$.

The results at $t = 8$ with 200 grid points are shown in Fig. 13. It can be seen that, present method can improve the accuracy not only near the discontinuities, but also for the peak of the half ellipse wave.

This example is also used to test the generalized finite compact(FC) scheme. From Fig. 14, it can be seen that, the present FC method can improve the accuracy not only for the continuous solution(Gaussians and sharp triangle wave), but also for the discontinuous waves(square wave). It needs to point out that the new method overshoots the exact solution at the tip of the half ellipse. However, with the increased grid points, the errors of both schemes at the tip of the half ellipse are reduced and the solution converges to the exact solution.

3.8.3. 1D Shock Wave Tube, Shu-Osher Problem

This problem is governed by the one-dimensional Euler equations:

$$\frac{\partial \mathbf{U}}{\partial t} + \frac{\partial \mathbf{F}}{\partial x} = 0 \tag{100}$$

where

$$\mathbf{U} = \begin{bmatrix} \rho \\ \rho u \\ \rho e \end{bmatrix}, \quad \mathbf{F} = \begin{bmatrix} \rho u \\ \rho u^2 + p \\ u(\rho e + p) \end{bmatrix}, \quad \mathbf{p = (\gamma - 1)(\rho e - \rho u^2/2)}, \quad \gamma = \mathbf{1.4}.$$

For solving the 1D Euler equations, in this paper, the first-order global Lax-Friedrichs flux(GLF), which is used as the low-order building block for the high-order reconstruction of WENO-Z5 and the present FC method for the characteristic variables[43, 52, 42, 70], is used.

The initial condition is

$$(\rho, u, p) = \begin{cases} (3.857143, 2.629369, 10.3333), & \text{when } x < -4, \\ (1 + \varepsilon \sin(5x), 0.0, 1.0), & \text{when } x \geq -4. \end{cases} \tag{101}$$

where, $\varepsilon = 0.2$. This case represents a Mach 3 shock wave interacting with a sine entropy wave[40]. The result at time $t = 1.8$ with mesh size of 300 is plotted in Fig. 15. The "exact" solutions are the numerical solutions of the original WENO-Z5 scheme with grid points of $N = 4000$. It can be seen that the present FC scheme is more accurate than the WENO-Z5 scheme.

3.8.4. 1D Shock Wave Tube, Interacting Blast Waves

The interacting blast wave example[77] is another difficult test of shock capturing schemes. The initial condition is

$$(\rho, u, p) = \begin{cases} (1, 0, 1000), & \text{if } 0 \leq x < 0.1, \\ (1, 0, 0.01), & \text{if } 0.1 \leq x < 0.9, \\ (1, 0, 100), & \text{if } 0.9 \leq x \leq 1, \end{cases} \tag{102}$$

The solutions at $t = 0.038$ are given in Fig.16. The "exact" solution is obtained using the original WENO-Z5 scheme with grid points of $N = 4000$. The various peaks and valleys are better resolved by the new scheme. The result demonstrates that the present FC scheme can capture strong shock structures very well.

3.8.5. Transonic Converging-Diverging Nozzle

To examine the performance of the new 7th order WENO scheme for capturing the weak shock waves that do not align with the mesh lines, the 2D Euler equations is solved for an inviscid transonic converging-diverging nozzle flow. The nozzle was designed and tested at NASA and was named as Nozzle A2[36]. The cell size is 175×80. The grid is clustered near the wall. The inlet Mach number is 0.22. CFL=3 is used.

Fig. 17 is the convergence histories. From this figure, it can be seen that, with $\varepsilon = 10^{-6}$, the residuals of both the present WENO-Z7 and the WENO-BS schemes are only around 0.2×10^{-1}. With $\varepsilon = 10^{-2}$, both schemes can achieve 10^{-12}, but WENO-BS converges faster than WENO-Z7.

Fig. 18 are the pressure contours obtained by WENO-BS and WENO-Z7 with $\varepsilon = 10^{-2}$, respectively. They all can capture the shock waves well.

Fig. 19 is the distribution of the wall pressure coefficient. It can be seen that the schemes with $\varepsilon = 10^{-2}$ resolve sharper shock waves with higher peak values. This also indicates that the not well converged solutions decrease the resolution of the numerical results.

3.8.6. Transonic RAE2822 Airfoil

The steady state solution of the transonic RAE2822 airfoil is calculated using the new 7th order WENO schemes with Reynolds averaged NS equation. The Baldwin-Lomax turbulent model is used. The computational region is divided into two blocks, and the mesh size of 128×55 is used for each block, the freestream Mach number M_∞ is 0.729, the Reynolds number based on chord is 6.5×10^6, and the angle of attack is 2.31^o. CFL=3 is used.

Fig. 20 is the convergence histories. Similarly, with $\varepsilon = 10^{-6}$, the residuals of WENO-BS and WENO-Z7 remain around 0.2×10^{-3}, and the amplitude of WENO-Z7 is smaller than WENO-BS. However, with $\varepsilon = 10^{-2}$, both schemes' residual can achieve 10^{-12}, even though with small oscillation.

Figs. 21 and 22 are the comparisons of the pressure coefficient and the skin friction coefficient, respectively. All results are in good agreement with the experiment. Figs. 23 and 24 are the pressure contours showing the shock captured on the suction surface. This case shows the robustness of the new 7th order WENO scheme to resolve a practical transonic flow with shock boundary layer interaction.

3.9. Conclusions

This section gives the following conclusions:

The analysis in this section indicates that the current 5th-order WENO schemes are not optimal at the transition points near discontinuities. Without using more grid points, a new method combined the forth-order fluxes with higher order smoothness indicator can overcome this drawback. Numerical examples show that the new scheme is effective to improve the accuracy near discontinuities.

The generalized finite compact scheme combines advantages of non-oscillation of the WENO scheme near shock wave and the compact accuracy of the compact scheme in

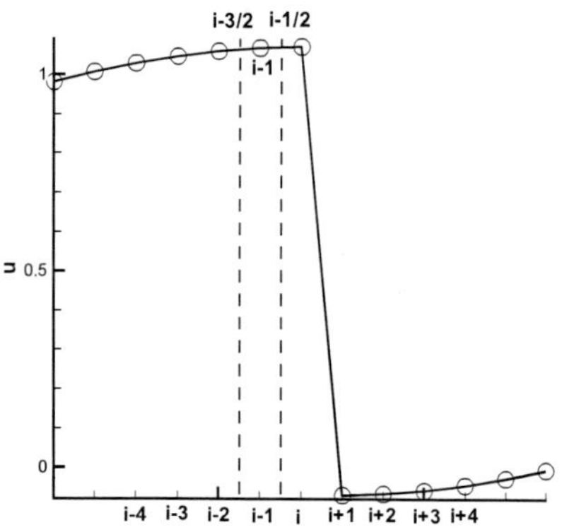

Figure 7. The sketch of transition point.

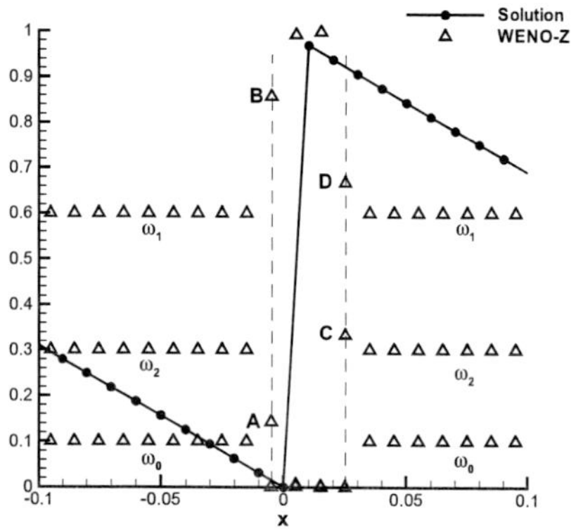

Figure 8. The distribution of weights of WENO-Z scheme.

smooth region. There is no free parameter introduced to switch the compact scheme and the WENO scheme.

A new parameter τ_7 is defined as the absolute value of a linear combination of original smoothness indicators β_k and is required to cover the whole stencil width and achieve maximum order of the truncated terms. The parameter τ_7 is used to construct a new seventh-order WENO (WENO-Z7) scheme, which satisfies the sufficient condition for seventh-order accuracy.

The role of the parameter ε is discussed. Since ε always appears with the smoothness indicator, it is naturally required that ε has the minimal effect on the weights, especially near shock waves. The view point of Henrick et al[41] is that the ε should be as small as possible. However, in smooth regions, it is crucial that the weights are not affected by the small disturbance of the smoothness indicator since the weights oscillation cause the convergence problem and decrease numerical accuracy. Hence, an enlarged and optimized[76] ε value of 10^{-2} is suggested to improve the convergence and accuracy of the seventh-order WENO scheme for practical applications.

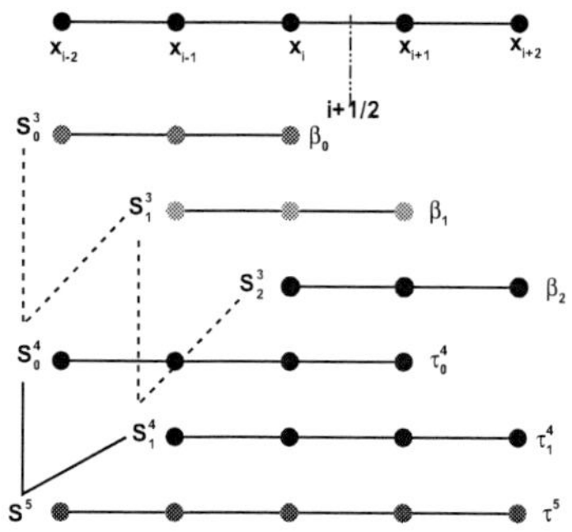

Figure 9. The sketch of reconstruction process.

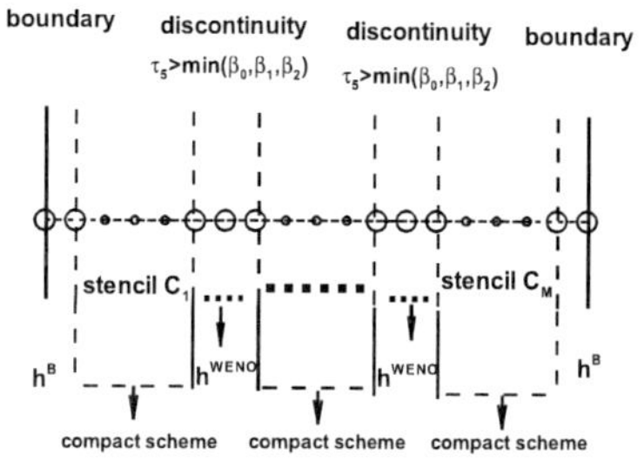

Figure 10. The sketch of the finite compact scheme.

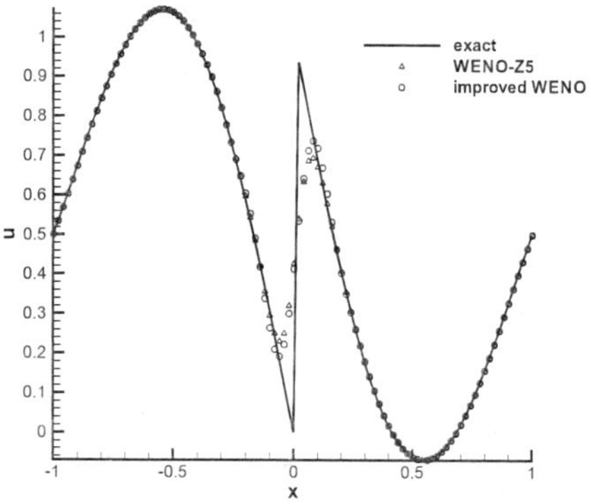

Figure 11. Numerical results of linear transport equation, initial condition (97). $t = 10$.

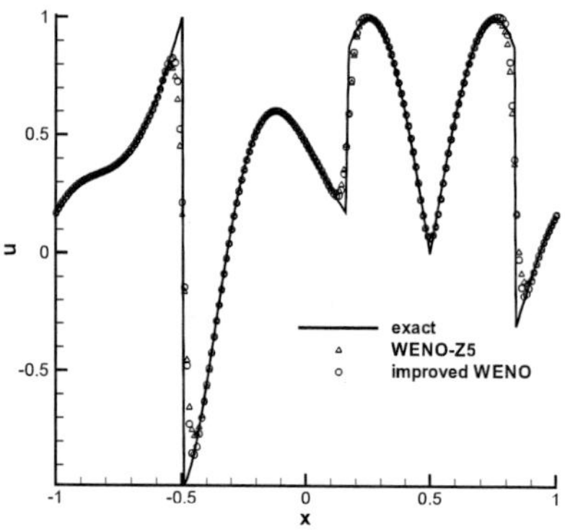

Figure 12. Numerical results of linear transport equation, initial condition (98). $t = 6$.

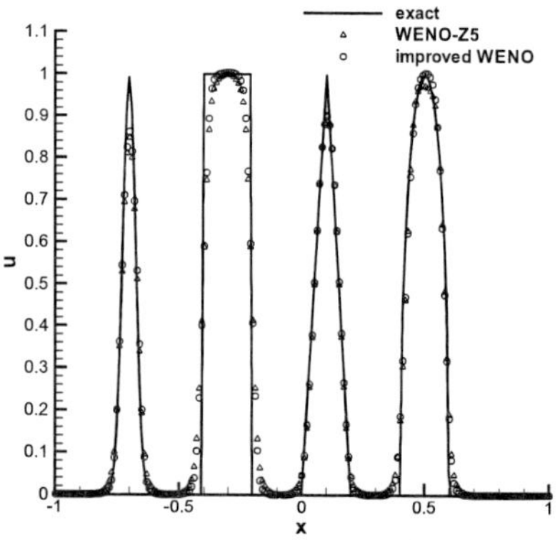

Figure 13. Numerical results of linear transport equation, initial condition (99). $t = 8$.

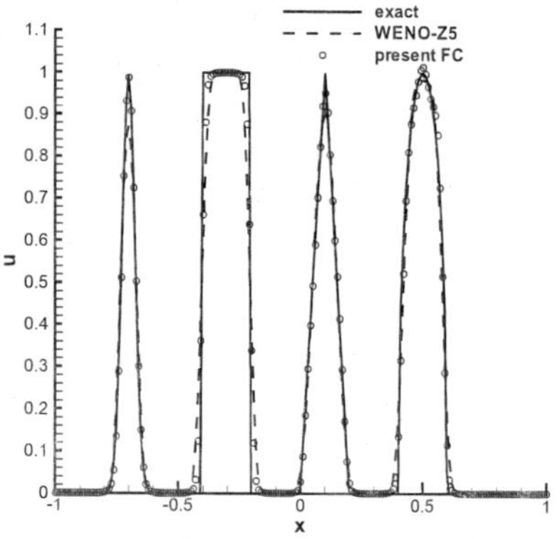

Figure 14. Numerical results of linear transport equation, initial condition (99). $t = 8$.

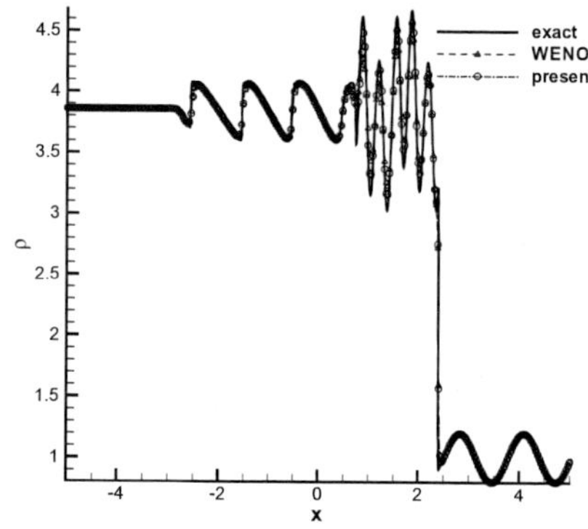

Figure 15. Density distribution, Shu-Osher problem, $N = 300$.

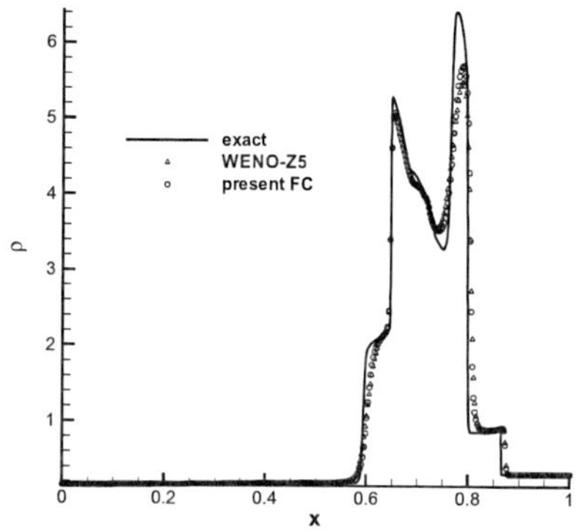

Figure 16. Density distribution, Interacting Blast Waves, $N = 400$.

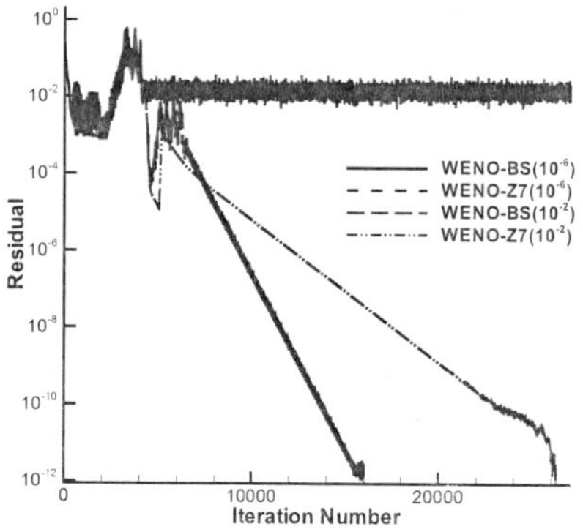

Figure 17. Convergence histories of the transonic converging-diverging nozzle flows.

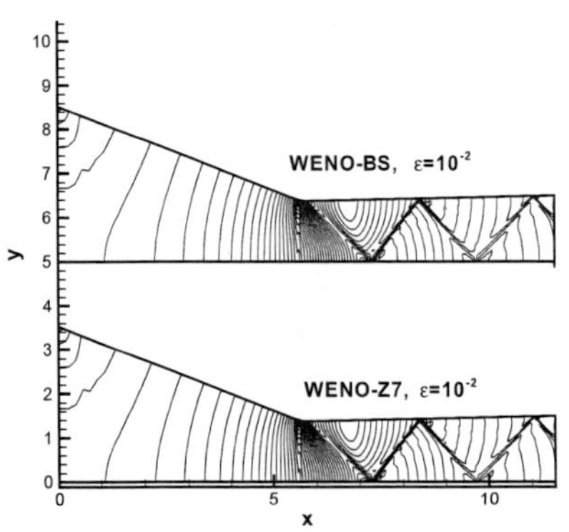

Figure 18. The pressure contours of the transonic converging-diverging nozzle flows, WENO-Z7.

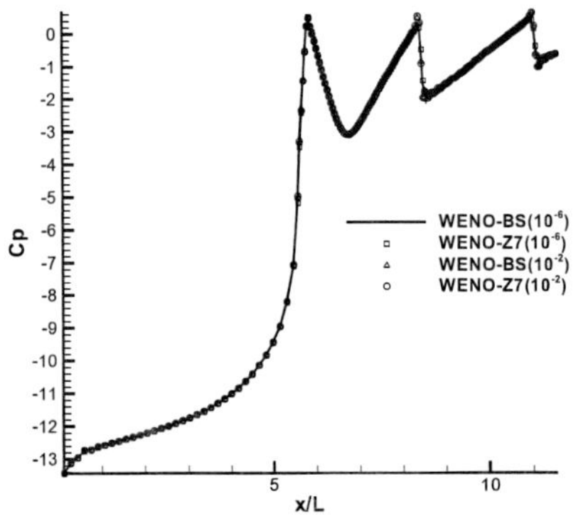

Figure 19. The pressure coefficients at the upper wall of the transonic converging-diverging nozzle flows.

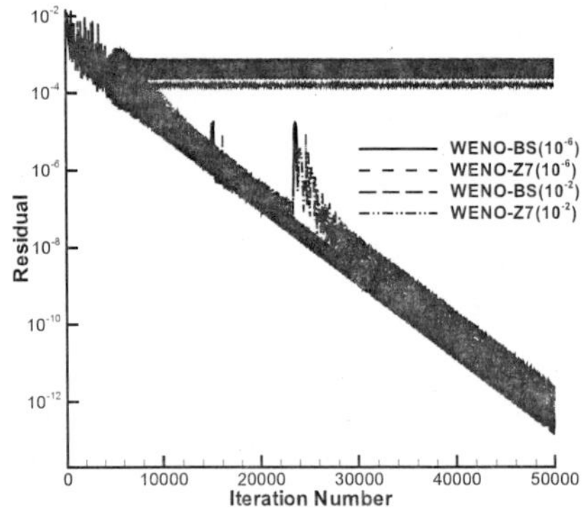

Figure 20. Convergence histories of the transonic flow for RAE2822 airfoil.

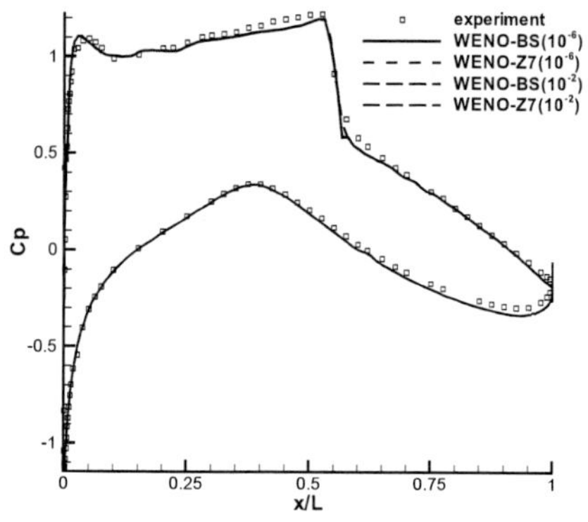

Figure 21. The pressure coefficients at the airfoil surface of the transonic flow for RAE2822 airfoil.

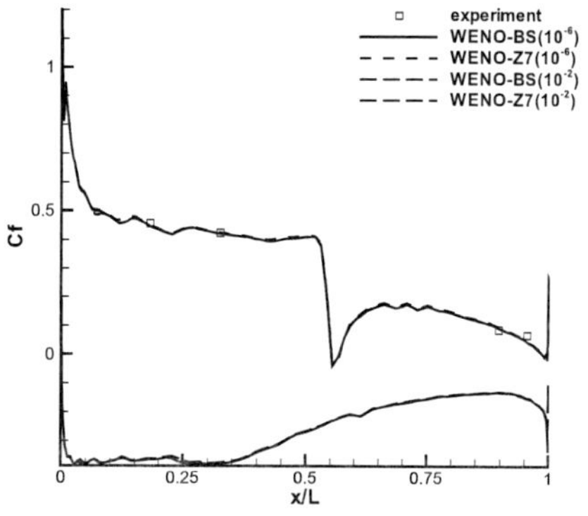

Figure 22. The skin friction coefficients at the airfoil surface of the transonic flow for RAE2822 airfoil.

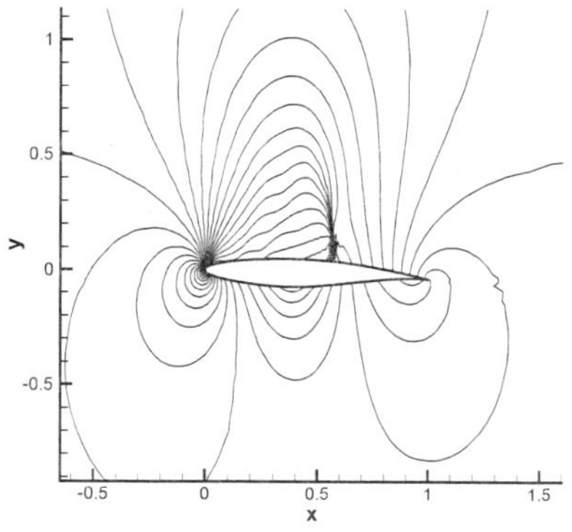

Figure 23. The pressure contours of the transonic flow for RAE2822 airfoil, WENO-BS, $\varepsilon = 10^{-2}$.

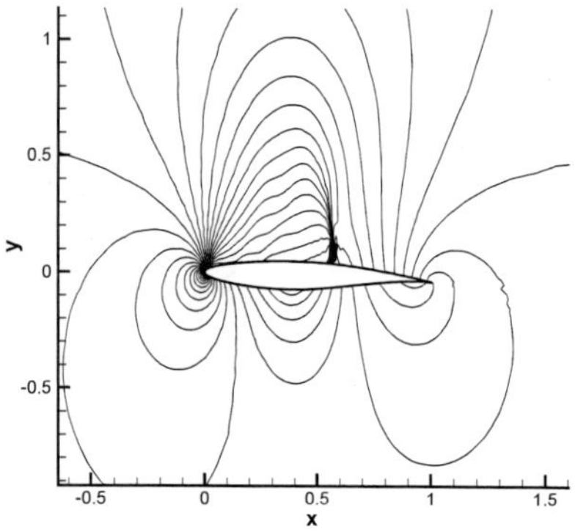

Figure 24. The pressure contours of the transonic flow for RAE2822 airfoil, WENO-Z7, $\varepsilon = 10^{-2}$.

4. High Order Conservative Central Differencing Schemes for Viscous Terms

4.1. Introduction

High-order accuracy requires high order evaluation of both the inviscid and viscous fluxes. However, most of research focus on the inviscid fluxes to resolve discontinuities. For example, the essentially non-oscillatory (ENO) schemes[39, 78] and weighted essentially non-oscillatory (WENO) schemes[38, 79, 43, 76, 80] are all aimed at resolving the inviscid fluxes with high order accuracy in smooth regions and achieving the capability to capture shock wave and contact discontinuities.

To capture discontinuities, the hyperbolic equations need to be solved in a conservative manner. A conservative numerical discretization is also essential to satisfy the conservation laws of fluid physics. A finite volume method based on the integral form of Navier-Stokes equations has the advantage to naturally obtain flux conservation. For finite differencing method, a partial derivative needs to be discretized using the interface location between two solution points in order to be conservative.

To discretize the viscous terms using finite differencing schemes, it is important that the stencil width of the viscous term discretization does not exceed the WENO stencil width. Otherwise, the overall stencil width will be wider than the WENO stencil and the advantage of the WENO scheme achieving a certain high order scheme within a compact stencil is lost. In addition, the narrower the stencil, the easier to treat boundary conditions.

For the viscous fluxes that contain 2nd order derivatives, such as

$$\frac{\partial}{\partial x}(\mu\frac{\partial f}{\partial x}) \quad or \quad \frac{\partial}{\partial x}(\mu\frac{\partial f}{\partial y}), \tag{103}$$

where μ is the viscosity coefficient, achieving a finite differencing scheme with both high order accuracy and flux conservation is not trivial. Usually, a 2nd order derivative is discretized first by a high order differencing of the first order derivative, and then the same differencing scheme is applied again to obtain high order discretization of the 2nd order derivative. These methods include the standard high order finite differencing schemes directly discretized on the node points as well as the compact central differencing schemes[71].

If the viscosity coefficient is constant, the conservative high-order schemes are straightforward, for example, a fourth-order discretization can be constructed by using five points. However, if the viscosity coefficient is variable as in compressible flows, the conservative high order finite differencing schemes for those viscous terms are not obvious. A conservative finite differencing scheme for the cross derivatives are even more complicated.

If a standard central difference scheme is used, the discretization for a 2nd order derivative will involve a large number of grid points with a wide stencil. For example, for $2r$th-order accuracy, the following discretization will have

$$\frac{\partial}{\partial x}\left(\mu\frac{\partial f}{\partial x}\right)\bigg|_i = \frac{1}{\Delta x}\sum_{k=-r}^{k=r}\beta_k\mu_{i+k}\frac{\partial f}{\partial x}\bigg|_{i+k} + O(\Delta x^{2r}) \tag{104}$$

and

$$\frac{\partial f}{\partial x}\bigg|_l = \frac{1}{\Delta x}\sum_{k=-r}^{k=r}\beta_k f_{l+k} + O(\Delta x^{2r}) \tag{105}$$

where β_k are the coefficients to make Eqs. (104) and (105) achieve $2r$th order accuracy. The stencil width of the viscous term discretization in Eq. (104) is hence $(f_{i+2r}, f_{i+2r-1}, \cdots, f_{i-2r})$. For example, the stencil $(f_{i+6}, f_{i+5}, \cdots, f_{i-6})$ will be used for constructing the 6th-order standard central schemes. The width of the used stencil is thus very large.

The following sub-sections introduce the high order conservative central finite difference schemes using the same stencil widths as those of corresponding WENO schemes.

4.2. High Order Conservative Central Schemes for Viscous Terms [33, 32]

A set of conservative high order accurate finite central differencing schemes for the viscous terms is suggested in [33, 32]. These central differencing schemes are constructed so that the stencil widths are within the stencil width of the corresponding WENO schemes for the inviscid fluxes and achieve their maximum order accuracy. We take the viscous flux derivative in ξ-direction as the example to explain how the schemes are constructed.

To conservatively discretize the viscous derivative term in Navier-Stokes equations Eq.(2), we have

$$\frac{\partial R}{\partial \xi}\bigg|_i = \frac{\tilde{R}_{i+1/2} - \tilde{R}_{i-1/2}}{\Delta\xi} \tag{106}$$

To obtain $2r$th order accuracy, $\tilde{R}$ needs to be reconstructed as

$$\tilde{R}_{i-1/2} = \sum_{I=i-r+1/2}^{i+r-3/2} \alpha_I R_I \tag{107}$$

$$R_I = [(\xi_x\tau_{xx}) + (\eta_y\tau_{xy}) + (\zeta_z\tau_{xz})]_I$$

and

$$(\tau_{xx}) = \mu\left\{\frac{4}{3}\left[\left(\xi_x\frac{\partial u}{\partial \xi}\right) + \left(\eta_x\frac{\partial u}{\partial \eta}\right) + \left(\zeta_x\frac{\partial u}{\partial \zeta}\right)\right]\right.$$
$$-\frac{2}{3}\left[\left(\xi_y\frac{\partial v}{\partial \xi}\right) + \left(\eta_y\frac{\partial v}{\partial \eta}\right) + \left(\zeta_y\frac{\partial v}{\partial \zeta}\right)\right.$$
$$\left.\left.\left(\xi_z\frac{\partial w}{\partial \xi}\right) + \left(\eta_z\frac{\partial w}{\partial \eta}\right) + \left(\zeta_z\frac{\partial w}{\partial \zeta}\right)\right]\right\} \tag{108}$$

are taken as an example to describe the high order formula. In order to achieve the highest order accuracy of R_I, the approximation of each term in Eq. (107) using the same stencil is given below,

$$\begin{cases} \mu_I = \sum_{l=m}^{n} C_l^I \mu_{i+l} + O(\Delta\xi^{(n-m+1)}), \\ \left.\frac{\partial u}{\partial \xi}\right|_I = \frac{1}{\Delta\xi}\sum_{l=r}^{s} D_l^I u_{i+l} + O(\Delta\xi^{(s-r)}), \\ \left.\frac{\partial u}{\partial \eta}\right|_I = \sum_{l=m}^{n} C_l^I \left.\frac{\partial u}{\partial \eta}\right|_{i+l,j} + O(\Delta\xi^{(n-m+1)}, \Delta\eta^{(n-m+1)}), \end{cases} \tag{109}$$

where

$$\left.\frac{\partial u}{\partial \eta}\right|_{i,j} = \frac{1}{\Delta\eta}\sum_{l=p}^{q} C_l^c u_{i,j+l} + O(\Delta\eta^{(q-p)}) \tag{110}$$

The other terms are determined similarly. By choosing different ranges for $(m,n), (r,s), (p,q)$ and using Taylor's series expansion of Eqs. (109)-(110) to calculate the coefficients C_l^I, D_l^I, C_l^c, one can obtain different order accuracy approximation to the viscous terms. The principle of choosing $(m,n), (r,s), (p,q)$ is to ensure that the approximation of $\frac{\partial R}{\partial \xi}|_i$ in Eq.(106) is a central differencing.

For example, for the fourth order schemes ($r = 2$),

$$(m,n) = (-2,1), \quad (r,s) = (-3,2) \quad \text{and} \quad (p,q) = (-2,2)$$

$$\alpha_{i-3/2} = -\frac{1}{24}, \; \alpha_{i-1/2} = \frac{26}{24}, \; \alpha_{i+1/2} = -\frac{1}{24} \tag{111}$$

and the coefficients C_l^I, D_l^I, C_l^c are listed in Tables 3-5.

Table 3. The coefficients C_l^I of the fourth order schemes

I	C_{-2}^I	C_{-1}^I	C_0^I	C_1^I
$i-3/2$	5/16	15/16	-5/16	1/16
$i-1/2$	-1/16	9/16	9/16	-1/16
$i+1/2$	1/16	-5/16	15/16	5/16

Table 4. The coefficients D_l^I of the fourth order schemes

I	D_{-3}^I	D_{-2}^I	D_{-1}^I	D_0^I	D_1^I	D_2^I
$i-3/2$	71/1920	-141/128	69/64	1/192	-3/128	3/640
$i-1/2$	-3/640	25/384	-75/64	75/64	-25/384	3/640
$i+1/2$	-3/640	3/128	-1/192	-69/64	141/128	-71/1920

Table 5. The coefficients C_l^c of the fourth order schemes

C_{-2}^c	C_{-1}^c	C_0^c	C_1^c	C_2^c
1/12	-8/12	0	8/12	-1/12

For the sixth order schemes ($r = 3$),

$$(m,n) = (-3,2), (r,s) = (-4,3), and\, (p,q) = (-3,3)$$

$$\alpha_{i-5/2} = -\frac{9}{1920}, \quad \alpha_{i-3/2} = -\frac{116}{1920}, \quad \alpha_{i-1/2} = \frac{2134}{1920}, \quad \alpha_{i+1/2} = -\frac{116}{1920}, \quad \alpha_{i+3/2} = \frac{9}{1920} \tag{112}$$

and the coefficients C_l^I, D_l^I, C_l^c are listed in Tables 6-8.

Table 6. The coefficients C_l^I of the sixth order schemes

I	C_{-3}^I	C_{-2}^I	C_{-1}^I	C_0^I	C_1^I	C_2^I
$i-5/2$	63/256	315/256	-105/128	63/128	-45/256	7/256
$i-3/2$	-7/256	105/256	105/128	-35/128	21/256	-3/256
$i-1/2$	3/256	-25/256	75/128	75/128	-25/256	3/256
$i+1/2$	-3/256	21/256	-35/128	105/128	105/256	-7/256
$i+3/2$	7/256	-45/256	63/128	-105/128	315/256	63/256

It can be proved below that Eq. (106) is $2r$th order accuracy and symmetric with respect to cell i. We will take a single term from Eq. (106) and the sixth order schemes as the example

$$\frac{1}{\Delta\xi}\left[\widetilde{\left(\mu\frac{\partial u}{\partial \xi}\right)}_{i+1/2} - \widetilde{\left(\mu\frac{\partial u}{\partial \xi}\right)}_{i-1/2}\right] \tag{113}$$

where

$$\widetilde{\left(\mu\frac{\partial u}{\partial \xi}\right)}_{i-1/2} = \sum_{I=i-5/2}^{i+3/2} \alpha_I \left(\mu\frac{\partial u}{\partial \xi}\right)_I$$

Both $\widetilde{(\mu\frac{\partial u}{\partial \xi})}_{i-1/2}$ and $\widetilde{(\mu\frac{\partial u}{\partial \xi})}_{i+1/2}$ are evaluated by Eq.(109)-(110). The coefficients C_l^I and D_l^I in Table 6 and 7 are based on interface $i-\frac{1}{2}$. For $\widetilde{(\mu\frac{\partial u}{\partial \xi})}_{i+1/2}$, the interfaces used

Table 7. The coefficients D_l^I of the sixth order schemes

I	D_{-4}^I	D_{-3}^I	D_{-2}^I	D_{-1}^I	D_0^I	D_1^I	D_2^I	D_3^I
$i-5/2$	3043/107520	-5353/5120	4731/5120	733/3072	-239/1024	597/5120	-167/5120	143/35840
$i-3/2$	-143/35840	185/3072	-1185/1024	1175/1024	-125/3072	-51/5120	5/1024	-5/7168
$i-1/2$	5/7168	-49/5120	245/3072	-1225/1024	1225/1024	-245/3072	49/5120	-5/7168
$i+1/2$	5/7168	-5/1024	51/5120	125/3072	-1175/1024	1185/1024	-185/3072	143/35840
$i+3/2$	-143/35840	167/5120	-597/5120	239/1024	-733/3072	-4731/5120	5353/5120	-3043/107520

Table 8. The coefficients C_l^c of the sixth order schemes

C_{-3}^c	C_{-2}^c	C_{-1}^c	C_0^c	C_1^c	C_2^c	C_3^c
-1/60	3/20	-3/4	0	3/4	-3/20	1/60

are off set by 1. To distinguish the coefficients from those for $\widetilde{(\mu\frac{\partial u}{\partial \xi})}_{i-1/2}$, $\tilde{C}_l^I$, $\tilde{D}_l^I$ and $\tilde{\alpha}_I$ are used for $\widetilde{(\mu\frac{\partial u}{\partial \xi})}_{i+1/2}$. Based on Table 6 and 7, we have

$$\tilde{C}_l^I = C_{l-1}^{I-1}, \quad \tilde{D}_l^I = D_{l-1}^{I-1},$$
$$\tilde{\alpha}_I = \alpha_{I-1}, \quad I = i-3/2, i-1/2, i+1/2, i+3/2, i+5/2$$

Substitute Eq. (109)-(110) to Eq. (113), we have

$$\frac{1}{\Delta\xi}\left[\widetilde{\left(\mu\frac{\partial u}{\partial \xi}\right)}_{i+1/2} - \widetilde{\left(\mu\frac{\partial u}{\partial \xi}\right)}_{i-1/2}\right] = \frac{1}{\Delta\xi^2}\sum_{m=-3}^{2}\sum_{r=-4}^{3} C_{i+m,i+r}\mu_{i+m}u_{i+r} \tag{114}$$

Using simple algebraic operation, it can be proved that

$$C_{i+m,i+r} = C_{i-m,i-r}, \quad m = -3,\ldots,2; \quad r = -4,\ldots,3. \tag{115}$$

This shows that Eq. (114) is symmetric about cell i as a central differencing.

Next, we prove that the 6th order accuracy given by Eq. (106) is satisfied. Again, take the term $T = \mu\frac{\partial u}{\partial \xi}$ from $\widetilde{(\mu\frac{\partial u}{\partial \xi})}_{i-1/2}$ in Eq. (113) as an example, which is located at $I = i-5/2$. Based on Eq. (109) and Taylor's series expansion, there is

$$\begin{aligned}
T_{i-5/2}^{-} &= \left(\sum_{l=m}^{n} C_l^I \mu_{i+l}\right)\left(\frac{1}{\Delta\xi}\sum_{l=r}^{s} D_l^I u_{i+l}\right) \\
&= \left[\mu_{i-5/2} + A_I\mu_{i-5/2}^{(6)}\Delta\xi^6 + O(\Delta\xi^7)\right]\left[\left.\frac{\partial u}{\partial \xi}\right|_{i-5/2} + O(\Delta\xi^7)\right] \\
&= \mu_{i-5/2}\left.\frac{\partial u}{\partial \xi}\right|_{i-5/2} + A_I\mu_{i-5/2}^{(6)}\left.\frac{\partial u}{\partial \xi}\right|_{i-5/2}\Delta\xi^6 + O(\Delta\xi^7)
\end{aligned} \tag{116}$$

where A_I is the coefficient of Taylor's series expansion, $\mu_I^{(6)}$ stands for the 6th order derivative of μ at interface I, i.e.,

$$\mu_I^{(6)} = \left.\frac{\partial^6 \mu}{\partial \xi^6}\right|_I$$

The corresponding symmetric term $T = \mu\frac{\partial u}{\partial \xi}$ from $\widetilde{(\mu\frac{\partial u}{\partial \xi})}_{i+1/2}$, which is located at $I = i - 3/2$, is

$$\begin{aligned}
T^{+}_{i-3/2} &= \left(\sum_{l=m}^{n} \tilde{C}_l^I \mu_{i+1+l}\right)\left(\frac{1}{\Delta\xi}\sum_{l=r}^{s} \tilde{D}_l^I u_{i+1+l}\right) \\
&= \left[\mu_{i-3/2} + \tilde{A}_I \mu^{(6)}_{i-3/2}\Delta\xi^6 + O(\Delta\xi^7)\right]\left[\left.\frac{\partial u}{\partial \xi}\right|_{i-3/2} + O(\Delta\xi^7)\right] \\
&= \mu_{i-3/2}\left.\frac{\partial u}{\partial \xi}\right|_{i-3/2} + \tilde{A}_I \mu^{(6)}_{i-3/2}\left.\frac{\partial u}{\partial \xi}\right|_{i-3/2}\Delta\xi^6 + O(\Delta\xi^7)
\end{aligned} \tag{117}$$

Note that $A_I = \tilde{A}_I$, and

$$\mu^{(6)}_{i-3/2}\left.\frac{\partial u}{\partial \xi}\right|_{i-3/2} = \mu^{(6)}_{i-5/2}\left.\frac{\partial u}{\partial \xi}\right|_{i-5/2} + O(\Delta\xi)$$

hence

$$T^{+}_{i-3/2} - T^{-}_{i-5/2} = \mu_{i-3/2}\left.\frac{\partial u}{\partial \xi}\right|_{i-3/2} - \mu_{i-5/2}\left.\frac{\partial u}{\partial \xi}\right|_{i-5/2} + O(\Delta\xi^7)$$

The other terms $(i - 3/2,\ i - 1/2,\ i + 1/2,\ i + 3/2)$ can be analyzed similarly as above, thus Eq.(106)

$$\frac{1}{\Delta\xi}(\tilde{R}_{i+1/2} - \tilde{R}_{i-1/2}) = R'(\xi_i) + O(\Delta\xi^6)$$

is proved, i.e. the constructed schemes are formally 6th order accuracy. More details of the 4th and 6th order central differencing schemes can be seen in [33, 32].

4.3. Numerical Results

4.3.1. Verification of Scheme Accuracy

This subsection is to verify the accuracy of the conservative central different schemes. In order to match the form of the viscous terms in compressible Navier-Stokes equations, the testing function is taken as

$$\frac{\partial}{\partial x}\left(\mu\frac{\partial f}{\partial x}\right)$$

and two cases are validated:

(1)

$$\mu = Ae^{-2x}, \quad f(x) = \frac{1 - e^{-Rx}}{1 - e^{-R}}, \quad 0 \le x \le 1$$

and $A = 0.01,\ R = 20$. The function $f(x)$ has the similar distribution as the velocity of a wall boundary layer.

(2)

$$\mu = Ae^{2x}, \quad f(x) = \sin(Bx), \quad 0 \le x \le 1$$

and $A = 0.1,\ B = 10$. This function $f(x)$ represents a high frequency wave.

Tables 9 and 10 show the errors and accuracy order. It can be seen that the scheme achieves the expected fourth order and sixth order accuracy.

Table 9. Accuracy comparison of case (1)

Scheme	N	L_∞ error	L_∞ order	L_1 error	L_1 order
4th	20	2.7556e-2	—	2.0653e-3	—
	40	1.8554e-3	3.893	1.0964e-4	4.236
	80	1.1753e-4	3.981	6.1103e-6	4.165
	160	7.3680e-6	3.996	3.5846e-7	4.091
	320	4.6085e-7	3.999	2.1676e-8	4.048
6th	20	5.2509e-3	—	3.9354e-4	—
	40	8.4762e-5	5.953	5.0090e-6	6.296
	80	1.3271e-6	5.997	6.8999e-8	6.182
	160	2.0741e-8	6.000	1.0091e-9	6.095
	320	3.2409e-10	6.000	1.5274e-11	6.046

Table 10. Accuracy comparison of case (2)

Scheme	N	L_∞ error	L_∞ order	L_1 error	L_1 order
4th	20	1.9647e-2	—	6.8659e-3	—
	40	1.2169e-3	4.013	4.2900e-4	4.000
	80	7.5868e-5	4.004	2.6950e-5	3.993
	160	4.7388e-6	4.001	1.6895e-6	3.996
	320	2.9628e-7	3.999	1.0575e-7	3.998
6th	20	7.9318e-4	—	2.5760e-4	—
	40	1.2556e-5	5.981	4.1441e-6	5.958
	80	1.9699e-7	5.994	6.5585e-8	5.982
	160	3.0843e-9	5.997	1.0316e-9	5.990
	320	5.1379e-11	5.908	1.6836e-11	5.937

4.3.2. Flow Past a Circular Cylinder with $Re = 3900$

The turbulent flow passing a circular cylinder at Mach number of 0.2 and Reynolds number of 3900 is calculated using the implicit LES, which does not employ an explicit SGS models and rely on numerical dissipation for SGS stress terms[10, 11, 12].

The large-eddy simulation for this problem has been performed extensively by different research groups[81, 82, 83, 84, 85]. Because periodic conditions are enforced in the spanwise direction, the span must be sufficient long such that large-scale structures are not artificially forced to become two-dimensional. Even though the spanwise size of $\pi D/2$ is appropriate for some analysis, it severely constrains the development of the spanwise vortex structures[83]. Hence, a spanwise size of πD is used in this paper.

Two sets of the spatial discretization schemes are used: 1) 7th order reconstruction for convective terms with 6th order central differencing for viscous terms(7-6) as described in this paper; 2) 5th order reconstruction for convective terms with 4th order central differencing for viscous terms(5-4) as described in Ref.[33]. The 2nd order accuracy implicit time marching scheme described in section 2.5 is used for the LES.

Three sets of O-type grid are used.

A. The baseline mesh size is $240 \times 160 \times 90$. Same as in Ref. [85], the grid points are spaced uniformly in the circumferential and spanwise direction, and are clustered in the radial direction with the first grid-point non-dimensional spacing equal to 7.7×10^{-5}. The mesh is clustered close to the wall by using a stretching function,

$$y(j) = d_1 \frac{1 - q^{j-1}}{1 - q}, \quad j = 1, 2, \ldots, N + 1 \tag{118}$$

with the first grid-point non-dimensional spacing $d_1 = 7.7 \times 10^{-5}$ and the stretching factor $q = 1.07220$ are used. The mesh sketch is shown in Fig. 25.

B. The second mesh size is $240 \times 160 \times 180$, the refined mesh has doubled in spanwise direction and other two directions are kept as mesh A.

C. The third mesh has the same size as the baseline mesh, but in the circumferential direction, the grid points is doubled uniformly within a $60°$ wake region, and then the grid is stretched from the uniform grid to the leading edge by using the same stretching transformation defined by Eq. (118) with a stretching factor of $q = 1.02065$.

The uniform initial flowfield is used to start the simulation and the transition period is up to dimensionless time $t = 100$. Then the statistical results are obtained from $t = 100$ to $t = 250$. The time step size in the computations is $\Delta t = 0.05$. Within each physical time step, twenty pseudo time steps are used with CFL number of 10 and the L_2 Norm residual reduced by $3 - 4$ orders of magnitude.

The shear Reynolds stress is given in Fig. 26. The results of 7-6 scheme with mesh A and C agree well with the results of Rizzetta et al.[84], who used a mesh size of $199 \times 197 \times 53$ and a sixth-order compact scheme with a tenth-order non-dispersive filter. However, both our computation and theirs significantly under-predicted the Reynolds stresses. It can be seen that the results of the scheme 7-6 agree with the experiment[86] much better than the scheme 5-4. The mesh refined near wake (mesh C) gives a better streamwise, shear and lateral Reynolds stress distribution than other meshes. This means a higher resolution grid system in wake region is beneficial for large eddy simulation to resolve the turbulence structures. With doubled grid points in spanwise direction(mesh B), both 7-6 and 5-4 schemes obtain lower Reynolds stress than that with the baseline mesh. This agrees with the conclusion of Kravchenko and Moin[83], in which, two sets of mesh for different spanwise size of $\pi D/2$ and πD are used to test the grid resolutions and domain size on the turbulent structures, the results shown that the Reynolds stress are noticeably higher in the coarse grid size.

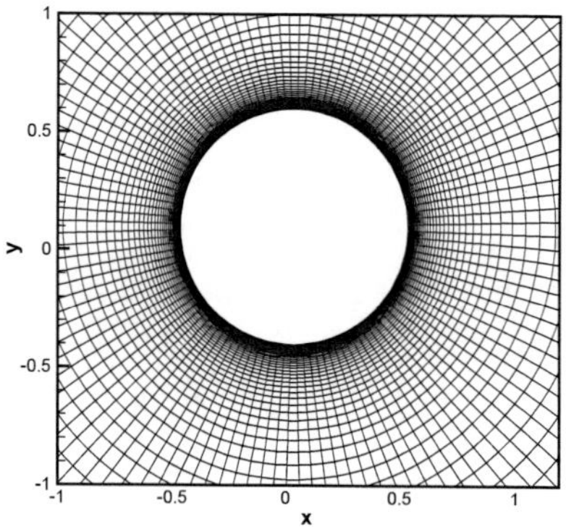

Figure 25. Sketch of baseline mesh in $x - y$ plane.

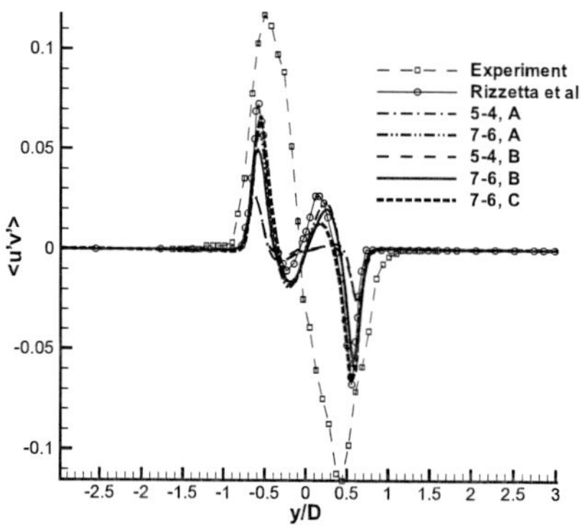

Figure 26. Shear Reynolds Stress in the wake at $x/D = 1.54$ Plane.

Fig. 27 shows a 3D instantaneous vorticity iso-surface with the refined grid of $240 \times 160 \times 180$ at $T = 250$ by using the 7-6 scheme. It can be seen that many small vortex structures are resolved.

More detailed LES results can be seen in [32].

4.4. Conclusions

A set of high order conservative central differencing schemes for the viscous terms of compressible Navier-Stokes equations is presented. The high order central differencing schemes have the same stencil widths as those used for inviscid fluxes by the corresponding high order WENO schemes. The accuracy of the central schemes are proved analytically and verified numerically. The LES simulation of the paper shows that the central differencing schemes are robust when used with high order WENO schemes.

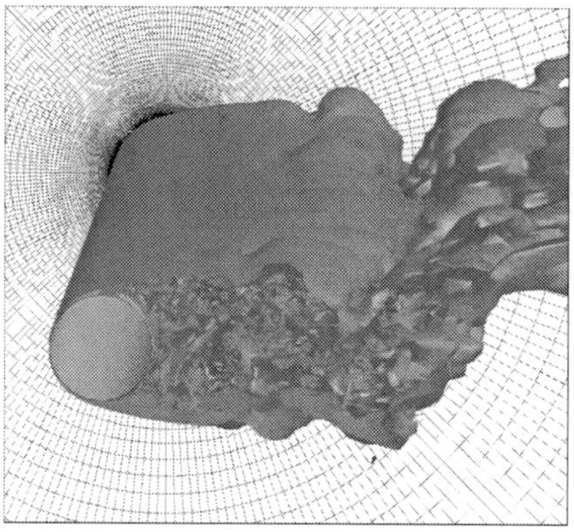

Figure 27. Instantaneous vorticity magnitude at $T = 250$, 7-6 scheme, mesh B.

5. Preconditioning Methods

5.1. Introduction

In recent years, there is a growing interest to develop an unified algorithm for compressible and incompressible flows in computational fluid dynamics (CFD) for two reasons: First, there exist flow problems of mixed compressible/incompressible type. Second, it is convenient to use the same CFD code for pure incompressible or compressible flows. For example, the design of an aircraft or a launch vehicle involves the calculation of flows during a full flight mission passing through various flow regimes[87]. The other example of co-existing incompressible and compressible flow is the ship-aircraft dynamic interface[88], where the flow near ship deck is low speed incompressible flow, and the flow around aircraft is compressible due to helicopter rotor tip speed and high speed engine nozzle jet.

The system of compressible flow governing equations at very low Mach numbers is stiff due to the large ratio of the acoustic and convective time scales, or the large disparity in acoustic wave speed, $u + c$, and the waves convected at fluid speed, u, where c is the speed of sound. That is, there exists a large difference in the eigenvalues of the convective flux Jacobians when the incompressible limit is approached. The largest eigenvalue tends to approach the speed of sound, whereas the smallest eigenvalue approaches zero. The large difference of the eigenvalues will exacerbate the condition numbers of the linearized system and create the analytical stiffness[89].

In addition to the stiffness problem, the other issue of direct applying compressible flow equation to incompressible flows is that the numerical dissipation is large at stagnation points. This is again because of the large difference between the speed of sound and flow speed. The large dissipation may distort the solution of a wall boundary layer.

Preconditioning is to change the eigenvalues of the compressible flow equations system in order to remove the large disparity of wave speeds. Usually, the system of compressible flow equations is preconditioned by multiplying the time derivatives with a suitable matrix[90, 91, 92, 93, 94]. In recent years, considerable progress has been made in the development of preconditioning methods for accelerating the convergence of Euler and Navier-Stokes solvers at low Mach numbers. An excellent review is given by Turkel in [95]. The preconditioning methods have been used to conduct the large-eddy simulations of turbulent channel flows and cylinder flows[96], the turbulent pipe flows[97], the turbulent flows over an airfoil and wing at subsonic and transonic conditions[98], nonequilibrium condensate flows in a nozzle[99].

For the spatial discretization with preconditioning, the 2nd order central differencing is adopted by Choi and Merkle [93, 100] and Bortoli[87]. The Roe-type flux-difference splitting (FDS) is used by Weiss and Smith [94, 101, 102]. The third-order MUSCL extrapolation is used by Briley et al [103]. The flux-vector splitting (FVS) is applied by Turkle et al [104]. Edwards and Liu[105] have extended the advective upwind splitting method (AUSM) to all flow speeds. Nigro et al[106] combined the preconditioning mass matrix[93] with an SUPG finite element formulation. Rossow[107] developed a blended pressure/density based approach, the MAPS+ flux scheme in terms of Mach number. All these preconditioned methods employ 2nd order schemes.

Many engineering applications may have discontinuities in the flows such as shock waves or contact surfaces. The essentially non-oscillatory (ENO) or weighted essentially non-oscillatory (WENO) schemes are attractive for their capability to capture discontinuities and achieve the consistent high order accuracy in smooth regions. By using a convex combination of all candidate stencils to replace the smoothest one in the ENO scheme, a WENO scheme has more advantages over its ENO counterpart. For example, it approaches certain high order accuracy in smooth regions and has better convergence due to the smoother numerical flux used. From its appearance [37, 38] to present, the WENO schemes have been extensively applied to different flow problems in many areas. Engineering problems often have the compressible and incompressible flow regimes existing simultaneously, including shock wave and contact surface (e.g. air/water interface). To improve the simulation accuracy, it is desirable to apply high order ENO/WENO schemes with preconditioning to treat the complicated flows at all speed.

Recently, Huang et al[108] developed a class of lower-upper symmetric Gauss-Seidel implicit weighted essentially non-oscillatory (WENO) scheme consisted of a first-order part and high-order part for solving the preconditioned Navier-Stokes equations with Spalart-Allmaras one-equation turbulence model. Numerical results show that the application of the preconditioned WENO schemes can improve accuracy and robustness. However, Shen et al[30] demonstrate that LU-SGS has significantly lower CPU efficiency than the unfactored Gauss-Seidel iteration due to the approximate factorization.

In recent years, the convective upwind and split pressure (CUSP) family schemes have achieved great success. The CUSP schemes can be basically categorized to two types, the H-CUSP and E-CUSP[109, 110, 111]. The H-CUSP schemes have the total enthalpy from the energy equation in their convective vector, while the E-CUSP schemes use the total energy in the convective vector. The Liou's AUSM family schemes[112, 113, 114, 115, 116], Van Leer-Hänel scheme[117], and Edwards's LDFSS schemes[118, 119] belong to the H-CUSP group. The schemes developed by Zha, et al.[120, 121, 122, 18, 123] belong to the E-CUSP group.

From the characteristic theory point of view, the H-CUSP schemes are not fully consistent with the disturbance propagation directions[124, 125], which may affect the stability and robustness of the schemes. By splitting the eigenvalues of the Jacobians to convection (velocity) and waves (speed of sound), one will find that the convection terms only contain the total energy[120], which will lead to the E-CUSP schemes.

In this section, combined with the unfactored implicit Gauss-Seidel relaxation scheme and the fifth order WENO scheme, the preconditioned Roe scheme and a low diffusion E-CUSP scheme are developed to simulate the flows at all speeds. Several flows at various speeds from low speed incompressible flows to supersonic flows are calculated to demonstrate the robustness, accuracy, and efficiency of these methods.

5.2. Preconditioning Methods

The preconditioned system for steady state flows in conservative form is obtained by multiplying the preconditioning matrix Γ to the time derivative terms of Eq.(2) to give

$$\Gamma\frac{\partial q}{\partial t}+\frac{\partial E}{\partial \xi}+\frac{\partial F}{\partial \eta}+\frac{\partial G}{\partial \zeta}=\frac{1}{Re}\left(\frac{\partial R}{\partial \xi}+\frac{\partial S}{\partial \eta}+\frac{\partial T}{\partial \zeta}\right) \tag{119}$$

The preconditioning matrix Γ has various forms[90, 92, 93, 94], and is dependent on the choice of primitive variables q. This paper adopts the method of Weiss and Smith described in Ref.[94]. The q and Γ are taken as the following,

$$q=(p,u,v,w,T)^T$$

$$\Gamma=\begin{bmatrix} \Theta & 0 & 0 & 0 & \rho_T \\ \Theta u & \rho & 0 & 0 & \rho_T u \\ \Theta v & 0 & \rho & 0 & \rho_T v \\ \Theta w & 0 & 0 & \rho & \rho_T w \\ \Theta H-1 & \rho u & \rho v & \rho w & \rho_T H+\rho C_p \end{bmatrix}$$

where Θ is given by

$$\Theta=\left(\frac{1}{U_r^2}-\frac{\rho_T}{\rho C_p}\right)$$

U_r is a reference velocity. In this paper, the reference velocity proposed by Edwards and Roy[105] is used:

$$U_r=\min[c,\max(|V|,k|V_\infty|)]$$

where, c is the speed of sound, $|V|=\sqrt{u^2+v^2+w^2}$ is the velocity magnitude, $|V_\infty|$ is a reference velocity. H is the total enthalpy, ρ_T stands for $\frac{\partial \rho}{\partial T}$, C_p is the specific heat at constant pressure. k is a constant and $k=0.5$ is used in this paper.

This methodology can be easily generalized to the time-dependent Navier-Stokes equations by using dual time stepping method as suggested by Weiss and Smith[94].

5.3. Preconditioned Flux Difference Splitting Method [126]

For preconditioned flux difference splitting scheme, the Roe's approximate Riemann solver[28] is used with the WENO scheme. For the rest of the paper, we will take the flux in ξ direction as the example to explain the numerical methodology. Other directions can be obtained following the symmetric rule.

The preconditioned Roe scheme can be expressed as the following[94]:

$$E_{i+\frac{1}{2}}=\frac{1}{2}[E(q^L)+E(q^R)-\tilde{A}(q^R-q^L)]_{i+\frac{1}{2}} \tag{120}$$

where

$$\tilde{A} = \Gamma M_\Gamma |\Lambda_\Gamma| M_\Gamma^{-1}$$

The subscript Γ denotes that the diagonal matrix of eigenvalues and the eigenvector matrix are derived from the preconditioned system. Here, the diagonal matrix of eigenvalues and the eigenvector matrix are calculated using Roe averaged variables $\tilde{\rho}$, $\tilde{u}$, et al.

$$\Lambda_\Gamma = diag(U, U, U, U' + C', U' - C')$$

where

$$U = \xi_x \tilde{u} + \xi_y \tilde{v} + \xi_z \tilde{w}$$

$$U' = U(1-\alpha),\ C' = \sqrt{\alpha^2 U^2 + (\xi_x^2 + \xi_y^2 + \xi_z^2) U_r^2}$$

$$\alpha = (1 - \beta U_r^2)/2,\ \beta = \rho_p + \frac{\rho_T}{\tilde{\rho} C_p}$$

and

$$M_\Gamma = \begin{bmatrix} 0 & 0 & 0 & 1 & 1 \\ 0 & \eta_x & \zeta_x & \frac{-\xi_x}{X_1} & \frac{-\xi_x}{X_2} \\ 0 & \eta_y & \zeta_y & \frac{-\xi_y}{X_1} & \frac{-\xi_y}{X_2} \\ 0 & \eta_z & \zeta_z & \frac{-\xi_z}{X_1} & \frac{-\xi_z}{X_2} \\ 1 & 0 & 0 & X_3 & X_4 \end{bmatrix}$$

$$M_\Gamma^{-1} = \begin{bmatrix} \frac{1}{X_5} & \frac{\xi_x X_6}{X_5} & \frac{\xi_x X_6}{X_5} & \frac{\xi_x X_6}{X_5} & 1 \\ 0 & \eta_x & \eta_y & \eta_z & 0 \\ 0 & \zeta_x & \zeta_y & \zeta_z & 0 \\ -\frac{\tilde{U}\alpha - \tilde{C}'}{2\tilde{C}'} & -\xi_x X_7 & -\xi_y X_7 & -\xi_z X_7 & 0 \\ \frac{\tilde{U}\alpha + \tilde{C}'}{2\tilde{C}'} & \xi_x X_7 & \xi_y X_7 & \xi_z X_7 & 0 \end{bmatrix}$$

where

$$X_1 = \tilde{\rho}(\tilde{U}\alpha - \tilde{C}')$$

$$X_2 = \tilde{\rho}(\tilde{U}\alpha + \tilde{C}')$$

$$X_3 = \frac{1 - (\rho_p - \Theta)\tilde{U}(\tilde{U}\alpha - \tilde{C}')}{(\rho_T + \tilde{\rho} C_p \Theta)(\tilde{U}\alpha - \tilde{C}')^2}$$

$$X_4 = \frac{1 - (\rho_p - \Theta)\tilde{U}(\tilde{U}\alpha + \tilde{C}')}{(\rho_T + \tilde{\rho} C_p \Theta)(\tilde{U}\alpha + \tilde{C}')^2}$$

$$X_5 = (\rho_T + \tilde{\rho} C_p \Theta)(\tilde{U}^2\alpha^2 - \tilde{C}'^2)$$

$$X_6 = \tilde{\rho}\tilde{U}[2\alpha - (\rho_p - \Theta)(\tilde{U}^2\alpha^2 - \tilde{C}'^2)]$$

$$X_7 = \frac{\tilde{\rho}(\tilde{U}^2\alpha^2 - \tilde{C}'^2)}{2\tilde{C}'}$$

and

$$\tilde{U} = \xi'_x \tilde{u} + \xi'_y \tilde{v} + \xi'_z \tilde{w},\ \ \tilde{C}' = \sqrt{\alpha^2 \tilde{U}^2 + U_r^2},\ \ \vec{\xi}' = \vec{\xi}/|\vec{\xi}|$$

The analysis in the generalized coordinates is similar to that in Cartesian coordinates as given in Ref.[94]. For an ideal gas, $\beta = (\gamma RT)^{-1} = 1/c^2$. Thus, when $U_r = c$, $\alpha = 0$, and the eigenvalues of the preconditioned system take their original form $U \pm c\sqrt{\xi_x^2 + \xi_y^2 + \xi_z^2}$ for compressible flows. At low speed, as $U_r \to 0$, $\alpha \to \frac{1}{2}$, all the eigenvalues become the same order of magnitude of U.

5.4. Preconditioned E-CUSP Scheme [127]

In [120, 121, 18, 123, 124], the characteristic analysis is given as the foundation to construct the E-CUSP scheme, which is a low diffusion approximate Riemann solver different from the Roe's scheme to avoid the matrix operation. The basic idea of E-CUSP scheme is to split the flux $\mathbf{E}$ to the convective flux $\mathbf{E}^c$ and the pressure flux $\mathbf{E}^p$. That is:

$$\mathbf{E} = E^c + E^p = \begin{pmatrix} \rho U \\ \rho u U \\ \rho v U \\ \rho w U \\ \rho e U \end{pmatrix} + \begin{pmatrix} 0 \\ l_x p \\ l_y p \\ l_z p \\ p U \end{pmatrix} \tag{121}$$

The flux at interface $\frac{1}{2}$, $E_{1/2}$ is evaluated as,

$$E_{1/2} = E^c_{1/2} + E^p_{1/2} \tag{122}$$

where

$$E^c_{1/2} = U^+ f^c_L + U^- f^c_R \tag{123}$$

and

$$E^p_{\frac{1}{2}} = p_{1/2} \begin{pmatrix} 0 \\ l_x \\ l_y \\ l_z \\ U_{1/2} \end{pmatrix}, \quad f^c = \begin{pmatrix} \rho \\ \rho u \\ \rho v \\ \rho w \\ \rho e \end{pmatrix} \tag{124}$$

$$p_{1/2} = P^+ p_L + P^- p_R \tag{125}$$

The different formulations for U^+, U^-, P^+ and P^- can be found in[120, 121, 18, 123] with different behavior.

The preconditioning of the the LDE scheme needs to satisfy two conditions when the flow velocity approaching zero: 1) the eigenvalues of the Jacobian matrices should be at the same order of magnitude of the velocity, 2) the numerical dissipation should diminish. The condition 1 is explained in the previous section. The condition 2 is described as the following.

The interface flux $E_{1/2}$ can be generally expressed as consisting of a central differencing plus a numerical dissipation $\mathbf{D}$ as[105]

$$E_{1/2} = \frac{1}{2}\left[E_L + E_R + \mathbf{D}(L, R)\right] \tag{126}$$

Similarly, the interface pressure $p_{1/2}$ of (125) can be written as[128]

$$\begin{aligned} p_{1/2} &= P^+ p_L + P^- p_R \\ &= \frac{1}{2}\left[(p_L + p_R) + (P^+ - P^-)(p_L - p_R) + (P^+ + P^- - 1)(p_L + p_R)\right] \end{aligned} \quad (127)$$

The last two terms can be regarded as the diffusion terms of $p_{1/2}$. A low Mach number results in large pressure value, hence the second diffusion term could be excessively large. A better scaling is found by replacing $(p_L + p_R)$ by $2\rho_{1/2}\tilde{c}^2_{1/2}$ or $2\rho_{1/2}U^2_{r,1/2}$[129, 128]. That is, Eq. (127) is replaced by

$$\begin{aligned} p_{1/2} &= P^+ p_L + P^- p_R \\ &= \frac{1}{2}\left[(p_L + p_R) + (P^+ - P^-)(p_L - p_R) + \rho_{1/2}U^2_{r,1/2}(P^+ + P^- - 1)\right] \end{aligned} \quad (128)$$

Similarly, the split velocities $U^{\pm}$ used in the low-diffusion flux splitting scheme can be written as

$$U^+ = \tilde{c}_{1/2}[M^+ - M^+_{1/2}]$$

$$U^- = \tilde{c}_{1/2}[M^- + M^-_{1/2}]$$

where

$$M^+_{1/2} = M_{1/2}\left(1 - \frac{p_L - p_R + \delta|p_L - p_R|}{p_L + p_R}\right) \quad (129)$$

$$M^-_{1/2} = M_{1/2}\left(1 + \frac{p_L - p_R - \delta|p_L - p_R|}{p_L + p_R}\right) \quad (130)$$

Using $2\rho_L U^2_{r,1/2}$ and $2\rho_R U^2_{r,1/2}$ to replace $(p_L + p_R)$ in Eqs. (129) and (130), respectively, the new split velocities are obtained as

$$U^+ = \tilde{c}_{1/2}\left[M^+ - M_{1/2}\left(1 - \frac{p_L - p_R + \delta|p_L - p_R|}{2\rho_L U^2_{r,1/2}}\right)\right] \quad (131)$$

$$U^- = \tilde{c}_{1/2}\left[M^- + M_{1/2}\left(1 + \frac{p_L - p_R - \delta|p_L - p_R|}{2\rho_R U^2_{r,1/2}}\right)\right] \quad (132)$$

$\delta = 1$ is used in [128], $\delta = 0$ is used in [130] and also in this paper.

Other quantities used in the preceding definitions are

$$M^+ = \alpha^+_L(1 + \beta_L)M_L + \beta_L M^+_L$$

$$M^- = \alpha^-_R(1 + \beta_R)M_R - \beta_R M^-_R$$

$$M^{\pm}_{L,R} = \pm\frac{1}{4}(M_{L,R} \pm 1)^2$$

$$\alpha^{\pm}_{L,R} = \frac{1}{2}[1 \pm sign(1.0, M_{L,R})]$$

$$\beta_{L,R} = -\max[0, 1 - int(|M_{L,R}|)]$$

$$M_{1/2} = \frac{1}{2}(M^+ - \alpha_L^+ M_L - M^- + \alpha_R^- M_R)$$

$$P^{\pm} = \alpha_{L,R}^{\pm}(1 + \beta_{L,R}) - \frac{\beta_{L,R}}{2}[1 \pm M_{L,R}]$$

$$\tilde{c}_{1/2} = \left. \frac{\sqrt{(1 - M_r^2)^2 U^2 + 4C^2 M_r^2}}{1 + M_r^2} \right|_{1/2}$$

and

$$M_{L,R} = \frac{U_{L,R}}{\tilde{c}_{1/2}}$$

Since the H-CUSP scheme uses total enthalpy and the E-CUSP scheme uses total energy in Eq.(121) for the convective vector, H-CUSP schemes[105, 128, 130] hence do not have the term of $p_{1/2}U_{1/2}$ in Eq. (124). In this paper, the interface velocity $U_{1/2}$ is evaluated as

$$U_{1/2} = U^+ + U^- \tag{133}$$

5.5. Numerical Results

For all cases in this paper, the fifth order WENO scheme with $\varepsilon = 10^{-2}$ is used for the reconstruction of conservative variables. The matrices of implicit part are calculated by using those formula in Section 2.2 with the preconditioned Roe's fluxes. The conservative fourth order central differencing scheme is used for viscous terms.

5.5.1. Lid Driven Cavity Flow

The lid driven cavity flow, which is a flow in a cavity with the lid moving at a constant speed, is a benchmark solution used to validate incompressible flow calculation[108, 131, 132].

The flowfiled with Reynolds of 3200, no-slip isothermal wall boundary condition and a Mach number of 10^{-3} for the moving lid is calculated. For the purpose of comparison, the uniform mesh systems of 100×100 and 200×200 are used.

Fig. 28 is the streamlines for the lid driven cavity flow calculated by the preconditioned Roe scheme with a mesh of 100×100. It exhibits a large primary vortex with two secondary vortices in the two bottom corners and a secondary vortex near the upper-left corner, which is the same as other researchers predict[108, 131, 132].

Figs. 29 and 30 give the convergence histories and show that the residuals are reduced by $8 \sim 9$ order of magnitude for both the preconditioned Roe scheme and E-CUSP scheme. Since the primitive pressure p is used in the present preconditioned method, the decreased Mach number results in increased machine round off errors, which increase proportionally with M^2[126, 93] and make the residual floating at the level of 10^{-5}. It needs to point out that the method without preconditioning can not get the correct solution for this case due to excessive dissipation.

Figs. 31 and 32 are the comparison of the velocity component in x-direction along the vertical centerline. The present results are in good agreement with that obtained by solving incompressible Navier-Stokes equations[131].

5.5.2. Wall Boundary Layer

A steady state laminar boundary layer flow with $M_\infty = 10^{-3}$ on an adiabatic flat plate is calculated. The Reynolds number based on the length of the flat plate is 4.0×10^4. The Prandtl number of 1.0 is used in order to compare with the analytical solution. The computation domain is taken to be $[0, 2] \times [0, 1.6]$. The mesh size is 180×80.

Figs. 33 and 34 show the convergence histories. It can be seen that the residual of preconditioning is reduced 2 and 4 order of magnitude lower than those without preconditioning for the Roe scheme and E-CUSP scheme, respectively. In addition, the residual of E-CUSP scheme without preconditioning oscillates at a high level with a large amplitude.

The velocity profiles in Figs. 35 and 36 demonstrate that the numerical solutions without preconditioning is greatly diffused due to the large numerical dissipation, whereas the preconditioned solvers accurately resolve the velocity profile.

5.5.3. Transonic Converging-Diverging Nozzle

This is the same case calculated in Section 2 to demonstrate the ability of the preconditioned system to deal with the flows at all speeds. Figs. 37 and 38 are the comparison of the convergence histories with and without precondition. The preconditioned convergence rates are about 30% and 20% faster than those without preconditioning for Roe scheme and E-CUSP scheme, respectively.

For this case, identical solutions are obtained by using preconditioning method and the method without preconditioning. The Mach contours given in Figs. 39 and 40 show that the preconditioned schemes get the right solutions.

5.5.4. Transonic RAE2822 Airfoil

The transonic compressible flows over RAE2822 airfoil in Section 2 is also computed using the preconditioned system to demonstrate its capability to deal with the flows at all speeds. From Figs. 41 and 42, we can see that both the preconditioned methods only need about half of the iterations of the original schemes to converge to machine zero. The pressure coefficients given in Figs. 43 and 44 show that the preconditioned systems obtain the same solutions.

6. Conclusions

A low diffusion E-CUSP (LDE) scheme is developed for preconditioned Navier-Stokes equations. Combined with the 5th-order WENO scheme for inviscid flux and the unfactored implicit Gauss-Seidel relaxation scheme for time integration, both the preconditioned E-CUSP scheme and Roe scheme are used to calculate flow fields from very low speed incompressible flows to transonic compressible flows. The numerical simulations show that the preconditioned methods are efficient, accurate and robust, not only for the low Mach number incompressible flows, but also for the subsonic and transonic compressible flows.

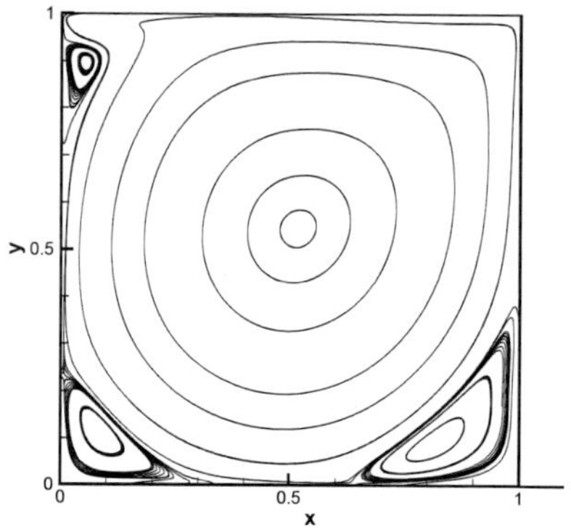

Figure 28. Streamlines for the lid driven cavity flow, Roe scheme, $Re = 3200$, $M = 10^{-3}$.

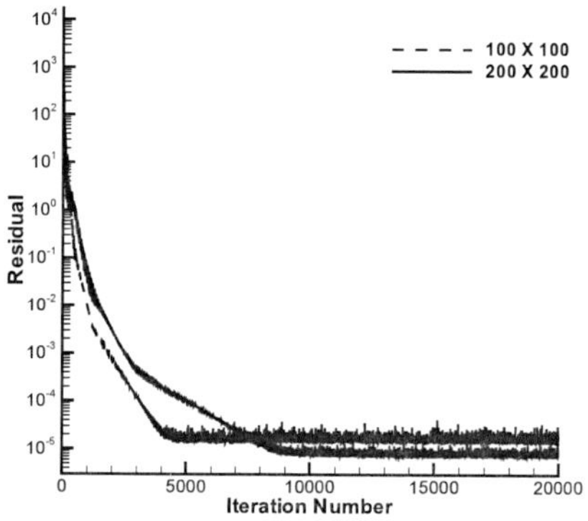

Figure 29. Convergence histories of the lid driven cavity flow, Roe scheme, $Re = 3200$, $M = 10^{-3}$.

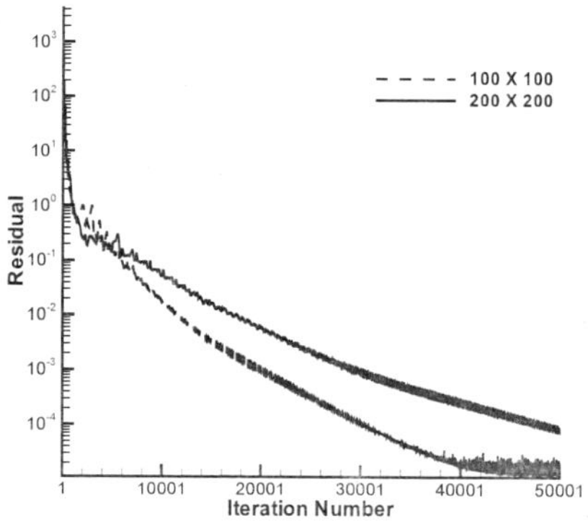

Figure 30. Convergence histories of the lid-driven cavity flow, E-CUSP scheme, $Re = 3200$, $M = 10^{-3}$.

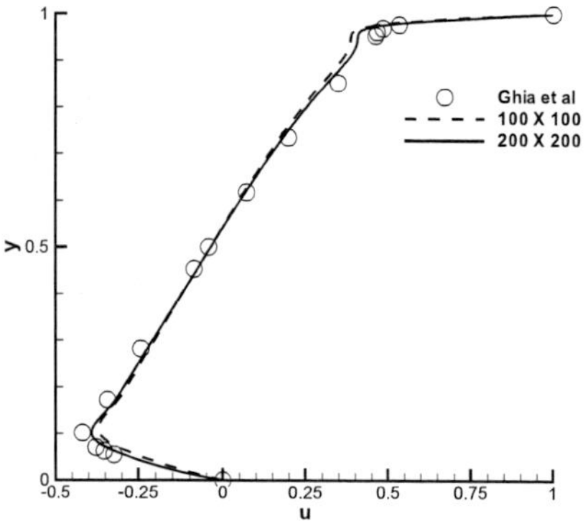

Figure 31. Comparison of u velocity along the vertical centerline for the lid driven cavity flow, Roe scheme, $Re = 3200,\ M = 10^{-3}$.

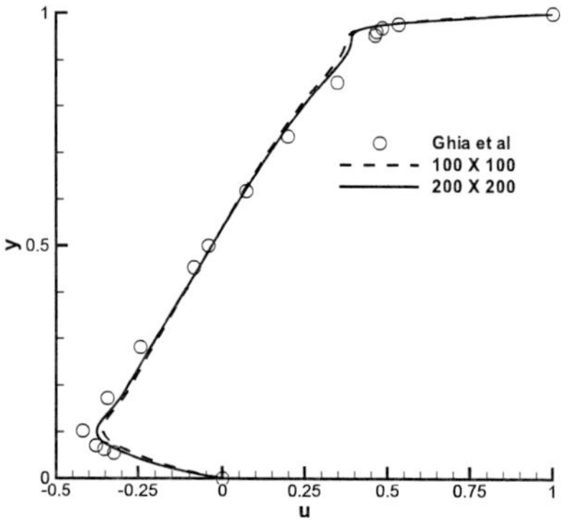

Figure 32. Velocity u along the vertical centerline for the lid driven cavity flow, E-CUSP scheme, $Re = 3200,\ M = 10^{-3}$.

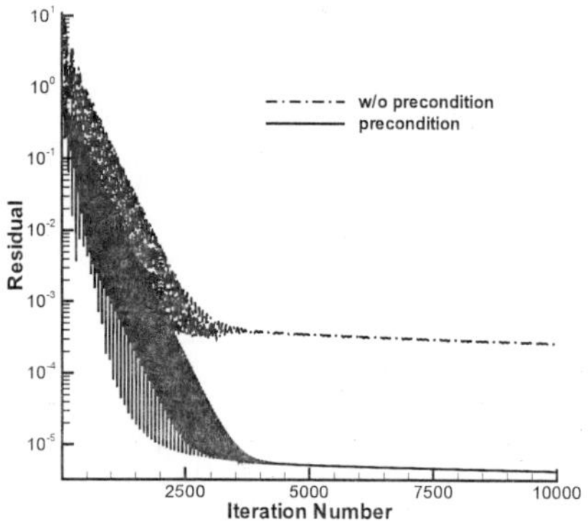

Figure 33. Convergence histories of the subsonic boundary layer flow, Roe scheme, $M = 10^{-3}$.

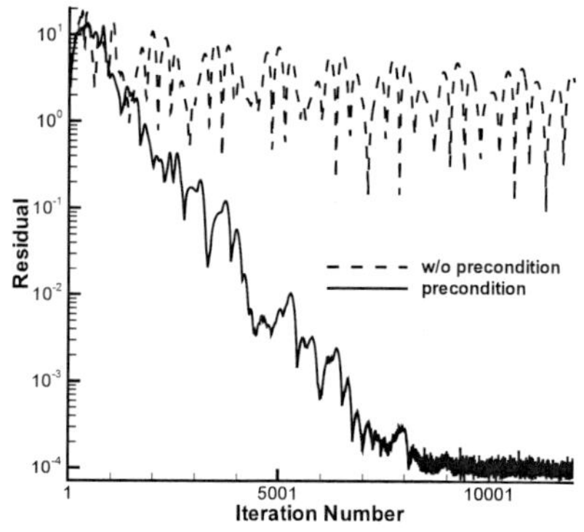

Figure 34. Convergence rate of the subsonic boundary layer flow, E-CUSP scheme, $M = 10^{-3}$.

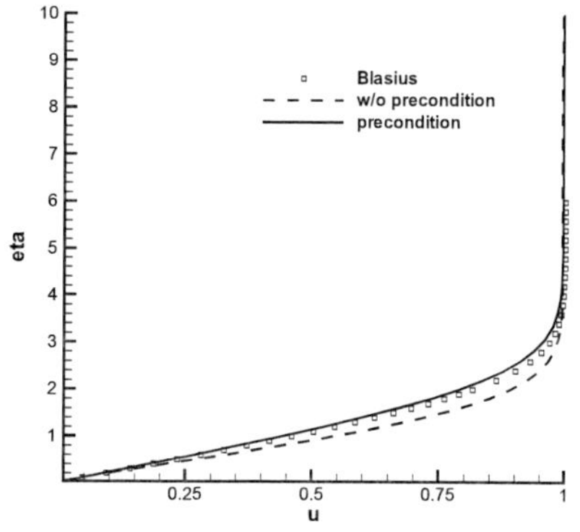

Figure 35. Velocity profile of the subsonic boundary layer flow, Roe scheme, $M = 10^{-3}$.

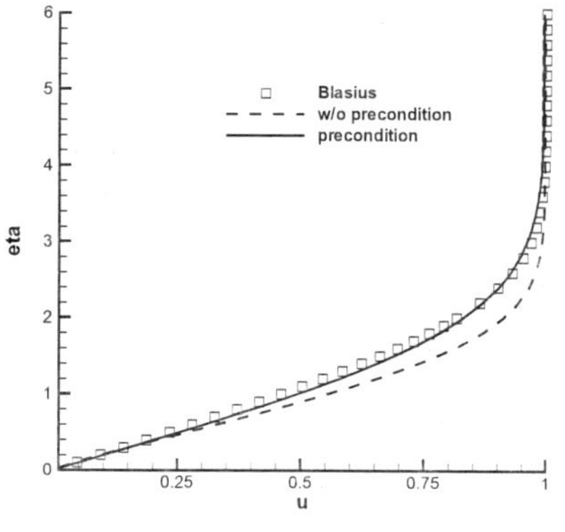

Figure 36. Velocity profile of the subsonic boundary layer flow, E-CUSP scheme, $M = 10^{-3}$.

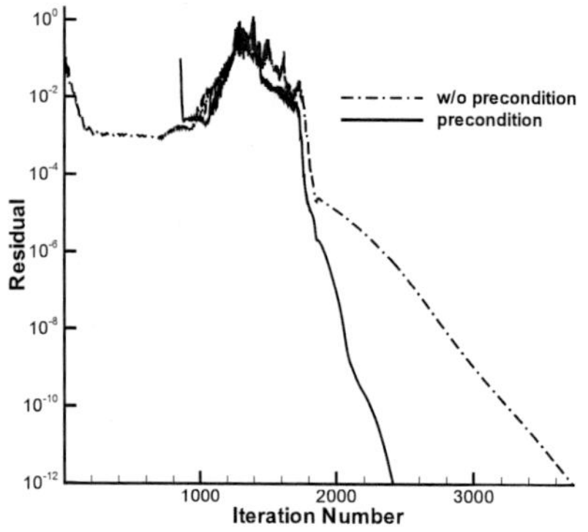

Figure 37. Convergence histories of the transonic converging-diverging nozzle flow, Roe scheme.

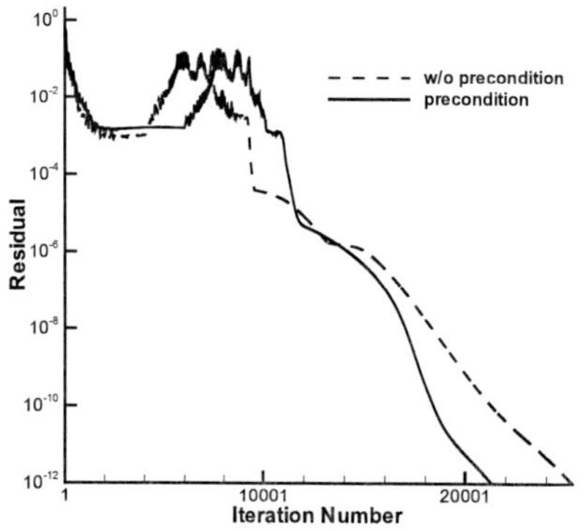

Figure 38. Convergence histories of the transonic converging-diverging nozzle flow, E-CUSP scheme.

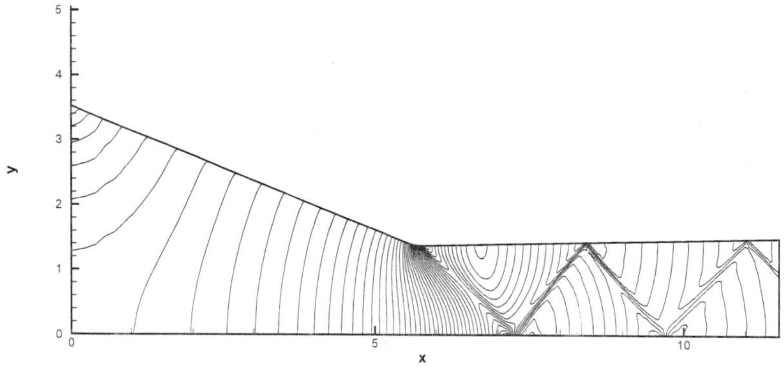

Figure 39. Mach number contours of the transonic converging-diverging nozzle flow, preconditioned Roe scheme.

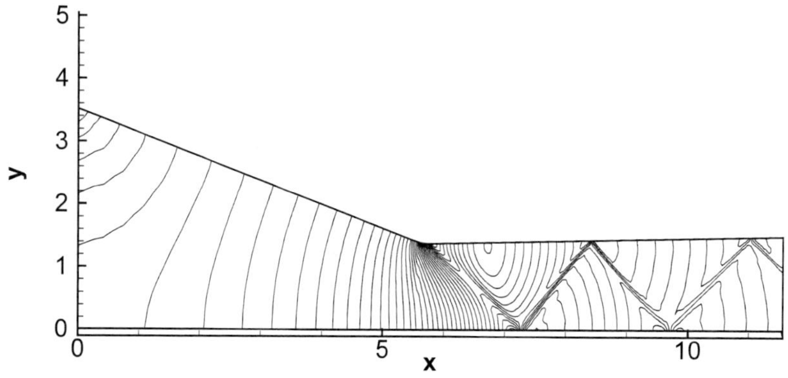

Figure 40. Mach number contours of the transonic converging-diverging nozzle flow, preconditioned E-CUSP scheme.

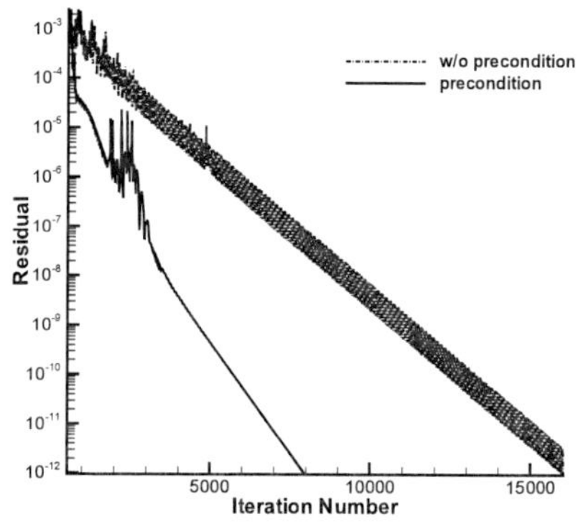

Figure 41. Convergence histories of the transonic flow over RAE2822 airfoil, Roe scheme.

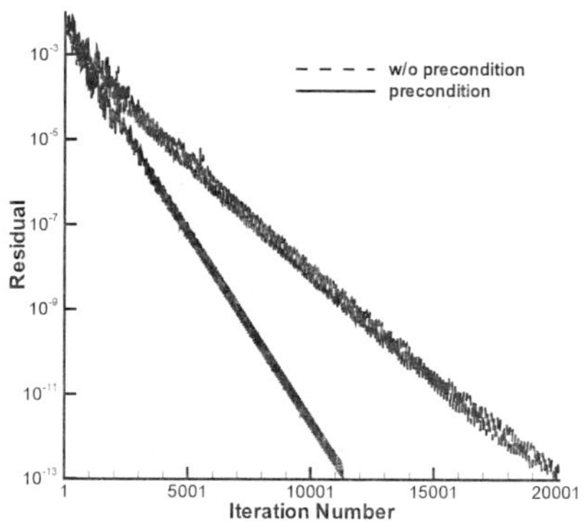

Figure 42. Convergence rate of the transonic flow over RAE2822 airfoil, E-CUSP scheme.

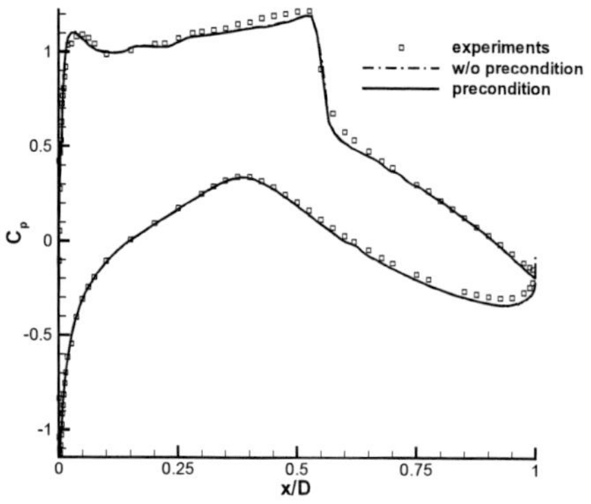

Figure 43. The pressure coefficients at the airfoil surface of the transonic flow over RAE2822 airfoil, Roe scheme.

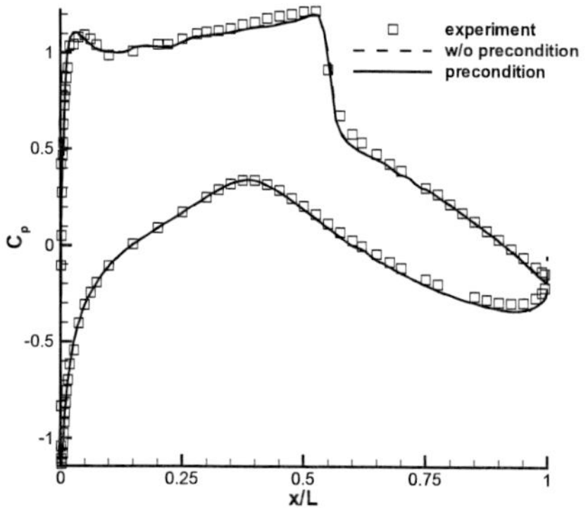

Figure 44. The pressure coefficients at the airfoil surface of the transonic flow over RAE2822 airfoil, E-CUSP scheme.

References

[1] J. C. Tannehill, D. A. Anderson, and R. H. Pletcher, *Computational Fluid Mechanics and Heat Transfer, 2nd Edition*. Taylor & Francis, 1997.

[2] B. Baldwin, H. Lomax, "Thin Layer Approximation and Algebraic Model for Separated Turbulent Flows." *AIAA Paper* 78-257, 1978.

[3] P. Spalart and S. Allmaras, "A One-equation Turbulence Model for Aerodynamic Flows." AIAA-92-0439, 1992.

[4] D. C. Wilcox, " Reassessment of the scale-determining equation for advanced turbulence models," *AIAA J.*, vol. 26, pp. 1299–1310, 1988.

[5] D. C. Wilcox, " Multi-scale model of turbulent flows," *AIAA J.*, vol. 26, pp. 1311–1320, 1988.

[6] C. G. Speziale, R. Abid, E. C. Anderson, " Critical evaluation of two-equation models for near-wall turbulence," *AIAA J.*, vol. 30, pp. 324–331, 1992.

[7] D. Knight, G. Zhou, N. Okong'o, V. Shukla, "Compressible large eddy simulation using unstructured grids." *AIAA Paper* 98-0535, 1998.

[8] J. Smagorinsky, "General Circulation Experiments with the Primitive Equations, I. The Basic Experiment," *Monthly Weather Review*, vol. 91, pp. 99–164, 1963.

[9] M. Germano, U. Piomeli, P. Moin, W. Cabot, "A Dynamic Subgrid-scale Eddy Viscosity Model," *Physics of Fluids A*, vol. 3, pp. 1760–1765, July 1991.

[10] J. P. Boris, "On large eddy simulation using subgrid turbulence models," *in Whither Turbulence? Turbulence at the Crossroads*, edited by J.L. Lumley, New York: Springer-Verlag, 1990.

[11] J. P. Boris, F. F. Grinstein, E. S. Oran, R. L. Kolbe , " New insights into large eddy simulation ," *Fluid Dynamics Research* , vol. 10, p. 199, 1992.

[12] J. P. Boris, "More for LES: A brief historical perspective of MILES." in: Implicit Large Eddy Simulation, Editors: F. F. Grinstein, L. G. Margolin, and W. J. Rider, Cambridge University Press, pp.9-38, 2007.

[13] R. Beam and R. Warming, "An Implicit Factored Scheme for the Compressible Navier-Stokes Equations ," *AIAA Journal*, vol. 16, No. 4, pp. 393–402, 1978.

[14] A.C. Taylor III, W.F. Ng, R.W. Walters, "An improved upwind finite volume relaxation method for high speed viscous flows," *J.Compu.Phys.*, vol. 99, pp. 159–168, 1992.

[15] S.E. Rogers, D. Kwak, "An upwind difference scheme for time accurate incompressible navier-stokes equations," *AIAA J.*, vol. 28, pp. 253–262, 1990.

[16] S.E. Rogers, F.R. Menter, N.N. Mansour, and P.A. Durbin, "A comparison of turbulence models in computing multi-element airfoil flows." *AIAA* 94-0291, Jan. 1994.

[17] G.-C. Zha and E. Bilgen, "Numerical Study of Three-Dimensional Transonic Flows Using Unfactored Upwind-Relaxation Sweeping Algorithm," *Journal of Computational Physics*, vol. 125, pp. 425–433, 1996.

[18] G.-C. Zha and Z.-J. Hu, "Calculation of Transonic Internal Flows Using an Efficient High Resolution Upwind Scheme," *AIAA Journal*, vol. 42, No. 2, pp. 205–214, 2004.

[19] X. Chen and G.-C. Zha, "Fully coupled fluid-structural interactions using an efficient high solution upwind scheme," *Journal of Fluid and Structure*, vol. 20, pp. 1105–1125, 2005.

[20] A. Jameson and S. Yoon, "Lower-upper implicit schemes with multiple grids for the Euler equations," *AIAA J.*, vol. 7, pp. 929–935, 1987.

[21] S. Yoon and A. Jameson, "Lower-upper symmetric-gauss-seidel method for the euler and navier-stokes equations," *AIAA J.*, vol. 26, pp. 1025–1026, 1988.

[22] G.H. Klopfer and S. Yoon, "Multizonal Navier-Stokes Code with LU-SGS Scheme." *AIAA* 93-2965, July 1993.

[23] M. Kandula and P.G. Buning, "Implementation of LU-SGS Algorithm and Roe Upwinding Scheme in OVERFLOW Thin-Layer Navier-Stokes Code." *AIAA* 94-2357, Jun. 1994.

[24] A. Jameson and D.A. Caughey, "How many steps are required to solve the Euler equations of steady, compressible flow: in search of a fast solution algorithm." *AIAA*-2001-2673, Jun. 2001.

[25] S.E. Rogers, "Comparison of implicit schemes for the incompressible navier-stokes equations," *AIAA J.*, vol. 33, pp. 2066–2072, 1995.

[26] L. Yuan, "Comparison of implicit multigrid schemes for three-dimensional incompressible flows," *J. Compu. Phys.*, vol. 177, pp. 134–155, 2002.

[27] S.E. Rogers, D. Kwak, C. Kiris, "Numerical solution of the incompressible navier-stokes equations for steady-state and time-dependent problems," *AIAA J.*, vol. 29, pp. 603–610, 1991.

[28] P. Roe, "Approximate Riemann solvers, parameter vectors, and difference schemes," *Journal of Computational Physics*, vol. 43, pp. 357–372, 1981.

[29] B.van Leer, "Towards the ultimate conservative difference scheme, v:a second-order sequel to godunov's method," *J. Compu. Phys.*, vol. 32, pp. 101–136, 1979.

[30] Y.-Q. Shen, B.-Y. Wang, G.-C. Zha, "Comparison study of implicit Gauss-Seidel line iteration method for transonic flows." *AIAA*-2007-4332, June 2007.

[31] A. Jameson, "Time Dependent Calculations Using Multigrid with Application to Unsteady Flows past Airfoils and Wings." *AIAA Paper* 91-1596, 1991.

[32] Y.-Q. Shen, G.-Z. Zha, "Large eddy simulation using a new set of sixth order schemes for compressible viscous terms," *Journal of Computational Physics*, vol. 229, pp. 8296–8312, 2010.

[33] Y.-Q. Shen, G.-Z. Zha, X.-Y. Chen , "High order conservative differencing for viscous terms and the application to vortex-induced vibration flows," *Journal of Computational Physics*, vol. 228, pp. 8283–8300, 2009.

[34] X. Chen, G. -C Zha, "Fully coupled fluid-structural interactions using an efficient high solution upwind scheme," *Journal of Fluid and Structure*, vol. 20, pp. 1105–1125, 2005.

[35] B.-Y. Wang, G.-Z. Zha, "Numerical simulation of transonic limit cycle oscillations using high-order low-diffusion schemes," *Journal of Fluids and Structures*, vol. 26, pp. 579–601, 2010.

[36] Mason, M. L. and Putnam, L. E. , "The Effect of Throat Contouring on Two-Dimensional Converging-Diverging Nozzles at Static Conditions ." *NASA Technical Paper* 1704, 1980.

[37] X.D. Liu, S. Osher, T. Chan, "Weighted essentially non-oscillatory schemes," *J.Comput.Phys.*, vol. 115, pp. 200–212, 1994.

[38] G.-S. Jiang, C.-W. Shu, "Efficient implementation of weighted ENO schemes," *J.Comput.Phys.*, vol. 126, pp. 202–228, 1996.

[39] A. Harten, B. Engquist, S. Osher, S. Chakravarthy, "Uniformly high order essentially non-oscillatory schemes, III," *Journal of Computational Physics*, vol. 71, pp. 231–303, 1987.

[40] C.-W. Shu, O. Osher, "Efficient Implementation of Essentially Non-Oscillatory Shock Capturing Schemes, II," *Journal of Computational Physics*, vol. 83, pp. 32–78, 1989.

[41] A.K. Henrick, T.D. Aslam, J.M. Powers, "Mapped weighted essentially non-oscillatory schemes:Achiving optimal order near critical points," *J.Comput.Phys.*, vol. 208, pp. 206–227, 2005.

[42] R. Borges, M. Carmona, B. Costa, W.S. Don, "An improved weighted essentially non-oscillatory scheme for hyperbolic conservation laws," *Journal of Computational Physics*, vol. 227, pp. 3191–3211, 2008.

[43] D.S. Balsara, C.-W. Shu, "Monotonicity preserving weighted essentially non-oscillatory schemes with increasingly high order of accuracy," *J.Comput.Phys.*, vol. 160, pp. 405–452, 2000.

[44] Z.J. Wang and R.F. Chen, "Optimized weighted essentially non-oscillatory schemes for linear waves with discontinuity," *J.Comput.Phys.*, vol. 174, pp. 381–404, 2001.

[45] M. P. Martin, E. M. Taylor, M. Wu, and V. G. Weirs, "A Bandwidth-Optimized WENO Scheme for the Direct Numerical Simulation of Compressible Turbulenc," *Journal of Computational Physics*, vol. 220, pp. 270–289, 2006.

[46] Y.-Q. Shen, R.-Q. Wang, H.-Z. Liao, " A fifth-order accurate weighted ENN difference scheme and its applications," *Journal of Computational Mathematics*, vol. 19, pp. 531–538, 2001.

[47] B. Cockburn, C.W. Shu, "Nonlinearly stable compact schemes for shock calculations," *SIAM Journal on Numerical Analysis*, vol. 31, pp. 607–627, 1994.

[48] K. S. Ravichandran, "Higher Order KFVS Algorithms Using Compact Upwind Difference Operators," *Journal of Computational Physics*, vol. 130, pp. 161–173, 1997.

[49] Guo-Hua Tu, Xiang-Jiang Yuan, "A characteristic-based shock-capturing scheme for hyperbolic problems," *Journal of Computational Physics*, vol. 225, pp. 2083–2097, 2007.

[50] N. A. Adams, K. Shariff, "A High-Resolution Hybrid Compact-ENO Scheme for Shock-Turbulence Interaction Problems," *Journal of Computational Physics*, vol. 127, pp. 27–51, 1996.

[51] S. Pirozzoli, "Conservative hybrid compact-WENO schemes for shock-turbulence interaction," *J.Comput.Phys.*, vol. 178, pp. 81–117, 2002.

[52] Yu-Xin Ren, Miao'er Liu, Hanxin Zhang, "A characteristic-wise hybrid compact-WENO scheme for solving hyperbolic conservation laws," *Journal of Computational Physics*, vol. 192, pp. 365–386, 2003.

[53] Qiang Zhou, Zhaohui Yao, Feng He, M.Y. Shen, "A new family of high-order compact upwind difference schemes with good spectral resolution," *Journal of Computational Physics*, vol. 227, pp. 1306–1339, 2007.

[54] Y.-Q. Shen, G.-W. Yang, Z. Gao, "High-resolution finite compact difference schemes for hyperbolic conservation laws," *J.Comput.Phys.*, vol. 216, pp. 114–137, 2006.

[55] Y.-Q. Shen, G.-W. Yang, "Hybrid finite compact-WENO schemes for shock calculation," *International Journal for Numerical Methods in Fluids* , vol. 53, pp. 531–560, 2007.

[56] Xiaogang Deng, Hanxin Zhang, "Developing High-Order Weighted Compact Nonlinear Schemes," *Journal of Computational Physics*, vol. 165, pp. 22–44, 2000.

[57] Shuhai Zhang, Shufen Jiang, Chi-Wang Shu, "Development of nonlinear weighted compact schemes with increasingly higher order accuracy," *Journal of Computational Physics*, vol. 227, pp. 7294–7321, 2008.

[58] Donald P. Rizzetta, Miguel R. Visbal, Philip E. Morgan, "A high-order compact finite-difference scheme for large-eddy simulation of active flow control," *Progress in Aerospace Sciences*, vol. 44, pp. 397–426, 2008.

[59] H. C. Yee, Sjogreen, B., "Adaptive filtering and limiting in compact high order methods for multiscale gas dynamics and MHD systems," *Computers & Fluids*, vol. 37, pp. 593–619, 2008.

[60] M. R. Visbal, D. Gaitonde, "High Order-Accurate Methods for Complex Unsteady Subsonic Flows," *AIAA Journal*, vol. 37, No. 10, pp. 1231–1239, 1999.

[61] A.W. Cook, W.H. Cabot, "A high-wavenumber viscosity for high resolution numerical method," *Journal of Computational Physics*, vol. 195, pp. 594–601, 2004.

[62] A.W. Cook, W.H. Cabot, "Hyperviscosity for shock-turbulence interactions," *Journal of Computational Physics*, vol. 203, pp. 379–385, 2005.

[63] B. Fiorina, S.K. Lele, "An artificial nonlinear diffusivity method for supersonic reacting flows with shocks," *Journal of Computational Physics*, vol. 222, pp. 246–264, 2007.

[64] S. Kawai, S.K. Lele, "Localized artificial diffusivity scheme for discontinuity capturing on curvilinear meshes," *Journal of Computational Physics*, vol. 227, pp. 9498–9526, 2008.

[65] C. Bogey, N. de Cacqueray, C. Bailly, "A shock-capturing methodology based on adaptative spatial filtering for high-order non-linear computations," *Journal of Computational Physics*, vol. 228, pp. 1447–1465, 2009.

[66] H. C. Yee, N. D. Sandham, M. J. Djomehri, "Low-Dissipative High-Order Shock-Capturing Methods Using Characteristic-Based Filters," *Journal of Computational Physics*, vol. 150, pp. 199–238, 1999.

[67] S.-C. Lo, G. A. Blaisdell, A. S. Lyrintzis, "High-order shock capturing schemes for turbulence calculations," *International Journal For Numerical Methods in Fluids*, vol. 62, pp. 473–498, 2010.

[68] G. A. Gerolymos, D. Senechal, I.Vallet, " Very-high-order WENO schemes," *Journal of Computational Physics*, vol. 228, pp. 8481–8524, 2009.

[69] Y.-Q. Shen, G.-C. Zha, "Improvement of weighted essentially non-oscillatory schemes near discontinuity." *AIAA paper* 2009-3655, 2009.

[70] Y.-Q Shen, G.-C Zha, "Generalized finite compact difference scheme for shock/complex flowfield interaction." accepted by *Journal of Computational Physics*, 2011.

[71] S. A. Lele, "Compact finite difference schemes with spectral-like resolution," *Journal of Computational Physics*, vol. 103, No. 1, pp. 16–42, 1992.

[72] Y.-Q. Shen, G.-C. Zha, " A robust seventh-order WENO scheme and its applications." *AIAA*-2008-0757, 2008.

[73] Y.-Q. Shen, G.-C. Zha, "Improved Seventh-Order WENO Scheme." 48th AIAA Aerospace Sciences Meeting Including the New Horizons Forum and Aerospace Exposition, AIAA-2010-1451, 4-7, Jan. 2010, Orlando, Florida.

[74] Y.-Q. Shen, G.-C. Zha, "A Seventh-Order WENO Scheme and Its Applications." submitted to *Journal of Computational Physics*, manuscript JCOMP D-10-00192, 26, Feb. 2010.

[75] M. Castro, B. Costa, W. S. Don, "High order weighted essentially non-oscillatory WENO-Z schemes for hyperbolic conservation laws," *Journal of Computational Physics*, vol. 230, pp. 1766–1792, 2011.

[76] Y.-Q. Shen, G.-C. Zha, B.-Y. Wang, "Improvement of stability and accuracy of implicit WENO scheme," *AIAA Journal*, vol. 47, pp. 331–344, 2009.

[77] P. Woodward, P. Colella, "The numerical simulation of two-dimensional fluid flow with strong shocks," *Journal of Computational Physics*, vol. 54, pp. 115–173, 1984.

[78] C.-W. Shu, O. Osher, "Efficient implementation of essentially non-oscillatory shock capturing schemes," *Journal of Computational Physics*, vol. 77, pp. 439–471, 1988.

[79] C.-W. Shu, "Essentially non-oscillatory and weighted essentially non-oscillatory schemes for hyperbolic conservation laws." NASA/CR-97-206253, *ICASE Report* No.97-65, Nov. 1997.

[80] Y.-Q. Shen, G.-Z. Zha, "Improvement of the WENO scheme smoothness estimator," *International Journal for Numerical Methods in Fluids*, vol. 64, pp. 653–675, 2010.

[81] P. Beaudan, P. Moin , " Numerical experiments on the flow past a circular cylinder at a sub-critical Reynolds number ." Report No.TF-62, Department of Mechanical Engineering, Stanford University, 1994.

[82] R. Mittal, P. Moin, "Suitability of upwind-biased schemes for large-eddy simulation of turbulent flows," *AIAA Journal*, vol. 36, pp. 1415–1417, 1997.

[83] G. Kravchenko, P. Moin , " Numerical studies of flow over a circular cylinder at $Re_D = 3900$," *Phys.Fluids* , vol. 12, pp. 403–417, 2000.

[84] D.P. Rizzetta, M.R. Visbal, G.A. Blaisdell, " A time-implicit high-order compact differencing and flitering scheme for large-eddy simulation ," *Int.J.Numer.Meth.Fluids* , vol. 42, pp. 665–693, 2003.

[85] A. Kasliwal, K. Ghia, U. Ghia , " Higher-order accurate solution for flow past a circular cylinder at $Re = 13,400$." *AIAA*-2005-1123, 43rd AIAA Aerospace Sciences Meeting and Exhibit, Reno, Nevada,10-13 Jan., 2005.

[86] L.M. Lourenco, C. Shih , " Characteristics of the plane turbulent near wake of a circular cylinder, A particle image velocity study ." private communication by Beaudan and Moin (data taken from[81]), 1993.

[87] A. L. De Bortoli, "Multigrid based aerodynamical simulations for the NACA 0012 airfoil," *Applied Numerical Mathematics*, vol. 40, pp. 337–349, 2002.

[88] S. Polsky, "Progress Towards Modeling Ship/Aircraft Dynamic Interface." hpcmp-ugc,pp.163-168, *HPCMP Users Group Conference (HPCMP-UGC'06)*, 2006.

[89] Niles A. Pierce, Michael B. Giles, "Preconditioned Multigrid Methods for Compressible Flow Calculations on Stretched Meshes," *Journal of Computational Physics*, vol. 136, pp. 425–445, 1997.

[90] E. Turkel, "Preconditioned methods for solving the incompressible and low speed compressible equations," *Journal of Computational Physics*, vol. 72, pp. 277–298, 1987.

[91] C.L. Merkle, Y.H. Choi, "Computation of low-speed compressible flows with time-marching procedures," *International Journal for Numerical Methods in Engineering* , vol. 25, pp. 293–311, 1988.

[92] E. Turkel, "Preconditioning techniques in computational fluid dynamics," *Annu. Rev. Fluid Mech.*, vol. 31, pp. 385–416, 1999.

[93] Y. H. Choi and C. L. Merkle, "The Application of Preconditioning in Viscous Flows," *Journal of Computational Physics* , vol. 105, pp. 207–223, 1993.

[94] J.M. Weiss, and W.A. Smith, "Preconditioning Applied to Variable and Constant Density Flows," *AIAA Journal*, vol. 33, pp. 2050–2057, 1995.

[95] E. Turkel, "Review of preconditioning methods for fluid dynamics," *Applied Numerical Mathematics*, vol. 12, pp. 257–284, 1993.

[96] N. Alkishriwi, M. Meinke, W. Schroder, "A large-eddy simulation method for low Mach number flows using preconditioning and multigrid," *Computers & Fluids*, vol. 35, pp. 1126–1136, 2006.

[97] Z.F. Xu and C.-W. Shu, "Anti-diffusive flux corrections for high order finite difference WENO schemes," *J.Comput.Phys.*, vol. 205, pp. 458–485, 2005.

[98] R.C. Swanson, E. Turkel, C.-C. Rossow, "Convergence acceleration of Runge-Kutta schemes for solving the Navier-Stokes equations," *Journal of Computational Physics*, vol. 224, pp. 365–388, 2007.

[99] S. Yamamoto, "Preconditioning method for condensate fluid and solid coupling problems in general curvilinear coordinates," *Journal of Computational Physics*, vol. 207, pp. 240–260, 2005.

[100] C.L. Merkle, J.Y. Sullivan, P.E.O. Buelow and S. Venkateswaran, "Computation of Flows with Arbitrary Equations of State," *AIAA Journal*, vol. 36, pp. 515–521, 1998.

[101] J.M. Weiss, J.P. Maruszewski, W.A. Smith, "Implicit solution of preconditioned Navier-Stokes equations using algebraic multigrid," *AIAA Journal*, vol. 37, pp. 29–36, 1999.

[102] C. -C. Rossow, "Efficient computation of compressible and incompressible flows," *Journal of Computational Physics* , vol. 220, pp. 879–899, 2007.

[103] W. R. Briley, L. K. Taylor, D. L. Whitfield, "High-resolution viscous flow simulations at arbitrary Mach number," *Journal of Computational Physics* , vol. 184, pp. 79–105, 2003.

[104] E. Turkel, R. Radespiel, N. Kroll, "Assessment of preconditioning methods for multidimensional aerodynamics," *Computers & Fluids*, vol. 26, pp. 613–634, 1997.

[105] J.R. Edwards,M.-S. Liou, "Low-Diffusion Flux-Splitting Methods for Flows at All Speeds," *AIAA Journal*, vol. 36, pp. 1610–1617, 1998.

[106] N. Nigro, M. Storti, S. Idelsohn, T. Tezduyar, "Physics based GMRES preconditioner for compressible and incompressible Navier-Stokes equations," *Computer Methods in Applied Mechanics and Engineering*, vol. 154, pp. 203–228, 1998.

[107] C. -C. Rossow, "A blended pressure/density based method for the computation of incompressible and compressible flows," *Journal of Computational Physics*, vol. 185, pp. 375–398, 2003.

[108] J.-C. Huang, H. Lin, and J.-Y. Yang, "Implicit preconditioned WENO scheme for steady viscous flow computation," *Journal of Computational Physics*, vol. 228, pp. 420–438, 2009.

[109] A. Jameson, "Analysis and Design of Numerical Schemes for Gas Dynamics I: Artificial Diffusion, Upwind Biasing, Limiters and Their Effect on Accuracy and Multigrid Convergence in Transonic and Hypersonic Flow." *AIAA Paper* 93-3359, July, 1993.

[110] A. Jameson, "Analysis and Design of Numerical Schemes for Gas Dynamics I: Artificial Diffusion, Upwind Biasing, Limiters and Their Effect on Accuracy and Multigrid Convergence in Transonic and Hypersonic Flow," *Journal of Computational Fluid Dynamics*, vol. 4, pp. 171–218, 1995.

[111] A. Jameson, "Analysis and Design of Numerical Schemes for Gas Dynamics II: Artificial Diffusion and Discrete Shock Structure," *Journal of Computational Fluid Dynamics*, vol. 5, pp. 1–38, 1995.

[112] M.-S. Liou and C. J. Steffen, "A New Flux Splitting Scheme," *Journal of Computational Physics*, vol. 107, pp. 1–23, 1993.

[113] Y. Wada and M.-S. Liou, "An Accurate and Robust Splitting Scheme for Shock and Contact Discontinuities." *AIAA Paper* 94-0083, 1994.

[114] M.-S. Liou, "Progress Towards an Improved CFD Methods: AUSM$^+$." *AIAA Paper* 95-1701-CP, June, 1995.

[115] M.-S. Liou, "A Sequel to AUSM: AUSM$^+$," *Journal of Computational Physics*, vol. 129, pp. 364–382, 1996.

[116] M.-S. Liou, "Ten Years in the Making-AUSM-Family." *AIAA* 2001-2521, 2001.

[117] D. Hänel, R. Schwane, G. Seider, "On the Accuracy of Upwind Schemes for the Solution of the Navier-Stokes Eqautions." *AIAA paper* 87-1105 CP, 1987.

[118] J. R. Edwards, "A Low-Diffusion Flux-Splitting Scheme for Navier-Stokes Calculations." *AIAA Paper* 95-1703-CP, June, 1995.

[119] J. R. Edwards, "A Low-Diffusion Flux-Splitting Scheme for Navier-Stokes Calculations," *Computer & Fluids*, vol. 6, pp. 635–659, 1997.

[120] G. -C. Zha, E. Bilgen, "Numerical Solutions of Euler Equations by Using a New Flux Vector Splitting Scheme ," *International Journal for Numerical Methods in Fluids* , vol. 17, pp. 115–144, 1993.

[121] G.-C. Zha, "Numerical Tests of Upwind Scheme Performance for Entropy Condition ," *AIAA Journal*, vol. 37, pp. 1005–1007, 1999.

[122] G.-C. Zha, "Comparative Study of Upwind Scheme Performance for Entropy Condition and Discontinuities." *AIAA Paper* 99-CP-3348, June 28- July 1, 1999.

[123] G.-C. Zha, Y.-Q. Shen, B.-Y. Wang, "Calculation of Transonic Flows Using WENO Method with a Low Diffusion E-CUSP Upwind Scheme." AIAA Paper 2008-0745, Jan 2008.

[124] G.-C. Zha, "A Low Diffusion Efficient Upwind Scheme ," *AIAA Journal*, vol. 43, pp. 1137–1140, 2005.

[125] G.-C. Zha, "A Low Diffusion E-CUSP Upwind Scheme for Transonic Flows." *AIAA Paper 2004-2707, to appear in AIAA Journal, 34th AIAA Fluid Dynamics Conference*, June 28 - July 1 2004.

[126] Y.-Q. Shen, G.-C. Zha, "Simulation of Flows at All Speeds with High-Order WENO Schemes and Preconditioning." *47th AIAA Aerospace Sciences Meeting*, AIAA-2009-1312, Jan 2009.

[127] Y.-Q. Shen, G.-C. Zha, "Low Diffusion E-CUSP Scheme with High Order WENO Scheme for Preconditioned Navier-Stokes Equations." *48th AIAA Aerospace Sciences Meeting Including the New Horizons Forum and Aerospace Exposition* , AIAA-2010-1452, 4-7, Jan. 2010, Orlando, Florida.

[128] Mao, D., and Edwards, J. R., and Kuznetsov, A. V., and Srivastava, R. K. , "Development of low-diffusion flux-splitting methods for dense gas-solid flows," *Journal of Computational Physics*, vol. 185, pp. 100–119, 2003.

[129] J. Edwards, R. Franklin, and M.-S. Liou, "Low Diffusion Flux-Splitting Methods for Real Fluid Flows with Phase Transitions," *AIAA Journal*, vol. 38, pp. 1624–1633, 2000.

[130] Neaves, M. D., and Edwards, J. R., "All-speed time-accurate underwater projectile calculations using a preconditioning algorithm," *Journal of fluids engineering*, vol. 128, pp. 284–296, 2006.

[131] U. Ghia and K. N. Ghia and C. T. Shin, "High-Re solutions for incompressible flow using the Navier-Stokes equations and a multigrid method," *Journal of Computational Physics*, vol. 48, pp. 387–411, 1982.

[132] Sheng-Tao Yu and Bo-Nan Jiang and Nan-Suey Liu and Jie Wu , "The least-squares finite element method for low-mach-number compressible viscous flows," *International Journal for Numerical Methods in Engineering* , vol. 38, pp. 3591–3610, 1995.

In: Navier-Stokes Equations
Editor: R. Younsi

ISBN: 978-1-61324-590-3

Chapter 8

The Phenomenon of the Effective Viscosity for the Flow in Inhomogeneous Granular Medium

A.V. Gavrilov and I.V. Shirko
Moscow Institute of Physics and Technology, Russia

Abstract

The chapter is expository and contains both recent and the latest results of the research devoted to investigation of the transfer process occurring under the fluid flow in inhomogeneous granular medium. When fluid flows around the particles of a granular medium in a longitudinal direction, pulsating components of the velocity occur in the transverse direction, which are ignored by Darcy's law. If the average flow velocity changes in a transverse direction, these pulsations transfer additional momenta from layer to layer, which leads to the occurrence of effective viscosity forces. It is assumed that the fluid flow in a porous space is described by the Navier–Stokes equations. Using the fundamental propositions of Reynolds' averaging theory and Prandtl's mixing path, the structure of the effective viscosity coefficient is determined and hypotheses are formulated which enable it to be assumed to be independent of the flow velocity. It is established by comparison with experimental data that the effective viscosity coefficient can exceed the viscosity coefficient of the flowing fluid by an order of magnitude. The equations of average motion are obtained, which in the case of an incompressible fluid have the form of the Navier–Stokes equations with body forces proportional to the velocity. It is shown that the solution of these equations can be found in boundary layer approximation. It is established that, in addition to the well-known dimensionless flow numbers, there is a new number So which characterizes the ratio of the Darcy porous drag forces to the effective viscosity forces. Structured packing of granular medium is shown to be an effective tool of elimination of velocity profile non-uniformity. It is shown that flow of a fluid in regularly packed granular medium is anisotropic and the medium permeability coefficients form a second-rank tensor. The values of these coefficients may vary in wide range what allows to control the field of velocities meeting certain given criteria of optimization and reduce flow porous drag by 5 – 8 times compared to irregular packing. The proposed equations are extended to the case of the flow of an aerated fluid. The solution of a new problem of the fluid flow in a plane channel with permeable walls is presented using three models: Darcy's law for an incompressible and aerated fluid, and also of an aerated fluid taking the effective viscosity into account. It is established that, for the same pressure drop, the maximum flow rate corresponds to Darcy's law. Compressibility leads to its reduction, but by simultaneously taking into account the compressibility and the effective viscosity one obtains minimum values of the flow rate. The effective viscosity and aeration of the fluid has a considerable effect on the flow parameters.

INTRODUCTION

Numerous experimental papers [[1], [2], [3], [4], [5], [6]] show that, for a fluid flow through a granular medium with variable porosity, when the flow field is inhomogeneous, the velocity profiles that occur differ considerably from the profiles predicted by Darcy's law. As a rule, experiments have been carried out in vertical circular tubes, containing a layer of granular material, peculated by a fluid with constant velocity distribution. At the layer exit considerable changes in the velocity profile were observed: the presence of a minimum in the centre of the tube, a maximum close to the wall, and zero velocity on the wall itself (i.e., so-called "ears" have been observed).

Characteristic experimental results [[1]] for a tube of radius D = 3 in, filled with spherical particles of diameter d = 0.25 in, are shown on the upper graph of Figure 1 by the points, where υ_m is the fluid velocity, averaged over the cross section. On the same graph, the dashed curve represents an attempt to process these results using Darcy's law, taking into account the experimentally determined change in the porosity ε over the radius (the continuous curve in the lower graph of Figure 1). The experimental results are in good agreement with Darcy's law in the central part of the tube but differ considerably in the boundary region.

This phenomenon has been mainly investigated in two directions. In the earlier investigations [[1], [2]] when processing the experimental results, Prandtl's formula was directly invoked for the turbulent shear stresses. Since, in Prandtl's theory, the shear stresses depend on the rate of deformation of the shear component according to a square law, the flow pattern must vary when the mean flow velocity changes. However, no such changes have been observed experimentally, and the authors have got round this contradiction by determining a missing length for each individual experiment.

Later [[5], [6], [7]] the inhomogeneity of the flow field was explained by deformation of the granular layer due to the action of the flowing fluid and buckling of the supporting net. A theory of the deformation of a granular medium was used in [7].

The view has been expressed in [[4], [5]] that inhomogeneities of the velocity profile only arise on the free boundary of the layer. The correctness of Darcy's law for describing flows with a variable velocity profile was not in question, despite its obvious breakdown in the boundary region.

Without excluding the influence of the above factors on the fluid flow field, we will show that the experimentally observed results can be explained by introducing the idea of a transferred effective viscosity, similar in its physical nature to additional Reynolds stresses. Note that Eqs (1.10) then obtained when inertial terms are neglected is formally similar to the equations proposed by Brinkman [[8], [9]] when investigating the flow of a solvent through tangled macromolecules consisting of long chains, considered as a porous medium. The usual derivation of Brinkman's equations consists of using Stokes' equations of a viscous fluid with external mass forces continuously distributed in the volume. The latter were assumed to be related to the velocity vector by Darcy's law. As Brenner has pointed out [[10]], procedures of this kind are logically unjustified, since they combine equations describing two different continua. We will show in Section 1 that Eqs (1.10) are free from this drawback.

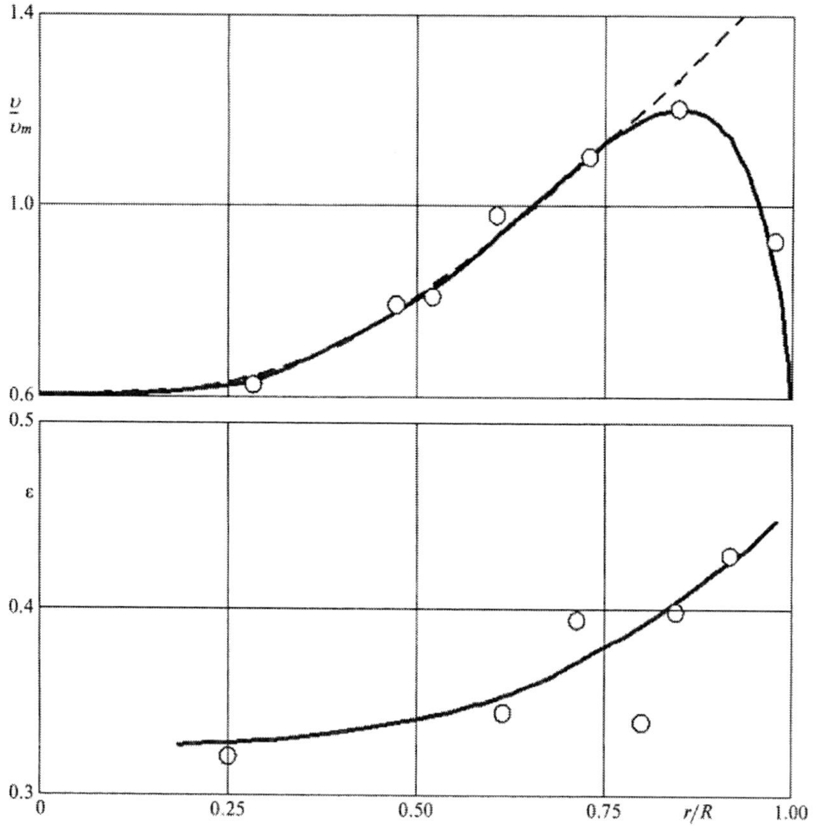

Figure 1.

1. The Transport Phenomenon and the Effective Viscosity for Flow in an Inhomogeneous Granular Medium

When a compressible fluid flows through a porous medium, Darcy's law and the continuity equations are usually employed to determine the flow field. In an orthogonal Cartesian system of coordinates x_i $(i = 1, 2, 3 \rightarrow x, y, z)$ these relations take the form

$$\mu\alpha\upsilon_i = -\frac{\partial P}{\partial x_i}, \quad \frac{\partial\rho}{\partial t} + \frac{\partial\rho\upsilon_i}{\partial x_i} = 0 \qquad (1.1)$$

In Eqs (1.1) υ_i are the components of the mean-velocity vector in an elementary area, normal to the corresponding coordinate axis, μ is the coefficient of viscosity of the flowing fluid, ρ is its density, P is the pressure and α is the hydraulic drag, which is the inverse of the Darcy seepage factor. Equations (1.1) must be supplemented by the equation of state, which defines the relation between the density ρ and the pressure P. This case will be considered in Section 5. In the second formula of (1.1) and elsewhere summation is assumed over repeated indices.

Although Darcy's law was obtained experimentally and was treated as an independent law of nature, the first formula of (1.1) can be obtained if, following Zhukovskii [[11]] we assume that the fluid, flowing in a porous space, is an ideal Euler fluid, on which fictitious porous drag mass forces $-\mu\alpha\upsilon_i$ act. Euler's equations can then be written as [[12]]

$$\rho\frac{\partial\upsilon_i}{\partial t}+\rho\upsilon_j\frac{\partial\upsilon_i}{\partial x_j}+\mu\alpha\upsilon_i=-\frac{\partial P}{\partial x_i} \tag{1.2}$$

If we carry out the simplest averaging of Eqs (1.2) assuming that υ_i is the average velocity on elementary areas, normal to the coordinate axis considered, and inertial forces are neglected, we obtain Darcy's law (1.1).

A justification of Zhukovskii's hypothesis regarding the structure of the bulk porous drag force can be obtained using the methods and approaches developed by Slichter, on the assumption that the granular medium, through which the incompressible viscous percolates, consists of spherical particles of the same diameter, and that the centres of each of the eight adjoining spheres are at the vertices of a rhombohedron. These assumptions enable the shape of the porous channel, in which the elementary fluid particles move, to be determined. The actual porous channel obtained was replaced by a fictitious cylindrical channel of definite dimensions and cross section. Further, using the well-known analytical solution for laminar flow in such a channel, the average velocity component was found, i.e., the coefficient α in the first formula of (1.1) was determined.

An independent statement of the results was obtained by Slichter [[12]] and there have also been numerous investigations by other researchers, devoted to a theoretical determination of the coefficient α. It follows from these that this coefficient can be represented in the form.

$$\alpha=f(\varepsilon)/d^2 \tag{1.3}$$

where d is the characteristic size of a granule or pore and $f(\varepsilon)$ is a dimensionless function of the porosity ε, which depends on the shape of the granules and the type of packing.

In a porous medium, described by Darcy's law, the full no-slip conditions on the boundary surfaces may not be satisfied. Hence, neither Darcy's law itself nor its well-known extensions (the non-linear dependence of the pressure gradient on the velocity vector, the presence of turbulence and stagnation zones at points of contact of the particles, etc.) cannot explain the phenomena in the boundary region described above.

When fluid flows around the particles of a granular medium in a longitudinal direction, pulsating components of the velocity occur in the transverse direction, which are ignored by Darcy's law. If the average flow velocity changes in a transverse direction, these pulsations transfer additional momenta from layer to layer, which leads to the occurrence of effective viscosity forces. This phenomenon can be investigated if it is assumed that a viscous fluid flows in the porous space, the equation of motion of which [[13]] can be represented in the form

$$\rho\frac{\partial \upsilon_i}{\partial t}+\rho\upsilon_j\frac{\partial \upsilon_i}{\partial x_j}+\frac{\partial \tau_{ij}}{\partial x_j}=-\frac{\partial P}{\partial x_i} \quad (1.4)$$

where τ_{ij} are the components of the viscous stress tensor.

We will write the components of the fluid flow velocity in the granular medium υ_i in the form of the sum of the average smoothed values $\overline{\upsilon}_i$ and pulsating components υ_i', the average values of which are equal to zero, i.e., $\upsilon_i=\overline{\upsilon}_i+\upsilon_i'$. Substituting these values into Eq. (1.4) and carrying out well-known averaging operations [[13]], we obtain the equations of averaged motion

$$\rho\frac{\partial \overline{\upsilon}_i}{\partial t}+\rho\overline{\upsilon}_j\frac{\partial \overline{\upsilon}_i}{\partial x_j}+\mu\alpha\overline{\upsilon}_i=-\frac{\partial \overline{P}}{\partial x_i}+\frac{\partial}{\partial x_j}\left(-\rho\overline{\upsilon_i'\upsilon_j'}\right) \quad (1.5)$$

which differ from the initial equations (1.4) in the fact that, in accordance with Zhukovskii's hypothesis, the stresses due to the fluid viscosity were averaged over the bulk force of the Darcy porous drag, while a consideration of the inertial forces led to the occurrence of six terms

$$\tau_{ij}^R=\rho\overline{\upsilon_i'\upsilon_j'} \quad (1.6)$$

called additional stresses.

When deriving Eqs (1.5) the physical nature of the velocity pulsations that occur was not stipulated; the fact that they existed was only important. In the case considered, unlike turbulent flows, the pulsations are due to the presence of granules in the flow field, and hence it is necessary to establish whether flow modes exist for which the stresses, due to the effective viscosity, are comparable with the other forces acting on the fluid.

To find the relation between the stresses (1.6) and the average velocity field in the flow (the closure problem) we will use the fundamental propositions of Prandtl's mixing path theory. Consider a steady plane translational flow parallel to the x axis, the mean velocity of which $\upsilon_x=u(y)$ depends on the transverse coordinate y. The pulsating component of the longitudinal velocity u' is taken in the form $u'\approx l'du/dy$, where l' is Prandtl's mixing length. The pulsating velocity component $\upsilon_y'=\upsilon'$ transverse to the average flow is taken as the transfer velocity. Substituting these quantities into formula (1.6), we obtain

$$\tau_{xy}^R=-\rho\overline{(\upsilon'u')}=\rho\overline{(\upsilon'l')}du/dy$$

It can be seen that $\overline{(\upsilon'l')}=\text{const}$, if we assume that l' will be shorter the greater the pulsating velocity υ'. Assuming that this assertion holds for the whole flow field (Boussinesq's hypothesis), we will have

$$\tau_{xy}^{R} = -\rho\overline{(\upsilon'u')} = \rho\overline{(\upsilon'l')}du/dy \tag{1.7}$$

The quantity μ_e can be treated as an effective (transfer) viscosity coefficient, which occurs when the fluid flow through the porous medium is spatially inhomogeneous. The effective viscosity coefficient can conveniently be represented in the form $\mu_e = \mu\beta(\varepsilon)$, where $\beta(\varepsilon)$ is a dimensionless porosity function. In the general case of an isotropic medium, the effective stresses form a symmetrical second-rank tensor, whose components, based on Eqs (1.7), can be expressed in terms of the average velocities in the form

$$\tau_{ij}^{R} = \mu\beta(\varepsilon)\left(\frac{\partial\bar{\upsilon}_i}{\partial x_j} + \frac{\partial\bar{\upsilon}_j}{\partial x_i}\right) \tag{1.8}$$

while the components of the complete stress tensor will have the form

$$\tau_{ij} = \mu(1+\beta(\varepsilon))\left(\frac{\partial\bar{\upsilon}_i}{\partial x_j} + \frac{\partial\bar{\upsilon}_j}{\partial x_i}\right) \tag{1.9}$$

By comparing the theoretical results with experimental data it will be shown below that the effective viscosity coefficient exceeds the viscosity coefficient μ by a factor of a hundred. Taking this into account and substituting expression (1.9) into Eq. (1.5), we obtain for an incompressible fluid.

$$\rho\left(\frac{\partial\upsilon_i}{\partial t} + \upsilon_j\frac{\partial\upsilon_i}{\partial x_j}\right) - \frac{\partial}{\partial x_j}\left[\mu\beta(\varepsilon)\left(\frac{\partial\upsilon_i}{\partial x_j} + \frac{\partial\upsilon_j}{\partial x_i}\right)\right] + \mu\alpha\upsilon_i = -\frac{\partial P}{\partial x_i}, \tag{1.10}$$

In formulae (1.10) and below we omit the bar over average quantities and assume that summation is carried out over repeated indices. Equations similar to (1.10) were obtained in a somewhat different form in [[14]] from phenomenological considerations as extensions of Stokes' equations. The important result that the occurrence of an effective (transfer) viscosity is practically independent of the natural viscosity of the fluid has not been previously mentioned.

Equations (1.10) can be reduced to dimensionless form by choosing certain constant quantities T, h, V, P characteristic for the flow, as scales of time, length, velocity and pressure. Denoting the corresponding dimensionless quantities by a prime and introducing, as is usually done, the dimensionless Strouhal, Euler and Reynolds similitude numbers

$$\mathrm{Sh} = \frac{h}{VT}, \quad \mathrm{Eu} = \frac{P}{\rho V^2}, \quad \mathrm{Re} = \frac{hV\rho}{\mu\beta}$$

we can write Eqs (1.10) in the form

$$\mathrm{Sh}\frac{\partial \upsilon_i'}{\partial t'}+\upsilon_j'\frac{\partial \upsilon_i'}{\partial x_i'}-\frac{1}{\mathrm{Re}}\left[\frac{\partial}{\partial x_j'}\left(\frac{\partial \upsilon_i'}{\partial x_j'}+\frac{\partial \upsilon_j'}{\partial x_i'}\right)-\mathrm{So}\,\upsilon_i'\right]=-\mathrm{Eu}\frac{\partial P'}{\partial x_i'} \tag{1.11}$$

The dimensionless number So represents the ratio of the Darcy porous drag forces to the effective viscosity forces. This was also first proposed in a somewhat different form in [[14]]. The effect of the So number on the flow pattern will be illustrated below using a number of examples.

2. The Flow in Plane and Axisymmetric Channels

We will consider, as an example, the fluid flow in a porous medium, situated in a long plane channel of height $2h$, assuming the porosity to be constant. It follows from this condition that the coefficients α and β are also constant. Directing the x axis along the middle line of the channel and substituting the values $\upsilon_y=\upsilon_z=0$, $\upsilon_x=u(y)$ into Eq (1.10), we obtain

$$\beta\frac{d^2u}{dy^2}-\alpha u=\frac{\theta}{\mu},\ \theta=\frac{dP}{dx}=\mathrm{const}$$

The solution of this equation, which satisfies the boundary conditions $u=0$ when $y=\pm h$, will be

$$u=\mu\frac{\theta}{\alpha}\left(1-\frac{\mathrm{ch}\chi\bar{y}}{\mathrm{ch}\chi}\right),\ \chi=\sqrt{\frac{\alpha}{\beta}}h=\sqrt{\frac{f(\varepsilon)}{\beta(\varepsilon)}}\frac{h}{d}=\sqrt{\mathrm{So}},\ \bar{y}=\frac{y}{h} \tag{1.12}$$

Graphs of the dimensionless velocity $\tilde{u}=u(\bar{y})/u(0)$ for a series of increasing values of the number So are shown in Figure 2, where the horizontal dashed line represents the solution in the case of Darcy's law. It follows from the graph that the solution becomes Darcy's solution $u_0=\theta/(\mu\alpha)$ when the So number increases without limit, i.e., when, for a fixed porosity ε, the size of the granules is much less than the channel width. Formulae (1.12) show that when $\mathrm{So}\to 0$ ($\alpha\to 0$), the solution becomes a Poiseuille parabola.

Knowing the velocity distribution along the channel height y, we obtain the fluid flow rate in an arbitrary cross section x in the form

$$Q=\frac{2h^3(1-\lambda)}{\mu\bar{\alpha}}\theta,\ \bar{\alpha}=\alpha h^2,\ \lambda=\frac{\mathrm{th}\chi}{\chi}=\frac{\mathrm{th}\sqrt{\mathrm{So}}}{\sqrt{\mathrm{So}}} \tag{1.13}$$

It follows from formulae (1.13) that, for a specified pressure drop, the flow depends considerably on the dimensionless So number and decreases from the maximum value

$Q_0 = 2h^3\theta/(\mu\bar{\alpha})$ corresponding to Darcy's law, as the So number decreases by virtue of the increase in the influence of the effective viscosity. This result has an important practical value since it enables one to judge, from the experimentally determined dependence of the flow rate on the pressure drop, the actual rheological properties of the flowing fluid.

A graph of the dimensionless flow rate $\bar{Q} = Q\mu/(h^3\theta) = 2(1-\lambda)/\bar{\alpha}$ as a function of the dimensionless parameters β and $\bar{\alpha}$ is shown in Figure 3. The curve in the $\beta = 0$ plane corresponds to Darcy's law $\bar{Q} = 2/\bar{\alpha}$, while in the plane $\bar{\alpha} = 0$ it corresponds to the Navier–Stokes equations $\bar{Q} = 2/(3\beta)$ with a viscosity coefficient of $\mu\beta$.

To determine how the effective viscosity coefficient depends on the porosity [[15]] the above-mentioned experimental results [[1]] for velocity profiles in a tube of radius R, filled with a granular medium with a characteristic particle size d, were used.

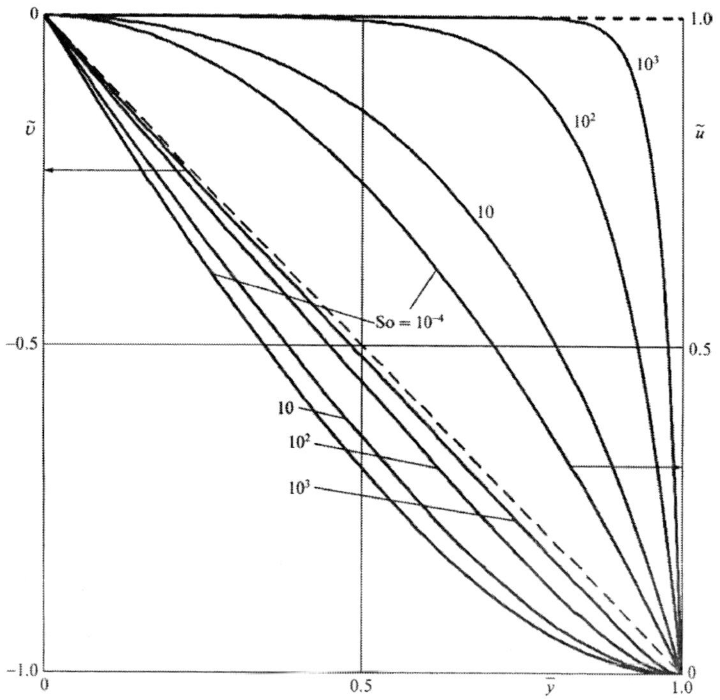

Figure 2.

Equation (1.10) has the following form in the axisymmetric case (the x coordinate is directed along the tube axis)

$$\beta\frac{d^2\upsilon}{d^2r} + \left(\frac{\beta}{r} + \frac{d\beta}{dr}\right)\frac{d\upsilon}{dr} - \alpha\upsilon = -\frac{1}{\mu}\frac{\partial P}{\partial x} \tag{1.14}$$

The coefficient α (1.3) is given by the Kozeny formula, in which the experimental dependence of the porosity on the radius is used (the lower graph in Figure 1). Starting from the mechanism by which the effective velocity occurs, the following form of the relation $\beta = \beta(\varepsilon)$ has been proposed

$$\beta(\varepsilon) = A/\varepsilon^{\lambda} \tag{1.15}$$

where A and λ are dimensionless coefficients, which are independent of the porosity of the medium and the size of the granules.

Equation (1.14) was solved numerically for several values of the tube diameter and the size of the packing granules. It was found that the value of the parameter A varies in the range from 50 to 60, while the value of the parameter λ varies in the range from 3.5 to 4. On the upper graph in Figure 1 the points represent experimental data for a tube of diameter D = 3 in, filled with particles of diameter d = 0.25 in, while the continuous curve represents the velocity profile obtained. The proposed relation (1.15) describes the experimental data quite well over a wide range of the parameters and shows that, for the usual values of the porosity of 0.3–0.45, the value of β is of the order of $10^2 - 10^3$, i.e., the value of the effective viscosity is many times greater than the value of the natural viscosity of the fluid, so the latter can be neglected.

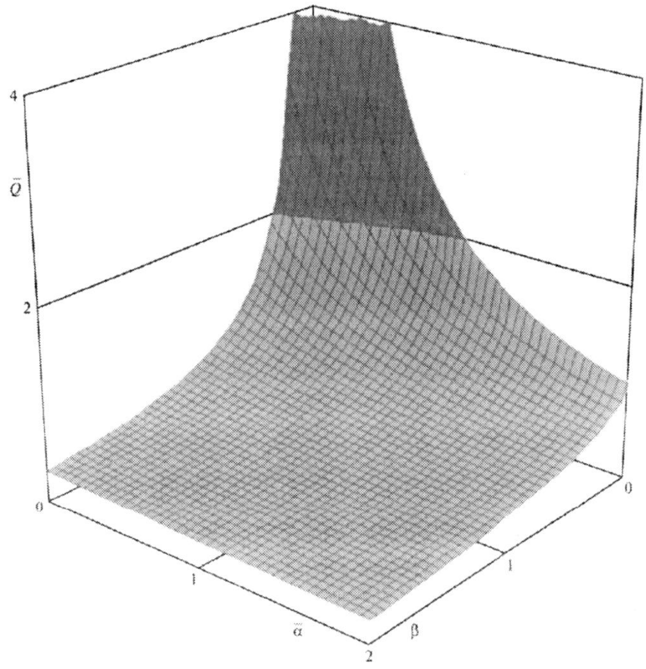

Figure 3.

As was noted above, a number of researchers have shown that non-uniformity of the velocity field only occurs in the outer surface layer of the granules. To check this assertion, we considered the problem of the flow in a long channel of circular cross section in which there was a layer of granular material [[16]]. The change in the porosity along the radius was also taken in the form shown on the lower graph of Figure 1.We used the Navier–Stokes equations in the parts of the channel that were free from granules. The numerical solution of the problem showed that perturbations of the velocity profile occur before the fluid enters the layer. They reach maximum values inside the layer and decrease somewhat in the exit section.

3. The Flow in Channels with Permeable Walls

The fluid flow in channels with permeable walls is encountered in many technological processes and is of particular interest for problems involving the extraction of hydrocarbon raw material, where the high-yield method has become widely used, involving drilling horizontal boreholes with subsequent use of the hydrofracturing procedure.

We will consider a plane channel of constant width $2h$ and length a. We will direct the x axis along the middle line of the channel so that the points $x=0$ and $x=a$ correspond to the beginning and end of the channel sections. At the boundaries of the channel, with $y=\pm h$, we will specify a constant velocity w, which is independent of the x coordinate: $\upsilon(x,\pm h)=\mp w=\text{const}$. We will denote the pressure at $x=a$ by P_w, and at $x=0$ by P_r ($P_w < P_r$).

Assuming, in Eqs (1.10), that the forces of inertia of the average motion are negligibly small compared with the porous drag and effective viscosity forces, we obtain

$$\beta\left(\frac{\partial^2 u}{\partial x^2}+\frac{\partial^2 u}{\partial y^2}\right)-\alpha u=\frac{1}{\mu}\frac{\partial P}{\partial x},\ \beta\left(\frac{\partial^2 \upsilon}{\partial x^2}+\frac{\partial^2 \upsilon}{\partial y^2}\right)-\alpha \upsilon=\frac{1}{\mu}\frac{\partial P}{\partial y},\ \frac{\partial u}{\partial x}+\frac{\partial \upsilon}{\partial y}=0 \qquad (1.16)$$

where u and υ are the components of the velocity vector along the x and y axes respectively.

We will consider the solution of the problem using Darcy's law. Assuming $\beta=0$ in Eqs (1.16) and integrating the system of equations obtained, we will have

$$u=wx/h,\ \upsilon=-wy/h,\ P(x)=P_r-w\bar{\alpha}\mu\left(x^2-y^2\right)/\left(2h^3\right) \qquad (1.17)$$

Hence it follows that the fluid flow rate in the section $x=a$ is related to the pressure drop $\Delta P=P_r-P_w$ by the equality $Q_w=4h^3\Delta P/(a\bar{\alpha}\mu)$. Note that this quantity, for the same pressure drop, is double the flow rate Q_0 corresponding to the flow in a channel with impermeable walls.

Returning to the complete equations (1.16), we can verify that the velocity components u and υ, and the pressure P, which satisfy these equations and the boundary conditions, have the form

$$u=\frac{wx}{h(1-\lambda)}\left(1-\frac{\text{ch}(\chi\bar{y})}{\text{ch}\chi}\right),\ \upsilon=-\frac{w}{1-\lambda}\left(\lambda\frac{\text{sh}(\chi\bar{y})}{\text{sh}\chi}-\bar{y}\right),$$

$$P=P_r-\frac{w\bar{\alpha}\mu}{2h^3(1-\lambda)}\left(x^2-y^2\right) \qquad (1.18)$$

where we have used the notation (1.12) and (1.13).

The first formula of (1.18) shows that the form of the relation $\tilde{u}(\bar{y}) = u(x,\bar{y})/u(x,0)$ is the same as in the case of a channel with impermeable walls, while the graph for $\tilde{u}(\bar{y})$ is identical with the graph for $\tilde{u} = \tilde{u}(\bar{y})$, shown in Figure 2. The relation $\tilde{\upsilon}(\bar{y}) = \upsilon(\bar{y})/w$ is shown in Figure 2 for a series of values of the So number. These graphs become graphs corresponding to Darcy's law as $\text{So} \to \infty$ (the dashed curves in Figure 2: for $\tilde{u} = \tilde{u}(\bar{y})$ the horizontal line and for $\tilde{\upsilon} = \tilde{\upsilon}(\bar{y})$ the diagonal line), and as $\text{So} \to 0$ they correspond to the Navier–Stokes equations with viscosity coefficient $\beta\mu$.

It follows from the last formula of (1.18) that the total fluid flow rate in the channel $Q_w = Q(a) = 2wa$ is related to the pressure drop $\Delta P = P_r - P_w$ and, in dimensionless form, the relation

$$\bar{Q} = Q_w \mu a / \left(2h^3 \Delta P\right) = 2(1-\lambda)/\bar{\alpha}$$

is identical with the relation $\bar{Q}(\bar{\alpha},\beta)$, shown in Figure 3.

4. The Flow Optimization in Non-homogenious Anisotropic Granular Medium

In this section we will extend Eqs (1.10) to anisotropic flows, which occur due to the natural or technogenic anisotropy of a porous space [[17]]. The source of technogenic anisotropy may be structured packing of the granular layer by granules of special shape, which enable the flow parameters to be directionally changed [[18]] and enable specified optimality criteria to be satisfied. Preliminary experiments [[19]] showed that that structured packing may reduce flow resistance by 5-8 times compared to irregular packing. In the present analysis two-dimensional flow fields are examined in cases of packing along a logarithmic spiral and spiral of Archimedes. The dependence of the field of flow on parameters of the spiral is determined.

If the pellets have geometrical anisotropy then the flow of a fluid in a regular packing is anisotropic. In this case the permeability coefficient in the Darcy law form a second-rank tensor α_{ij} and effective viscosity coefficients form a fourth-rank tensor β_{ijkm}. Assuming that the forces of inertia of the average motion are negligibly small compared with the porous drag and effective viscosity forces the equation (1.10) in an arbitrary orthogonal Cartesian coordinates system xyz may be written as:

$$\frac{\partial}{\partial x_i}\left[\beta_{ijkm}\left(\frac{\partial \upsilon_k}{\partial x_m} + \frac{\partial \upsilon_m}{\partial x_k}\right)\right] - \beta_{ij}\upsilon_i = \frac{1}{\mu}\frac{\partial P}{\partial x_j} \tag{1.19}$$

As an example of Eq (1.19) application a flow in a 2-dimensional canal of height $2h$ is considered. XY plane is situated in the middle plane of the canal and the pressure gradient is directed along axis x, i.e. $\partial P/\partial y = 0$. The pellets are assumed to be the rings of Raschig type (pieces of tube) and are placed in rows at angle of γ to the axis x (Figure 4). The equation (1.19) is written in an orthogonal coordinate system the axis 1 of which is parallel to the lines of packing and the axis 2 lies in the plane of flow. Under these assumptions: $\upsilon_3 = 0$, $\upsilon_1 = \upsilon_1(z)$, $\upsilon_2 = \upsilon_2(z)$ and in Eq (1.19) only the following anisotropic coefficients are to be taken into account: $\alpha_{11} = \alpha_1$, $\alpha_{22} = \alpha_2$, $\beta_{3131} = \beta_1$, $\beta_{3232} = \beta_2$. Therefore Eq (1.19) split into two equations, each of them containing only one of two velocity components:

$$\begin{aligned} \beta_1 \frac{\partial^2 \upsilon_1}{\partial z^2} - \alpha_1 \upsilon_1 &= \frac{1}{\mu}\frac{\partial P}{\partial x}\cos\gamma \\ \beta_2 \frac{\partial^2 \upsilon_2}{\partial z^2} - \alpha_2 \upsilon_2 &= \frac{1}{\mu}\frac{\partial P}{\partial x}\sin\gamma \end{aligned} \qquad (1.20)$$

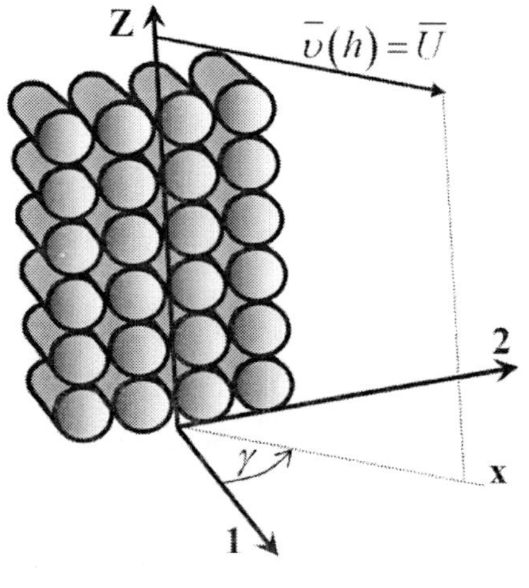

Figure 4.

and the equation of continuity is satisfied. The solution of Eqs (1.20) meeting the boundary conditions $z = \pm h$ and $\upsilon_1 = \upsilon_2 = 0$ is:

$$\begin{aligned} \upsilon_1 &= -\frac{\partial P}{\partial x}\frac{\cos\gamma}{\mu\alpha_1}\left(1 - \frac{ch\,\chi_1 z/h}{ch\,\chi_1}\right), \quad \chi_1 = h\sqrt{\frac{\alpha_1}{\beta_1}} \\ \upsilon_2 &= -\frac{\partial P}{\partial x}\frac{\sin\gamma}{\mu\alpha_2}\left(1 - \frac{ch\,\chi_2 z/h}{ch\,\chi_2}\right), \quad \chi_2 = h\sqrt{\frac{\alpha_2}{\beta_2}} \end{aligned} \qquad (1.21)$$

It may be easily shown that equations (1.19) have boundary layer solutions and for a flow near a wall they are given by expressions:

$$\upsilon_1 = \upsilon_1^0\left(1 - e^{-\chi_1 z/\delta}\right),\ \upsilon_1^0 = -\frac{\partial P}{\partial x}\frac{\cos\gamma}{\mu\alpha_1}$$
$$\upsilon_2 = \upsilon_2^0\left(1 - e^{-\chi_2 z/\delta}\right),\ \upsilon_2^0 = -\frac{\partial P}{\partial x}\frac{\sin\gamma}{\mu\alpha_2} \tag{1.22}$$

where υ_1^0 and υ_2^0 are the total flow velocity components determined by the Darcy law, δ is a boundary layer thickness. In we consider a pipe filled with a granulated medium of pellets packed along screw lines the expression (1.22) may also be applied if δ is much less than the pipe radius R.

In numerous technological processes (adsorption, chemical catalysis, etc.) the condition of flow optimization is that it takes the same time for any elementary volume of fluid to pass the packing. As mentioned in the introduction the velocity profiles of flows described above reach their maximum near a wall and the optimization condition does not hold true. But provided that screw-line racking in the proximity of the wall is organized, certain parameters (inclination angle, diameter of a pellet etc.) may be calculated from (1.22) in order to satisfy the optimization condition described above.

Equation (1.19) may also be solved for a flow represented in Figure 4. The space between two planes is filled with anisotropic granulated medium with constant fractional void volume and the pellets are packed so that the axes of anisotropy are directed along axes X and Y. The lower plane is fixed and it corresponds to the XY coordinate plane. The upper plane is parallel to the lower one and moves at a velocity U in the direction at angle γ to X axis (Figure 4). The distance between planes is equal to h. Eqs (1.19) in main axes of anisotropy ($\alpha_{12} = \alpha_{21} = 0$) may be written as:

$$\begin{cases} \dfrac{\partial}{\partial z}\left[\beta_{3131}\dfrac{\partial \upsilon_x}{\partial z}\right] - \alpha_{11}\upsilon_1 = 0 \\ \dfrac{\partial}{\partial z}\left[\beta_{3232}\dfrac{\partial \upsilon_y}{\partial z}\right] - \alpha_{22}\upsilon_2 = 0 \\ \upsilon_{1,2}(0) = 0 \\ \upsilon_1(h) = U\cos\gamma;\quad \upsilon_2(h) = U\sin\gamma \end{cases} \tag{1.23}$$

The solution of these differential equations can be obtained analytically:

$$\upsilon_1 = 2A_1\sinh\left(\delta_1\frac{z}{h}\right),\ \delta_1^2 = \frac{\alpha_1 h^2}{\beta_1} \tag{1.24}$$

$$\upsilon_1 = 2A_2\sinh\left(\delta_2\frac{z}{h}\right),\ \delta_2^2 = \frac{\alpha_2 h^2}{\beta_2} \tag{1.25}$$

$$A_i = \frac{U \sin \gamma}{2 \sinh \delta_i} \quad (1.26)$$

where δ_i is the determining parameter of flow similar to χ_i in (1.21). The plot of the flow profile described by Eqs (1.24) and (1.25) is shown in Figure 5. Apparently, lines of the flow are curved and diverge from U direction due to the difference in physical properties of the medium in the directions of anisotropy. Velocity profiles for a series of δ values are plotted in Figure 6.

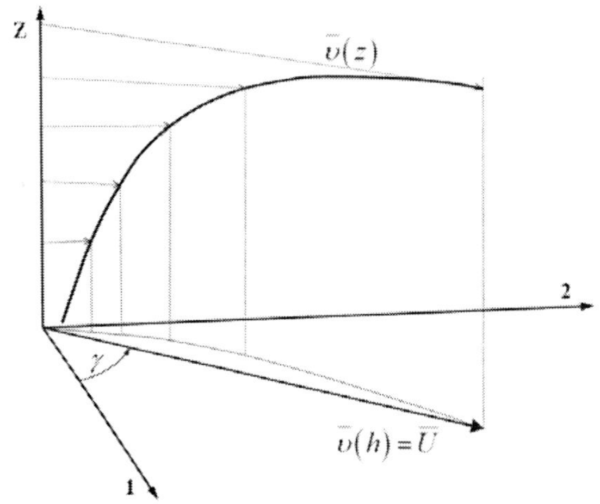

Figure 5.

There are two extreme regimes of flow: when $\delta \to 0$ the velocity profile tends to a linear profile characteristic of a standard Quett flow in an empty canal. On the other hand, at bigger values of δ the flow exists only in the proximity of the upper moving plane and the velocity in the lower region tends to zero. Provided that the value of fractional void volume and the type of packing do not vary, then $\delta \sim (h/d)^2$ and therefore, the flow profile tends to the linear profile when the diameter of a pellet and the characteristic size of the region of flow are comparable. When this ratio and the relative size of a pellet decrease the so-called "dead flow space" appears where the fluid is not affected by the moving plate.

In numerous technological processes radial-type devices are used, consisting of two coaxial cylinders filled between them with granular material under constant pressure difference. The resulting flow of a fluid through the inner region of this sort of reactor is considered in cylindrical coordinates system $r\varphi z$.

Again, we consider a regular packing of Raschig-type cylindrical pellets along 2-dimensional spirals. For such packing there exist two main directions of filtration: along the pellet's axis and perpendicular to it. Then the tensor of hydraulic drag coefficients α_{ij} in the Darcy law is symmetrical and may be written in a polar coordinates system in case of an axial-symmetric flow as follows:

$$\alpha_{rr}\upsilon_r + \alpha_{r\varphi}\upsilon_\varphi = -\frac{\partial P}{\partial r}, \quad \alpha_{\varphi\varphi}\upsilon_\varphi + \alpha_{\varphi r}\upsilon_r = 0 \tag{1.27}$$

$$\frac{\partial \upsilon_r}{\partial r} + \frac{\upsilon_r}{r} = 0 \tag{1.28}$$

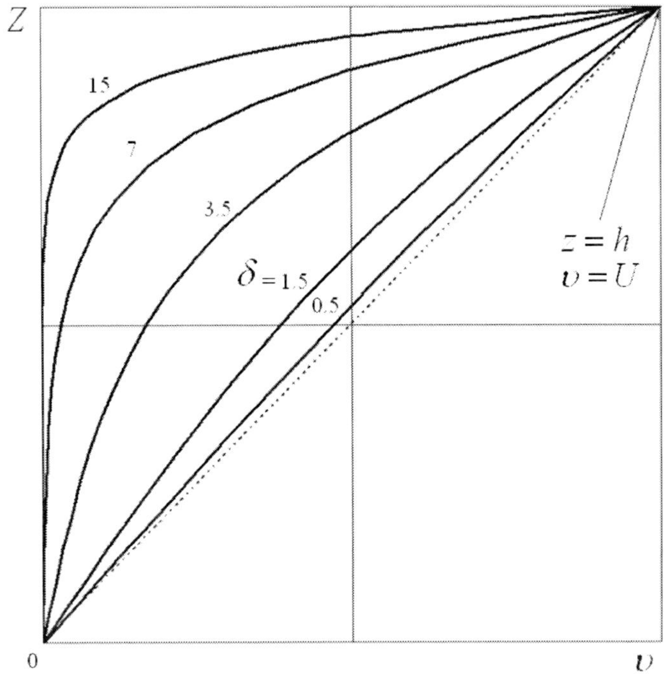

Figure 6.

The equation of continuity (1.28) is analogous to the isotropic flow and the expression for radial velocity is:

$$\upsilon_r = \frac{C}{r} \tag{1.29}$$

and does not depend on anisotropic coefficients.

Velocity components υ_φ are determined from the second equation of (1.27) and may vary in wide range with variation of anisotropic coefficients. Therefore, the values of the latter may be chosen so that the dependence of the scalar of velocity from radial position $\upsilon = \upsilon_r$ will meet the necessary physicochemical criteria of the flow.

Hydraulic drag coefficients in Eqs. (1.27) may be expressed in terms of their values in main axes: α_1 – along the pellet's axis and α_2 – perpendicular to it:

$$\left.\begin{array}{l}\alpha_{rr}\\ \alpha_{\varphi\varphi}\end{array}\right\}=\frac{1}{2}(\alpha_1+\alpha_2)\pm\frac{1}{2}(\alpha_1-\alpha_2)\cos 2\eta$$
$$\alpha_{r\varphi}=\frac{1}{2}(\alpha_1-\alpha_2)\sin 2\eta \tag{1.30}$$

where η is the angle between the radial direction and the pellet's axis.

$$tg\eta=\frac{dr}{rd\varphi} \tag{1.31}$$

In (1.31) the relationship $r=r(\varphi)$ is the expression for a curve of packing. Here we will consider three curves: spiral of Archimedes: $r=a_1\varphi$, logarithmic spiral: $r=\exp(a_2\varphi)$, and the following spiral:

$$\varphi=\sqrt{(r/a_3)^2-1}-\arccos(a_3/r) \tag{1.32}$$

corresponding to the most dense packing. For the each type of packing the field of flow, the field of pressure and the fluid flow rate are obtained. Raschig-type rings are assumed to be of 5×5×1 mm size, for which the ratio of hydraulic drag coefficients is $\alpha_1/\alpha_2=10$. The lines of flow and corresponding lines of packing are shown on Figure 7.

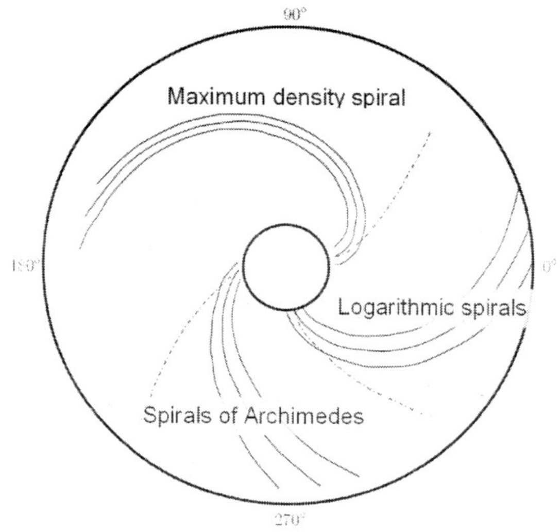

Figure 7.

The dependences of scalar velocity on radial coordinate for a series of packing angles β_0 on the inner cylinder of the reactor in case of the densest packing is shown in Figure 8. If $0<\beta_0<\pi/3$ the dependences above are significantly different and this allows one to select the value of β_0 so that certain optimization criteria are met. Noticeably, scalar velocities corresponding to different values of $\beta_0 > \pi/3$ are almost equal to those obtained in an isotropic flow $\upsilon = \upsilon_0/r$.

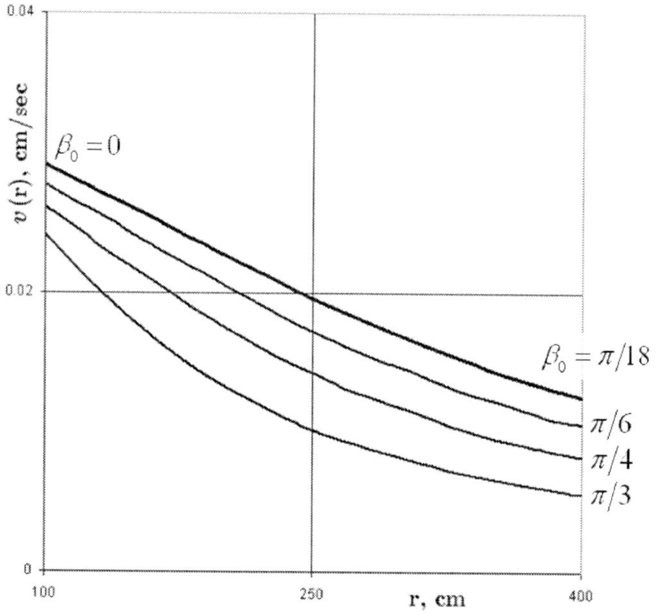

Figure 8.

5. The Flow of an Aerated Liquid

Consider the steady flow of an aerated liquid in an undeformed granular medium. We will introduce the momentum vector $\boldsymbol{K} = \rho \boldsymbol{V}$ with components $k_i = \rho \upsilon_i$. We will assume that the flow in the porous space is described by Eqs (1.4) and the continuity equation, the second formula of (1.1). In this case they take the form

$$k_j \frac{\partial \upsilon_i}{\partial x_j} + \rho \frac{\partial \tau_{ij}}{\partial x_j} = -\rho \frac{\partial P}{\partial x_i}, \ \frac{\partial k_j}{\partial x_j} = 0 \tag{2.1}$$

Using the new function q, such that

$$dq = \rho \, dP \tag{2.2}$$

and taking into account the last equation of (2.1),we obtain

$$\frac{\partial k_j \upsilon_i}{\partial x_j} + \rho \frac{\partial \tau_{ij}}{\partial x_j} = -\frac{\partial q}{\partial x_i} \tag{2.3}$$

From the same considerations as in Section 1 when analysing the velocity field, we can represent the components of the momentum vector k_i in the form

$$k_i = \bar{k}_i + k_i' \tag{2.4}$$

where $\bar{k}_i$ are the average values of k_i, while k_i' is their pulsating components. Substituting (2.4) into Eqs (2.3) and into the last equation of (2.1), we obtain after averaging

$$\bar{k}_j \frac{\partial \bar{k}_i}{\partial x_j} + \frac{\partial \overline{k_i' k_j'}}{\partial x_j} - \mu\alpha \bar{k}_i = \frac{\partial \bar{q}}{\partial x_i}, \quad \frac{\partial \bar{k}_j}{\partial x_j} = 0 \tag{2.5}$$

In deriving the averaged equations (2.5) we took into account the fact that, as a result of the averaging of Eqs (2.3), terms containing viscous stresses due to the Zhukovskii hypothesis [[11]] are converted into the mass Darcy porous drag force. We have also assumed that the compressibility of the fluid can be neglected on the left-hand sides of Eqs (2.3), and we have only taken it into account in the averaged continuity equation. This approach is widely used [[12]] in the theory of the compressible fluid flow in porous media.

The four equations (2.5) contain three components of the momentum vector $\bar{k}_i$ and the quantity $\bar{q}$ as the desired functions. However, this system of equations is not closed, since it contains the terms $\overline{k_i' k_j'}$. To establish their dependence on the average values $\bar{k}_i$ we will use, as previously, the Prandtl mixing path theory. As in Section 1, for this purpose we will consider a plane translational flow, parallel to the x axis, the average momentum of which $k_x = k_x(y)$ depends on the transverse coordinate y. Repeating the discussion of Section 1, we obtain for the quantity $\psi_{ij} = \overline{k_i' k_j'}$, for $i = x$ and $j = y$,

$$\psi_{xy} = \overline{k_y' l'} \frac{d\bar{k}_x}{dy} = A \frac{d\bar{k}_x}{dy} \tag{2.6}$$

The coefficient of proportionality A can be taken to be a constant quantity if we assume that the pulsations l' will be less the greater the pulsations k_y'. Assuming this assumption to hold for all flow fields, equality (2.6) can be extended to the case of any three-dimensional motion and represented in the form

$$\psi_{ij} = \mu\eta \left(\frac{\partial \bar{k}_i}{\partial x_j} + \frac{\partial \bar{k}_j}{\partial x_i} \right), \quad \eta = \frac{A}{\mu} \tag{2.7}$$

Introducing these quantities into equations of motion (2.5) we obtain

$$k_j \frac{\partial k_i}{\partial x_j} + \alpha\mu k_i - \frac{\partial}{\partial x_j}\left[\mu\eta\left(\frac{\partial k_i}{\partial x_j} + \frac{\partial k_j}{\partial x_i}\right)\right] = -\frac{\partial q}{\partial x_i} \tag{2.8}$$

Here and below the bar above average quantities will be omitted. Equations (2.8), like Eqs (1.10), describe the change in the average momentum for the fluid flow through a granular medium. However, the required functions in Eqs (1.10) are the components of the velocity vector, whereas in Eqs (2.8) they are the components of the momentum vector.

The system consisting of the last equation of (2.5) and the three equations of (2.8) contain four required functions and enable us to determine the velocity field and the pressures of the compressible fluid, using Eq. (2.2), if we know how the density ρ depends on the pressure P.

Assuming the flow to be isothermal, this relation takes the form [[12]]

$$\bar{\rho} = \frac{\bar{P}}{\theta + (1-\theta)\bar{P}},\ \bar{\rho} = \frac{\rho}{\rho_0},\ \bar{P} = \frac{P}{P_0} \tag{2.9}$$

where P_0 and ρ_0 are the pressure and density for which the fluid phase (the liquid with the gas dissolved in it) and the gaseous phase (a gas in the form of fine dust particles, freely moving in the pores) is in an equilibrium state, and θ is an experimentally determined parameter. When $P < P_0$, by Henry's law, the gas separates out from the liquid phase, and the overall density per unit volume of the liquid phase is reduced, i.e., when $P < P_0$ we have $\rho < \rho_0$.

Substituting expressions (2.9) into Eq. (2.2), we obtain

$$\bar{q} = \frac{q}{\rho_0 P_0} = b\bar{P} - \theta b^2 \ln\left(\bar{P} + \theta b\right) + C,\ b = \frac{1}{1-\theta} \tag{2.10}$$

where C is an integration constant.

As an example of the use of the equations obtained, we will again consider the solution of the above problem on the fluid flow in a plane channel. Unlike the discussion in Section 3, on the boundary of the channel $y = \pm h$ we will specify a constant mass flow rate ω independent of the x coordinate, i.e., $k_y(\pm h) = \mp\omega = \text{const}$, and the value of the pressure function (2.2) at the points $x = 0$, $y = 0$ and $x = a$, y = 0 will be denoted by q_r and q_w respectively. Henceforth we will assume

$P_0 = P_r$, $\rho_0 = \rho_r$, $q_0 = q_r$

Then expression (2.10) then takes the form

$$\bar{q} = \frac{q_r}{P_r \rho_r} + b\bar{P} - 1 - \theta b^2 \ln\left(\frac{\bar{P} + \theta b}{1 + \theta b}\right) \tag{2.11}$$

If the flow field is such that the average inertial forces and effective viscosity forces are much less than the porous drag forces, the first two terms in the first equation of (2.5) can be neglected, and we can obtain, together with the continuity equation of (the second formula of (2.5)) a system of four equations in the four required functions k_i and q. This system describes the flow of a compressible fluid in a granular medium in the Darcy law approximation, and its solution reduces to solving Laplace's equation for the function q [[12]]. In this case the solution of the above problem on the flow in a channel with permeable walls takes the form [[20]]

$$k_x = \omega x/h,\ k_y = -\omega y/h,\ q = q_r - \bar{\omega}\left(x^2 - y^2\right),\ \bar{\omega} = \mu\bar{\alpha}\omega/\left(2h^3\right) \tag{2.12}$$

Formulae (2.11) and (2.12) show that the mass flow rate through the channel is given by the relation

$$Q_\rho = 2\omega a = 4h^3\Delta q/\left(\mu\bar{\alpha}a\right);\ \Delta q = q_r - q_w \tag{2.13}$$

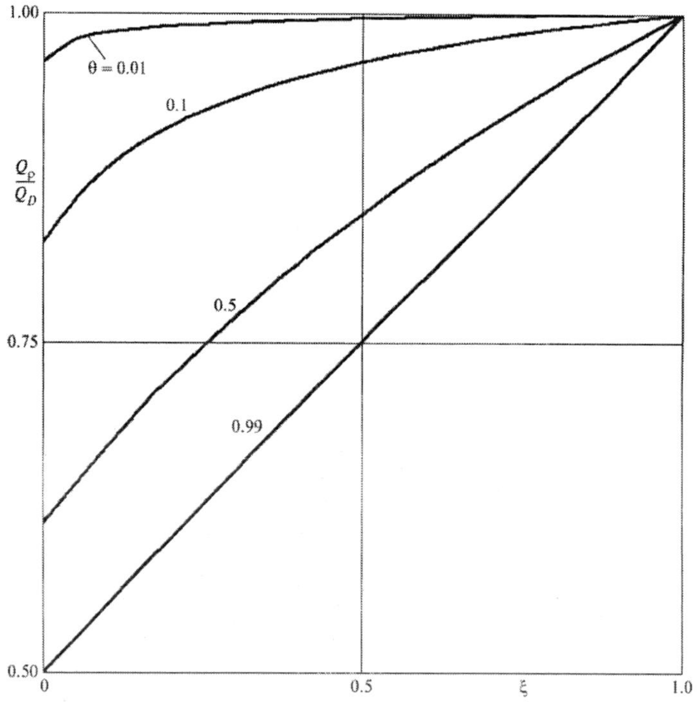

Figure 9.

It follows from formula (2.11) that $\Delta q < \rho\Delta P$ always, and consequently for a specified pressure drop, the mass flow rate (2.13) obtained will always be less than the mass flow rate

$$Q_D = 4h^3\Delta P\rho_r/\left(\mu\bar{\alpha}a\right) \tag{2.14}$$

corresponding to Darcy's law for a fluid with a constant density ρ_r. The dependence of the ratio Q_ρ/Q_D on the value of the pressure-difference parameter $\xi = P_w/P_r$ is shown in Figure 9 for a series of values of θ, where, when $\theta = 0$ (an incompressible fluid), we have $Q_\rho = Q_D$. Note that, in dimensionless form, the relation

$$\bar{Q} = Q_\rho \mu a/(2h^3 \Delta q) = 2/\bar{\alpha}$$

coincides with the curve in the $\beta = 0$ plane, Figure 3.

Formula (2.11) and the last of formulae (2.12) enable us to obtain an equation which defines the pressure distribution $\bar{P} = \bar{P}(x, y)$, $\bar{P} = P/P_r$ in the form

$$b(1-\bar{P}) + \theta b^2 \ln \frac{\bar{P} + \theta b}{1 + \theta b} = \tilde{x}^2 - \tilde{y}^2, \quad \frac{\tilde{x}}{x} = \frac{\tilde{y}}{y} = \sqrt{\frac{\bar{\omega}}{2h^3 \rho_r P_r}} \tag{2.15}$$

In the space $\bar{P}$, $\tilde{x}$, $\tilde{y}$ the first formula of (2.15) defines a family of surfaces as a function of the parameter θ. Knowing the value of the pressure P_w in the exit section of the channel we can find the coordinate $\tilde{x}_w$ corresponding to it and obtain the relation $\bar{P} = \bar{P}(x/a)$, since $x/a = \tilde{x}/\tilde{x}_w$. A series of curves $\bar{P} = \bar{P}(\tilde{x}, 0)$ for a number of values of the parameter θ is shown in the upper part of Figure 10. For example, for $\theta = 0.5$ and $P_w = 0.2P_r$ it turned out that $\tilde{x}_w = 0.76$. A series of curves $\bar{\rho} = \bar{\rho}(\tilde{x}, 0)$ for the same values of θ are given in the lower part of Figure 10, and we obtained a value $\bar{\rho}_w = \rho(\tilde{x}_w)/\rho_r = 0.34$ for the dimensionless density in the exit section of the channel. The corresponding points in both parts of Figure 10 are shown by the light circles.

Bearing in mind that $y/a = \tilde{y}/\tilde{x}_w$, and specifying the ratio h/a, for each section x we can construct the relation $P = P(\bar{y})$, which has a minimum when $\bar{y} = 0$, i.e. on the axial line of the channel. Knowing the pressure field $\bar{P} = \bar{P}(\bar{x}, \bar{y})$, from formula (2.9) we can obtain the density distribution $\rho = \rho(x, y)$. In all the sections x the density has a minimum at the centre of the channel cross section and a maximum on the walls. Consequently, the concentration of the gaseous phase is a maximum along the axial line of the channel and a minimum on its walls. However, these differences are slight.

The graphs in Figure 10 show that changes in the pressure and density along the channel axis depend very much on the compressibility parameter θ. The pressure and density are practically constant in the transverse direction.

If, in Eqs (2.8), the term in the square brackets is retained and the averaged inertia forces are neglected, the corresponding system of equations for the plane case takes the form

$$\eta\left(\frac{\partial^2 k_x}{\partial x^2}+\frac{\partial^2 k_x}{\partial y^2}\right)-\alpha k_x=\frac{1}{\mu}\frac{\partial q}{\partial x},\ \eta\left(\frac{\partial^2 k_y}{\partial x^2}+\frac{\partial^2 k_y}{\partial y^2}\right)-\alpha k_y=\frac{1}{\mu}\frac{\partial q}{\partial y},$$

$$\frac{\partial k_x}{\partial x}+\frac{\partial k_y}{\partial y}=0 \tag{2.16}$$

Equations (2.16) and the boundary conditions are identical, apart from the notation, with Eqs (1.16) and the boundary conditions considered in Section 3, and consequently, their solution, by analogy with formulae (1.18), can be written as

$$k_x=\frac{1-\mathrm{ch}(\chi\bar{y})/\mathrm{ch}\chi}{1-\lambda}\frac{\omega x}{h},\ k_y=\frac{\lambda\,\mathrm{sh}(\chi\bar{y})/\mathrm{sh}\chi-\bar{y}}{1-\lambda}\omega,$$

$$q=q_r-\frac{\omega\bar{\alpha}\mu}{2h^3(1-\lambda)}\left(x^2-y^2\right) \tag{2.17}$$

where we have introduced the notation

$$\chi=\sqrt{\mathrm{So}}=\sqrt{\frac{\bar{\alpha}}{\eta}},\ \lambda=\frac{\mathrm{th}\chi}{\chi}$$

If we assume that the variables $\tilde{x}$, $\tilde{y}$ in (2.15) have the form

$$\frac{\tilde{x}}{x}=\frac{\tilde{y}}{y}=\sqrt{\frac{\overline{\omega}}{2h^3\rho_r P_r(1-\lambda)}}$$

the method of constructing the velocity and pressure field for the case considered is identical with the method described above for the case of Darcy's law. The difference between the graphs shown in Figure 10 and the graphs for the case considered reduces solely to a change of scale along the corresponding axes.

Formulae (2.17) show that

$$\tilde{k}_x=k_x(x,\bar{y})/k_x(x,0)=\tilde{u},\ \tilde{k}_y=k_y(\bar{y})/\omega=\tilde{\upsilon}$$

and the graphs of $\tilde{k}_x$ and $\tilde{k}_y$ against the dimensionless coordinate $\bar{y}$ coincide with the graphs of $\tilde{u}$ and $\tilde{\upsilon}$, shown in Figure 2, respectively.

It follows from the last formula of (2.17) that the overall mass flow rate of the fluid through the channel Q_η differs from the similar flow rate ignoring the effective viscosity (2.13) by the factor $(1-\lambda)$. Consequently, the graph of the ratio $Q_\eta/((1-\lambda)Q_D)$ against ξ coincides with the one shown in Figure 9.

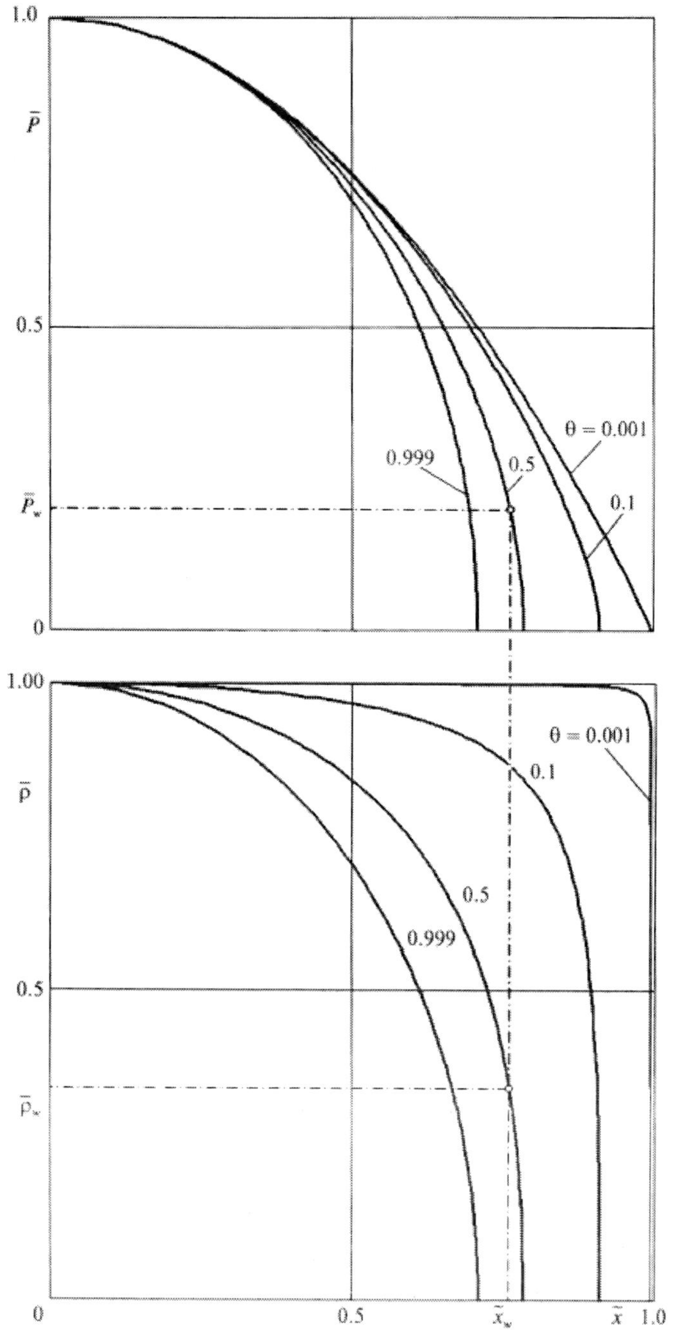

Figure 10.

The value of λ varies from 0 to 1; $\lambda \to 0$ as $\chi \to \infty$ (the condition $\alpha \to \infty$ corresponds to Darcy's law), and $\lambda \to 1$ as $\chi \to 0$ (the condition $\alpha = 0$ corresponds to the absence of porous drag). For a specified value of the pressure difference $\Delta P = P_r - P_w$ the mass flow rate Q_ρ (2.13), corresponding to a compressible fluid, turns out to be less than the flow rate Q_D (2.14), corresponding to Darcy's law for an incompressible fluid, while the flow rate $Q_\eta = (1-\lambda)Q_\rho$ when both the compressibility and the effective viscosity are taken into account simultaneously turns out to be less than the flow rate Q_ρ, i.e., $Q_\eta < Q_\rho < Q_D$. Consequently, if both the compressibility and the effective viscosity are taken into account, there is a considerable increase in the overall hydraulic drag of the channel, which is important in practice.

REFERENCES

[1] Schwartz C.E., Smith J.M.; Flow distribution in packed beds. *Ind Eng Chem* 1953; 45(6):1209–18.

[2] Sato K., Akehata T.; Flow distribution in packed beds. *Chem Eng* 1958; 22(7):430–6.

[3] Popov Ye. K., Smirnova Ye. V., Abayev G.M., Shtern P.G., et al. Problems of investigating reactors with a fixed layer of catalyst. *Aerodynamics of Chemical Reactors* 1976:65–70.

[4] Kirillov V.A., Kuz'min V.A., P'yanov V.I., Khanayev V.M.; The velocity profile in a fixed granular layer. *Dokl Akad Nauk SSSR* 1979; 245(1):159–62.

[5] Gol'dshtik M.A.; Transfer Processes in a Granular Layer. Novosibirsk: *Izd SO Akad Nauk SSSR*; 1984.

[6] Vaisman A.M., Gol'dshtik M.A.; A dynamical model of the fluid flow in a porous medium. *Izv Akad Nauk SSSR* MZhG 1978; 6:89–95.

[7] Grigoryan S.S., Dao Min Ngok.; *Hydrodynamic Problems of Chemical Technologies*. Moscow: Izd MGU; 1979.

[8] Brinkman H.C.; A calculation of the viscosity and the sedimentation constant for solutions of large chain molecules taking into account the hampered flow of the solvent through macromolecules. *Physica* 1947; 13(8):447–8.

[9] Brinkman H.C.; Problems of fluid flow through swarms of particles and through macromolecules in solution. *Research* (London) 1949; 2:190–4.

[10] Brenner H.; Rheology of two-phase systems. *Ann Rev Fluid Mech* 1970; 2:137–76.

[11] Zhukovskii N.Ye.; A theoretical investigation of the flow of subsurface water. *Zh Russ Fiz-Khim Obshch* 1889; 21(1). Moscow: Gostekhizdat; 1949.

[12] Leibenzon L.S.; *The Flow of Natural Fluids in a Porous Medium*; Gostekhizdat: Moscow, 1947;

[13] Loitsyanskii L.T.; *Fluid Mechanics*; Drofa: Moscow, 2003;

[14] Shirko I.V.; Viscous properties of hydrodynamic flow in porous media. *Proc 4th National Congr Theoret Appl Mechanics*, Varna, Bulgaria 1981; 6:912–7.

[15] Shirko I.V., Parfus V.O.; Viscous properties of the flows of an incompressible fluid in an anisotropic and inhomogeneous granular medium. *Teor. Osznovy Khim Tekh* 2004; 6:630–42.

[16] Shirko I.V., Neginsky M.Yu.; Numerical investigation of hydrodynamic flow pattern in porous media. In: *Proc. XX JAHR Congr. Subject C*. 1983. p. 537–48.

[17] Shirko I.V., Parfus V.O.; Flow distribution in nonhomogenious anisotropic packed beds. In: *Proc. 2nd BIOT Conf. Poromechanics*. 2002. p. 539–45.

[18] Shirko I.V., Sobakin S.V.; Flow optimization in nonhomogeneous anisotropic granular medium. In: *Proc. 2nd BIOT Conf. Poromechanics*. 2002. p. 547–8.
[19] Matushenko V.K.; 1998, Effect of regular packing of catalyst in pipe reactor. Spb. Hydrodynamical problems of technological processes, M. *Science*.
[20] Gavrilov A.V.; The effect of aeration of a liquid on the seepage in a porous channel. *Proc 50th Scientific Conf of MFTI Moscow: MFTI* 2007; 2(Pt. 3):63–6.
[21] Gavrilov A.V., Shirko I.V.; Fluid flow in an inhomogeneous granular medium. *Prikl. Mat. Mekh.* 2010; 3(74): 375-90.

In: Navier-Stokes Equations
Editor: R. Younsi

ISBN: 978-1-61324-590-3

Chapter 9

TRANSIENT FLOW OF FLUIDS: SOME APPLICATIONS OF THE NAVIER-STOKES EQUATIONS

***Z. Ouchiha*[1], *A. Ghezal*[1] *and S.M. Ghiaasiaan*[2]**
[1]USTHB, faculté de physique, Bp 32, 16111, El Alia, BEZ, Algiers, Algeria
[2]G. W. Woodruff School of Mechanical Engineering, Georgia Institute of Technology, Atlanta, GA, U. S.

1. INTRODUCTION

By definition, a fluid is a substance that deforms continuously under the action of a shearing stress. The Navier-Stokes equations govern the motion of fluids in general, and are applicable to Newtonian as well as non-Newtonian fluids. Fluids are classified based on the rate at which they deform in response to an imposed shear stress. In Newtonian fluids there is a linear relation between the sear stress and the strain rate, whereas in non-Newtonian fluids this relation is non-linear. When the complete set of conservation equations governing the flow of a pure fluid, along with its equation of state, are used, the velocity, the pressure, the density, and the temperature (or, equivalently, the enthalpy or internal energy) are the unknown parameters. In the case of a three – dimensional flow, for example, the pressure, three velocity components, the density and temperature constitute the six unknowns in a set of six equations that include the three components of the momentum conservation equation, the mass continuity, the energy conservation, and the fluid equation of state.

The derivation of the conservation equations for fluids can be found in a multitude of book, including

Mass Continuity: For a compressible fluid is in unsteady flow, we have:

$$\frac{D\rho}{Dt} = \frac{\partial \rho}{\partial t} + \nabla.\left(\rho\vec{V}\right)$$

or, equivalently,

$$\frac{D\rho}{Dt} + \rho\nabla.(\vec{V}) = 0, \tag{1}$$

where, in an Eulerian frame, the total (substantial) derivative $\frac{D\rho}{Dt}$ represents the combination of the local contribution term $\frac{\partial\rho}{\partial t}$ and the advective contribution term (due to translation) $\vec{V}.\nabla\rho$, namely:

$$\frac{D\rho}{Dt} = \frac{\partial\rho}{\partial t} + \vec{V}.\nabla\rho$$

2. Momentum

The basic equations governing the fluid motion are derived by applying Newton's second law to an infinitesimally small fluid volume. The law states that the center-of-mass of a fluid particle of mass m is accelerated under the action of two categories of forces. The first category represents the body forces (gravitational force when gravity is the only force field), whereas the second category represents surface forces resulting from pressure and viscous stresses that act on the fluid particle's boundaries. Thus, the momentum equation in generic form contains three terms:

$$\rho\frac{D\vec{V}}{Dt} = \vec{B} + \vec{F} \tag{2}$$

where $\frac{D\vec{V}}{Dt} = \frac{\partial\vec{V}}{\partial t} + (\vec{V}.\nabla)\vec{V}$ represents the fluid acceleration. The body forces are external forces which are imposed on the fluid without physical contact, and are distributed over the volume of the fluid. The body force, as well as the surface force, can evidently be represented in terms of their components in a coordinate system. In the Cartesian coordinate system, for example,

$$\vec{B} = B_x\vec{\imath} + B_y\vec{\jmath} + B_z\vec{k}$$

$$\vec{F} = F_x\vec{\imath} + F_y\vec{\jmath} + F_z\vec{k}$$

When gravity is the only cause of body force, furthermore, we have

$$\vec{B} = \rho\vec{g}$$

The surface forces depend on the rate at which the fluid is strained by its velocity field. Fluids in which the shear stress is linearly related to the rate of deformation are termed Newtonian. We will focus our attention on isotropic and Newtonian fluids. All gases and many liquids, in particular water, belong to the class of Newtonian fluids. The linearity of the stress – strain rate also implies isotropy of the fluid which obeys Stokes's law of friction.

The forces at the surfaces of a fluid particle are the results of the *stress tensor*; a symmetric, second-rank dyadic tensor that in Cartesian coordinate can be represented as:

$$\underline{\underline{\tau}} = \begin{bmatrix} \sigma_x \vec{i}\vec{i} & \tau_{xy}\vec{i}\vec{j} & \tau_{xz}\vec{i}\vec{k} \\ \tau_{xy}\vec{j}\vec{i} & \sigma_y \vec{j}\vec{j} & \tau_{yz}\vec{j}\vec{k} \\ \tau_{xz}\vec{k}\vec{i} & \tau_{yz}\vec{k}\vec{j} & \sigma_z \vec{k}\vec{k} \end{bmatrix}$$

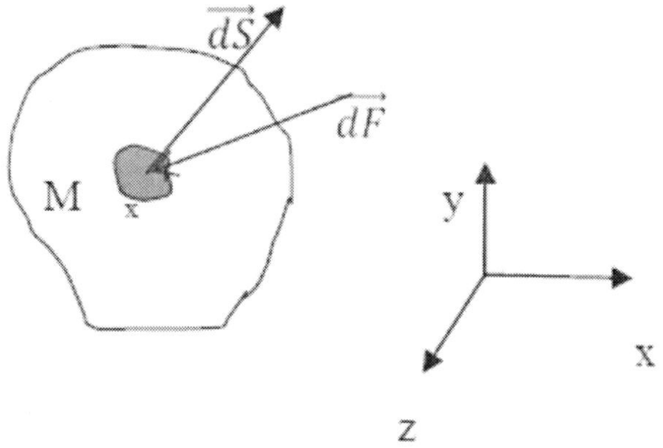

Figure 1.

The surface force $\overrightarrow{dF}$ acting upon the element of area $\overrightarrow{dS}$ at point M, as shown in Fig. 1, can then be found from.

$$d\vec{F} = \underline{\underline{\tau}} \cdot d\vec{S} \tag{3}$$

The magnitude of $\overrightarrow{dS}$ is the area of the element, and the direction is normal to the surface.

Each of the different components of the surface force dF_x, dF_y and dF_z has two components: one due to the normal stress to which pressure is a contributor, and the other due to shear stresses. It can be shown that the net surface force acting on the control volume depicted in Fig. 2 will be

$$\vec{F} = \nabla \cdot \underline{\underline{\tau}}$$

In Cartesian coordinates this will give, for x direction:

$$F_x = \frac{\partial \sigma_x}{\partial x} + \frac{\partial \tau_{xy}}{\partial y} + \frac{\partial \tau_{xz}}{\partial z} \tag{4}$$

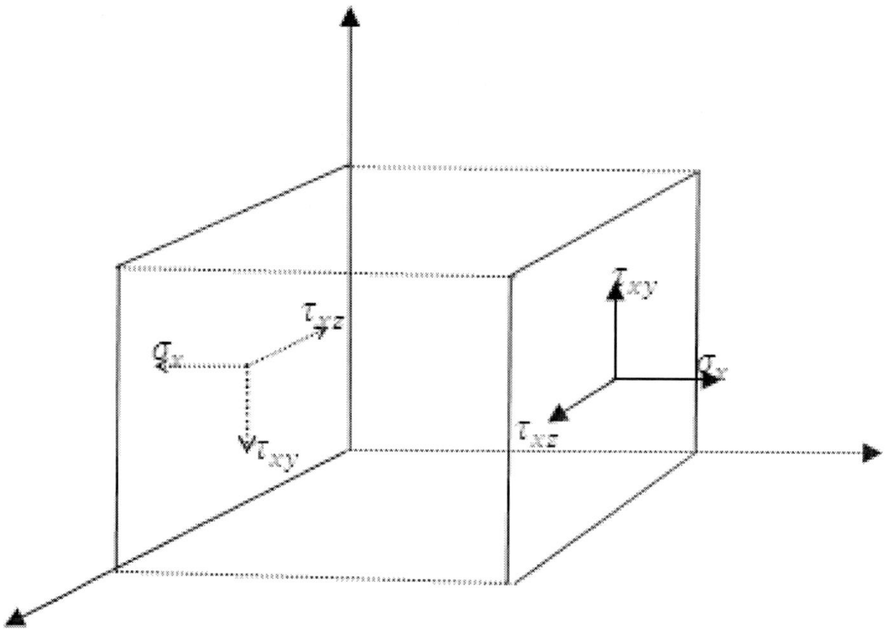

Figure 2. Sketch for stresses impact.

When the fluid is at rest, the normal stress is the same as the hydrostatic pressure, which is a the same as –p, where p is the thermodynamic pressure. When the fluid is in motion, however, the equation of state along with the conservation principles determine the pressure at every point (hence the "principle of local state"). The normal stresses will then have an additional viscous component. In the x direction, for example, the normal viscous stress will be $\sigma_x' = \sigma_x + p$. Similar expressions can be written for y and z directions.

The assumptions that the viscous stress varies linearly with strain rate, and that the fluid is isotropic, lead to the following general formulation for viscous stresses, $\sigma_x' = \lambda div\vec{V} + 2\mu\frac{\partial u}{\partial x}$;

$$\sigma_y' = \lambda div\vec{V} + 2\mu\frac{\partial v}{\partial y}; \sigma_z' = \lambda div\vec{V} + 2\mu\frac{\partial w}{\partial z} \qquad (5)$$

$$\tau_{xy} = \tau_{yx} = \mu\left(\frac{\partial v}{\partial x} + \frac{\partial u}{\partial y}\right); \tau_{yz} = \tau_{zy} = \mu\left(\frac{\partial w}{\partial y} + \frac{\partial v}{\partial z}\right); \tau_{zx} = \tau_{xz} = \mu\left(\frac{\partial u}{\partial z} + \frac{\partial w}{\partial x}\right) \qquad (6)$$

In these equations μ is the viscosity of the fluid and λ represents the second coefficient of viscosity. The deviatoric components represent the normal stresses which cause viscous dissipation. For monatomic gases the gas kinetic theory predicts that $\lambda = -\frac{2}{3}\mu$. The same expression is often assumed to apply to other fluids as well, however (Stokes' hypothesis).

Substitution from eqs. (xx-1) and (xx-2) in eq. (2) leads the following equation for the x-direction:

$$\rho \frac{Du}{Dt} = B_x + F_x;\ F_x = -\frac{\partial p}{\partial x} + \left(\frac{\partial \sigma_x'}{\partial x} + \frac{\partial \tau_{xy}}{\partial y} + \frac{\partial \tau_{xz}}{\partial z}\right) \quad (7)$$

Furthermore, substitution of eqs. (5) and (6) leads to:

$$F_x = -\frac{\partial p}{\partial x} + \frac{\partial}{\partial x}(2\mu\frac{\partial u}{\partial x} - \frac{2}{3}\mu div\vec{V}) + \frac{\partial}{\partial y}\left[\mu\left(\frac{\partial u}{\partial y} + \frac{\partial v}{\partial x}\right)\right] + \frac{\partial}{\partial z}\left[\mu\left(\frac{\partial w}{\partial x} + \frac{\partial u}{\partial z}\right)\right] \quad (8)$$

Similar equations can be written for the momentum conservation in the y and z directions.

The mass continuity equation [eq. (1a)] in Cartesin coordinates for compressible flow gives:

$$\left(\frac{\partial \rho}{\partial t} + \frac{\partial(\rho u)}{\partial x} + \frac{\partial(\rho v)}{\partial y} + \frac{\partial(\rho w)}{\partial z}\right) = 0 \quad (9)$$

The solution of the momentum equation for a compressible fluid requires an equation of state because temperature, pressure and density are coupled. For an ideal gas the equation of state is $p - \rho RT = 0$ in which R is the gas constant and T is the temperature in absolute scale.

3. The Energy Equation

Unsteady flows occur in a vast variety of engineering applications. For incompressible and constant-property fluids the momentum and energy conservation equations are uncoupled. The flow field for these fluids can be analyzed using only the mass continuity and momentum equations, even when heat transfer or strong viscous dissipation are encountered. However, in many unsteady flow problems the momentum and energy conservation equations are coupled and an analysis of the flow must include the solution of the energy conservation equation as well. The energy conservation equation for a fluid can be derived by applying the first law of thermodynamics to a fluid particle in a Lagrangian frame. Considering an elementary control volume $\Delta V = dxdydz$ that contains $\Delta M = \rho\Delta V$ of a flowing fluid mass. The control volume receives dQ heat during the infinitesimally short time dt contributes with total energy dE_T and elementary work dW. Globally, the energy equation is:

$$\frac{dQ}{dt} = \frac{dE_T}{dt} + \frac{dW}{dt} \quad (10)$$

With assumptions that the radiation transfer is neglected then the heat can be transferred only through conduction, according to Fourier's law which is involving the heat flux: $q = \frac{1}{A}\frac{dQ}{dt} = -k\frac{\partial T}{\partial n}$. In SI units the area A is in m^2, T is the temperature in K and the thermal conductivity k is in W/(m.K).

The heat added by conduction to a volume ΔV can be written as:

$$\frac{dQ}{dt} = \Delta V \left\{ \frac{\partial}{\partial x}\left(k\frac{\partial T}{\partial x}\right) + \cdots \right\} \tag{11}$$

$\frac{dE_T}{dt}$ is a derivative which consists of a local contribution represented by the variation of internal energy $\frac{de}{dt}$, and a convection effect which is defined as a change in kinetic energy. Definitely the total energy is expressed as:

$$\frac{dE_T}{dt} = \rho \Delta V \left\{ \frac{de}{dt} + \frac{d}{dt}\left(\frac{1}{2}u^2 + gz \ldots\right) \right\} \tag{12}$$

The total work performed by the normal and shearing stresses is:

$$\frac{dW}{dt} = -\Delta V \left\{ \left(\frac{\partial}{\partial x}(u\sigma_x + v\tau_{xy} + w\tau_{xz} \right) + \cdots \right\} \tag{13}$$

After some simplifications, the equation becomes

$$\rho \frac{de}{dt} + \rho div\vec{V} = \frac{\partial}{\partial x}\left(k\frac{\partial T}{\partial x}\right) + \frac{\partial}{\partial y}\left(k\frac{\partial T}{\partial y}\right) + \frac{\partial}{\partial z}\left(k\frac{\partial T}{\partial z}\right) + \mu\emptyset \tag{14}$$

$\emptyset$ denotes the function of dissipation:

$$\emptyset = 2\left\{ \left(\frac{\partial u}{\partial x}\right)^2 + \left(\frac{\partial v}{\partial y}\right)^2 + \left(\frac{\partial w}{\partial z}\right)^2 \right\} + \left(\frac{\partial v}{\partial x} + \frac{\partial u}{\partial y}\right)^2 + \left(\frac{\partial w}{\partial y} + \frac{\partial v}{\partial z}\right)^2 + \left(\frac{\partial u}{\partial z} + \frac{\partial w}{\partial x}\right)^2 - \frac{2}{3}\left(\frac{\partial u}{\partial x} + \frac{\partial v}{\partial y} + \frac{\partial w}{\partial z}\right)^2$$

Considering the continuity equation (9) with both of the variation in the internal energy of a perfect gas de = C_vdT and the enthalpy variation dh=C_pdT, the energy equation (14) takes the final form as below:

$$\rho C_p \frac{dT}{dt} = \frac{dp}{dt} + \left\{ \frac{\partial}{\partial x}\left(k\frac{\partial T}{\partial x}\right) + \frac{\partial}{\partial y}\left(k\frac{\partial T}{\partial y}\right) + \frac{\partial}{\partial z}\left(k\frac{\partial T}{\partial z}\right) \right\} + \mu\emptyset \tag{15}$$

In: Navier-Stokes Equations
Editor: R. Younsi

ISBN: 978-1-61324-590-3

Application A

Transient Flow of Highly Pressurized Fluids in Pipelines

***Zohra Ouchiha**[*,1], **Jean C. Loraud**[2] and **Abderahmane Ghezal**[1]*
[1]USTHB, Faculté de Physique, Bp 32, 16111, El Alia, BEZ, Algiers, Algeria
[2]IUSTI, Polytech' Marseille, France

The solution of the Navier-Stokes equations, which are partial and nonlinear differential equations, presents many mathematical difficulties. With simplifications coming from physical insight, however, it is often possible to derive useful solutions for some important problems. In this application, our attention is devoted to the presentation of the mathematical model of transient gas flow in a pipe based on a set of partial differential equation such as those advanced by Osiadac[2] and Zucrow[3]. The form of these equations varies with the assumptions made as regards to the conditions of operation of the gas flow within the pipeline. The equations may be linear or, quite generally, nonlinear. They may be parabolic or hyperbolic of the first or second order. The basic equations describing the transient gas flow in a pipe are derived below and the simplified model is discussed.

Assuming one-dimensional flow, the conservation equations are cast in the form of a closed set of couple partial differential equations (PDEs). The PDEs, which represent a compressible fluid in transient flow, are hyperbolic, and are solved by the method of characteristics (MOC). The discretization for the unsteady numerical analysis is based on the finite difference technique with constant time and spatial steps.

In the present paper we analyze the pressure and flow rate variations inside a horizontal gas transmission system that is typical of pipelines that supply power to facilities in a region. Because of the likelihood of large swings in the daily load requirements, such a pipeline should be analysed to ensure that the swing can be accommodated by the existing capacity of the pipe without the need for additional storage.

A.1. Basic Formulation for One-dimensional Compressible Flow

When density variations within a flow are not negligible, the fluid must be treated as compressible. We may note that most liquid flows are essentially incompressible. In the gas flow case, however, the significance of compressibility effect depends on *Mach number, M,* which is defined as the ratio between flow velocity *v* and the velocity of sound in the fluid, *a.* It is commonly known that when $M < 0.3$, changes in density are limited to around 2 percent of the mean density value. Such gas flows are therefore treated as incompressible. However, compressible flows (namely, flows with $M > 0.3$) occur frequently in engineering applications, for example in high pressure transmission pipelines and dental drills.

Following Mekbel[5], the one-dimensional transient flow in a pipe is represented by the forthcoming set of coupled PDEs. The system of PDEs describes the compressible unsteady natural gas flow in a horizontal pipeline with length L. The continuity equation is derived from recasting equation (1) as:

$$\frac{\partial \rho}{\partial t} + \frac{\partial (\rho v)}{\partial t} = 0 \tag{16}$$

The friction force δF always act in opposite direction to the mean flow. In the one-dimensional momentum equation we have $\delta F = \frac{\rho v^2}{2}\frac{\lambda dx}{d}A$, where λ is the Darcy friction factor, and from there the momentum equation can be cast as:

$$\frac{\partial v}{\partial t} + v\frac{\partial v}{\partial x} + \frac{1}{\rho}\frac{\partial p}{\partial x} + \frac{\lambda}{2d}v^2 = 0 \tag{17}$$

The one-dimensional energy equation can be written as:

$$v\frac{\partial u}{\partial t} + v^2\frac{\partial v}{\partial x} + \delta v C_p\frac{\partial p}{\partial x} - \beta C_p\frac{\partial p}{\partial t} + vC_p\frac{\partial T}{\partial x} + C_p\frac{\partial T}{\partial t} - \frac{1}{\rho}\frac{4k(\theta - T_0)}{d} = 0 \tag{18}$$

In these equations δ, β and γ are defined as:

$$\delta = \frac{1}{C_p}\left[\frac{1}{\rho} - T\left(\frac{\partial(1/\rho)}{\partial T}\right)\right]_p$$

$$\beta = \frac{T}{C_p}\left(\frac{\partial}{\partial T}(1/\rho)\right)_p$$

$$\gamma = \frac{C_p}{C_v} = \frac{1}{\left(1 - \frac{p}{T}\beta\right)}$$

The equation of state for the fluid is $\rho = \frac{p}{zRT}$ and following Mekbel[5] it is assumed that z = 0.99, and the speed of sound is found from:

$$A\,a = \sqrt{\frac{\gamma P}{\rho}} = \sqrt{\gamma z R T_0}$$

The set of PDEs represented by equations (10 through (3) is now transformed into the following set of equations in accordance with the method of characteristics (MOC) (see Zucrow[3] and Hoffman[3], for a discussion),

Characteristic C^0

$$\text{(II)} \begin{cases} \left[\dfrac{dx}{dt}\right] = v & (19) \\ \left[\dfrac{dp}{dt}\right]_{(0)} - \dfrac{\gamma}{\gamma - 1}\dfrac{p}{T}\left[\dfrac{dT}{dt}\right]_{(0)} = -E_0 & (20) \end{cases}$$

Forward characteristic C^+

$$\text{(III)} \begin{cases} \left[\dfrac{dx}{dt}\right] = v + a & (21) \\ \left[\dfrac{dp}{dt}\right]_{(1)} + a\rho\left[\dfrac{dv}{dt}\right]_{(1)} = E_1 & (22) \end{cases}$$

Backward characteristic C^-

$$\text{(IV)} \begin{cases} \left[\dfrac{dx}{dt}\right] = v - a & (23) \\ \left[\dfrac{dp}{dt}\right]_{(2)} - a\rho\left[\dfrac{dv}{dt}\right]_{(2)} = E_2 & (24) \end{cases}$$

where: $E_0 = \dfrac{4h(\theta - T_{CM})}{d} + \rho_{CM}\dfrac{\lambda}{2d}v_{CM}^3$, and

$$E_1 = (\gamma - 1)\rho_{AM}\left(\frac{4k(\theta - T_{AM})}{d\rho_{AM}} + \frac{\lambda}{2d}v_{AM}^3\right) - \rho_{AM}a\frac{\lambda}{2d}v_{AM}^2 \quad (25)$$

$$E_2 = (\gamma - 1)\rho_{BM}\left(\frac{4k(\theta - T_{BM})}{d\rho_{BM}} + \frac{\lambda}{2d}v_{BM}^3\right) + \rho_{BM}a\frac{\lambda}{2d}v_{BM}^2 \quad (26)$$

where $\rho_{AM} = \dfrac{p_{AM}}{zRT_{AM}}$, $\rho_{BM} = \dfrac{p_{BM}}{zRT_{BM}}$, $p_{AM} = \dfrac{p_A + p_M}{2}$, and $p_{BM} = \dfrac{p_B + p_M}{2}$.

A.2. Determination of the Different Parameters by Transformation

To derive the mathematical model for transient analysis, we assume that steady state friction factor applies to transient flow. In view of Figure 3, the equation systems (III) and (IV) are then recast as follows.

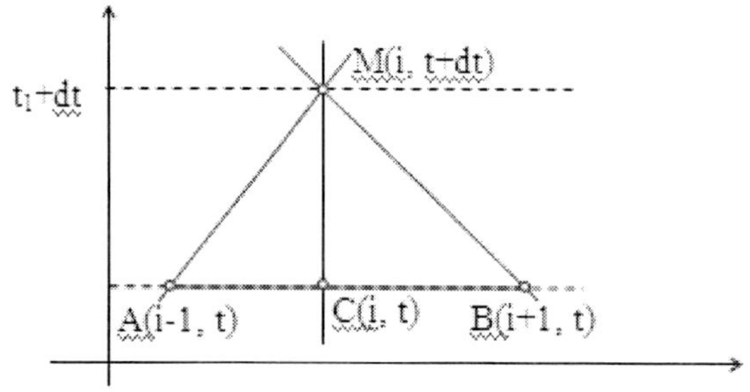

Figure 3. Schematic representation of MOC.

First, we consider an internal grid point (i.e., a point in the bulk of fluid inside pipeline). We then write:

$$\frac{(p_M - p_A)}{dt} + a\rho_{AM}\frac{(v_M - v_A)}{dt} = E_1 \quad (27)$$

$$\frac{(p_M - p_B)}{dt} - a\rho_{BM}\frac{(v_M - v_B)}{dt} = E_2 \quad (28)$$

By subtracting eq. (27) from eq. (28) one obtains the following equation for v_M, the mixture velocity:

$$v_M = \frac{p_A - p_B + a(\rho_{BM}v_B + \rho_{AM}v_A) + (E_1 - E_2)dt}{a(\rho_{AM} + \rho_{BM})} \quad (29)$$

The above expression is inserted into eq. (27), for example, to derive an equation for pressure. It leads to:

$$p_M = p_A + a\rho_{AM}(v_A - v_M) + E_1 dt \tag{30}$$

In discretizing these equations, a uniform mesh distribution ($dx = const.$), and constant time step ($dt = const.$) are used. The time step is chosen small enough to satisfy the Courant-Friedriech-Lewy (CFL) stability criterion $dt \le \left|\frac{dx}{v_0 \pm a}\right|$.

We now address the development of equations for the pipe inlet. Figure 4 depicts the conditions of the fluid upstream from the inlet. The known parameters at point B(2, t) at time t, along with the backward characteristic (C^-), are used to calculate the parameters at point M(1, t+dt). Since $v_M = v_0$ at point M, eq. (28) leads to the following expression for pressure:

$$p_M = p_B + a\rho_{BM}(v_M - v_B) + E_2 dt \tag{31}$$

where $\rho_{BM} = \frac{P_{BM}}{ZRT_0}$, $P_{BM} = (P_B + P_M)/2$, and E_2 is to be found from eq. (26).

With respect to the pipe exit, similar to the aforementioned analysis, the boundary downstream from the pipe exit is treated by utilizing the positive characteristic (C^+), as displayed in Figure 5. Thus, the unknown parameters at point M (n+1, t+dt) are calculated by considering the parameters at the known point A(n, t). The transient flow effect in very long pipes vanishes (i.e., $u_M = 0$) at the exit is approached. Therefore, using eq. (27), we can write for the pressure p_M at point M(n+1,t+dt):

$$p_M = p_A - a\rho_{AM}(v_M - v_A) + E_1 dt \tag{32}$$

where $\rho_{AM} = \frac{p_{AM}}{zRT_0}$, $p_{AM} = \frac{p_A + p_M}{2}$, and E_1 is defined as in eq. (10).

A.3. Results and Discussions

This study considered the phenomenology of gas pressurization which commonly occurs during gaseous flow in pipelines. Natural gas transportation and distribution often take place in piping systems which are designed purely for constant supply and demand flow conditions, whereby the transmission to any specific point is to take place at a specific pressure and flow rate.

Pipelines in the ideal situation would have constant flow rates, and the volumetric flow rates of gases delivered at the end of a pipeline would not vary with time. In practice, however, the consumption rates frequently rise and fall in pipeline networks. In fact, the flow rates at boundaries of a pipeline occasionally occur in periodic or pseudo periodic form during a day in response to consumer demand. Highly pressurised flows also occur in pipelines during transients caused by pumps and valves and can cause vibration problems. These flow disturbances, which often generate pressure and mass flow fluctuations, can be slow or rapid. Gato and Henriques[6] associate the slow disturbances to the cyclic variations in the demand for natural gas during the day. The gas is alternatively compressed and expanded in the pipeline. Thus, because in practice the global boundary conditions for a pipeline are always changing with time, an accurate prediction of flow rate, pressure drop, and temperature distribution during transient flows is essential for the optimum design of pipelines, online network control and monitoring, as well as economics. Tentis et al[4] developed a system of Euler equations for describing the slow or rapid gas transients in pipelines, and applied the Method of Lines (MOL) along with an adaptive grid algorithm. Our solution method is similar. Typical results are presented below.

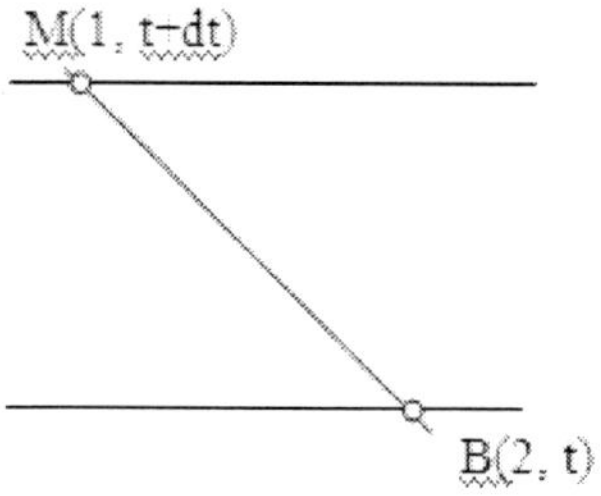

Figure 4. Conditions at the pipe inlet.

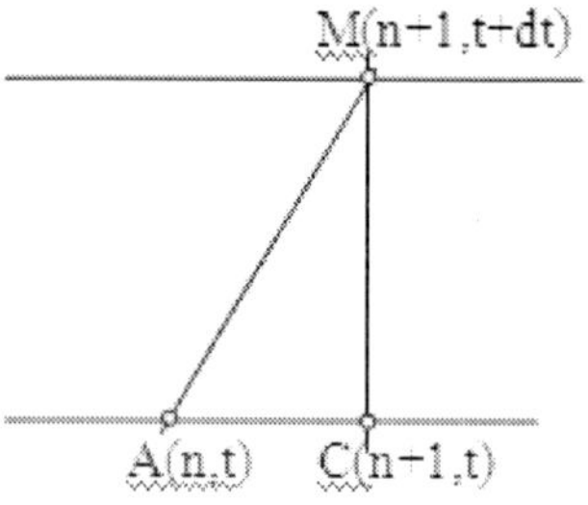

Figure 5. Conditions at the pipe outlet.

In pipeline networks, the perturbation processes resulting from flow disturbances propagate upstream.

In fact, any event that occurs at any point along the pipeline is reverberated in other points at different sections of the pipeline as reported by Mekbel[5]. The initial regime of gas flow in a pipe line may thus be disturbed as a result of an upstream event, leading to the slowing down of flow, or even an abrupt stoppage of the flow, a downstream event, and a pressure drop whose magnitude essentially depends on the flow rate and the magnitude and intensity of the disruption.

Figs. 6 and 7 illustrate respectively the velocity and the pressure distributions at different instants within the pipe sections for an inlet velocity $v=25 m.s^{-1}$. The propagation of the wave front can be seen in the profiles and is captured at chronological progression, t'_1, t'_2,..., until t'_{11}. As noted, at any instant the fluid velocity monotonically decreases along the pipe in the flow direction. All parts of the pipe start to be filled with gas uniformly once the wave front has reached the downstream end of the pipe. Also, up to t_9' the velocity at any location in the pipe monotonically increases with time, while an opposite trend is observed from t_9' to t_{12}'. This trend is the outcome of flow oscillations which take place after the wave front has reached the end of the pipe.

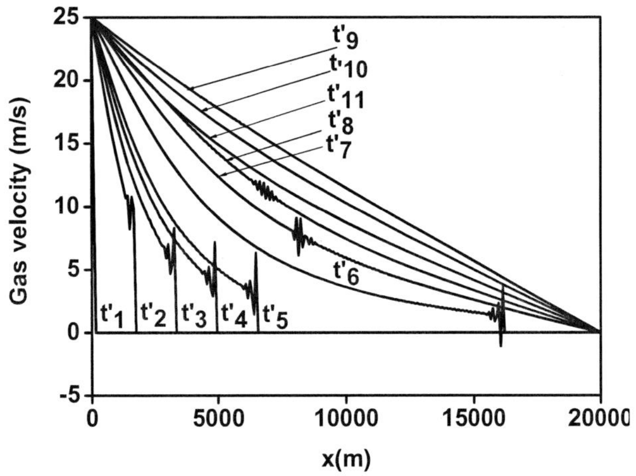

Figure 6. Propagation of the wave front through gaseous flow.

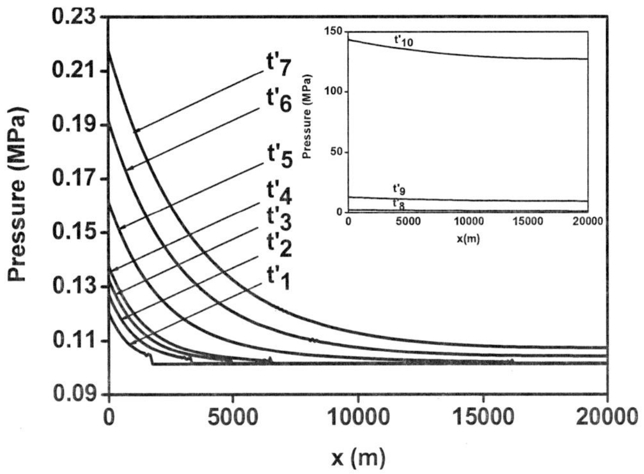

Figure 7. Propagation of the pressure wave through gaseous flow.

References

[1] H. Schlichting, *Boundary-layer theory*, McGraw-Hill book company, New-York, 1979.

[2] Osiadacz A.J. *Simulation and Analysis of Gas Networks*. Gulf Publishing Company: Houston, 1987.

[3] Zucrow M.J., Hoffman J.D. The method of Characteristics applied to unsteady one-dimensional flow. *Gas Dynamics: Multidimensional flow*. New York: John Wiley & Sons; 1976; II, 295-311.

[4] Tentis E., Margaris D., Papanikas D. Transient Gas Flow Simulation Using an Adaptive Method of Lines. *C. R. Mec.* 2003, 331, 481-487.

[5] Mekbel S. *Contribution à l'Etude de l'Ecoulement Variable dans les Conduites de Transport de Gaz à Haute Pression.* Algiers University, Algiers, Algeria, 1978.

[6] Gato L.M.C., Henriques, J.C.C. Dynamic Behaviour of High Pressure Natural Gas Flow in Pipelines. *Int. J. Heat and Fluid Flow* 2005, 26, 817-825.

In: Navier-Stokes Equations
Editor: R. Younsi

ISBN: 978-1-61324-590-3

Application B

Periodic Flow

***A. Ghezal*[1], *Jean C. Loraud*[2] *and Z. Ouchiha*[1]**
[1]USTHB, Faculté de Physique, Bp 32, 16111, El Alia, BEZ, Algiers, Algeria
[2]IUSTI, Polytech' Marseille, France

B.1. Analytical Study

B.1.1. General Remarks and Definitions

An oscillatory flow can be a pulsatile flow or inverse flow. In the first case the velocity oscillation amplitude is smaller than the mean velocity and therefore the flow direction remains unchanged. In the second case, the oscillatory amplitude velocity is greater than the mean amplitude and the flow direction is inversed periodically.

Similitude parameters:

-inversed flow

We consider an incompressible inversed flow in a cylindrical conduct (D=2a) while the inversed flow is caused by a sinusoidal displacement of compressor for example is given by,

$$x_m = \frac{x_{\max}}{2}(1 - \cos \omega t)$$

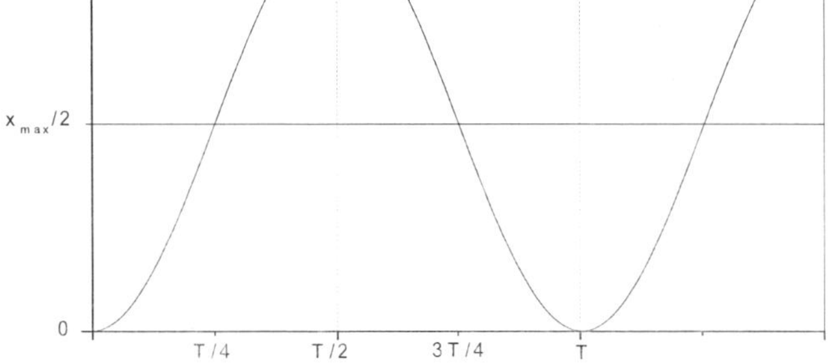

The velocity is defined by:

$$U_m = \frac{dx_m}{dt} = \frac{\omega x_{max}}{2} \sin \omega t$$

$$= U_{max} \sin \omega t \quad \Rightarrow \quad U_{max} = \frac{\omega x_{max}}{2}$$

$$U_{max} = \omega \frac{x_{max}}{2} = \frac{2\pi}{T} \frac{x_{max}}{2} \Rightarrow U_{max} = \frac{\pi x_{max}}{T}$$

We have: $U_m = \frac{\omega x_{max}}{2} \sin \omega t = \frac{\omega x_{max}}{2} \sin\left\{\left(\frac{2\pi}{T}(t) + \Phi\right)\right\}$

We pose $t/T = i - 1$, while i is the number of periods

$$\langle U_m \rangle = \frac{\omega x_{max}}{2} \sin\Phi$$

$$\langle U_m \rangle = U_{max} \sin\Phi$$

$$U_{max} = \frac{\omega x_{max}}{2}$$

The dimensionless numbers are defined as bellow:

$$\tau = \omega t$$

$$V = V / U \max$$

$$P = p/\rho_f U^2{}_{max}$$

$$\theta = \begin{cases} \dfrac{T - T_r}{T_w - T_r} (\text{case } T_{Wall} = C^{st}),\ T_r = T_{reference} \\ k_f T / q_w D \ (\text{case } q_W : \text{Heat flux at wall} = C^{st}) \end{cases}$$

The dimensionless equations are:

$$\nabla \vec{V} = 0$$

$$\frac{\partial \vec{V}}{\partial \tau} + \frac{A_0}{2}\left[\left(\vec{V}.\nabla\right)\vec{V} + \nabla P\right] = \frac{1}{R_{ew}} \nabla^2 \vec{V}$$

$$\frac{\partial \theta}{\partial \tau} + \frac{A_0}{2}\left(\vec{V}.\nabla\right)\theta = \frac{1}{R_{ew} P_r} \nabla^2 \theta$$

While: $R_{ew} = \frac{\omega D^2}{\upsilon}$ is the kinetic Reynolds number.

$A_0 = \frac{x_{max}}{D}$: is the dimensionless amplitude of oscillation

$Pr = \frac{\upsilon}{\alpha_f}$: is the Prandtl number

For the conduct length the ratio L/D must be taken as a parameter in the mathematic problem across the inlet and outlet conditions.

The Reynolds number associated with the maximum velocity, Re_{max} , is defined as:

$$R_{emax} = \frac{U_{max} D}{\upsilon}$$

The dependence of Re_{max} on the frequency can be shown by writing:

$$R_{emax} = \frac{\omega x_{max}}{2} \frac{D}{\upsilon} = \frac{x_{max}}{D} \frac{\omega D^2}{2\upsilon}$$

$$\Rightarrow \quad R_{emax} = \frac{A_0}{2} R_{ew}$$

We observe that the maximum Reynolds number Re_{max} depends on the product of the amplitude and the kinetic Reynolds number Re_w . This indicates that the nature of oscillatory flow cannot be defined by Re_w alone, and the importance of the amplitude $A_0/2$ should be borne in mind.

B.1.2. Pulsatile Flow

The velocity throughout the conduct can be expressed as bellow:

$$U(r,t) = U_s(r) + U_t(r,t)$$

While: U_s is the velocity of stationary flow.

U_t is the velocity of oscillatory flow.

The mean velocity at the cross section can be obtained by integrating U(r, t)

$$\langle U \rangle = \langle U_s \rangle + \langle U_t \rangle$$

U_s is the velocity of stationary flow

U_t is the velocity of oscillatory flow

If $U_t \approx \sin \omega t \Rightarrow \langle U_t \rangle = U_{max} \sin \varphi$

$$\Rightarrow \langle U \rangle = \langle U_s \rangle + U_{max} \sin \omega t$$

The dimensionless equations are:

$$\nabla \vec{V} = 0$$

$$\frac{\partial \vec{V}}{\partial \tau} + (\vec{V}.\nabla)\vec{V} + \vec{\nabla} P = \frac{1}{R_e} \nabla^2 \vec{V}$$

$$\frac{\partial \theta}{\partial \tau} + (\vec{V}.\nabla)\theta = \frac{1}{R_e P_r} \nabla^2 \theta$$

where $R_e = \frac{\langle U_s \rangle}{\upsilon} D$ is the Reynolds number of the mean flow.

If we consider a pulsatile flow; the inlet dimensionless velocity is given by:

$$\langle U \rangle (t) = \langle U_s \rangle + U_{max} \sin \omega t \ ,$$

We can write

$$U = \frac{\langle U_s \rangle}{\langle U_s \rangle} + \frac{U_{max}}{\langle U_s \rangle} \sin \omega t = 1 + \frac{A_0}{2} \frac{R_{ew}}{R_e} \sin \omega t$$

$$\Rightarrow U = 1 + \frac{A_0}{2} \frac{R_{ew}}{R_e} \sin \omega t$$

We compare the expression above which is used in literature with our expression below:

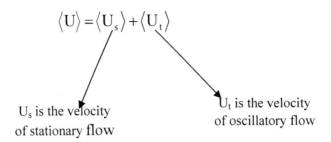

$$U = U_{inlet}\,(1+\varepsilon_W \sin \omega t)$$

$$\Rightarrow U_{inlet} = 1\ (\text{referrence velocity}\ , \varepsilon_W = \frac{A_0}{2}\frac{R_{ew}}{R_e}\ )$$

We can now derive a relation between ω and Re_w by noting that:

$$R_{ew} = \frac{uD^2}{\upsilon} \quad \Rightarrow \quad u = \frac{\upsilon}{D^2} R_{ew}$$

We define $u_{ref} = \dfrac{\nu}{D^2}$ as the characteristic frequency of the flow

$$\Rightarrow \omega = \omega_{ref}\ R_{ew}$$

The dimensionless frequency is defined by: $\omega_{add} = \dfrac{\omega}{\omega_{ref}} = R_{ew}$

We see that the kinetic Reynolds number Re_w in fact represent also the dimensionless frequency.

$$\omega_d = Re_w$$

The term $A \cos\left(\omega t + \phi\right)$ is thus equivalent to $A \cos\left(Re_w(nk) + \phi\right)$ where nk is the dimensionless time.

B.2. Pulsatile Flow in Rectangular Conduct

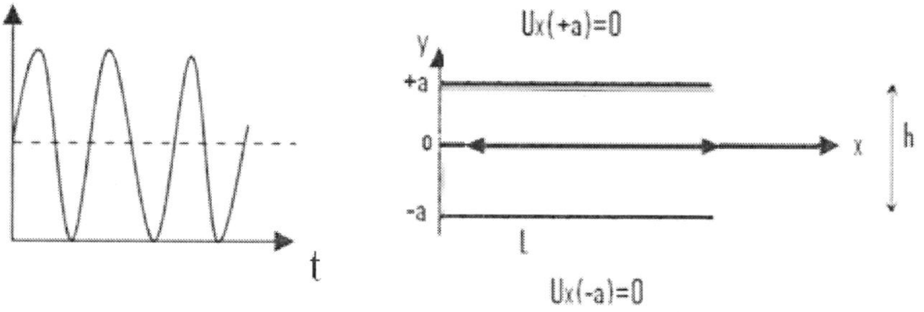

We consider a unidirectional flow. The mass continuity and momentum conservation equations will be:

Equation of continuity:

$$\frac{\partial u}{\partial x}+\frac{\partial v}{\partial y}=0$$

Momentum equation:

$$\frac{\partial \mathrm{U}}{\partial \mathrm{t}}+(\mathrm{U}.\nabla)\mathrm{U}=-\frac{\mathbf{1}}{\boldsymbol{\rho}}\nabla \mathrm{P}+\nu\nabla^2\mathrm{U}$$

Because the flow is two-dimensional we can write:

$$\frac{\partial u}{\partial t}+u\frac{\partial u}{\partial x}+v\frac{\partial u}{\partial y}=-\frac{1}{\rho}\frac{\partial p}{\partial x}+\nu\left(\frac{\partial^2 u}{\partial x^2}+\frac{\partial^2 u}{\partial y^2}\right)$$

$$\frac{\partial v}{\partial t}+u\frac{\partial v}{\partial x}+v\frac{\partial v}{\partial y}=-\frac{1}{\rho}\frac{\partial p}{\partial y}+\nu\left(\frac{\partial^2 v}{\partial x^2}+\frac{\partial^2 v}{\partial y^2}\right)$$

However, we have

$$v=0\rightarrow\frac{\partial v}{\partial y}=0\rightarrow\frac{\partial u}{\partial x}=0$$

The Navier-Stokes equation in the *x* direction then simplifies to:

$$\frac{\partial u}{\partial t}=-\frac{1}{\rho}\frac{\partial p}{\partial x}+\nu\frac{\partial^2 u}{\partial y^2} \tag{1}$$

In the y directional the equation is: $0=-\frac{1}{\rho}\frac{\partial p}{\partial y}\rightarrow p\neq p(y)$

The equation (1) indicate that the pressure gradient must be invariant in x direction when the flow is fully developed

$$p=\ p(x,t)$$

We find: $\frac{\partial p}{\partial x}=\frac{\partial p(t)}{\partial x}$, and from there:

$$u=u(y,t)\rightarrow\frac{\partial p}{\partial y}=0\rightarrow\frac{\partial p}{\partial x}=\frac{\partial p(t)}{\partial y}$$

The Fourier expansion proposed by Majdalani (2002) is:

$$-\frac{1}{\rho}\frac{\partial p}{\partial x}=p_0+\sum_{n=1}^{\infty}p_{cos}\cos(\omega nt)+p_{sin}\sin(\omega nt)=p_0+\sum_{n=1}^{\infty}p_n\,e^{i\omega nt}$$

where p_0 represents the stationary solution, and p_{cos} and p_{sin} represent the sinus and cosinus amplitudes of the harmonic function, respectively. Furthermore,

$$p_n = p_{cos} - ip_{sin}$$

and:

$$-\frac{1}{\rho}\frac{\partial p}{\partial x} = p_0 + \sum_{n=1}^{\infty} p_n\, e^{i\omega nt} \tag{2}$$

In this case we also have:

$$u = u_0 + \sum_{n=1}^{\infty} u_{cos} \cos(\omega nt) + \sum_{n=1}^{\infty} u_{sin} \sin(\omega nt)$$

Equivalently we can write:

$$u = u_0 + \sum_{n=1}^{\infty} u_n e^{i\omega nt} \tag{3}$$

where $u_n = u_{cos} - ip_{sin}$

B.3. Exact Solution

Consider the case where n=1, which corresponds to the first harmonic. We can then write:

$$\frac{\partial u}{\partial t} = -\frac{1}{\rho}\frac{\partial p}{\partial x} + \nu\frac{\partial^2 u}{\partial y^2}$$

$$-\frac{1}{\rho}\frac{\partial p}{\partial x} = p_0 + p_1 e^{i\omega t}$$

$$u = u_0 + u_1 e^{i\omega t}$$

$$\frac{\nu\,\partial^2 u}{\partial y^2} = \nu\frac{\partial^2 u_0}{\partial y^2} + \nu\frac{\partial^2 u_1}{\partial y^2} e^{i\omega t}$$

$$\frac{\partial u}{\partial t} = i\omega u_1 e^{i\omega t}$$

Substitution into eq. (1), and further manipulation then give:

$$\frac{\partial u}{\partial t} = p_0 + p_1 e^{i\omega t} + \nu\frac{\partial^2 u_0}{\partial y^2} + \nu\frac{\partial^2 u_1}{\partial y^2} e^{i\omega t}$$

$$i\omega u_1 e^{i\omega t} = p_0 + p_1 e^{i\omega t} + \nu\frac{\partial^2 u_0}{\partial y^2} + \nu\frac{\partial^2 u_1}{\partial y^2}e^{i\omega t}$$

$$0 = p_0 + \nu\frac{\partial^2 u_0}{\partial y^2} + (p_1 + \nu\frac{\partial^2 u_1}{\partial y^2} - i\omega u_1)e^{i\omega t}$$

$$p_0 + \nu\frac{\partial^2 u_0}{\partial y^2} + (p_1 + \nu\frac{\partial^2 u_1}{\partial y^2} - i\omega u_1)e^{i\omega t} = 0$$

For this equation to be valid we must have:

$$\begin{cases} \dfrac{d^2 u_0}{dy^2} + \dfrac{p_0}{\nu} = 0 \\ \text{et} \\ \dfrac{d^2 u_1}{dy^2} - \dfrac{i\omega u_1}{\nu} + \dfrac{p_1}{\nu} = 0 \end{cases}$$

The solution is:

$$\begin{cases} u_0 = -\dfrac{p_0}{2v}y^2 + c_1 y + c_2 \\ u_1 = -\dfrac{i}{\omega}p_1 + c_3 sh(y\sqrt{\dfrac{i\omega}{v}}) + c_4 ch(y\sqrt{\dfrac{i\omega}{v}}) \end{cases}$$

Thus

$$u = u_0 + e^{i\omega t}u_1$$

This leads to:

$$u = -\frac{p_0}{2v}y^2 + c_1 y + c_2 + \left(-\frac{ip_1}{\omega} + c_3 sinh(y\sqrt{\frac{i\omega}{v}}) + c_4 cosh(y\sqrt{\frac{i\omega}{v}})\right)e^{i\omega t}$$

The boundary conditions for this equation are as follows. Symmetry with respect to the x axis requires that at $y = 0$

$$\frac{\partial u(0,t)}{\partial y} = 0$$

Now

$$\left.\frac{\partial u}{\partial y}\right)_{y=0} = 0 \Rightarrow -\frac{p_0}{v}(0) + c_1 + + \left(c_3\sqrt{\frac{i\omega}{v}}ch(0\sqrt{\frac{i\omega}{v}}) + c_4\sqrt{\frac{i\omega}{v}}sh(0\sqrt{\frac{i\omega}{v}})\right)e^{i\omega t} = 0$$

$$\Rightarrow c_1 + + \left(c_3 \sqrt{\frac{i\omega}{\nu}}(1) + c_4 \sqrt{\frac{i\omega}{\nu}}(0) \right) e^{i\omega t} = 0$$

$$\Rightarrow c_1 + c_3 \sqrt{\frac{i\omega}{\nu}} e^{i\omega t} = 0$$

$$\Rightarrow c_1 + c_3 \sqrt{\frac{i\omega}{\nu}} e^{i\omega t} = 0 \ \forall \, t \text{ ou } c_1 = c_3 = 0$$

Furthermore, the no-slip conditions at the upper and lower walls require that:

$$u(a,t) = u(-a,t) = 0$$

This leads to:

$$-\frac{p_0}{2\nu} a^2 + c_2 + \left(-\frac{ip_1}{\omega} + c_4 cosh(a \sqrt{\frac{i\omega}{\nu}}) \right) e^{i\omega t} = 0$$

We thus find:

$$-\frac{p_0}{2\nu} a^2 + c_2 = 0 \ et \ -\frac{ip_1}{\omega} + c_4 cosh(a \sqrt{\frac{i\omega}{\nu}}) = 0$$

$$c_2 = \frac{p_0}{2\nu} a^2 \text{ et } + c_4 = \frac{ip_1}{\omega \cosh\left(a \sqrt{\frac{i\omega}{\nu}} \right)}$$

The solution is therefore:

$$u = \frac{p_0}{2\nu}(a^2 - y^2) + i \frac{p_1}{\omega} \left(-1 + \frac{cosh(y \sqrt{\frac{i\omega}{\nu}})}{cosh\left(a \sqrt{\frac{i\omega}{\nu}} \right)} \right) e^{i\omega t}$$

Oscillatory case:

$$u(y,t) = \frac{k}{i\omega} \left(1 - \frac{cosh(y \sqrt{\frac{i\omega}{\nu}})}{cosh\left(a \sqrt{\frac{i\omega}{\nu}} \right)} \right) e^{i\omega t}$$

B.4. Exercises

B.4.1. Exercise 1

The velocity distribution for pulsatile flow in a rectangular duct that has a height of 2a and a width of b (see the figure below) is.

$$u = \frac{p_0}{2\nu}(a^2 - y^2) + i\frac{p_1}{\omega}\left(-1 + \frac{\cosh\left(y\sqrt{\frac{i\omega}{\nu}}\right)}{\cosh\left(a\sqrt{\frac{i\omega}{\nu}}\right)}\right)e^{i\omega t} \quad \text{(a)}$$

This solution can be cast as:

$$u = u_0(y) + u_1(y, t)$$

1. Specify the terms $u_0(y)$ and $u_1(y, t)$ and explain what they represent.
2. Derive an expression for the mean velocity by using:

$$u_m = \frac{1}{2a}\int_{-a}^{+a} u\,dy$$

3. By using the following dimensionless parameters

$$\eta = \frac{y}{a}, u^* = \frac{u}{u_m}, t^* = \omega t, \varphi = \frac{P_1}{P_0}$$

show that the dimensionless velocity expression can be cast as:

$$u^*(\eta, t^*) = \frac{3}{2}(1 - \eta^2) + 3i\frac{\varphi}{Re_w}\left[-1 + \frac{\cosh(\eta\sqrt{iRe_w})}{\cosh(\sqrt{iRe_w})}\right]e^{it^*} \quad \text{(b)}$$

where $Re_w = \frac{\omega a^2}{\nu}$

4. Using the above expression determining the positions of the maximum of velocity.
5. Simplify the above expression for the following limits:
   (a) small frequencies
   (b) large frequencies

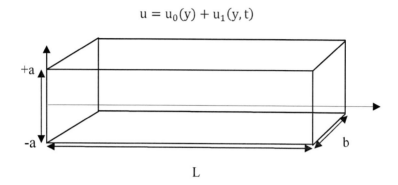

B.4.2. Exercise 2

Consider the dimensionless velocity expressed in eq. (b) of the previous exercise. Find the wall stress expression according to the parameters of the problem.

1. Simplify the expression you have derived for the two following limiting cases:
   (a) very small frequencies
   (b) very large frequencies

B.5. Oscillatory Flow

We define the movement of the flow generated by sinusoidal variation of the pressure gradient which is given by the next:

$$-\frac{1}{\rho}\frac{\partial p}{\partial x} = A_0 \cos \omega t$$

B.5.1. Velocity Profile

Uchida obtain an exact solution for velocity profile in the case of fully developed flow for the cylindrical conduct given by the next expression:

$$u(r,t) = \frac{A_0}{\omega}[B\cos\omega t + (1-A)\sin\omega t] \qquad (1)$$

A and B are a Bessel functions.

The dimensionless velocity is given by:

$$u(R,\tau') = \frac{32}{\sigma R_{ew}}\left[B\cos\tau' + (1-A)\sin\tau'\right] \qquad (2)$$

If R_{ew} <16, (Small frequencies):

$$u(R,\tau') \cong \frac{2}{\sigma}\left[1-\left(\frac{r}{2}\right)^2\right]\cos\tau' + \frac{1}{32\sigma}R_{ew}\left[\left(\frac{r}{a}\right)^4 + 4\left(\frac{r}{a}\right)^2 - 5\right]\sin\tau' \qquad (3)$$

If R_{ew} >16, (great frequencies):

$$u(R,\tau') \cong \frac{32}{\sigma R_{ew}}\left[\sin\tau' - \frac{e^{-E}}{\sqrt{r/a}}\sin(\tau' - E)\right] \qquad (4)$$

where: $E = \left[1-\left(\frac{r}{a}\right)^2\right]\sqrt{\frac{R_{ew}}{8}}$

We check that the expressions vanishes for r = a (at the wall, the dimensionless length is r/a)

If R_{ew} <16 (small frequencies):

$$u(R,\tau') \cong 0.\cos\tau' + \frac{1}{32\sigma} R_{ew} \left[\left(\frac{a}{a} \right)^4 + 4\left(\frac{a}{a} \right)^2 - 5 \right] \sin\tau'$$

$$u(R,\tau') = 0$$

If R_{ew} >16, (great frequencies):

$$u(R,\tau') \cong \frac{32}{\sigma R_{ew}} \left[\sin\tau' - \frac{e^{-0}}{\sqrt{1}} \sin\tau' \right] = 0$$

The relation above confirm the boundary conditions.

B.5.2. Velocity Expression for the Limit Case of the Frequency

The velocity in the cylindrical tube is given by:

$$U(r,t) = U_s(r) + U_t(r,t) \tag{5}$$

where: $U_s(r)$ is the velocity of steady flow and $U_t(r)$ is the velocity of unsteady flow.

The mean velocity is obtained by integrating the expression (5) across a section:

$$\langle U\rangle(t) = \langle U_s\rangle + \langle U_t\rangle(t) \tag{6}$$

In the sinusoidal flow case:

$$\langle U\rangle(t) = \langle U_s\rangle + U_{max} sin\omega t \tag{7}$$

Generally the dimensionless variables are defined as bellow:

$$U = \frac{U}{\langle U_s\rangle} \quad V = \frac{V}{\langle V_s\rangle} \text{ and } P = \frac{P}{\rho\langle U_s\rangle^2}$$

with $\langle U_s\rangle$ characteristic velocity.

The dimensionless equations are:

$$\vec{\nabla}.\vec{V} = 0$$

$$\frac{\partial\vec{V}}{\partial t} + (\vec{V}.\vec{\nabla})\vec{V} + \vec{\nabla}P = \frac{1}{Re}\nabla^2\vec{V}$$

$$\frac{\partial\theta}{\partial t} + (\vec{V}.\vec{\nabla})\theta = \frac{1}{RePr}\nabla^2\theta$$

where: $Re = \frac{\langle U_s \rangle D}{\nu}$ is the Reynolds number according to the inlet condition.

$$U(t) = 1 + \Delta \sin\omega t = 1 + \frac{A_0}{2}\frac{Re_w}{Re}\sin \quad (8)$$

where:

$$\Delta = \frac{U_{max}}{\langle U_s \rangle} = \frac{A_0}{2}\frac{Re_w}{Re} \quad (9)$$

We observe that the pulsatile flow associated with heat transfer must be characterized by the next dimensionless numbers:

Re, A_0(or Δ), Re_ω (ou λ), Pr, L/D

with $A_0 = x_{max}/D$, $\lambda = a\sqrt{\frac{\omega}{\upsilon}}$ (Stokes or Womorsely number)

The state of flow:

The state of turbulence in the oscillatory flow is characterized by the dimensionless displacement A_0 and the kinetic Reynolds number Re_w .

For the pulsatile flow we must add into account the effect of the mean flow.

Experimental results give these next values:

Re=1500, λ (Womorshey) =10 $A_0 \leq 5.8$.

B.5.3. Wall Friction τ_p

We have:

$$U(R,t') = \frac{u(r,t')}{\langle u_s \rangle} = 2\left(1 - \frac{r^2}{a^2}\right) + \frac{15A_0}{Re\sigma}\left[B\cos\tau' + (1-A)\sin\tau'\right] \quad (10)$$

with: $u_s = \frac{A_0}{4\upsilon}\left[a^2 - r^2\right]$

and: $u_m = \frac{u_{max}}{2} = \frac{A_0 a^2}{8\upsilon}$

The friction coefficient is defined by:

$$\tau_p = \frac{2\mu \frac{\partial u}{\partial r}\Big|_{r=a}}{\rho u_m^2} \quad (11)$$

We have: $\frac{\partial u}{\partial r} = u\left[2\left(-\frac{2r}{a^2}\right) + \frac{16A_0}{Re\sigma}(B'\cos\tau' - A'\sin\tau')\right]$

The steps in computing progression are developed as follow:

$$\tau_P = \frac{\mu}{u_m^2}u_m\left[-4 + \frac{16A_0}{R_e\sigma}(B'(r=a)\cos\tau' - A'(r=a)\sin\tau')\right]$$

$$= \frac{2\mu}{u_m}\left[-4 + \frac{16A_0}{R_e\sigma}(B'(r=a)cos\tau' - A'(r=a)sin\tau')\right]$$

$$= 2\frac{\mu}{u_m D}D\left[-4 + \frac{16A_0}{R_e\sigma}(B'(r=a)cos\tau' - A'(r=a)sin\tau')\right]$$

$$= \frac{4a}{R_e}\left[+4 - \frac{16A_0}{R_e\sigma}(B'(r=a)cos\tau' - A'(r=a)sin\tau')\right]$$

Finally,

$$\tau_p = \frac{16}{R_e}\left[1 + \frac{4A_0}{R_e\sigma}(cos(\tau' - \emptyset)\right] \quad (12)$$

with,

$$\frac{A_0 Re_w}{2R_e} = \Delta, \sigma_3 = \frac{2}{\lambda}\sqrt{c_1^2 + c_2^2}$$

where:

$$\frac{4A_0}{R_e} = 4\frac{2\Delta}{Re_w}$$

Also:

$$\Delta\frac{\sigma_3}{\sigma} = \frac{A_0 Re_w}{2Re}\frac{\sigma_3}{\sigma}$$

with:

$$\lambda = \frac{1}{2}\sqrt{Re_w}$$

$$\sigma_3 = \frac{2}{\lambda}\sqrt{c_1^2 + c_2^2} = \frac{2}{{}^1/_2\sqrt{Re_w}}\sqrt{c_1^2 + c_2^2} = \frac{4}{\sqrt{Re_w}}\sqrt{c_1^2 + c_2^2}$$

$$\xrightarrow{yields} \frac{\Delta\sigma_3}{\sigma} = \frac{1}{\sigma}\frac{A_0 Re_w}{2Re}\frac{4}{\sqrt{Re_w}}\sqrt{c_1^2 + c_2^2} = \frac{2A_0\sqrt{Re_w}}{\sigma Re}\sqrt{c_1^2 + c_2^2}$$

We find the final expression:

$$\tau_p = \left[1 + \frac{A_0 Re_w}{2Re}\frac{\sigma_3}{\sigma}(cos(\tau' - \emptyset_s)\right] \quad (13)$$

with: $\emptyset_s = Artg\, {}^{C_1}/_{C_2}$

B.5.4. Calculate τ_p from the Zhao Expression

The velocity is given by:

$$u(r,\tau') = \frac{2}{a\sigma}\left[(1-({}^{r}/_{a})^2)\cos\tau' + \frac{Re_\omega}{16}\left\{\left(\frac{r}{a}\right)^4 + 4\left(\frac{r}{a}\right)^2 - 5\right\}\sin\omega t\right] \quad (14)$$

We have, $u(R,\tau') = \frac{u(r,t)}{u_{max}}$

$$\frac{\partial u}{\partial R} = \frac{2}{a\sigma}\left[-\frac{2r}{a^2}\cos\tau' + \frac{Re_\omega}{16}\left\{\frac{4r^3}{a^4} + 8\frac{r}{a^2}\right\}\sin\tau'\right] \quad (15)$$

with: $u_{max} = \frac{A_0 D^2 \sigma}{32\upsilon}$ and $A_0 = \left|\frac{1}{\rho}\frac{\partial P}{\partial x}\right|$

We find:

$$\tau_p \equiv \frac{1}{\sigma^2}\left[-\cos\,\tau' + \frac{3}{8}Re_\omega \sin\tau'\right] \quad (16)$$

For the Small frequencies

$$\tau_p \equiv Re_\omega^2$$

– **Great frequencies,** $R_i = a$ et $B = E$

$$\frac{u(r,\tau')}{u_{max}} = \frac{32}{\sigma\, Re_\omega}\left[\sin\tau' - \frac{e^{-E}}{\sqrt{{}^{r}/_{a}}}\sin(\tau' - E)\right] \quad (17)$$

$$\frac{\partial u}{\partial r} = \frac{32}{a\sigma Re_\omega}\left[\left(-\frac{2r}{a^2}\right)\sqrt{\frac{Re_\omega}{8}}\ \frac{e^{-E}(\sin(\tau'-E)}{\sqrt{{}^{r}/_{a}}} - e^{-E}\frac{d}{dr}\left(\frac{\sin(\tau'-E)}{\sqrt{{}^{r}/_{a}}}\right)\right] \quad (18)$$

Derivation: $$\frac{d}{dr}\left(\frac{\sin(\tau'-E)}{\sqrt{{}^{r}/_{a}}}\right) = \frac{\frac{2r}{a^2}\sqrt{\frac{Re_\omega}{8}}\cos(\tau'-E)}{\left(\sqrt{{}^{r}/_{a}}\right)^2} - \sin(\tau'-E)\left(-\frac{1}{2}\left(\frac{r}{a}\right)^{-3/2}\frac{1}{a}\right)$$

$$\Rightarrow \frac{\partial u}{\partial r} = \frac{32}{a\sigma Re_\omega}\left[\frac{-2r}{a^2}\sqrt{\frac{Re_\omega}{8}}e^{-E}\frac{\sin(\tau'-E)}{\sqrt{r/a}} - e^{-E}\left\{\frac{d_x\cos(\tau'-E) - d_z\sin(\tau'-E)}{r/a}\right\}\right] \quad (19)$$

In the wall:

$$\left.\frac{\partial u}{\partial r}\right|_{r=a} = \frac{32}{a\sigma \mathrm{Re}_{\omega}}\left[\frac{-2}{a^2}\sqrt{\frac{\mathrm{Re}_{\omega}}{8}}e^{0}\frac{\sin\tau'}{\sqrt{r/a}} - e^{0}\left\{\frac{2\sqrt{\frac{\mathrm{Re}_{\omega}}{8}}\cos\tau' - \frac{1}{2}\sin\tau'}{r/a}\right\}\right] \tag{20}$$

$$= \frac{32}{a^2\sigma \mathrm{Re}_{\omega}}\left[-2\sqrt{\frac{\mathrm{Re}_{\omega}}{8}}\sin\tau' + \frac{1}{2}\sin\tau' - 2\sqrt{\frac{\mathrm{Re}_{\omega}}{8}}\cos\tau'\right]$$

$$\tau_p = \frac{\mu\frac{\partial u}{\partial y}}{\rho u_{\max}^2}$$

$$= \frac{\mu u_{\max}\frac{32}{a\sigma \mathrm{Re}_{\omega}}\left[\frac{-2r}{a^2}\sqrt{\frac{\mathrm{Re}_{\omega}}{8}}e^{-E}\frac{\sin(\tau'-E)}{\sqrt{r/a}} - e^{-E}\left\{\frac{d_x\cos(\tau'-E) - d_z\sin(\tau'-E)}{r/a}\right\}\right]}{\rho u_{\max}^2}$$

$$= \frac{C^{ste}}{\sigma \mathrm{Re}_{\omega}}\left[\frac{-2r}{a^2}\sqrt{\frac{\mathrm{Re}_{\omega}}{8}}e^{-E}\frac{\sin(\tau'-E)}{\sqrt{r/a}} - e^{-E}\left\{\frac{d_x\cos(\tau'-E) - d_z\sin(\tau'-E)}{r/a}\right\}\right]$$

$$= \frac{C^{ste}}{\frac{1}{\mathrm{Re}_{\omega}^{2}}\mathrm{Re}_{\omega}}\left[\frac{-2r}{a^2}\sqrt{\frac{\mathrm{Re}_{\omega}}{8}}e^{-E}\frac{\sin(\tau'-E)}{\sqrt{r/a}} - e^{-E}\left\{\frac{d_x\cos(\tau'-E) - d_z\sin(\tau'-E)}{r/a}\right\}\right]$$

$$\equiv (\mathrm{Re}_{\omega})^{3/2}$$

- **At Small frequencies** $\mathrm{Re}_{\omega} < 16$

$$u(r,\tau') = \frac{2}{a\sigma}\left[(1-(r/a)^2)\cos\tau' + \frac{\mathrm{Re}_{\omega}}{16}\left\{\left(\frac{r}{a}\right)^4 + 4\left(\frac{r}{a}\right)^2 - 5\right\}\sin\omega t\right] \tag{21}$$

$$u(R,\tau') = \frac{2}{a\sigma}\left[\left(1-\left(\frac{r}{a}\right)^2\right)\cos\tau'\right] \tag{22}$$

We note that if $\mathrm{Re}_{\omega} \to 0$

$$u(R,\tau') \to \frac{2}{a\sigma}\left(1-\left(\frac{r}{a}\right)^2\right)\cos\tau' \tag{23}$$

This expression can be formulated as shown below:

$$u(R,\tau') = A_u \cos\tau'$$

$$A_u = \frac{2}{a\sigma}\left(1-\left(\frac{r}{a}\right)^2\right) \tag{24}$$

This is identical to the Poiseuille profile.

2nd case R_{ew} >16

At the wall r = a

$$\left.\frac{\partial u}{\partial r}\right|_{r=a} = \frac{2}{a\sigma}\left[\frac{-2}{a}\cos\tau' + \frac{Re_\omega}{16}\left[\frac{4}{a}+\frac{8}{a}\right]\sin\tau'\right]$$
$$= \frac{2}{a\sigma}\left[\frac{-2\cos\tau'}{a} + \frac{Re_\omega}{16}\frac{12}{a}\sin\tau'\right] = \frac{4}{a^2\sigma}\left[-\cos\tau' + \frac{6Re_\omega}{16}\sin\tau'\right] \tag{25}$$

B.5.5. Determination of the Position of the Maximum Velocity

We have $u = U_{max}$ if $\frac{\partial u}{\partial r} = 0$

1. **Case of small frequencies** Re_ω **<4**

$$u(r,t) = \frac{k R_i}{4\,\upsilon}\left[\left(1-\left(\frac{r}{R_i}\right)^2\cos\omega t + \frac{Re_\omega}{16}\left[\left(\frac{r}{R_i}\right)^4 + 4\left(\frac{r}{R_i}\right)^2 - 5\right]\sin\omega t\right)\right]$$

$$\frac{\partial u}{\partial r} = \frac{k\,R_i}{4\,\upsilon}\left(\frac{-2r}{R_i}\cos\omega t + \frac{Re_\omega}{16}\left[\frac{4r^3}{{R_i}^4} + 8\frac{r}{{R_i}^2}\right]\sin\omega t\right)$$

$\frac{\partial u}{\partial r} = 0 \Rightarrow r = 0 \Rightarrow u = U_{max}$ in the center

2. **Case of great frequencies:** Re_ω **>4**

$$u = \frac{k\;R_i}{4\,Re_w}\left[\sin wt - \frac{e^{-B}}{\sqrt{\frac{r}{R_i}}}\sin(\omega t - B)\right]$$

$$\frac{\partial u}{\partial r} \equiv \frac{\partial}{\partial r}\left[-\frac{e^{-B}}{\sqrt{\frac{r}{R_i}}}\sin(\omega t - B)\right]$$

=

Then, $\frac{\partial u}{\partial r} = \sin(\omega t - B)\,e^{-B}\left[\frac{\sqrt{\frac{Re_\omega}{2}}}{R_i} - \frac{1}{2\left(\frac{r}{R_i}\right)^{5/2}}\right] + \frac{\cos(\omega t - B)}{\sqrt{\frac{r}{R_i}}}\sqrt{\frac{Re_\omega}{2}}$

In the condition where $\frac{\partial u}{\partial r} = 0 \Rightarrow \sqrt{\frac{Re_\omega}{2}} - \frac{1}{2\left(\frac{r}{R_i}\right)^{5/2}} = 0$

If sin (ωt - B) = cos (ωt - B) ⇒ sin B = cos B ⇒ B = π/4

$$\left(1 - \frac{r}{R_i}\right) = \frac{\pi}{4} \Rightarrow \frac{r}{R_i} = 1 - \frac{\pi}{4} \Rightarrow r = \frac{4-\pi}{4} R_i = 0.215 R_i$$

We note that r is in the vicinity of the wall.

Particulars cases:

- Si cos B= 0 ⇒ B= π/2 ⇒ $1 - \frac{r}{R_i} = \frac{\pi}{2} \Rightarrow \frac{r}{R_i} = \frac{2-\pi}{2} < 0$

-Si cos B = 1 ⇒ B=0 ⇒ $1 - \frac{r}{R_i} = 0 \Rightarrow r = R_i$

General case

$$\frac{\cos B}{\sqrt{\frac{r}{R_i}}} = \frac{1 - \frac{\left(\left(1 - \frac{r}{R_i}\right)\sqrt{\frac{Re_\omega}{2}}\right)^2}{2}}{\sqrt{\frac{r}{R_i}}}$$

$$\frac{\partial u}{\partial r} = 0 \Rightarrow 2 - \left(\left(1 - \frac{r}{R_i}\right)\sqrt{\frac{Re_\omega}{2}}\right)^2 = 0 \Rightarrow \left(1 - \frac{r}{R_i}\right)\sqrt{Re_\omega} = 2$$

The final solution is $r = R_i\left(1 - \frac{2}{\sqrt{\frac{1}{Re_\omega}}}\right)$

In the vicinity of the wall the next approximation is introduced: B→0 ($B = (1-r/Ri)(Re_\omega/2)^{1/2}$ ⇒ sinB=0, $\cos B = 1 - B^2/2$

The previous approximation is injected into the equation $\frac{\partial u}{\partial r}$

Then this formulation is obtained:

$$\frac{1}{R_i}\sqrt{\frac{r}{R_i}}\sqrt{\frac{Re_\omega}{2}}\left(1 - \frac{1}{2}\left(\left(1 - \frac{r}{R_i}\right)\sqrt{\frac{Re_\omega}{2}}\right)^2\right) = 0$$

The term between brackets is in the form $1-\frac{B^2}{2}=0 \Rightarrow B=\sqrt{2} \Rightarrow \left(1-\frac{r}{R_i}\right)\sqrt{\frac{Re_\omega}{2}}=\sqrt{2}$

The radius is: $r = R_i\left(1-\frac{2}{\sqrt{Re_\omega}}\right)$

B.5.6. Calculation of the Lag Shift between the Gradient Pressure, the Velocity and the Wall Stress Space Phase

The lag phase between the pressure and the velocity can be determined on the space phase as bellow:

In the case while the pressure varies periodically its expression is:

$$\Delta P(t) = \Delta P_0(1+\gamma_p sin\omega t) \tag{26}$$

We find that the flow rates can be written as:

$$Q_t = Q_0(1+\gamma_a \sin(\omega t + \emptyset_0)) \tag{27}$$

The ellipse of phase lag.

The trajectory is

$$\begin{cases} x(t) = \sin(\omega t + \emptyset_x) \\ y(t) = \sin(\omega t + \emptyset_y) \end{cases} \tag{28}$$

In the space phase it is an ellipse with equation:

$$\frac{x'^2}{a^2}+\frac{y'^2}{b^2}=1 \tag{29}$$

$$\sin\emptyset = \frac{Oa}{OA}=\frac{Ob}{OB}=\frac{aa'}{AA'}=\frac{bb'}{BB'}$$

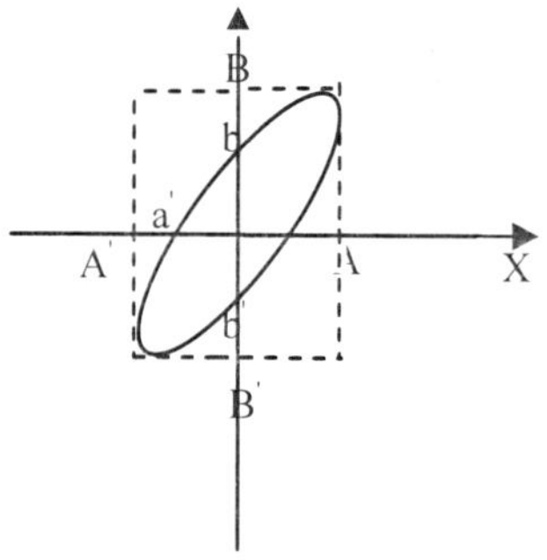

Particular case:

1st – if Oa = OA $\Rightarrow \sin\emptyset = 1 \Rightarrow \emptyset = \frac{\pi}{2} \Rightarrow$ OA=OB (it is a circle)

2nd - if Oa = 0 $\Rightarrow$ Ob $= 0 \Rightarrow$ Ob = 0 $\Rightarrow$ a=b=0 $\Rightarrow \sin\emptyset = 0 \Rightarrow \emptyset = 0$

$$a = \sqrt{2}cos\left[\frac{(\emptyset_y-\emptyset_x)}{2}\right],\ b = \sqrt{2}sin\left[\frac{(\emptyset_y-\emptyset_x)}{2}\right]$$

and

$$x' = asin\left[\omega t + \frac{(\emptyset_y - \emptyset_x)}{2}\right] = \frac{\sqrt{2}}{2}(x+y)$$

$$y' = bcos\left[\omega t + \left(\frac{(\emptyset_y - \emptyset_x)}{2}\right)\right] = \frac{\sqrt{2}}{2}(-x+y)$$

We have:

$$\mathrm{x = sin\ (\omega t + \emptyset_x)\ and}$$

$$\mathrm{y = sin(\omega t + \emptyset_y) = cos\left(\frac{\pi}{2} - \omega t - \emptyset_y\right) = cos\ (\omega t + \emptyset_y - \frac{\pi}{2})}$$

$$x^2 + y^2 = 1\ then\ R^2 = 1\ \mathrm{if}\ \emptyset_x = \emptyset_y - \frac{\pi}{2} \Rightarrow \emptyset_y - \emptyset_x = \frac{\pi}{2}$$

Consequently: $x' = \mathrm{asin}\ (\omega t + \frac{\pi}{4})$ and $y' = bsin(\omega t + \frac{\pi}{4})$
where, $a = \sqrt{2}cos\frac{\pi}{4}$ and $b = \sqrt{2}sin\frac{\pi}{4}$ then a = b = 1.

$x'^2 + y'^2 = 1$, it is an equation of circle with radius R=1. The trajectory is the same in the reference (xoy)

In the case while: $\emptyset_y - \emptyset_x = \mathrm{cste} = \emptyset \Rightarrow \begin{cases} \mathrm{x' = asin}\left(\omega t + \frac{(\emptyset_y-\emptyset_x)}{2}\right) = \mathrm{asin}\ (\omega t + \frac{\emptyset}{2}) \\ \mathrm{y' = bcos}\left(\omega t + \frac{(\emptyset_y-\emptyset_x)}{2}\right) = \mathrm{bsin}\ (\omega t + \frac{\emptyset}{2}) \end{cases}$

with: $\mathrm{a} = \sqrt{2}\cos\frac{\emptyset}{2}$ and $\mathrm{b} = \sqrt{2}\sin\frac{\emptyset}{2} \Rightarrow {}^{b}/_{a} = \frac{\sin\frac{\emptyset}{2}}{\cos\frac{\emptyset}{2}} = \mathrm{tg}({}^{\emptyset}/_{2})$

The rotational angle is defined by: $\psi = (x', x) = (y', y)$
It is the rotational angle of axes x', y' regarding the axes x, y

$$\mathrm{tg}(2\psi) = \frac{2a_{xa_y}\cos\phi}{a_x^2 - a_y^2}$$

if $a_x = a_y \Rightarrow \mathrm{tg}(2\psi) \to \infty \Rightarrow \psi = \frac{\pi}{4} \Rightarrow \begin{cases} \mathrm{Ox', is\ the\ first\ biocectrice} \\ \mathrm{Oy', is\ the\ 2nd\ biocectrice} \end{cases}$

In the case of small frequencies ($Re_\omega < 200$)
If $a_x \neq a_y \Rightarrow \psi = \frac{\pi}{4}$
The angle of the ellipse is defined by: $\mathrm{tg}(\eta) = \pm b/a$
Linear polarization b=0 $\Rightarrow \eta = 0$
Linear polarization b = a $\Rightarrow \eta = \pm\pi/4$
The general case: $-{}^{\pi}/_{4} \leq \eta \leq {}^{\pi}/_{4}$, $sin\eta = {}^{b}/_{\sqrt{a^2+b^2}}$, $cos\eta = {}^{a}/_{\sqrt{a^2+b^2}}$
The lag shift is defined from the next expression of wall friction:

$$\tau = CDe^{j\omega t}e^{j\phi\tau}$$

$$\phi_\tau = Artg\frac{\cos(-\phi_p)}{\omega^A/_C + \sin(-\phi_p)} = Artg\frac{\cos\phi_p}{\omega^A/_C - \sin\phi_p}$$

$$\Rightarrow \phi_\tau = \mathrm{Artg}\frac{1}{\omega^A/_{C\cos\phi_p} - \mathrm{tg}\phi_p}$$

For the small frequencies τ is in phase with $^{\partial p}/_{\partial x}$

* Increases the lag shift ($\emptyset_\tau$ - $\emptyset_p$) increases also.
* very great the lag shift between $^{\partial p}/_{\partial x}$ and τ_p tends to 90° ;

Demonstration:

$$* \text{ faibles} \Rightarrow \varphi_\tau = Artg\frac{1}{0 - tg\varphi_p} = Artg\left(\frac{-1}{tg\varphi_p}\right) \Rightarrow tg\varphi_\tau = \frac{-1}{tg\varphi_p}$$

$$\Rightarrow \varphi_\tau = \varphi_p - \frac{\pi}{2}$$

We have: $\mathrm{tg}\emptyset_\tau = \frac{\sin\left(\emptyset_p + \frac{\pi}{2}\right)}{\omega^A/_C + \cos\left(\emptyset_p + \frac{\pi}{2}\right)}$

If $\omega \to 0 \Rightarrow \mathrm{tg}\emptyset_\tau = \frac{\sin\left(\emptyset_p + \frac{\pi}{2}\right)}{0 + \cos\left(\emptyset_p + \frac{\pi}{2}\right)} \Rightarrow \mathrm{tg}\emptyset_\tau = \frac{\cos\emptyset_p}{-\sin\emptyset_p}$

Where: $\mathrm{tg}\emptyset'_\tau = \frac{\sin\emptyset_\tau}{\cos\emptyset_\tau} = \frac{\sin\left(\emptyset_p + \frac{\pi}{2}\right)}{\cos\left(\emptyset_p + \frac{\pi}{2}\right)} \Rightarrow \emptyset'_\tau = \emptyset_p + \frac{\pi}{2}$

If $\omega \to \infty$ $\mathrm{tg}\emptyset_\tau \to \frac{\sin\left(\emptyset_p + \frac{\pi}{2}\right)}{\omega^A/_C} \to 0 \Rightarrow \emptyset_\tau \to 0$

We have $\emptyset_\tau = \emptyset'_p - \frac{\pi}{2}$

If $\omega \to 0 \Rightarrow \emptyset_\tau = \emptyset_p + \frac{\pi}{2} - \frac{\pi}{2} = \emptyset_p$

If $\omega \to \infty \Rightarrow \emptyset_\tau = \emptyset'_\tau - \frac{\pi}{2} = -\frac{\pi}{2}$

General case:

In the case : $0 < \omega < \infty$, $\emptyset_\tau = \emptyset'_\tau - \frac{\pi}{2} \Rightarrow \emptyset'_\tau = \mathrm{Artg}\frac{\sin\left(\emptyset_p + \frac{\pi}{2}\right)}{\frac{\omega A}{C} + \cos\left(\emptyset_p + \frac{\pi}{2}\right)} = \mathrm{Artg}\frac{\cos\emptyset_p}{\frac{\omega A}{C} - \sin\emptyset_p}$

$$\Rightarrow \emptyset_\tau = \mathrm{Artg}\frac{\cos\emptyset_p}{\frac{\omega A}{C} - \sin\emptyset_p} - \frac{\pi}{2}$$

B.5.7. Pulsatile Flow throughout Channel with Arbitrary Flow Rate

1. Introduction

Pulsating flow rate involves the pulsating gradient pressure. This is because of easiness in measuring the dependence between the flow rate with time and the pressure gradient. In this study the Navier – Stokes equation and the boundary conditions are solved with the Laplacian transformation coupled with the residual method. This method was used recently by Das and Araberi.

2. Mathematical Formulation

The problem concerns a two dimensional flow in a rectangular channel with the height 2a and a length b. We use the Cartesian coordinates x, y. The axial velocity is u(y, t).

The continuity equation is: $\frac{\partial(\rho u)}{\partial x} = 0$ where $u \neq u(x)$ (26)

The momentum equation become:

$$\frac{\partial u}{\partial t} = -\frac{1}{\rho}\frac{\partial P}{\partial x} + \nu\frac{\partial^2 u}{\partial y^2} \quad (27)$$

At the y axe the momentum is simplified to:

$$0 = -\frac{1}{\rho}\frac{\partial P}{\partial y} \Rightarrow \; P \neq P(y) \quad (28)$$

Then:

$$\frac{\partial P}{\partial x} = \frac{\partial P(t)}{\partial x} \quad (29)$$

The boundary conditions are:

- On the wall at the Top and at the bottom:

$$u(a,t)=u(-a,t)=0 \quad (30)$$

- At the axis:

$$\frac{\partial u(0,t)}{\partial y} = 0 \quad (31)$$

The initial condition is determined by a given flow rate:

We can write it:

$$Q(t) = 2b\int_0^a u(y,t)dy = 2\,a\,b\,u_p(t) \quad (32)$$

where: $u_p(t)$ represents the piston velocity that causes the periodic variation of flow rate Q(t).

By using the condition given by the equation (28) it is possible to resolve equation (27) by Laplacian technique.

The Laplacian transformation is given by:

$$u(y,t)=\frac{1}{2\pi i}\int_{\gamma-i\alpha}^{\gamma+i\alpha}\overline{u}(y,s)\,\mathrm{e}^{st}\,\mathrm{ds}$$

Effectively the Laplacian transformation of equation (27) is:

$$\frac{\partial^2\overline{u}(y,s)}{\partial y^2}-\frac{s}{2}\overline{u}(y,s)=\frac{1}{\mu}\frac{\partial\overline{P}(x,s)}{\partial x}-\frac{1}{2}u(y,0) \tag{33}$$

$$\overline{u}(a,s)=\overline{u}(-a,s)=0 \tag{34}$$

$$\frac{\partial\overline{u}(0,s)}{\partial y}=0 \tag{35}$$

$$2b\int_0^a\overline{u}(y,s)\,\mathrm{dy}=2\,\mathrm{a}\,\mathrm{b}\,\overline{\mathrm{u}}_{\mathrm{p}}(s) \tag{36}$$

The equation is a differential equation of 2nd order. The next equation is the solution:

$$\overline{u}(y,s)=\overline{u}_h(y,s)+\phi_p=C_1\,\mathrm{e}^{-y\sqrt{s/2}}+C_2\,\mathrm{e}^{y\sqrt{s/2}}+\phi_P$$

where: ϕ_p is a particular solution and the constants C_1, C_2 are determined by using the boundary conditions at. $y=a$ And $y=0$, we find:

$$C_1=C_2=\frac{-\phi_p}{e^{pa}+e^{-pa}}\text{, avec } p=\sqrt{s/2}$$

We have: $\overline{u}_s(y,s)=\phi_p\left(1-\frac{\cosh(p\,y)}{\cosh(p\,a)}\right)$

ϕ_p is determined by using the equation (7).

We find:

$$\phi_p=-\frac{\overline{u}_p(y,s)}{\tanh(p\,\mathrm{a})}$$

The result is:

$$\bar{u}(y,s) = \frac{-\bar{u}_p(s)\,a\sqrt{s/2}}{\tanh\left(a\sqrt{s/2}\right)}\left(1 - \frac{\cosh\left(y\sqrt{s/2}\right)}{\cosh\left(a\sqrt{s/2}\right)}\right)$$

This solution is dependent of the piston motion. Generally:

$$u_p(t) = u_0 + \sum_{n=1}^{\infty} u_{cn}\cos(\omega nt) + \sum_{n=1}^{\infty} u_{sn}\sin(\omega nt)$$

This can be normalized as below:

$$u_p^*(T) = 1 + \sum_{n=1}^{\infty}\left(\frac{u_{cn}}{u_0}\cos(nT) + \frac{u_{sn}}{u_0}\sin(nT)\right)$$

The solution of the velocity profile is:

$$u(\eta,t) = u_m(\eta) + u_{n,+}(\eta,t) + u_{n,-}(\eta,t)$$

with:

$u_m(\eta) = \frac{s}{2}u_0(1-\eta^2)$, is the mean velocity.

$$u_{n,+} = \sum_{n=1}^{\infty} \frac{kn\left(\cosh\left(\sqrt{i\,k\,n}\right) - \cosh\left(\eta\sqrt{i\,k\,n}\right)\right)}{i\,k\,n\cosh\left(\sqrt{i\,k\,n}\right) - \sqrt{i\,k\,n}\sinh\left(\sqrt{i\,k\,n}\right)}\frac{1}{2}e^{(i\omega nt)}\left(u_{cn}i + u_{sn}\right)$$

$$u_{n,-} = \sum_{n=1}^{\infty} \frac{kn\left(\cosh\left(\sqrt{-i\,k\,n}\right) - \cosh\left(\eta\sqrt{-i\,k\,n}\right)\right)}{k\,n\cosh\left(\sqrt{-i\,k\,n}\right) + \sqrt{i\,k\,n}\sinh\left(\sqrt{-i\,k\,n}\right)}\frac{1}{2}e^{(-i\omega nt)}\left(u_{cn} + iu_{sn}\right)$$

with $k = \frac{\omega a^2}{2}$ is the kinetic Reynolds number.

Important particular case

This is the case of a first order development (fundamental n=1).

$$u_{1,+} = \frac{k\left(\cosh\left(\sqrt{i\,k}\right) - \cosh\left(\eta\sqrt{i\,k}\right)\right)}{i\,k\cosh\left(\sqrt{i\,k}\right) - \sqrt{i\,k}\sinh\left(\sqrt{i\,k}\right)}\frac{1}{2}e^{(i\omega t)}\left(u_{c1}.i + u_{s1}\right)$$

$$u_{1,-} = \frac{k\left(\cosh\left(\sqrt{-i\,\mathrm{k}}\right) - \cosh\left(\eta\sqrt{-i\,\mathrm{k}}\right)\right)}{\mathrm{k}\cosh\left(\sqrt{-\mathrm{i}\,\mathrm{k}}\right) + \sqrt{i\,\mathrm{k}}\sinh(-i\,\mathrm{k})}\frac{1}{2}e^{(-i\omega t)}\left(\mathrm{u}_{C1} + i\,u_{S1}\right)$$

with:

$$u_{1,+} + u_{1,-} = 0$$

$$u_{C1} = u_{S1}$$

$$\sqrt{i} = i^{1/2} = e^{(i\pi/2)^{1/2}} = e^{i\pi/4} = \frac{\sqrt{2}}{2}(1+i)$$

$$\cosh\sqrt{i\,\mathrm{k}} = \cosh(\sqrt{i}\sqrt{k}) = \cosh\sqrt{k}\left(\frac{\sqrt{2}}{2}(i+1)\right) = \cosh\frac{\sqrt{2}}{2}\sqrt{k}(1+i)$$

$$= \frac{1}{2}\left(e^{\frac{\sqrt{2}}{2}\sqrt{k}(1+i)} + e^{-\frac{\sqrt{2}}{2}\sqrt{k}(1+i)}\right)$$

$$\cosh\sqrt{-i\,\mathrm{k}} = \cosh\left(\sqrt{2}\frac{\sqrt{2}}{2}(1-i)\right)$$

$$-i = e^{-i\frac{\pi}{2}}, (-i)^{1/2} = e^{\left(-i\frac{\pi}{2}\right)^{1/2}} = e^{-i\pi/4}$$

$$\cosh\sqrt{-i\,\mathrm{k}} = \cosh\frac{\sqrt{2}}{2}\sqrt{2}(1-i) = \cos\left(\frac{-\pi}{4}\right) + i\sin\left(-\frac{\pi}{4}\right)$$

$$= \cos\left(\frac{\pi}{4}\right) - i\sin\left(\frac{\pi}{4}\right) = \frac{\sqrt{2}}{2}(1-i)$$

$$\cosh\frac{\sqrt{2}}{2}\sqrt{k}(1+i) = \frac{1}{2}\left(e^{\frac{\sqrt{2}}{2}\sqrt{k}(1+i)} + e^{-\frac{\sqrt{2}}{2}\sqrt{k}(1+i)}\right) = \frac{1}{2}\left(e^{\frac{\sqrt{2}}{2}\sqrt{k}}e^{\frac{\sqrt{2}}{2}\sqrt{k}\,i} + e^{-\frac{\sqrt{2}}{2}\sqrt{k}}e^{-\frac{\sqrt{2}}{2}\sqrt{k}\,i}\right)$$

$$= \frac{1}{2}\left(e^{\frac{\sqrt{2}}{2}\sqrt{k}}\left(\cos\frac{\sqrt{2}}{2}\sqrt{k} + i\sin\frac{\sqrt{2}}{2}\sqrt{k}\right) + e^{-\frac{\sqrt{2}}{2}\sqrt{k}}\left(\cos\frac{-\sqrt{2}}{2}\sqrt{k} - i\sin\frac{\sqrt{2}}{2}\sqrt{k}\right)\right)$$

$$= \frac{1}{2}\left(\cos\frac{\sqrt{2}}{2}\sqrt{k}\left(e^{\frac{\sqrt{2}}{2}\sqrt{\mathrm{k}}} + e^{-\frac{\sqrt{2}}{2}\sqrt{\mathrm{k}}}\right) + i\sin\frac{\sqrt{2}}{2}\sqrt{k}\left(e^{\frac{\sqrt{2}}{2}\sqrt{\mathrm{k}}} - e^{-\frac{\sqrt{2}}{2}\sqrt{\mathrm{k}}}\right)\right)$$

$$= \cos\left(\frac{\sqrt{2}}{2}\sqrt{k}\right)\cosh\left(\frac{\sqrt{2}}{2}\sqrt{\mathrm{k}}\right) + i\sin\left(\frac{\sqrt{2}}{2}\sqrt{k}\right)\sinh\left(\frac{\sqrt{2}}{2}\sqrt{\mathrm{k}}\right)$$

By setting:

$$A = \cos\left(\frac{\sqrt{2}}{2}\sqrt{k}\right)\cosh\left(\frac{\sqrt{2}}{2}\sqrt{\mathrm{k}}\right)$$

$$B = \sin\left(\frac{\sqrt{2}}{2}\sqrt{k}\right)\sinh\left(\frac{\sqrt{2}}{2}\sqrt{\mathrm{k}}\right)$$

We have also:

$$\cosh\left(\frac{\sqrt{2}}{2}\sqrt{\mathrm{k}}\,(1-\mathrm{i})\right) = \frac{1}{2}\left(\mathrm{e}^{\frac{\sqrt{2}}{2}\sqrt{\mathrm{k}}(1-\mathrm{i})} + \mathrm{e}^{-\frac{\sqrt{2}}{2}\sqrt{\mathrm{k}}(1-\mathrm{i})}\right)$$

$$= \frac{1}{2}\left(\mathrm{e}^{\frac{\sqrt{2}}{2}\sqrt{\mathrm{k}}}\left(\cos\left(\frac{\sqrt{2}}{2}\sqrt{\mathrm{k}}\right) - \mathrm{i}\sin\left(\frac{\sqrt{2}}{2}\sqrt{\mathrm{k}}\right)\right) + \mathrm{e}^{-\frac{\sqrt{2}}{2}\sqrt{\mathrm{k}}}\left(\cos\left(\frac{\sqrt{2}}{2}\sqrt{\mathrm{k}}\right) + \mathrm{i}\sin\left(\frac{\sqrt{2}}{2}\sqrt{\mathrm{k}}\right)\right)\right)$$

$$= \cos\left(\frac{\sqrt{2}}{2}\sqrt{\mathrm{k}}\right)\cosh\left(\frac{\sqrt{2}}{2}\sqrt{\mathrm{k}}\right) - \mathrm{i}\sin\left(\frac{\sqrt{2}}{2}\sqrt{\mathrm{k}}\right)\sinh\left(\frac{\sqrt{2}}{2}\sqrt{\mathrm{k}}\right)$$

$$\cosh\left(\frac{\sqrt{2}}{2}\sqrt{\mathrm{k}}(\mathrm{i}-1)\right) = \frac{1}{2}\left(\mathrm{e}^{\frac{\sqrt{2}}{2}\sqrt{\mathrm{k}}(\mathrm{i}-1)} + \mathrm{e}^{-\frac{\sqrt{2}}{2}\sqrt{\mathrm{k}}(\mathrm{i}-1)}\right)$$

$$\cosh\left(\frac{\sqrt{2}}{2}\sqrt{\mathrm{k}}(\mathrm{i}-1)\right)$$

$$= \frac{1}{2}\left[\mathrm{e}^{\frac{\sqrt{2}}{2}\sqrt{\mathrm{k}}}\left(\cos\left(\frac{\sqrt{2}}{2}\sqrt{\mathrm{k}}\right) - \mathrm{isin}\left(\frac{\sqrt{2}}{2}\sqrt{\mathrm{k}}\right)\right) + \mathrm{e}^{-\frac{\sqrt{2}}{2}\sqrt{\mathrm{k}}}\left(\cos\left(\frac{\sqrt{2}}{2}\sqrt{\mathrm{k}}\right)\mathrm{isin}\left(\frac{\sqrt{2}}{2}\sqrt{\mathrm{k}}\right)\right)\right]$$

$$= \cos\left(\frac{\sqrt{2}}{2}\sqrt{\mathrm{k}}\right)\cosh\left(\frac{\sqrt{2}}{2}\sqrt{\mathrm{k}}\right) - \mathrm{isinh}\left(\frac{\sqrt{2}}{2}\sqrt{\mathrm{k}}\right)\sinh\left(\frac{\sqrt{2}}{2}\sqrt{\mathrm{k}}\right)$$

Then:
$$\cosh\left(\sqrt{i\,\mathrm{k}}\right) = A + i\,\mathrm{B}$$
$$\cosh\left(\sqrt{-i\,\mathrm{k}}\right) = A - i\,\mathrm{B}$$

$$h$$

$$\sinh(\sqrt{ik}) = \frac{1}{2}\left(\mathrm{e}^{\frac{\sqrt{2}}{2}\sqrt{\mathrm{k}}(\mathrm{i}+1)} - \mathrm{e}^{-\frac{\sqrt{2}}{2}\sqrt{\mathrm{k}}(\mathrm{i}+1)}\right)$$

$$= \cos\left(\frac{\sqrt{2}}{2}\sqrt{\mathrm{k}}\right)\sinh\left(\frac{\sqrt{2}}{2}\sqrt{\mathrm{k}}\right) + \mathrm{isin}\left(\frac{\sqrt{2}}{2}\sqrt{\mathrm{k}}\right)\cosh\left(\frac{\sqrt{2}}{2}\sqrt{\mathrm{k}}\right)$$

$$\sin\sqrt{-i\,\mathrm{k}} = \frac{1}{2}\left(e^{\frac{\sqrt{2}}{2}\sqrt{k}}\left(\cos\frac{\sqrt{2}}{2}\sqrt{k} - i\sin\frac{\sqrt{2}}{2}\sqrt{k}\right) - e^{-\frac{\sqrt{2}}{2}\sqrt{k}}\left(\cos\frac{\sqrt{2}}{2}\sqrt{k} + i\sin\frac{\sqrt{2}}{2}\sqrt{k}\right)\right)$$

$$= \cos\frac{\sqrt{2}}{2}\sqrt{k}\sinh\frac{\sqrt{2}}{2}\sqrt{k} - i\sin\frac{\sqrt{2}}{2}\sqrt{k}\cosh\frac{\sqrt{2}}{2}\sqrt{k}$$

$$\sinh\sqrt{i\,\mathrm{k}} = C + i\,\mathrm{D}$$

$$\sinh\sqrt{-\mathrm{i}\,\mathrm{k}} = C - i\,\mathrm{D}$$

Calculation of $\mathrm{u}_{1,+}$

$$u_{1,+} = \frac{k\left(A + i\,\mathrm{B} - \cos\eta\sqrt{\mathrm{i}\,\mathrm{k}}\right)}{\mathrm{i}\,\mathrm{k}(\mathrm{A} + \mathrm{i}\,\mathbf{B}) - \sqrt{i\,\mathrm{k}}(C + i\,\mathrm{B})}\frac{1}{2}\mathrm{e}^{(\mathrm{i}\omega \mathrm{t})}\left(\mathrm{u}_{\mathrm{c}1}\,\mathrm{i} + \mathrm{u}_{\mathrm{s}1}\right)$$

$$\sqrt{\mathrm{i}} = \frac{\sqrt{2}}{2}(1 + i)$$

$$\sqrt{i\,\mathrm{k}} = \sqrt{k}\frac{\sqrt{2}}{2}(1 + i)$$

$$\sqrt{i\,\mathrm{k}}(C + i\,\mathrm{D}) = \frac{\sqrt{2}}{2}\sqrt{k}(1 + i)(C + i\,\mathrm{D}) = \frac{\sqrt{2}}{2}\sqrt{k}(C - D + i(C + D))$$

$$\mathrm{i}\,\mathrm{k}(\mathrm{A} + \mathrm{i}\,\mathrm{B}) = i\,\mathrm{k}\,\mathrm{A} - \mathrm{k}\,\mathrm{B}$$

$$\text{we pose}: \mathrm{u}_{1,+} = \frac{Num}{Denom}$$

$$Denom = i\,\mathrm{k}(\mathrm{A} + \mathrm{i}\,\mathrm{B}) - \sqrt{-i\,\mathrm{k}}(C + D)$$

$$= \mathrm{i}\,\mathrm{k}\,\mathrm{A} - \mathrm{k}\,\mathrm{B} - \frac{\sqrt{2}}{2}\mathrm{k}(\mathrm{C} - \mathrm{D} + \mathrm{i}(\mathrm{C} + \mathrm{D}))$$

$$= -\mathrm{k}\,\mathrm{B} - \frac{\sqrt{2}}{2}\sqrt{k}(C - D) + i\left(k\,\mathrm{A} - \frac{\sqrt{2}}{2}\sqrt{k}(C + D)\right)$$

$$= \mathrm{M} + \mathrm{i}\,\mathrm{N}$$

$$\mathrm{u}_{1,+} = \frac{k\left(A + i\,\mathrm{B} - \cosh\left(\eta\sqrt{i\,\mathrm{k}}\right)\right)}{\mathrm{M} + \mathrm{i}\,\mathrm{N}}\frac{1}{2}e^{(i\omega t)}\left(u_{c1}\,\mathrm{i} + \mathrm{u}_{\mathrm{s}1}\right)$$

Calculation of $\mathrm{u}_{1,-}$

We have:

$$u_{1,-} = \frac{k\left(\cosh\sqrt{-ik} - \cosh\eta\sqrt{-ik}\right)}{k\cosh\sqrt{-\mathrm{ik}} + \sqrt{ik}\sinh\sqrt{-\mathrm{ik}}}\frac{1}{2}\mathrm{e}^{(-\mathrm{i}\omega t)}\left(\mathrm{u}_{\mathrm{c}1} + i\,\mathrm{u}_{\mathrm{s}1}\right)$$

$$\mathrm{Num} = \mathrm{k}\left(\mathrm{A} - \mathrm{i}\,\mathrm{B} - \cosh\eta\sqrt{-\mathrm{ik}}\right)$$

$$Denom = k\left(A - i\,B\right) + \sqrt{ik}\left(C - i\,\mathrm{D}\right)$$

$$= \mathrm{k}\,(\mathrm{A} - \mathrm{i}\,\mathrm{B}) + \frac{\sqrt{2}}{2}\sqrt{k}(1+i)(\mathrm{C} - \mathrm{i}\,\mathrm{D})$$

$$= \mathrm{k}\,\mathrm{A} + \frac{\sqrt{2}}{2}\sqrt{k}(C+D) + \frac{\sqrt{2}}{2}\sqrt{k}\,\mathrm{i}\,(\mathrm{C} - \mathrm{D}) - \mathrm{k}\,\mathrm{i}^2\,\mathrm{B}$$

$$= \mathrm{M}' + \mathrm{i}\,\mathrm{N}'$$

We note that $u_{1,+}$ and $u_{1,-}$ are different.

Study of Particular Cases

1. **Small frequencies ($k \to 0$)**
   We have the next values

$$\left.\begin{array}{l} A \to 1 \\ B \to 0 \\ C \to 0 \\ D \to 0 \end{array}\right\} \Rightarrow \left\{\begin{array}{l} M \to 0 \\ N \to k\,\mathrm{A} = \mathrm{k} \times 1 \\ \mathrm{M}' \to \mathrm{k}\,\mathrm{A} = \mathrm{k} \times 1 \\ \mathrm{N}' \to 0 \end{array}\right.$$

$$u_{1,+} = \frac{k\left(1 - \cosh\eta\sqrt{i\,\mathrm{k}}\right)}{\mathrm{i}\,\mathrm{k}}\frac{1}{2}e^{(i\,\omega\,\mathrm{t})}\left(\mathrm{u}_{\mathrm{c}1}\,\mathrm{i} + \mathrm{u}_{\mathrm{s}1}\right)$$

$$= \frac{\left(1 - \cosh\eta\sqrt{ik}\right)}{i}\frac{1}{2}e^{(i\,\omega\,\mathrm{t})}\left(u_{c1}\,\mathrm{i} + \mathrm{u}_{\mathrm{s}1}\right)$$

$$= -i\left(1 - \cosh\eta\sqrt{ik}\right)\frac{1}{2}e^{(i\omega t)}\left(u_{c1}\,\mathrm{i} + \mathrm{u}_{\mathrm{s}1}\right)$$

$$u_{1,-} = \frac{k\left(1 - \cosh\eta\sqrt{-ik}\right)}{k}\frac{1}{2}\mathrm{e}^{(-\mathrm{i}\,\omega\,\mathrm{t})}\left(\mathrm{u}_{\mathrm{c}1} + i\,\mathrm{u}_{\mathrm{s}1}\right)$$

$$= \frac{k\left(1 - \cosh\eta\sqrt{-\mathrm{ik}}\right)}{k}\frac{1}{2}\mathrm{e}^{(-\mathrm{i}\,\omega\,\mathrm{t})}\left(\mathrm{u}_{\mathrm{c}1} + i\,\mathrm{u}_{\mathrm{s}1}\right)$$

$$= (1 - \cosh\eta\sqrt{-\mathrm{i}\,\mathrm{k}}\,)\frac{1}{2}\mathrm{e}^{(-\mathrm{i}\,\omega\,\mathrm{t})}\left(\mathrm{u}_{\mathrm{c}1} + i\,\mathrm{u}_{\mathrm{s}1}\right)$$

We have:
$$\cosh\eta\sqrt{ik} = A' + iB'$$
$$\cosh\eta\sqrt{-ik} = A' - iB'$$

If $k \to 0$ (small frequencies)

$$A' = \cos\frac{\sqrt{2}}{2}\sqrt{k}\eta \ \cosh\frac{\sqrt{2}}{2}\sqrt{k}\eta$$
$$B' = \sin\frac{\sqrt{2}}{2}\sqrt{k}\eta \ \sinh\frac{\sqrt{2}}{2}\sqrt{k}\eta$$

$$u_{1,+} = (1-(A'+i\,\mathrm{B}'))\frac{1}{2}e^{i\omega t}(u_{c1}-u_{s1}\,\mathrm{i})$$
$$u_{1,-} = (1-(A'-i\,\mathrm{B}'))\frac{1}{2}e^{-i\omega t}(u_{c1}+u_{s1}\,\mathrm{i})$$

$$u_{1,+} = \frac{1}{2}e^{i\omega t}(u_{c1}-i\,\mathrm{u}_{s1}-A'\,\mathrm{u}_{c1}+A'\,\mathrm{u}_{s1}\,\mathrm{i}-\mathrm{i}\,\mathrm{B}'\,\mathrm{u}_{c1}-B'\,\mathrm{u}_{s1})$$
$$= \frac{1}{2}e^{i\omega t}(u_{c1}-A'\,\mathrm{u}_{c1}-B'\,\mathrm{u}_{s1}+i(-u_{s1}+A'u_{s1}-B'\,\mathrm{u}_{c1}))$$
$$u_{1,-} = \frac{1}{2}e^{-i\omega t}(u_{c1}+i\,\mathrm{u}_{s1}-A'\,\mathrm{u}_{c1}-i\,\mathrm{A}'\,\mathrm{u}_{s1}+i\,\mathrm{B}'\,\mathrm{u}_{c1}-B'\,\mathrm{u}_{s1})$$
$$= \frac{1}{2}e^{-i\omega t}(u_{c1}-B'\,\mathrm{u}_{s1}-A'\,\mathrm{u}_{c1}+i(-A'\,\mathrm{u}_{s1}+B'\,\mathrm{u}_{c1}+u_{s1}))$$

We pose:

$$R = u_{c1}(1-A')-B'\,\mathrm{u}_{s1}$$
$$S = u_{s1}(1-A')+B'\,\mathrm{u}_{c1}$$

We find:

$$u_{1,+} = \frac{1}{2}e^{i\omega t}(R+i\,\mathrm{S})$$
$$u_{1,-} = \frac{1}{2}e^{-i\omega t}(R-i\,\mathrm{S})$$
$$u_{1,+}+u_{1,-} = \frac{1}{2}e^{i\omega t}(R+i\,\mathrm{S})+\frac{1}{2}e^{-i\omega t}(R-i\,\mathrm{S})$$
$$= \frac{1}{2}R(e^{i\omega t}+e^{-i\omega t})+\frac{1}{2}i\,\mathrm{S}(\mathrm{e}^{i\omega t}-e^{-i\omega t})$$
$$= \mathrm{R}\cos(\omega \mathrm{t})+i\,\mathrm{S}\sin(\omega \mathrm{t})$$

With:

$$R = U_{c1}\left(1-\cos\frac{\sqrt{2}}{2}\sqrt{k}\eta\cosh\frac{\sqrt{2}}{2}\sqrt{k}\eta\right)-U_{s1}\left(\sin\frac{\sqrt{2}}{2}\sqrt{k}\eta\sinh\frac{\sqrt{2}}{2}\sqrt{k}\eta\right)$$
$$S = U_{s1}\left(1-\cos\frac{\sqrt{2}}{2}\sqrt{k}\eta\cosh\frac{\sqrt{2}}{2}\sqrt{k}\eta\right)+U_{c1}\left(\sin\frac{\sqrt{2}}{2}\sqrt{k}\eta\sinh\frac{\sqrt{2}}{2}\sqrt{k}\eta\right)$$

and:

$$\text{Real}\left(u_{1,+}+u_{1,-}\right)=R\cos(\omega t)$$

Particular case: $u_{C1}=u_{S1}=u_0$

$$R=u_0\left(1-A'\right)-B'u_0=u_0\left(1-A'-B'\right)$$

$$S=u_0\left(1-A'\right)+B'u_0=u_0\left(1-A'+B'\right)$$

If $B'=0 \Rightarrow R=S=u_0\left(1-A'\right)$

In this case:

$$u_{1,+}=f(k)\frac{1}{2}e^{i\omega t}\left(i\,u_{c1}+u_{s1}\right)$$

$$u_{1,-}=f(k)\frac{1}{2}e^{-i\omega t}\left(u_{c1}+i\,u_{s1}\right)$$

if $\mathrm{u}_{c1}=u_{s1}=u_0$

$$u_{1,+}=\frac{1}{2}f(k)\,\mathrm{u}_0\,\mathrm{e}^{i\omega t}(i+1)$$

$$=\frac{1}{2}f(k)\,\mathrm{u}_0\,(\cos\omega t+\mathrm{i}\sin\omega t)(i+1)$$

$$\mathrm{Re}al\left(\mathrm{u}_{1,+}\right)=\frac{1}{2}u_0\,\mathrm{f(k)}\left(-\sin\omega t+\cos\omega t\right)$$

$$\mathrm{Re}al\left(\mathrm{u}_{1,-}\right)=\frac{1}{2}u_0\,\mathrm{f(k)}\left(\sin\omega t+\cos\omega t\right)$$

We note in the absence of reflexion ($u_{1,-}$=0), (i.e. infinite length), the solution is;

$$u=\mathrm{Re}al\left(u_{1,+}\right)=\frac{1}{2}u_0\,f(k)\left(\cos\omega t-\sin\omega t\right)$$

In the presence of reflexion the solution is:

$$u=u_{1,+}+u_{1,-}=\frac{1}{2}\mathrm{u}_0\,\mathrm{f(k)}\left(2\cos\omega t\right)=u_0\,\mathrm{f(k)}\cos\omega t$$

The real part that gives the velocity profile for the small frequencies:

$$u=u_m+u_0\cos(\omega t)\left[1-\cos\left(\frac{\sqrt{2}}{2}\sqrt{k\eta}\right)\cosh\left(\frac{\sqrt{2}}{2}\sqrt{k\eta}\right)-\sin\left(\frac{\sqrt{2}}{2}\sqrt{k\eta}\right)\sinh\left(\frac{\sqrt{2}}{2}\sqrt{k\eta}\right)\right]$$

If $\eta\rightarrow 0\,(center)$

$$u = u_m + u_0 \cos(\omega t)\left[1 - \cos\left(\frac{\sqrt{2}\sqrt{k}}{2}\eta\right)\cosh\left(\frac{\sqrt{2}\sqrt{k}}{2}\eta\right)\right]$$

$$u_m = \frac{3}{2}u_0\left(1-\eta^2\right)$$

Velocity Profile for Different Kinetic Reynolds Numbers and Different Phases on Channel and Cylinder

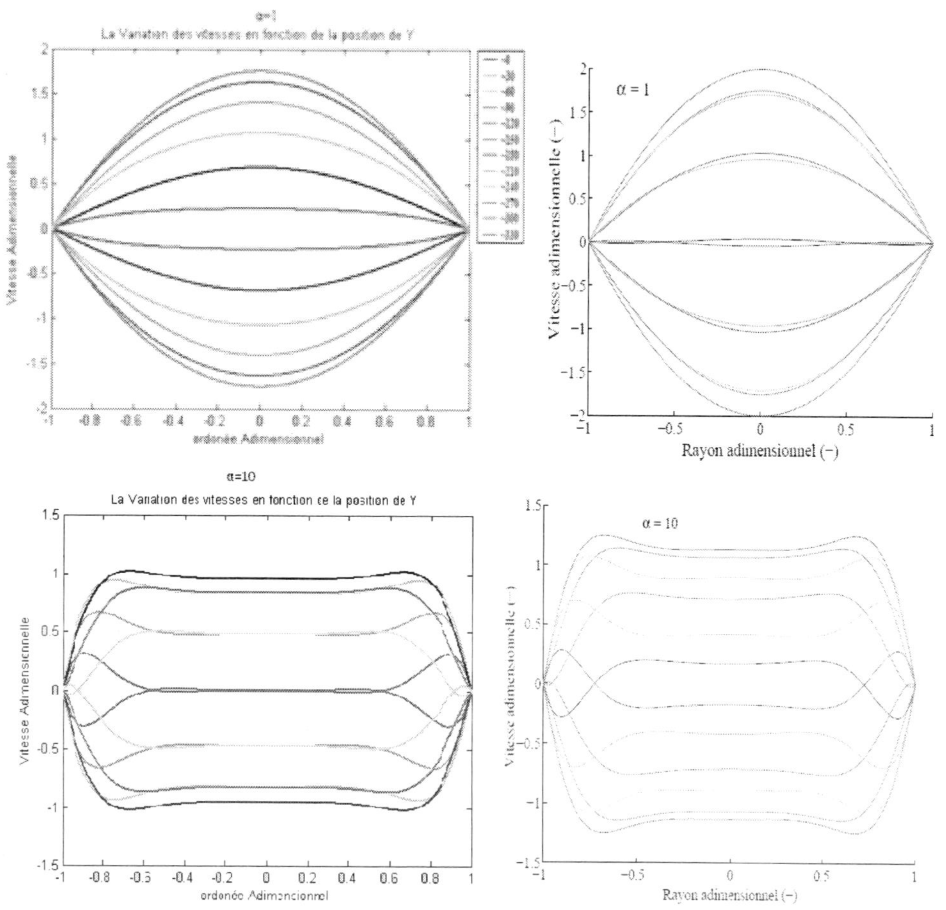

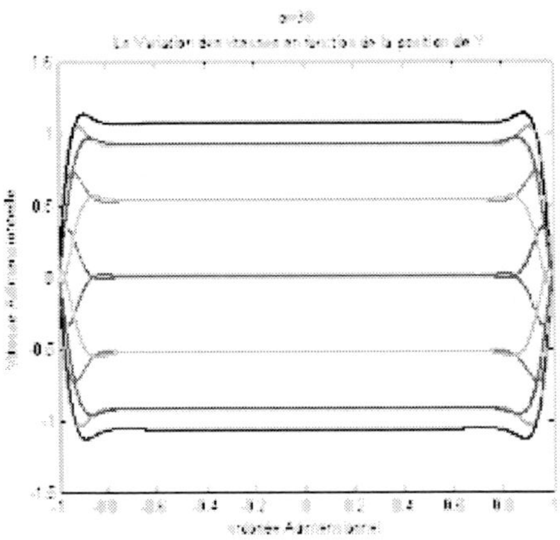

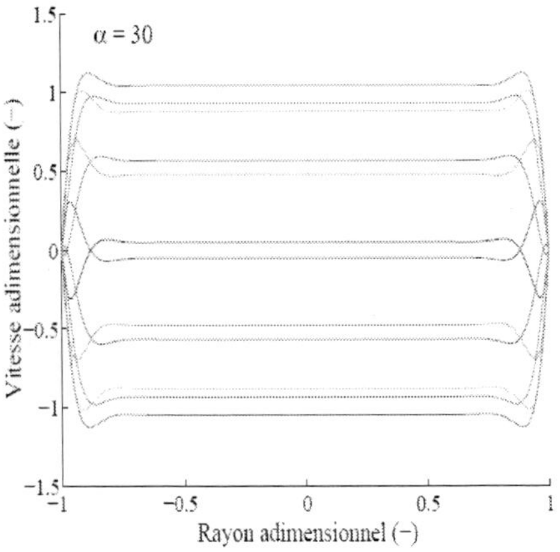

NOMENCLATURE

A_0 Amplitude of pressure gradient oscillations

$E=(1\text{-}r/a)\sqrt{(Re_\omega/2)}=B$

L Conduct Length ....(m)

P Pressure...................................(Nm^{-2})

R Duct radius...............................(m)

r Radial coordinate

Re Reynolds number.........................($w_m b/v$)

Re_ω Kinetic Reynolds number...........($\omega b^2/v$)

t Time

u, w Radial and axial velocity components

w_m Time average centreline velocity.....(ms^{-1})

z Axial coordinate

Greek Symbols

ε Amplitude of pressure gradient oscillations

ν Kinematic viscosity ....(m^2s^{-1})

ω Dimensionless frequency

REFERENCES

[1] Atabek,Oscillatory flow near the entry f circular tube, zamp 12, pages 403-422, 1961

[2] Drake, On the flow in a channel due to a periodic pressure gradient. *Quart J.Mech and Applied Math*, Vol 18, pt.11965

[3] D. Gedeon, Mean-parameter modeling of oscillating flow, *J. of Heat Transfe*r 108, (1986) 513-518.

[4] S.R. Huebner, A finite difference analysis of heat transfer in periodic cavity flows. *Numerical Methods in laminar and turbulent* flow, vol 6, part 2 (Swansea 1989).

[5] P. Andre et R.Creff, Etude des conditions particulières de fréquence favorisant les transferts thermiques en écoulements pulsés en canalisation cylindrique. *Int. J. Heat Mass Trans.*, vol.24, n°7 pp 1211-1219, 1981.

[6] A.Yakhot, L.Grinberg, Phase shift ellipses for pulsating flows *Physics of Fluids*. Vol 15, Number 7. July 2003

[7] R.Sevirino, De Farias Neto, Simulation numérique du transfert de matière global dans un écoulement tourbillonnaire non entretenu – Cas des forts nombres de Schmidt, *14 Congrès Français de mécanique Toulouse* 99 France.

[8] S. Kakac, Y. Yenner, Exact solution of transient forced convection energy equation for time, with variation of inlet temperature. *Int. J. Heat Mass Trans* vol.11, 2205, 1973

[9] J. Suces, An improved quasi-study approach for transient conjugated forced convection problems. *Int. J. Heat. Mass. Trans.*, vol.24, n°10, pp 1711-1722, 1981.

[10] M.T. Acker, B. Fourcher, Analyse en régime thermique périodique du couplage conduction convection entre un fluide en écoulement laminaire et une paroi de stockage. *Int. J. Heat. Mass Trans.* vol.24, n°7, pp 1201-1210, 1981.

[11] P. Singh, V. Radhkrishnan, Fluctuation flow due to unsteady rotation of a disk. *AIAAA Journal*, vol 27, n°2, pp 150-154, 1988.

[12] C.T. Aplet, M.A. Ledwich, Heat transfer in transient and unsteady flow past a heated circular cylinder in the range $1 \leq R \leq 40$. *J. Fluid. Mech.*, vo.95, part 4, 761-777 (1979)

[13] Huw G. Davies, Fluctuating heat transfer from hot wires in low Reynolds number. *J. Fluid. Mech.,* vol.73, Part 1, pp 49-51, 1976

[14] M. Lachi, G. Polidori, M. Rebay, J. Padet, Convention forcée Instationnaire sur une plaque soumise à une perturbation de flux périodique. *14ème Congrès Français de Mécanique*, Toulouse 99 France

[15] J. Majdalani, Pulsatory Channel Flows with Arbitrary Pressure Gradients. *3rd aiaa.* June, 2002

[16] T. Zhao, A Numerical Solution of Laminar flow convection in a heated pipe subjected to a reciprocating flow *.Int J Heat and Mass Trans*, vol 38 n° 16, pages 3011,3022 Nov 1995.

[17] Byunng-Hun Kim, Modelling pulsed blowing systems for active flow control. *PHD thesis Chicago*. Illinois May 2003.

[18] F. Fedel, D.Hitt, R.D. Prabhu, Revisiting the stability of pulsatile pipe flow. *European Journal of Mechanics.*

[19] Hadj Ali, Contribution à l'étude dynamique et thermique d'un écoulement pulsé dans une conduite horizontale en présence d'un obstacle cylindrique chauffé, *thèse de Magister*, université de USTHB, 2006.

[20] A. Ghezal, J. C. Loraud, Ecoulement confiné d'un fluide visqueux incompressible autour d'un obstacle cylindro-conique. *Mechanics Research New York* 1984.

[21] A.Ghezal, J. C. Loraud, Ecoulement confiné d'un fluide visqueux incompressible autour d'un obstacle cylindro- conique en mouvement hélicoïdal. *Mechanics. Research.* NewYork1989.

[22] A. Ghezal, J.C. Loraud, Helicoidally motion of a body through confined viscous flow. *Euro Mech* 245. CambridgeUK1989.

[23] A. Ghezal, B. Porterie, J.C. Loraud, Modélisation du transfert de chaleur, avec couplage conduction convection, entre un obstacle en mouvement hélicoïdal est un fluide visqueux en écoulement confiné. *Int. J. Heat. Mass. Transfer*. Jan.1991.
[24] A. Ghezal, N. Ait Moussa, Z. Ouchiha, J. C. Loraud, Frequency influence on thermal exchange in spiral flow 4th international conference on Heat Transfer, *Fluid Mechanics and Thermodynamics*. Cairo. HEFAT 2005. Egypt, paper number GA
[25] Y. Benakcha, Contribution a l'étude du transfert thermique entre un écoulement pulsé et un solide chauffé. *Magister USTHB*.2008.

In: Navier-Stokes Equations
Editor: R. Younsi

ISBN: 978-1-61324-590-3

Chapter 10

FIXED GRID NUMERICAL SIMULATION OF A PHASE CHANGE MATERIAL IN A RECTANGULAR ENCLOSURE HEATED FROM ONE SIDE

Annabelle Joulin[1], Zohir Younsi[2], Stéphane Lassue[1] and Laurent Zalewski[1]

[1]Univ Lille Nord de France – F59000 Lille, France
UArtois, LGCgE, F-62400 Béthune Cedex, France
[2]HEI, rue de Toul, 59000 Lille, France

ABSTRACT

The storage of thermal energy as latent heat of fusion presents advantages over sensible heat due to its high storage density and to the isothermal nature of the storage process at melting temperature. Latent heat thermal energy storage systems find application in space craft, solar energy system, greenhouses, heating and cooling of buildings and so on.

The use of phase change material (PCM) for thermal storage in buildings was one of the first applications considered for such materials along with typical sensible heat storage reservoirs and enclosures. Our general objective is to study the thermal behaviour of phase change materials so as to incorporate "bricks" of such materials into passive solar components.

This chapter presents a formulation and an implementation of a numerical method in order to optimize the design of solar passive walls involving phase change materials (PCMs). Particularly, it explores numerically the melting and solidification processes of a PCM. Indeed, the fusion and the solidification of the commercially available PCM (hydrated salts, designed to melt at *27 °C*), engineered by Cristopia™ are investigated.

The mathematical model for the numerical simulations is based on the enthalpy-porosity method, which is traditionally used to track the motion of the liquid-solid front and to obtain the temperature and velocity profiles in the liquid-phase. The governing equations are discretized on a fixed grid by means of a finite volume method. Numerical predictions are obtained with custom one-dimensional and two-dimensional Fortran codes. Several simulation runs were conducted to provide the heat storage during the melting process, as well as the heat recovery during the solidification process. Moreover, these results were compared to experimental data. Numerical simulations provided in 2D important features such as the streamlines in the liquid phase, and the temperature profiles inside the cavity along with the time evolution of the melted fraction.

Nomenclature

A	Matrix coefficients
B	Porosity function (Carman Kozeny)
b	Constant (*b=0.0003*)
C	Morphological constant in the Carman-Kozeny relation ($kg/m^3.s$)
c_p	Specific heat (J/kg°C)
f	Liquid fraction
g	Acceleration of gravity (m/s^2)
h	Sensible volumetric enthalpy (J/m^3)
k	Thermal conductivity (W/mK)
L_x	Horizontal domain length (m)
L_y	Vertical domain length (m)
L_f	Latent heat of fusion (kJ/kg)
P	Pressure (Pa)
S	Source term
t	Time (s)
T	Temperature (°C)
u	Velocity in the x direction (m/s)
v	Velocity in the y direction (m/s)
x	Component of x direction (m)
y	Component of y direction (m)

Greek Symbols

α	Thermal diffusivity (m^2/s)
β	Expansion coefficient (K^{-1})
Δt	Time step (s)
ΔV	Control volume
Γ	Diffusion coefficient
μ	Viscosity (kg/m.s)
ρ	Density (kg/m^3)
ϕ	Generalized variable
ω	Relaxation factor

Subscripts

0	Initial temperature
E	East node
F	Melting temperature
P	Centre node
w	Final temperature
W	West node
k	Relative to phase k

Superscripts

k	Iteration level
l	Liquid phase
0	Old value
s	Solid phase

Non dimensional Numbers

Pr	Prandtl number
Ra	Rayleigh number
R	Aspect ratio
Ste	Stefan number
Θ	Temperature

Introduction

Large heat-storage capacity of phase-change materials (PCMs) makes them attractive for use in various thermal energy storage systems where their latent heat is utilized. Indeed, the thermal latent energy storage provides is more interesting than sensible energy storage. This is due to its high energy storage density and to the isothermal nature of the storage process at the melting temperature of the material. A wide range of applications exists for such systems, such as energy storage in buildings [1, 2], electronics cooling [3], material processing, thermal management of aircraft [4], solar energy system [5, 6], greenhouses, and also in heating and cooling of buildings [7].

Thermal energy storage is important for the efficient generation and use of energy, because it can store energy during the day (during slack demand times at night) and provide energy during high demand times. It can also store energy during the summer to use it during the winter (inter-seasonal storage). Phase change materials (PCMs) were first used for thermal storage in buildings along with typical sensible heat storage reservoirs and enclosures. Recently, PCMs become a topic more and more abundant in the literature. For example, encapsulated materials used in plasterboard or packed beds are reviewed in a paper by [8]; PCM in buildings applications are the subject of the syntheses by [9] and Zhang *et al.*[10] while general storage applications are reviewed in that of Sharma *et al.* [11] and Farid *et al.* [12]. Dutil's work [13] provides an overview of the models employed in many applications of PCMs.

Our overall objective is to insert PCM bricks in a passive solar component such as the «composite Trombe wall» investigated by our laboratory [14]. This chapter deals with the investigation of the fusion process of a specific PCM. It presents the formulation and the implementation of numerical methods (1D and 2D) to model the melting and the solidification of this PCM.

The numerical codes used in this chapter were validated in different previous studies by comparing the obtained results with 1D analytical, 2D numerical (Fluent® software) and experimental results [15, 16, and 17].

Phase change problems, known as Stefan problems, occur in many physical domains [18]. In general, numerical simulations usually used for phase change problems are classified into two different approaches: the fixed grid and the transformed grid methods [19]. The transformed grid method employs the governing equations based on the classical Stefan formulation. The interface conditions, therefore, are accounted for differently according to the method incorporated in solving the phase change problem. In the transformed grid method, they are easily imposed because the interface is explicitly solved. However, in the fixed-grid methods, the interface conditions are described as suitable source terms in the governing equations. Numerical solutions of the governing equations are generally proposed. Some of these formulations are temperature based methods while others base their mathematical model upon enthalpy. The first ones rely on the prediction of the front of fusion/solidification while the other, much easier to implement, does not.

The enthalpy formulation is one of the most popular fixed-domain methods for solving the Stefan problem because of its easy formulation. One of the advantages of the fixed grid method is that a single set of conservation equations and boundary conditions is used for both solid and liquid phases. It allows avoiding the problem of tracking the solid/liquid interface. The enthalpy formulation involves the solution within a mushy zone, involving both solid and liquid material, between the two standard phases.

1. Physical Problem

The studied material (hydrated salts, mineral, containing potassium and calcium chlorides) has a melting point provided by the manufacturer equal to *27 °C* [15]. Its thermophysical properties are summarized in Table 1. The phase change material samples are $210 \times 140 \times 25\ mm^3$ parallelipedic elements. Figure 1 shows a rectangular two-dimensional cavity (brick) of length L_x and height L_y with adiabatic horizontal walls and isothermal vertical planes.

The principle of the use of phase change materials (PCMs) is simple: as the temperature increases, the material changes phase from solid to liquid. As the reaction is endothermic, the PCM absorbs heat. Similarly, as the temperature decreases, the material changes phase from liquid to solid and the PCM releases heat.

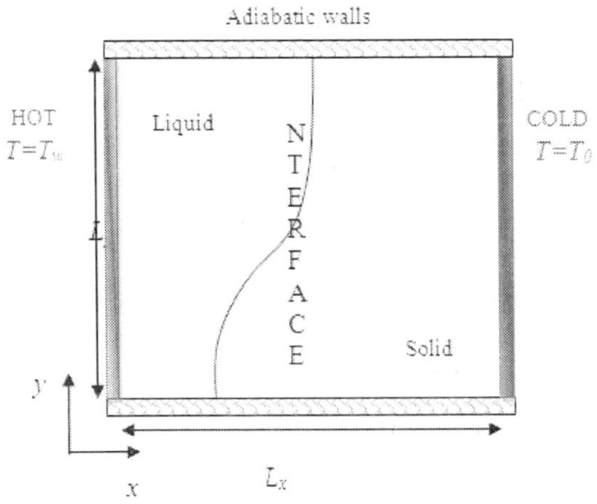

Figure 1. Schematic of the problem.

Table 1. Thermophysical properties of PCM 27 [16]

Variable	PCM 27	
	Solid	Liquid
k (W/m.K)	0.577	0.813
c_p (J/kg.K)	1751.5	2225
α (m^2/s)	1.93×10^{-7}	2.39×10^{-7}
ρ (kg/m^3)	1710	1530
L_f(kJ/kg)	172.42	
μ (Pa.s)	0.094	

Several difficulties appear when solving phase change problems. The use of analytical solutions is generally difficult because of the nonlinearity of the governing equations. Another issue pertains to the time-dependence of location, shape, and orientation of the solid-liquid interface as well as that of the presence of a time-dependent «mushy» zone that may involve solid particles surrounded by the liquid phase next to the interface.

One more important problem to solve often takes place during PCM solidification. Such problems become even more complex because of the supercooling phenomenon [18]. Experimental studies have been carried out in our laboratory and have revealed that the studied PCM, when solidifying, undergoes supercooling phenomenon [16].

2. Case 1: 1D Numerical Simulations

In this section, the thermal storage capacity of a phase change material used for energy conservation in buildings, and in particular in solar walls, is analyzed and discussed. For the numerical tests, the length of the domain was 1 m. Simulations were carried out using a 1D finite difference code (namely FDM code) written in Fortran [16].

2.1. Boundary Conditions

Initially (at $t=0$), the PCM was entirely solid at the uniform temperature $T_0=288.15$ K, lower than the melting temperature T_F. At time $t>0$ and at the outer wall ($x=0$), the temperature of the PCM was raised to a temperature $T_w=323.15$ K, and the PCM melting started, whereas the temperature at $x=L_x$ $(=1$ $m)$ was maintained at $T_0=288.15K$ (Table 2).

Thus, the boundary conditions were:

$$At\begin{cases} t=0 & 0\le x\le L_x & T=T_0<T_F \\ t>0 & x=0 & T=T_W>T_F \\ t>0 & x=L_x & T=T_0<T_F \end{cases}$$

Table 2. PCM test

Test conditions	
L (m)	1
T_0 (K)	288.15
T_F (K)	300.15
T_w (K)	323.15

2.2. Assumptions

The following assumptions were used to model the phase change material:

- The problem is *one-dimensional*
- *Radiative heat transfer is neglected*
- *Thermophysical properties are constant and different in solid and liquid phases*

2.3. Governing Equations

Thus, the governing partial differential equation for the phase change process can be written as:

$$\frac{\partial h}{\partial t} = \alpha \frac{\partial^2 h}{\partial x^2} - \rho L_f \frac{\partial f}{\partial t} \tag{1}$$

where

$$h = \int_{T_F}^{T} \rho c_p dT \text{ and } \alpha = \frac{k}{\rho c_P} \tag{2}$$

The liquid fraction f defined as the ratio of the liquid mass to the total mass in a given computational cell is given by:

$$f = \begin{cases} 0 & if \quad T < T_F \quad (solid) \\ 0-1 & if \quad T = T_F \quad (mushy) \\ 1 & if \quad T > T_F \quad (liquid) \end{cases} \tag{3}$$

Equation (1) can be solved using a fully implicit finite difference solution method. The discretization of this equation leads to the scheme:

$$h_P = h_P^0 + \alpha R(h_E + h_W - 2h_P) + \rho L_f (f_P^0 - f_P^k) \tag{4}$$

$$a_P h_P + a_E h_E + a_W h_W = Q \tag{5}$$

$$\text{where} \begin{cases} a_P = 1 + 2\alpha R \\ a_E = a_W = -\alpha R \\ Q = h_P^0 + \rho L_f (f_P^0 - f_P^k) \\ R = \dfrac{\Delta t}{(\Delta x)^2} \end{cases}$$

The quantities h_P^0 and f_P^0 represent the enthalpy and the liquid fraction from the previous time step. Source term Q keeps track of the heat evolution and superscript k means the *kth* iteration of f at node P.

Equation (4) has been solved using a tri-diagonal matrix algorithm (TDMA) and the liquid fraction update method [20, 21]. When the phase change was occurring $(0<f<1)$, the $(k+1)th$ estimate of the melt fraction needed to be updated such that $h_P=0$ in equation (4)

$$f_P^{k+1} = \frac{-a_E h_E - a_W h_W + h_P^0}{\rho L} + f_P^0 \tag{6}$$

Equation (6) was applied at each node along with under/over correction:

$$f = \begin{cases} 0 & if \quad f_P^{k+1} \leq 0 \\ 1 & if \quad f_P^{k+1} \geq 1 \end{cases} \tag{7}$$

Convergence at a given time step was obtained when the difference between the total enthalpy fields was below than a given tolerance (10^{-4}).

$$\frac{ABS(H^k - H^{k+1})}{\rho_k c_k} \leq 10^{-4} \tag{8}$$

Algorithm

Different steps were required to solve algebraic equations [22, 23]:

1. Solve equation (4) (TDMA algorithm);
2. Update the liquid fraction at every nodes;
3. Apply the correction (7);
4. Check convergence criterion (8);
5. If the convergence criterion is not satisfied, go back at step 1.

Several simulations have been carried out [16]. Different factors, namely the time step, the under-relaxation factor and the grid size (number of nodes) influenced the stability of the solution. Three different time steps (*1 s, 10 s* and *100 s*) have been studied on results, and a time step equal to *10 s* seemed to be appropriate. The influence of the mesh size on these results was evaluated (*40, 80* and *800* nodes); a grid of *80* nodes with a finer mesh near to the cold wall was found to be suitable. Taking into account a grid of *80* nodes and a time step of *10 s*, it was acceptable to choose an under-relaxation coefficient ω of *0.9*. The validation of the FDM code was performed by comparing results of the literature such as ice melting and gallium melting, the detailed results are presented in [16].

3. Case 2: 2D Numerical Simulations

The problem to be considered was that of a rectangular enclosure (figure 1) that initially held a solid PCM at temperature (T_0), lower than the melting temperature (T_F). This temperature was maintained until thermal equilibrium. The whole material was entirely solid. At time $t > 0$, the temperature of the outer left surface of the brick was increased to a constant temperature (T_W), such that $T_W > T_F$ (fusion process) The values of T_0, T_F and T_W are the same as the ones employed in the 1D simulations written in Table 2. Heat was conducted through the wall of the PCM brick, causing the PCM to start to melt next to the inner surface. The details of the mathematical model that is used to predict this melting process are outlined next.

3.1. Boundary Conditions

The rectangular two-dimensional cavity of length L_x and height L_y was subjected to the following boundary conditions:

- Heated surface (left surface, $x= 0, y, t$)

$$u = v = 0 T = T_w$$

- Solid / Liquid interface ($x= x_c, y = y_c, t$)

$$u = v = 0\ T = T_F$$

- Cooling surface (right surface, $x= L_x, y, t$)

$$u = v = 0\ T = T_0$$

- Top ($x, y=0, t$) and bottom ($x, y=L_y, t$) surfaces

$$u = v = 0 \frac{\partial T}{\partial y} = 0$$

where $T_w > T_F > T_0$

3.2. Assumptions

To obtain a solution of the governing equations [24, 25], several assumptions were essential to understand the limits of the validity of the selected method:

- The problem is *two-dimensional* (natural convection occurs in the liquid phase and conduction occurs in both phases)
- *Radiative heat transfer is neglected*

- The flow in the liquid phase is *Newtonian, laminar, and incompressible*. Buoyancy effects due to temperature variations are taken into account by invoking *Boussinesq's assumption*
- *Thermophysical properties* are assumed *constant* and are *different in both phases*
- The PCM material is considered *pure, homogeneous*, and *isotropic*
- *Volumetric expansion is neglected upon melting*

3.3. Governing Equations

Based on the aforementioned assumptions, the governing equations for the heat transfer melting process are as follows:

- Solid region:

$$\frac{\partial T}{\partial t} = \alpha_s \left[\frac{\partial^2 T}{\partial x^2} + \frac{\partial^2 T}{\partial y^2} \right] \tag{9}$$

- Liquid region:

$$\frac{\partial u}{\partial x} + \frac{\partial v}{\partial y} = 0 \tag{10}$$

$$\rho_l \left(\frac{\partial u}{\partial t} + u \frac{\partial u}{\partial x} + v \frac{\partial u}{\partial y} \right) = -\frac{\partial p}{\partial x} + \mu \left(\frac{\partial^2 u}{\partial x^2} + \frac{\partial^2 u}{\partial y^2} \right) + Bu \tag{11}$$

$$\rho_l \left(\frac{\partial v}{\partial t} + u \frac{\partial v}{\partial x} + v \frac{\partial v}{\partial y} \right) = -\frac{\partial p}{\partial y} + \mu \left(\frac{\partial^2 v}{\partial x^2} + \frac{\partial^2 v}{\partial y^2} \right) + \rho g \beta (T - T_{ref}) + Bv \tag{12}$$

$$\left(\frac{\partial T}{\partial t} + u \frac{\partial T}{\partial x} + v \frac{\partial T}{\partial y} \right) = \alpha_l \left(\frac{\partial^2 T}{\partial x^2} + \frac{\partial^2 T}{\partial y^2} \right) - \frac{L}{c_l} \frac{\partial f}{\partial t} \tag{13}$$

In the enthalpy-porosity approach, the condition that sets the velocities in the solid regions is provided by appropriately defining a parameter B in the momentum equations [26]. During the solution process, the velocity at the computational cell located in the solid phase should be suppressed while the velocity in the liquid phase remains unaffected. This can be achieved by assuming that such cells behave like a porous medium with porosity equal to the liquid fraction. In order to achieve this behavior, an appropriate definition of B is:

$$B = -\frac{C(1-f)^2}{(f^3 + b)} \tag{14}$$

which is the Carman-Kozeny relation. In this model, $f=1$ in the liquid region, $f=0$ in the solid region, while it takes a value between 0 and 1 in the mushy zone. The constant C has a large value to suppress the velocity as the cell becomes solid and b is a small constant used to avoid a division by zero when a cell is fully located in the solid region, namely $f = 0$. The choice of the constants is arbitrary. However, the constants should ensure sufficient suppression of the velocity in the solid region and should not influence the numerical results significantly. In this work, $C = 1\times10^9\ kg/m^3\ s$ and $b = 0.0003$ are used [27].

As recommended by Patankar [25], a single algorithm can be used where all the equations can be casted into a general form. Here, equations (9-13) can be formulated such that:

$$\frac{\partial(\rho\varphi)}{\partial t} + \nabla\cdot(\rho V\varphi) = \nabla\cdot(\Gamma\nabla\varphi) + S \tag{15}$$

where the appropriate variables for this problem are specified in Table 3:

Table 3. Coefficients for the general equations

Equation	φ	u	Γ	S
9	T	0	$[k/c_p]_s$	0
10	1	u	0	0
11	u	u		$S_u = -\frac{\partial P}{\partial x} + Bu$
12	v	u		$S_v = -\frac{\partial P}{\partial y} + Bv + \rho g\beta(T - Tref)$
13	T	u	$[k/c_p]_l$	$S_h = -L\frac{\rho}{c_l}\frac{\partial f}{\partial t}$

The phase change was handled using a single-domain with an enthalpy porosity technique [26]. In this method, the absorption of latent heat during melting is included as a source term S_h in the energy equation (Table 3). Latent heat content of each control volume in the PCM is evaluated after each energy equation iteration cycle. Based on the latent heat content, a liquid fraction for each control volume is determined. For control volumes containing a liquid phase of PCM, f is set to 1, and for control volumes containing solid phase, f is set to 0. The control volumes with values of f between 0 and 1 are treated as mushy. Even though the phase change is assumed to be isothermal, the idea of mushy zone is introduced to gradually switch off the velocities from liquid to solid at the interface. The switching off is controlled by the source terms S_u, S_v in the momentum equations (Table 3).

3.4. Numerical Procedure

The conservation governing equations were solved using a control volume based finite difference method employing a uniform Cartesian grid. The coupled energy-liquid fraction, resulting from the use of an enthalpy formulation, was handled by the procedure suggested by Brent et *al.* [26]. The standard SIMPLER algorithm was used to solve the coupled continuity and momentum equations. This code is fully implicit in time, for transient computations, and the convection-diffusion terms were treated by the hybrid scheme [25]. A line by line solver based on the TDMA (tri-diagonal matrix algorithm) was used to solve iteratively the algebraic discretized equations.

The general form of the discretized equation for any variable φ is given by:

$$A_p\varphi_p = \sum_{i=W,E,S,N} A_i\varphi_i + S \tag{16}$$

For each iteration, the solid-liquid interface has to be determined. In this work, it is based on the value of the liquid fraction. The liquid fraction is updated using the equation:

$$f_P^{k+1} = f_P^k + \omega\frac{\Delta t A_P^k h_P^k}{\rho L_f \Delta V} \tag{17}$$

where ω is an appropriate under-relaxation factor [26].

To prevent the calculation of unrealistic values for f, variations of f_p from one iteration to the next are limited by use of:

$$f_P^{k+1} = \begin{cases} 1 & if \;\; f_P^{k+1} > 1 \\ 0 & if \;\; f_P^{k+1} < 0 \end{cases} \tag{18}$$

Hence, despite the particular additions to the momentum and energy equations with respect to a standard mathematical description, the procedure proposed by Patankar [25] can readily be used to obtain discretized algebraic equations, to incorporate boundary conditions, to solve, and to provide relevant dependent variables via an adequate post-processing. Suffice it to say that for a given time step, when a converged solution is obtained for the discretized algebraic equations, the dimensionless liquid fraction is updated until no further change is detected in the solution for this particular time step. Then, the procedure proceeds with the next time step until t_{max} is reached. The convergence is reached when all the following conditions are satisfied simultaneously: for all the governing equations, iterations are terminated when the residual drop below 10^{-7}, and the liquid fraction field remains unchanged [15].

Several domain discretizations were studied. Mesh deployment has been used to refine the discretization along the heated wall (at $x=0$). A 80×100 grid was found to be satisfactory to obtain converged solutions, with $\omega_h=\omega_u=\omega_v=0.1$. The time step $\Delta t = 1s$ was used for all calculations [15].

4. RESULTS

4.1. Heat Storage and Recovery (1D Numerical Simulation)

Initially (at $t=0$), the material was isothermal ($T_0=15$ °C) and, at $t>0$, the material was evolving from temperature T_0 to temperature $T_W=50$ °C. Between these two permanent states, the material stored energy (sensible heat and latent heat). The second step consisted in decreasing the temperature of the PCM from $T_W=50$ °C down to the initial temperature $T_0=15$ °C. During this second phase, the material released heat. In Figure 2, the amount of energy is calculated versus temperature as the material underwent heating from *15 °C* to *50 °C* (complete melting), and then cooling from *50 °C* to *15 °C* (solidification). The cycle concerned the variations of the heat storage and the heat recovery of the sample.

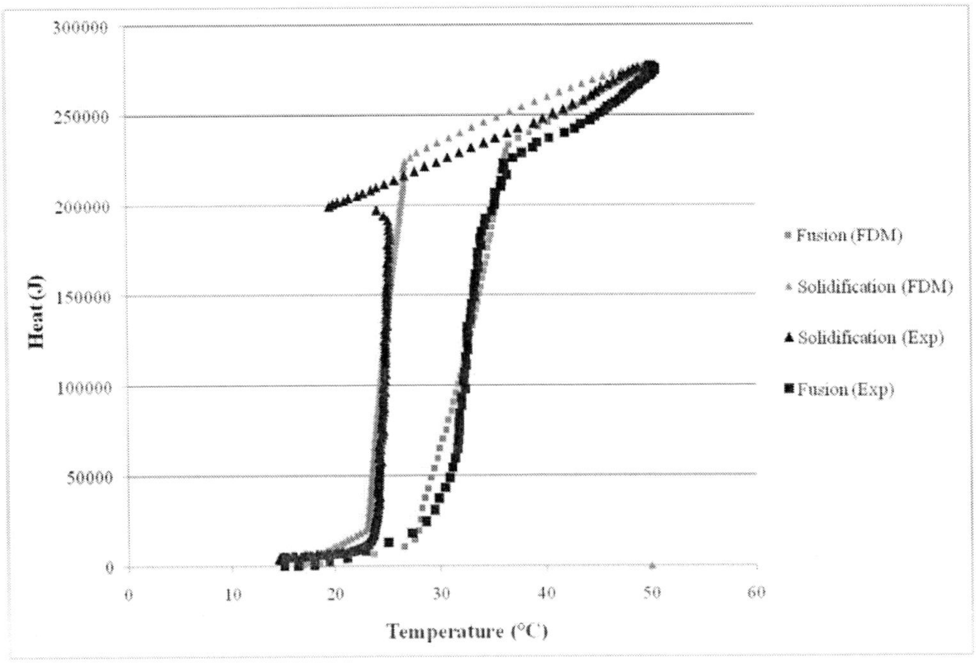

Figure 2. Heat stored and released on temperature cycle (15 °C · 50 °C and 50 °C · 15 °C).

Experimental results are presented thoroughly in papers [16, 28] along with the details that pertain to the experimental apparatus used for thermophysical property measurements and PCM bricks behavior determination. An experimental device was developed, based on the measurement of temperatures and heat fluxes exchanged between both lateral sides of the PCM samples, providing the total heat stored during the phase change process. The heat storage and recovery periods calculated by the FDM code were compared to those experimental measurements (figure 2). A small difference occurred between experimental data (black symbols) and numerical predictions (red symbols) in the case of sensible heat storage as well as latent heat storage during the fusion process.

However, the results are quite different during solidification, when the PCM released heat. Indeed, numerical simulations did not take into account the important effect of supercooling phenomenon [16]. Consequently, this figure reveals that the numerical simulations failed to predict correctly the thermal behavior of the PCM when solidifying.

Moreover, during the fusion process, the heat storage began to take place at a temperature (about *27 °C*) both in the numerical and experimental results, whereas the followed path during the solidification process was different. The melting started at a temperature higher than for the solidification process. The solidification began at *27 °C*. At this stage, the first solid layer appeared. Indeed, in the numerical data, the temperature of the material was decreasing down to *24 °C*, corresponding at the end of the release of latent heat. Heat recovery occurred in numerical simulations as well as in experimental results at a quasi constant temperature about *24 °C*.

In addition, there exists a supercooling phenomenon in the experimental solidification path. The temperature of the material is decreasing down to *19* °C while releasing sensible energy, corresponding at the beginning of the release of latent heat.

Furthermore, figure 2 also shows that the heat stored was much more important than sensible heat transfer when a phase change occurs. This confirms the advantage of latent heat storage.

4.2. Melting of PCM (2D Numerical Simulations)

The problem of gallium melting in a rectangular cavity heated from the side, simulating solar irradiation impinging on a vertical wall, is frequently used as a comparison exercise in the phase change community. In particular, both the experimental results of Gau and Viskanta [29] and the numerical results of Brent *et al.* [30] are often considered for that purpose. The validation of our numerical FVM code (2D) has been achieved both for melting of ice and gallium in [15]: satisfactory results were obtained from these comparisons. Several domain discretizations were studied for the 2D simulations. Mesh redeployment has been used to refine the discretization along the active wall (at $x=0$). A 80×40 grid was found to be satisfactory to obtain converged solutions, insensitive to further refinements. The time step used was found to be problem dependent; different temperature scales (differences between the hot and the cold surfaces) were found to require different time steps. A time step of $t=1$ s was found to be appropriate for all cases.

Heat transfer in a building wall is mostly one-dimensional, however, in vertical enclosures filled with liquid that undergoes a temperature difference between the vertical surfaces, natural convection occurs and two-dimensional effects have to be accounted for.

In this section, PCM was considered for simulation using the proposed FVM method (2D). The relevant dimensionless parameters are:

$$Ra = \frac{\rho^2 c_l g \beta L_y^3 (T_W - T_F)}{\mu k_l}, Ste = \frac{c_l (T_W - T_F)}{L}, Pr = \frac{c_l \mu}{k_l}$$

which are respectively the Rayleigh, Stefan, and Prandtl numbers. For the problem, the magnitudes of the non-dimensional parameters were: $R=0.12$; $Ra=1.34\times10^8$; $Pr=273.85$; $Ste=0.2968$; $c_s/c_l = 0.79$; and $k_s/k_l = 0.71$.

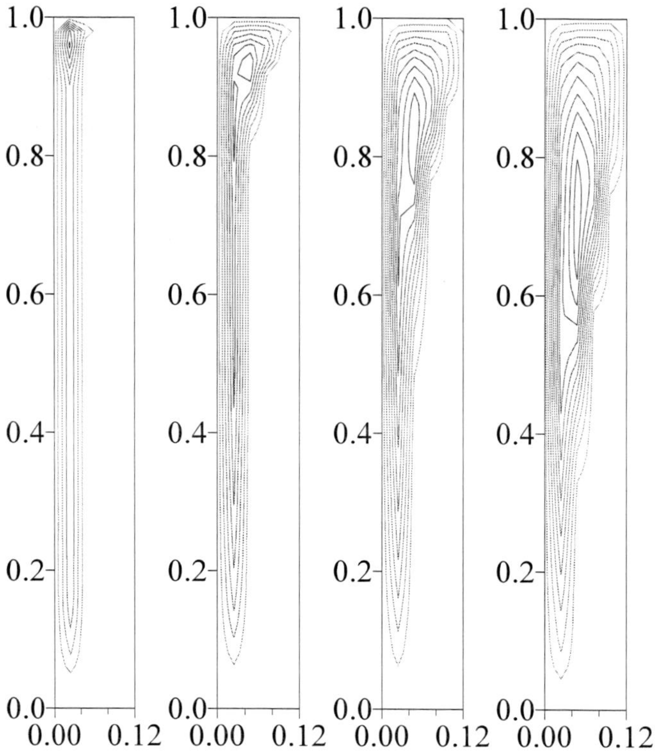

Figure 3. Evolution of the streamlines in the cavity (*R=0.12*) for several times: from left to right: *10 000 s*, *20 000 s*, *30 000 s* and *40 000 s*.

Figure 3 shows the calculated streamlines for four different times ($t=1.10^4$ s, $t=2.10^4$ s, $t=3.10^4$ s, and $t=4.10^4$ s) after the onset of melting, while figure 4 presents the predicted temperature profiles of the PCM at the same instants. The aspect ratio is constant (*R=0.12*). In figure 4, the temperature has been set non dimensional such as: $\theta = \frac{T-T_F}{T_W-T_F}$

In this way, on the hot plate ($T=T_W$), the non-dimensional temperature is =1 (red color in the figure 4), whereas the cold isothermal wall (in blue in figure 3) has a temperature equal to $T=T_0$ and $\theta = \frac{T_0-T_F}{T_W-T_F}$.

At the beginning of the melting process ($t=1.10^4$ s), conduction was almost the dominant mode of heat transfer even within the liquid phase. The isotherms were almost vertical, suggesting the dominance of the conduction mode of heat transfer (figure 4). During the next stages, the temperature difference in the melt gave rise to the natural convection. Natural convection became more and more important as the melt volume increased (figure 3). However, a phase transition for which both conduction and natural convection occurred as the melting process continued. As time increased, this buoyancy induced the motion of the fluid due to the temperature gradient, and then caused the melt volume at the top to move at a faster rate compared to the fluid at the bottom. This was obvious for the streamlines and the temperature profiles at times $t=2.10^4$-4.10^4 s, which were curved due to the effect of the natural convection, thus augmenting the overall melting process.

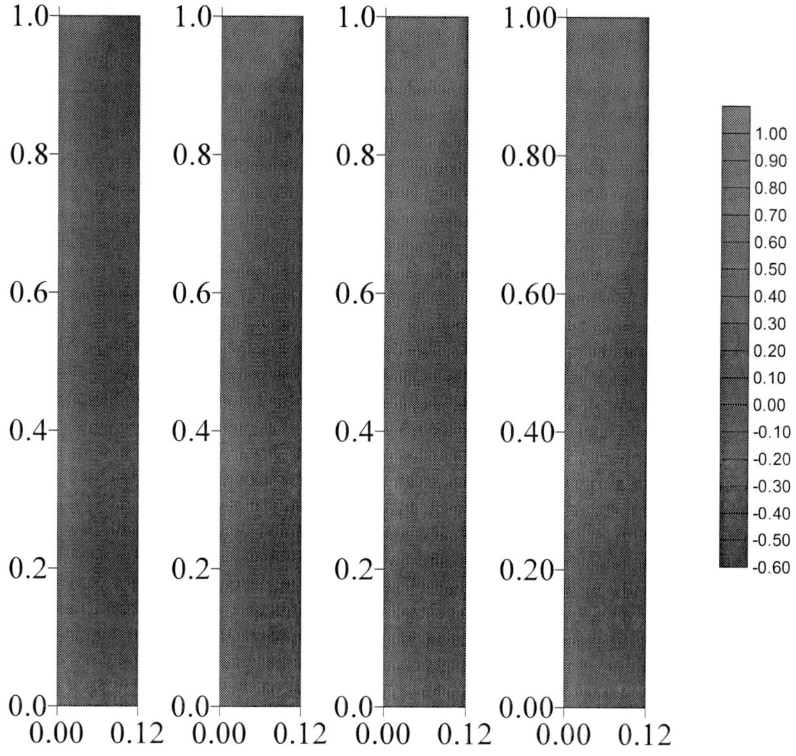

Figure 4. Dimensionless temperature profiles in the cavity ($R=0.12$) for several times: from left to right: *10 000 s*, *20 000 s*, *30 000 s* and *40 000 s*.

Indeed, the flow structure could significantly affect the phase change process. The convection must be taken into account as it greatly influenced the morphology of the streamlines (figure 3) by changing the flow structure in the melt. Also, the effects of buoyancy driven natural convection play a key role in the amount of enhancement produced by the PCM.

CONCLUSION

In this chapter, a study of the PCM melting problem in a cavity heated from the side was presented. A numerical investigation was conducted to analyze a convection phase change heat transfer of a PCM on the basis of the fixed-grid formulation. The phase change process and the velocity suppression were modeled by the enthalpy-porosity method. The mathematical model was developed to describe the phase change. The solution was obtained by use of a control volume-based finite difference method.

The model has been applied to predict the thermal behavior of this PCM, a material designed to melt at *27 °C*. The proposed parallelepipedic-shaped brick of such PCM was heated from the lateral side: this simulated the solar irradiation impinging on a vertical wall.

The energy storage and the energy recovery were studied and compared to experimental data. Results of the proposed model (1D) showed a good agreement with these experimental data, except during the solidification process where the influence of supercooling phenomenon has been clearly exhibited with the experimental curve. These numerical simulations 1D did not gave satisfaction because results were very far away from the experimental ones as the code 1D did not represent the supercooling phenomenon. It has to be taken into account in the modelling in order to correctly predict the PCM thermal behaviour.

A second result was presented in this chapter provided by the 2D numerical code. The melting driven by conduction as well as by convection in the rectangular cavity was investigated numerically. Natural convection played an important role after the first stages of melting that were dominated by conduction. Buoyancy effects became more and more important at later stages of the fusion process.

Further investigations are needed to have a better numerical description of this special phase change feature.

REFERENCES

[1] Fath H.E.S. Assessment of solar thermal energy storage technologies, *Renew. Energy,* 1998, 14, 35–40.

[2] Kurklu A., Energy storage applications in greenhouses by means of phase change materials (PCMs): a review, *Renew. Energy*, 1998, 13, 89–103.

[3] Evans A.G.; He M.Y.; Hutchinson J.W.; Shaw M., Temperature distribution in advanced power electronics systems and the effect of phase change materials on temperature suppression during power pulses, *J. Electron. Packag.—Trans. ASME*, 2001, 123, 211–217.

[4] Swanson T.D.; Birur G.C., NASA thermal control technologies for robotic spacecraft, *Appl. Therm. Eng.*, 2003, 23, 1055–1065.

[5] Cheng K.C.; Seki C., *Freezing and melting heat transfer in engineering*, Hemisphere Publishing Corporation, New York, U.S.A., 1991.

[6] Medrano M.; Yilmaz M.O.; Nogues M.; Martorell I.; Roca J.; Cabeza L.F., Experimental evaluation of commercial heat exchangers for use as PCM thermal storage systems, Applied Energy, 2009, 86, 10, 2047-2055.

[7] Agyenim F.; Hewitt N.; Eames P.; Smyth M., A review of materials, heat transfer and phase change problem formulation for latent heat energy storage systems (LHTESS), *Renewable and sustainable energy reviews*, 2010, 14, 615-628.

[8] Regin A.F.; Solanki S.C.; Saini, J.S., Heat transfer characteristics of thermal energy storage system using PCM capsules: A review, *Renewable and Sustainable Energy Reviews*, 2008, 12, 9, 2438-2458.

[9] Tyagi V.V.; Buddhi D., PCM thermal storage in buildings: A state of art, Renewable and *Sustainable Energy Reviews*, 2007, 11, 6, 1146-1166.

[10] Zhang Y.; Zhou G.; Lin K.; Zhang Q.; Di H., Application of latent heat thermal energy storage in buildings: State-of-the-art and outlook, *Building and Environment*, 2007, 42, 6, 2197-2209.

[11] Sharma A.; Tyagi V.V.; Chen C.R.; Buddhi V, Review on thermal energy storage with phase change materials and applications, *Renewable and Sustainable Energy Reviews*, 2009, 13, 2, 318-345.

[12] Farid M.M.; Khudhair A.M.; Razack S.A.K.; Al-Hallaj S., A review on phase change energy storage: materials and applications, *Energy Conversion and Management*, 2004, 45, 9-10,1597-1615.

[13] Dutil Y.; Rousse D.R.; Ben Salah N.; Lassue S.; Zalewski L., A review on phase-change materials : Mathematical modeling and simulations, *Ren. Sust. Energy Rev.*, 2011, 15, 1, 112-130.

[14] Zalewski L.; Lassue S.; Duthoit B.; Butez M., Study of solar walls validating a simulation model, *Building and Environment*, 2002, 37, 1, 109-121.

[15] Younsi Z.; Joulin A.; Zalewski L.; Rousse D.R.; Lassue S., A numerical study of PCM melting heated from a vertical wall of a rectangular enclosure, *Int. J. Computational Fluid Dynamics*, 2009, 23, 7, 553-566.

[16] Joulin A.; Younsi Z.; Zalewski L.; Lassue S.; Rousse D.R.; Cavrot J.P., Experimental and numerical investigation of a phase change material : thermal energy storage and release, *Applied Energy*, In Press, 2011.

[17] Younsi Z.; Joulin A.; Zalewski L.; Lassue S.; Rousse D., Phase change material: Experimental Measurements of Thermophysical properties, *Proc. ICTEA*, 2009.

[18] Ehmimed J.A.; Zeraouli Y.; Dumas J.P.; Mimet A., Heat transfer model during the crystallization of a dispersed binary solution, *International Journal of Thermal sciences*, 2003, 42, 1, 33-46.

[19] Lacroix M.; Voller V.R., Finite difference solutions of solidification phase change problems transformed versus fixed grids, *Numer. Heat Transfer*, 1990, 17(B), 24-41.

[20] Shaikh S.; Lafdi K., Effect of multiple phase change materials (PCMs) slab configurations on thermal energy storage, *Energy Conv. Management*, 2007, 47, 2103-2117.

[21] Voller V.R., Fast implicit finite-difference method for the analysis of phase change problems, *Num. Heat Transfer - Part B*, 1990, 17, 155-169.

[22] Costa M.; Buddhi D.; Oliva A., Numerical simulation of a latent heat thermal energy storage system with enhanced heat conduction, *Energy Conv. Management*, 1997, 39, 319-330.

[23] Sharma A.; Tyagy V.V.; Chen C.R.; Buddhi D., Review on thermal energy storage with phase change materials and applications, *Renew. Sustain Energy Rev*, 2009, 13, 2, 318-345.

[24] Voller V.R., An Overview of Numerical Methods for Solving Phase Change Problems, *Advances in Numerical Heat Transfer*, 1997, 1, edited by W.J. Minkowycz and E.M. Sparrow, Taylor & Francis.

[25] Patankar S.V.; *Numerical Heat Transfer and Fluid Flow*, Taylor & Francis, 1980.

[26] Brent A.D.; Voller V.R.; Reid K.J., Enthalpy-porosity technique for modeling convection-diffusion phase change: application to the melting of a pure metal, *Numerical Heat Transfer*, 1988, 13, 297-318.

[27] Viswanath R.; Jaluria Y., A Comparison of Different Solution Methodologies for Melting and Solidification Problems in Enclosures, *Num. Heat Transfer Part B*, 1993, 24, 77-105.

[28] Younsi Z.; Zalewski L.; Lassue S.; Rousse D.R.; Joulin A., A novel technique for the experimental thermophysical characterization of phase change materials, *Int. J. Thermophysics*, 2010, 31.

[29] Gau C.; Viskanta R., Melting and solidification of a pure metal from a vertical wall, *Journal of Heat Transfer*, 1986, 108, 171–174.

[30] Brent A.D.; Voller V.R.; Reid K.J., Enthalpy-porosity technique for modeling convection–diffusion phase change: application to the melting of a pure metal, *Numerical Heat Transfer*, 1988, 13:297–318.

INDEX

A

B

C

D

E

F

N

O

P

Q

R

S

T

U

V

W

Y